Right Triangle

$c^2 = a^2 + b^2$ [Pythagorean Theorem]

area $= \frac{1}{2}ab$

$\sin\theta = \dfrac{b}{c} = \dfrac{\text{opp}}{\text{hyp}}$

$\cos\theta = \dfrac{a}{c} = \dfrac{\text{adj}}{\text{hyp}}$

$\tan\theta = \dfrac{\sin\theta}{\cos\theta} = \dfrac{b}{a} = \dfrac{\text{opp}}{\text{adj}}$

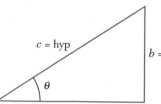

Any Triangle

area $= \frac{1}{2}bh = \frac{1}{2}ab\sin C$

$a^2 = b^2 + c^2 - 2bc\cos A$

$b^2 = a^2 + c^2 - 2ac\cos B$

$c^2 = a^2 + b^2 - 2ab\cos C$

$\dfrac{\sin A}{a} = \dfrac{\sin B}{b} = \dfrac{\sin C}{c}$

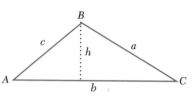

Trapezoid

area $= \frac{1}{2}b(h_1 + h_2)$

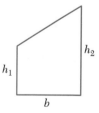

Circle

area $= \pi r^2$

circumference $= 2\pi r$

Circular Sector

area $= \frac{1}{2}r^2\theta$

arclength $= s = r\theta$

Circular Cylinder

volume $= \pi r^2 h$

Circular Cone

volume $= \frac{1}{3}\pi r^2 h$

Sphere

volume $= \frac{4}{3}\pi r^3$

surface area $= 4\pi r^2$

Trigonometric Identities

$\sin^2 x + \cos^2 x = 1$

$\sec^2 x = 1 + \tan^2 x$

$\sin(2x) = 2\sin x \cos x$

$\cos(2x) = \cos^2 x - \sin^2 x = 2\cos^2 x - 1 = 1 - 2\sin^2 x$

$\sin^2 x = \frac{1}{2}(1 - \cos(2x))$

$\cos^2 x = \frac{1}{2}(1 + \cos(2x))$

$\sin(x \pm y) = \sin x \cos y \pm \cos x \sin y$

$\cos(x \pm y) = \cos x \cos y \mp \sin x \sin y$

$\tan(x \pm y) = \dfrac{\tan x \pm \tan y}{1 \mp \tan x \tan y}$

$\sin x \sin y = \frac{1}{2}\big(\cos(x - y) - \cos(x + y)\big)$

$\cos x \cos y = \frac{1}{2}\big(\cos(x - y) + \cos(x + y)\big)$

$\sin x \cos y = \frac{1}{2}\big(\sin(x + y) + \sin(x - y)\big)$

Algebraic Identities

$ax^2 + bx + c = 0 \iff x = \dfrac{-b \pm \sqrt{b^2 - 4ac}}{2a}$ [Quadratic Formula]

$a^2 - b^2 = (a + b)(a - b)$

$a^3 + b^3 = (a + b)(a^2 - ab + b^2)$

$a^3 - b^3 = (a - b)(a^2 + ab + b^2)$

$(a + b)^n = a^n + na^{n-1}b + \cdots + \binom{n}{k}a^{n-k}b^k + \cdots + b^n;\ \binom{n}{k} = \dfrac{n!}{(n - k)!\,k!}$ [Binomial Theorem]

Derivative Formulas

$\dfrac{d}{dx}c = (c)' = 0$

$\dfrac{d}{dx}x = (x)' = 1$

$\dfrac{d}{dx}x^n = (x^n)' = nx^{n-1}$

$\dfrac{d}{dx}e^x = (e^x)' = e^x$

$\dfrac{d}{dx}\ln x = (\ln x)' = \dfrac{1}{x}$

$\dfrac{d}{dx}\sin x = (\sin x)' = \cos x$

$\dfrac{d}{dx}\cos x = (\cos x)' = -\sin x$

$\dfrac{d}{dx}\tan x = (\tan x)' = \sec^2 x$

$\dfrac{d}{dx}\sec x = (\sec x)' = \sec x \tan x$

$\dfrac{d}{dx}\arcsin x = (\arcsin x)' = \dfrac{1}{\sqrt{1 - x^2}}$

$\dfrac{d}{dx}\arctan x = (\arctan x)' = \dfrac{1}{1 + x^2}$

Derivative Rules

$\big(f(x) + g(x)\big)' = f'(x) + g'(x)$

$\big(f(x)g(x)\big)' = f'(x)g(x) + f(x)g'(x)$

$\left(\dfrac{f'(x)}{g(x)}\right)' = \dfrac{f'(x)g(x) - f(x)g'(x)}{\big(g(x)\big)^2}$

$\big((f \circ g)(x)\big)' = \big(f(g(x))\big)' = f'\big(g(x)\big)g'(x)$

CALCULUS

From Graphical, Numerical, and Symbolic Points of View

Volume I

- **ARNOLD OSTEBEE**

  St. Olaf College

- **PAUL ZORN**

  St. Olaf College

SAUNDERS COLLEGE PUBLISHING

Harcourt Brace College Publishers

Fort Worth   Philadelphia   San Diego   New York
Orlando   Austin   San Antonio   Toronto
Montreal   London   Sydney   Tokyo

This material is based upon work supported by the National Science Foundation under Grant No. USE-9053363

Any opinions, findings, and conclusions or recommendations expressed in this material are those of the authors and do not necessarily reflect the views of the National Science Foundation.

Derive is a registered trademark of Soft Warehouse, Inc.
HP-35 and HP-48 graphics calculators are registered trademarks of Hewlett-Packard Company.
IBM is registered trademark of International Business Machines Corporation.
Macintosh is a registered trademark of Apple Computer, Inc.
Maple is a registered trademark of Waterloo Maple Software.
Mathematica is a registered trademark of Wolfram Research, Inc.
TI-81, TI-82, TI-83, TI-85, and TI-92 graphics calculators are trademarks of Texas Instruments, Inc.
Windows is a trademark of Microsoft Corporation.

Text Typeface: New Caledonia
Compositor: ETP/Harrison
Senior Associate Editor: Alexa Epstein
Managing Editor: Carol Field
Senior Project Editor: Nancy Lubars
Copy Editor: Mary Patton
Manager of Art and Design: Carol Bleistine
Art Director: Caroline McGowan
Art & Design Coordinator: Kathleen Flanigan
Cover Designer: Chazz Bjanes
Text Artwork: Techsetters, Inc.
Vice President of EDP: Tim Frelick
Manager of Production: Joanne Cassetti
Production Manager: Alicia Jackson
Product Manager: Nick Agnew

Cover Credit: Art Wolfe/© Tony Stone Images
Printed in the United States of America

ISBN 0-03-096578-0
Library of Congress Catalog Card Number: 96-67893

6789012345 069 10 987654321

To Kay, Kristin, and Paul
To Janet, Anne, and Libby

About This Book: Notes for Instructors

This book aims to do exactly what its title suggests: present calculus from graphical, numerical, and symbolic points of view. In this preface we elaborate briefly on what this means.

Philosophy

Several common threads run through *Toward a Lean and Lively Calculus*, *Calculus for a New Century*, discussions on calculus at many professional meetings, and the NCTM *Curriculum and Evaluation Standards for School Mathematics*. First, there is a consistent call for *leaner and more conceptual courses*, driven by and focused on central ideas. Second, there is a realization that courses should *reflect modern technology both in content and in pedagogy*.

A Diagnosis. Many calculus courses, we believe, slight the conceptual foundations of the subject and overemphasize routine techniques—formal differentiation, antidifferentiation, and convergence testing, for example. Analytic objects (such as integral, derivative, and convergence) are represented and manipulated only algebraically (i.e., via symbolic manipulation of explicit elementary functions). For example, textbooks often treat limits, derivatives, and integrals—all *analytic* objects—only as *algebraic* operations on *algebraic* functions. We try to take a broader view.

Whether one views calculus as an introduction to pure mathematics or as a foundation for applications (or both!), the conclusion is the same—concepts, not techniques, are truly fundamental to the course. Whatever uses they make of the calculus, students need more than a compendium of manipulative techniques. The *sine qua non* for a useful command of the calculus is a conceptual understanding that is deep and flexible enough to accommodate diverse applications.

A Prescription. Our key strategy for improving conceptual understanding is combining, comparing, and moving among graphical, numerical, and algebraic "representations" of central concepts. This strategy pervades and unifies our exposition. Bringing graphical and numerical, as well as algebraic, viewpoints to bear on calculus ideas is the philosophical foundation of our text. By representing and manipulating calculus ideas and objects graphically, numerically, and algebraically, we believe that students gain a better, deeper, and more useful understanding.

Audience and Prerequisites

Our main strategy—combining graphical, numerical, and algebraic viewpoints to clarify concepts and make them concrete—aims explicitly at mainstream students, who especially need such help. We regard a more conceptual calculus as also more applicable. To use calculus ideas and techniques effectively, students must know what they are doing and why, not merely how. Thus, our text is appropriate for a general audience: mathematics majors, science and engineering majors, and non-science majors.

We expect that students have the "usual" precalculus preparation, including basic algebra and trigonometry and the rudiments of logarithm and exponential functions. However, we include reasonably complete reviews of all these subjects, with routine exercises, in the Appendices. Appendices are written in a style that supports independent reading and self-study; many instructors should be able to avoid explicit "coverage" of precalculus material in class time.

Volume 1 and Volume 2

We treat single-variable calculus in two volumes, each suitable for a semester-length course. To accommodate varying calendars and preferences, Chapters 5 and 6 appear in *both* volumes. In addition, a brief introduction to multivariable calculus topics appears at the end of Volume 2; some instructors may prefer it as an alternative to the topics of infinite sequences and series. (See also: Volume 3, *Multivariable Calculus from Graphical, Numerical, and Symbolic Points of View.*)

Annotated Table of Contents, Volume 1

Here is brief chapter-by-chapter information on Volume 1; for many more details, see the *Notes for Instructors*, described below.

Chapter 1 – Functions in Calculus This chapter sets the stage by emphasizing graphical and numerical points of view, especially on functions. Functions are described by algebraic formulas, but also by graphs, by tables, and in words. The standard elementary functions are reviewed briefly, along with a variety of ways of producing new functions from old. The chapter reviews key precalculus concepts and foreshadows important calculus ideas, such as slope, concavity, and rates of change. (Appendices provide still more precalculus review for students who need it.)

Chapter 2 – The Derivative The derivative function is first treated intuitively, in the parallel languages of slope and of rates of change. Next, a careful geometric study is made of functions and their derivatives, emphasizing relationships between the two. The second half of the chapter treats the derivative symbolically, via limits. Systematic symbolic calculation of derivatives begins in the next chapter.

Chapter 3 – Derivatives of Elementary Functions   Derivatives of the standard (algebraic and transcendental) elementary functions are first calculated, and their properties are studied. Antiderivatives are also introduced and used in simple calculations and applications. The second half of the chapter emphasizes combinatorial properties of functions and their derivatives—how the derivative of a built-up function (e.g., by composition) is formed from derivatives of its constituents. A final section treats inverse trigonometric functions.

Chapter 4 – Applications of the Derivative   A flexible and varied collection of applications of the derivative, designed to allow easy rearrangements or omissions. Some topics (e.g., optimization) are standard; others (e.g., early differential equations, quadratic splines) are less usual. The chapter ends with some topics of a more analytic character, exploring continuity, differentiability, and their most important consequences, such as the mean value theorem. *Important Note: Chapter 4 is designed as a sampler; most instructors will want to cover only parts of it.*

Chapter 5 – The Integral   The treatment generally mirrors that of the derivative in Chapter 2. The integral is first presented geometrically, as signed area. Analytic viewpoints follow, including a formal definition via approximating sums and a careful treatment of the fundamental theorem of calculus. Interpretations and applications of approximating sums foreshadow topics of Calculus II.

Chapter 6 – Finding Antiderivatives   Students began finding antiderivatives early, in Chapter 3. Combined with that experience, this brief chapter completes a short introduction to rudiments of antidifferentiation. The last section refers both to printed tables and to technological aids.

Annotated Table of Contents, Volume 2

Here is brief chapter-by-chapter information on Volume 2. For an examination copy, contact Saunders College Publishing, or see our World Wide Web site, at
http://www.stolaf.edu/people/zorn/ozcalc/

Chapters 5 and 6   See above; these chapters appear in both volumes.

Chapter 7 – Numerical Integration   This chapter introduces, briefly but systematically, the idea of approximating integrals numerically—and of estimating the errors committed in doing so. Several alternative methods are presented and compared.

Chapter 8 – Using the Definite Integral   Like Chapter 4, Chapter 8 offers a flexible variety of applications, this time of the definite integral. Geometric, physical, and economic applications are included. Choices and reordering are possible. An introductory section emphasizes what various applications have in common.

Chapter 9 – More Antidifferentiation Techniques This chapter continues and extends the themes of Chapter 6. Some instructors may choose to cover it before Chapter 8.

Chapter 10 – Improper Integrals This chapter both extends the idea of integral and foreshadows the treatment of infinite series in the next chapter.

Chapter 11 – Infinite Series Stressing analogies between improper integrals and infinite series, the chapter examines convergence and divergence numerically and graphically, as well as symbolically. The topic is traditionally difficult; technology is used to illustrate elusive ideas and make them concrete.

Chapter 12 – Differential Equations Differential equations are treated first graphically (using slope fields), then numerically (using Euler's method), and then symbolically (via separation of variables).

Chapter 13 – Polar Coordinates This brief, readily transportable chapter stresses geometric understanding of an alternative coordinate system.

Chapter 14 – A First Look at Multivariable Calculus The chapter offers a brief overview of derivatives and integrals of functions of several variables, stressing functions of two variables. The theme of linear approximation is stressed; technology is used to help students visualize the new objects.

More on Content and Treatment

Several special features of our treatment cut across chapters, and deserve emphasis. For many more details and specific examples, see the *Notes for Instructors*, described below.

Foreshadowing Important concepts and associated vocabulary often appear more than once—first informally and later in more detail. The geometric idea of concavity, for instance, is mentioned in Section 1.1, and then revisited several times. Instructors should know that it is *not* necessary to cover these ideas in full rigor at their first appearance.

Early Differential Equations Differential equations are mentioned early in Chapter 2, introduced informally (but at some length) in Chapter 4, and then revisited more formally in Chapter 12. We believe that DE's are so basic and so important that students should see them early and often. Calculus I emphasizes the *idea* of a DE and its solutions; the (harder) question of finding solutions systematically is deferred to Calculus II.

Graphs, then Symbols Derivatives and integrals are introduced in a parallel manner, with the graphical approach preceding the symbolic viewpoint. Thus Chapter 2 stresses the geometry of the derivative; the symbolic approach enters mainly in Chapter 3. Chapter 5 begins with a graphical definition of integral, and proceeds to the Riemann sum definition.

The Role of Technology

Combining graphical, numerical, and symbolic viewpoints in calculus can be forbiddingly time-consuming and distracting without technological assistance. With technology, these viewpoints become practically accessible; hence, we use technology freely.

Having said this, we emphasize that we do *not* intend to "automate" or "computerize" the calculus. We regard computers and graphing calculators strictly as instructional *tools*—albeit very powerful ones—to facilitate the crucial combination of graphical, numerical, and algebraic viewpoints. Our text exploits the capabilities of these tools to help students focus on the ideas that lie at the heart of the calculus.

We do not assume that students have any previous experience with using technology.

How Much Technology? What Kinds?   This text freely uses and refers to numerical and graphical computations, but it is independent of any particular technology. Any of the familiar high-level products—*Mathematica*®, *Maple*®, *Derive*®—are certainly sufficient, but so are many other forms of technology. Many graphing calculators (e.g., the TI-8X and 92, the HP-35 and 48) and special-purpose microcomputer software packages are also adequate.

Our use of technology is most conveniently described in terms of *functionalities*. The requirements for Calculus I and Calculus II differ somewhat:

Calculus I: Chapters 1–5 (the traditional content of Calculus I) draw freely on machine graphics; almost any up-to-date form will do. Most graphing calculators would suffice; so would almost any flexible microcomputer graphing program.

Calculus II: To make the best use of the remaining (Calculus II—not covered in this volume) material, students will require access to a modest level of numerical computation (mainly for estimating analytic quantities: integrals, series, etc.). Many microcomputer software packages provide the necessary functionality; so do programmable graphing calculators. In addition, access to simple symbolic operations (e.g., formal differentiation, Taylor series expansion) is desirable, but not strictly necessary.

Symbol Manipulation and Hand Computation

Although we emphasize concepts and use technology freely, we by no means ignore symbol manipulations in general, or hand computations in particular. We take the symbolic points of view of our title fully as seriously as the graphical and the numerical. We cover, for instance, the usual techniques for formal differentiation and antidifferentiation. Why do we do so?

That machines can do calculus manipulations does not render by-hand operations obsolete. Some manipulative practice and skill builds and supports conceptual understanding. Hand computation illustrates concepts concretely, builds "symbol sense" (the algebraic counterpart of numerical intuition) and an ability to estimate, and gives students a sense of mastery. It does *not* follow, though, that harder, more baroque computational problems are necessarily better or more useful; we deemphasize them. Both research results and our own experience suggest that diverting some time and attention to concepts does little, if anything, to reduce students' hand computational facility.

Distinctive Features of the Text

Our text differs significantly from standard treatments. Here are some of these differences, together with some of our assumptions, goals, and strategies.

Combining Graphical, Numerical, and Algebraic Viewpoints   Throughout the text we insist that students manipulate and compare graphical, numerical, and algebraic representations of mathematical objects. In studying functions, for example, students manipulate not only elementary functions but also functions presented graphically and tabularly. In the context of formal differentiation, exercises ask students to apply the chain rule to combinations of functions presented in various ways. Graphical and numerical techniques, with error estimates, complement algebraic antidifferentiation methods. For series, routine convergence tests are emphasized less than finding—and defending—numerical limit estimates.

Concepts vs. Rigor   Proving theorems in full generality is less valuable, we think, than helping students understand concretely what theorems say, why they're reasonable, and why they matter. Too often, fully rigorous proofs address questions that students are unprepared to ask.

Still, we believe that introducing calculus students to the idea of proof—and to some especially important classical proofs—is essential. We prove major results, but emphasize only those that we believe contribute significantly to understanding calculus concepts. In examples and problems, too, we pay attention to developing analytic skills and synthesizing mathematical ideas.

Exercises   However clear its exposition, a textbook's problems generate most of students' mathematical activity and occupy most of their time. Through the problems we assign, we tell students concretely what we think they should know and do.

Routine drills and challenging theoretical problems are standard in calculus texts; ours includes many of the former and some of the latter. More distinctive, perhaps, are problems that fall between these poles:

- Problems that combine and compare algebraic, graphical, and numerical viewpoints and techniques.
- Problems that require "translation" among various representations and interpretations of calculus ideas (e.g., to interpret derivative information in terms of either slope or rate of change).
- Problems that use calculus as a *language* for interpreting and solving problems. Students are asked to translate problems into mathematical terms, solve these problems using the tools of calculus, and reinterpret mathematical results in the context of the original problem.

"Basic" vs. "Further" Exercises   For instructors' convenience, exercises are classified as either "Basic Exercises" or "Further Exercises." The former are, typically, (1) relatively straightforward; (2) conceptual in nature; and (3) focused on some important, basic principle. More routine exercises (e.g., symbol manipulation problems) are usually *not* named as "basic"—even though instructors should assign some of them. For more details and advice on assigning exercises, see the Instructor's Notes.

Strategies for Better Problem Solving   Emphasizing problem solving is nowadays *de rigueur* in calculus textbooks; ours is no exception. What, concretely, does such an emphasis entail in content and strategy?

Conceptual understanding, we believe, is the weakest link in students' ability to solve nontrivial problems. Thus, *the goal of better problem-solving skills is implicit in the goal of deeper conceptual understanding.* To that end:

- We use numerical and graphical methods both to improve students' understanding of concepts and to enlarge their kit of tools to solve problems. Students, we find, are surprisingly quick to master and apply elementary numerical and graphical methods, even error estimation.

- When technology is available, students are freed, indeed forced, to analyze the structure of a problem and plan a solution strategy. We emphasize general problem-solving strategies (reduction to more tractable subproblems, estimation, search for patterns, etc.) explicitly wherever we can.

- We provide a greater qualitative variety of exercises including problems that are posed more generally, problems that call for more synthesis, and problems that rely on a larger set of solution techniques. In this richer environment, we hope students will come to see mathematics as an open-ended, creative activity, not a rigid collection of recipes.

Teaching from the Text; Notes on Pedagogy

Our approach to the subject is non-traditional in some respects, e.g., in emphasizing graphs and geometry and in the nature of its exercises. Most teachers will find that using such a text successfully calls for some corresponding changes and adaptations in pedagogy. Many strategies have been used, including group homework assignments, takehome exams, writing assignments, computer labs, "gateway" tests for proficiency in differentiation and integration, and many others. What combination, if any, works best will depend on instructors' individual preferences and practical logistics.

No particular combination is ideal for everyone, but we believe that almost all instructors will benefit from looking closely at our *Notes for Instructors*, which is available either in physical form from Saunders College Publishing or electronically, in PostScript format, from our World Wide Web site: `http://www.stolaf.edu/ people/zorn/ozcalc`. The *Notes for Instructors* includes such features as these:

- Overviews of philosophy, strategies, and use of technology.

- Extensive pedagogical notes and suggestions, treating such topics as reading and writing mathematics, exercises, testing, pace and coverage, suggestions for use of classroom time, etc.

- Sample lesson plans.

- Section-by-section notes and comments, with suggested "basic" homework exercises from each section.

Supplements to the Text

The following aids for students and instructors are available from the publisher:

Student Answer Booklets.   These booklets, prepared by the authors, are available in two volumes, either separately or shrinkwrapped with the text. They contain answers to selected exercises.

Student Solutions Manuals.   Prepared by the authors, these manuals are available in two volumes, and contain complete solutions to all odd-numbered exercises.

Lecture Guide and Student Notes.   Prepared by Stephen M. Kokoska of Bloomsburg University of Pennsylvania, available in two volumes, these notes offer effective teaching aids that allow instructors and students to use identical material in class. Instructors can use these key examples, exercises, definitions, and theorems to prepare overhead transparencies. Ample room on each page allows students to add notes about the discussion, additional examples, and justification. Thus, instructors can prepare classroom materials with ease and students spend class time in discussion and group work rather than taking notes.

Calculus and Graphing Calculators.   Written by Joe May of North Hennepin Community College and Glen Van Brummelen of the King's University College. Part I introduces basic commands and features of the TI-82 and TI-85 graphing calculators, with numerous examples and exercises. Exploratory activities in Part II build on various applications and historical perspectives.

Calculus Explorations.   Three volumes are available: *Calculus Explorations Using Mathematica*®, by Allen Hibbard (Central College); *Calculus Explorations Using Maple*®, by Phoebe Judson (Trinity College); and *Calculus Explorations Using Derive*®, by Mary Kay Abbey (Montgomery College–Takoma Park).

These three technology supplements introduce students to a computer algebra system through meaningful calculus explorations. As students complete each lab activity, they learn the basic commands and discover (or reinforce) key (single-variable) calculus concepts through the numerous examples. The questions and examples in each lab encourage students to interpret, explain, and communicate their results. Each supplement offers over 20 lab activities to give instructors variety; the activities are ideal for group work or in-class discussion. Each lab begins with clearly stated goals; progressive questions help students check their understanding as they work. No previous knowledge of *Mathematica*, *Maple*, or *Derive* is required.

Instructors who adopt this text may also receive, free of charge, the following items.

Instructor's Resource Manual   The manual offers a variety of teaching tips and suggestions. Part I includes (i) advice on using technology, group work, homework, and testing, etc.; (ii) sample "lecture" plans; and (iii) section-by-section notes. Part II contains essays from several class-testers about their experiences. Part III includes several writing assignments.

Instructor's Solutions Manual   Available in two volumes, the manual contains complete solutions, prepared by the authors, for all exercises. (Answers to selected exercises are also available in the Student Answer Booklets.)

Test Bank   Prepared by Scott Inch of Bloomsburg University of Pennsylvania, this resource contains approximately 700 test questions, covering multiple viewpoints and levels of difficulty. Answers to all questions are included.

EXAMASTER+™ **Computerized Test Bank**   Available in Windows, Macintosh, and IBM formats, this resource combines the items of the Test Bank with a convenient features that allow instructors to prepare quizzes and examinations quickly and easily. Instructors can edit questions, add their own, administer tests over a computer network, and record student grades with the gradebook software.

Overhead Transparencies and Masters   These are available in two volumes and contain selected figures from the exposition and exercises. A selection of important figures is already printed on acetate, for direct viewing in the classroom. Additional useful figures are given on the masters, from which instructors may prepare their own transparencies.

Saunders College Publishing may provide complimentary instructional aids and supplements or supplement packages to those adopters qualified under our adoption policy. Please contact your sales representative for more information. If as an adopter or potential user you receive supplements you do not need, please return them to your sales representative or send them to

> Attn: Returns Department
> Troy Warehouse
> 465 South Lincoln Drive
> Troy, MO 63379

Advice from You

We appreciate hearing instructors' comments, suggestions, and advice on this edition. Many of the suggestions received from users of earlier versions of this text have been incorporated in this version. Our physical and e-mail addresses are below.

Arnold Ostebee and Paul Zorn
Department of Mathematics
St. Olaf College
1520 St. Olaf Avenue
Northfield, Minnesota 55057-1098

e-mail: ostebee@stolaf.edu   zorn@stolaf.edu

Acknowledgments

This text owes its existence to (literally) countless professors, students, publishing company professionals, friends, advisors, critics, "competitors," family, and others. (These categories are not mutually exclusive!) It is a pleasure to acknowledge by name some—but, necessarily, only some—of the people who attended this text through its long gestation and birth.

Focus group participants, who offered useful early advice, include Wade Ellis, Jr., West Valley College; Gregory D. Foley, Sam Houston State College; Richard O. Hill, Jr., Michigan State University; Roger B. Nelsen, Lewis and Clark College; Patricia Roecklein, Montgomery College; Wayne Roberts, Macalester College; and Audrey Rose, Tulsa Junior College.

Various versions of the manuscript were meticulously reviewed, and many errors caught, by Mary Kay Abbey, Montgomery College–Takoma Park; Janet Andersen, Hope College; Michael J. Bonanno, Suffolk Community College–Ammerman Campus; Matthew Brahm, University of Minnesota; Susanna Epp, DePaul University; Stephen Kuhn, University of Tennessee–Chattanooga; Rebecca Lee, Bowie State University; Herbert A. Medina, Loyola Marymount University; Edward S. Miller, Lewis–Clark State College; Jeffrey Ondich, Carleton College; John Polking, Rice University; John C. Peterson, Chattanooga State Technical Community College; Karen Saxe, Macalester College; Marsha Schoonover, Chattanooga State Technical Community College; Anita E. Solow, Grinnell College; and Larry Thomas, University of Virginia.

Accuracy of examples and exercises was carefully checked during production of preliminary and first editions by Eric Bibelnieks, University of Minnesota; Tracy Bibelnieks, University of Minnesota; Dan Kemp, South Dakota State University; Norm Loomer, Ripon College; Lewis Lum, University of Portland; and Kerry Wyckoff, Brigham Young University. All surviving errors are, of course, our own.

The Test Bank was checked for accuracy by Susan E. Fettes, SUNY–Oswego; and Art Richert, Southern College of Seventh-Day Adventists.

The overhead transparencies and masters were chosen by Tracy Bibelnieks, University of Minnesota; Peter Collins, Chattanooga State Technical Community College; and Jane Serbousek, Northern Virginia Community College–Loudoun Campus.

We are indebted, for many and different reasons, to members of Saunders College Publishing's professional staff, including Nancy Lubars, Senior Project Editor; Alicia Jackson, Production Manager; Caroline McGowan, Art Director; Nick Agnew, Product Manager; and Liz Wilchacky, Marketing Coordinator. To Alexa Epstein, Senior Associate Editor, our most special thanks are due, for her tireless support (extending even to attending several week-long summer workshops), unflagging encouragement, practical help, concrete advice, common sense, generous friendship, and (from time to time) exquisitely gentle prodding. Alexa's help has been simply indispensable and we thank her very much.

We have learned much from professional colleagues, here and around the country. Some are our teachers; some are departmental colleagues; some course-tested preliminary versions of these volumes at their own schools; some reviewed our grant proposals; some reviewed our project; some invited us to review theirs; some invited us to speak on panels or at workshops; some spoke on panels we organized; some helped us run our workshops; some are "calculus reform competitors"; some are advisers; some are critics. We list a small sample below, in alphabetical order. Regardless of category, we thank them all. Janet Andersen, Tracy Bibelnieks, Judith Cederberg, Caspar Curjel, Tom Dick, Ed Dubinsky, Wade Ellis, Bob Foote, Bonnie Gold, Michael Henle, Deborah Hughes Hallett, Paul Humke, Zaven Karian, John Kenelly, Harvey Keynes, Steve Kuhn, Richard Mercer, Lang Moore, John Peterson, Art Richert, Wayne Roberts, Don Small, David Smith, Keith Stroyan, Tom Tucker, Jerry Uhl, and Ted Vessey.

We thank St. Olaf College in general, and our departmental colleagues in particular, for their generous advice, support, good humor, and (when needed) forbearance during the many years of this project's development. Students (here and at many other institutions where preliminary versions of this book were course-tested) offered generous and useful advice.

We are grateful, too, for grants from the National Science Foundation (Grants CSI-8650912 and USE-9053363) and from the Fund for Improvement for Postsecondary Education (Grant G008642079). This support, extending over several years, has been essential to completing this project.

Our families, finally, deserve our deepest thanks. They have coped cheerfully with peculiar hours, extended absences, mental distraction, blizzards of paper, missed meals, and every other vagary that such a project entails. Without their love and sacrifice, we would never have begun—let alone completed—this project.

How to Use This Book: Notes for Students

All authors want their books to be *used*: read, studied, thought about, puzzled over, reread, underlined, disputed, understood, and, ultimately, enjoyed. So do we.

That might go without saying for some books—beach novels, user manuals, field guides, etc.—but it may need repeating for a calculus textbook. We know as teachers (and remember as students) that mathematics textbooks are too often read *backwards*: faced with Exercise 231(b) on page 1638, we've all shuffled backwards through the pages in search of something similar. (Very often, moreover, our searches were rewarded.)

A calculus textbook isn't a novel. It's a peculiar hybrid of encyclopedia, dictionary, atlas, anthology, daily newspaper, shop manual, *and* novel—not exactly light reading, but essential reading nevertheless. Ideally, a calculus book should be read in *all* directions: left to right, top to bottom, back to front, and even front to back. That's a tall order. Here are some suggestions for coping with it.

Read the Narrative   Each section's narrative is designed to be read from beginning to end. The examples, in particular, are supposed to illustrate ideas and make them concrete—not just serve as templates for homework exercises.

Read the Examples   Examples are, if anything, more important than theorems, remarks, and other "talk." We use examples both to show already-familiar calculus ideas "in action," and to set the stage for new ideas.

Read the Pictures   We're serious about the "graphical points of view" mentioned in our title. The pictures in this book are not "illustrations" or "decorations." Pictures like this one—

Local but not global maxima and minima:
$$f(x) = x/2 + \sin x$$

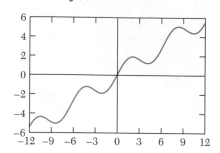

are everywhere, even in the middle of sentences. That's intentional: graphs are an important part of the language of calculus. An ability to think "pictorially"—as well as symbolically and numerically—about mathematical ideas may be the most important benefit calculus can offer.

Often they're put in margin notes, like this one.

Read with a Calculator and Pencil   This book is full of requests ◄ to check a calculation, sketch a graph, or "convince yourself" that something makes sense. Take these "requests" seriously. Mastering mathematical ideas takes more than reading; it takes doing, drawing, and thinking.

Read the Language   Mathematics is not a "natural language" like English or French, but it has its own vocabulary and usage rules. Calculus, especially, relies on careful use of technical language. Words like **rate**, **amount**, **concave**, **stationary point**, and **root** have precise, agreed-upon mathematical meanings. Understanding such words goes a long way toward understanding the mathematics they convey; misunderstanding the words leads inevitably to confusion. Whenever in doubt, consult the index.

Read the Appendices   The human appendix generally lies unnoticed—unless trouble starts, when it's taken out and thrown away. Don't treat *our* appendices that way. Though perhaps slightly enlarged, they're full of healthy matter: reviews of precalculus topics, help with "story problems," proofs of various kinds, even a graphical "atlas" of functions. Used as directed the appendices will help appreciably in digesting the material.

Read the Instructors' Preface (If You Like)   Get a jump on your teacher.

In short: *Read the book.*

A Last Note

Why study calculus at all? There are plenty of good practical and "educational" reasons: because it's good for applications; because higher mathematics requires it; because it's good mental training; because other majors require it; because jobs require it. Here's another reason to study calculus: because calculus is among our species' deepest, richest, farthest-reaching, and most beautiful intellectual achievements. We hope this book will help you see it in that spirit.

A Last Request

Last, a request. We sincerely appreciate—and take very seriously—students' opinions, suggestions, and advice on this book. We invite you to offer your advice, either through your teacher or by writing us directly. Our addresses appear below.

Arnold Ostebee and Paul Zorn
Department of Mathematics
St. Olaf College
1520 St. Olaf Avenue
Northfield, Minnesota 55057-1098

Contents

CALCULUS

From Graphical, Numerical, and Symbolic Points of View

Volume I

Functions in Calculus

1.1 Functions, Calculus Style

Calculus is a branch of mathematical analysis—the study of functions. Functions describe relationships among varying quantities; calculus helps us discover and quantify these relationships. But we're getting ahead of ourselves.

Let's start from the beginning. What is a function? What kinds of functions are there? What can functions do? Informally speaking,

> A *function is a procedure* ▶ *for assigning a unique output to any acceptable input.*

I.e., a recipe.

Functions With Formulas

The most familiar functions are given by explicit algebraic formulas, such as these:

$$f(x) = x^3 - 4x + 3; \qquad g(x) = \frac{x^2 - 4}{x + 3} .$$

Functions like f and g are easy to write down, easy to plug numbers into, easy to draw graphs of, and (as we'll see later) easy to differentiate and antidifferentiate. ▶

We'll define these technical words soon.

Functions With or Without Formulas: A Sampler

Not every useful function comes equipped with, or even has, a convenient algebraic formula. Working with functions that are described in many ways—by graphs, by tables, in words, and so on—is an important theme of this course. In this section we sample some of the variety of possible functions and their uses.

EXAMPLE 1 Following are world population figures, in millions, over 20 centuries. [The data are from the Population Division, United Nations Secretariat.]

Human Population													
Year (C.E.)	0	1000	1500	1750	1800	1850	1900	1930	1940	1950	1960	1970	1980	1990
Number of people (millions)	300	310	500	790	980	1260	1650	2070	2300	2520	3020	3700	4450	5300

The table shows what we expect: Population grows faster and faster over time. *Plotting* the data shows this even more clearly:

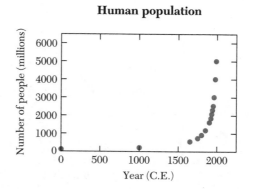

The preceding graph is a **scatter plot**, with one point per datum. Notice carefully how each point is plotted and how each axis is used.

Times are inputs; populations are outputs.

World population is certainly a function; corresponding to any time is a certain population. ◄   But does the population function have some convenient formula? An answer would be well worth having! ◄

What good would an explicit population formula be?

Later we'll describe ways of choosing mathematical formulas that "fit" observed data with reasonable accuracy. The following picture, for instance, is the graph of a mathematical function p, with formula

$$p(t) = 906e^{0.008(t-1800)}$$

superimposed on the data. The function appears to fit the data reasonably well.

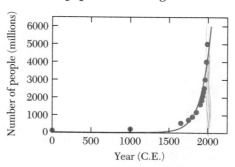

How and why we chose this particular formula for p doesn't matter here. ▶ The various constants—906, e, 0.008, 1800—involved in the formula are called **parameters**. Choosing parameter values appropriately is part of the process called **curve fitting**. ■

We'll return to such matters at the end of this chapter.

EXAMPLE 2 For any real number t, let $l(t)$ be the outside temperature, in degrees Celsius, t hours after midnight, January 1, 1995, in Northfield, Minnesota. This rule defines a function: To an input t (a time), l assigns an output $l(t)$ (a temperature). We expect l to be continuous ▶ because temperature varies smoothly, not by leaps and bounds. However, we don't expect l to be given *exactly* by a simple algebraic formula that is valid for all time. ▶ Nevertheless, we might reasonably try to *estimate* l in some approximate or average sense. ■

We'll define "continuous" rigorously later; here it's understood informally.

With such a formula we could predict the weather with perfect certainty.

No Formulas? Functions like l, without explicit formulas, may seem strange or artificial. In one sense the opposite is true: Most real-world functions are like the weather—messy, unpredictable, and imperfectly understood. The tame, well-understood functions we use to model natural phenomena are, from this point of view, comparatively artificial. The real surprise is how successfully the basic mathematical functions model, and even predict, real-world phenomena.

EXAMPLE 3 Information about a function n is given in the following table.

x	-3	-2	-1	0	1	2	3	4
$n(x)$	-4	-9	-1	0	1	8	3	-2

Our knowledge of n is, to say the least, incomplete. What's $n(100)$? For what inputs is n defined at all? Is the graph of n a smooth curve? Does n have some secret formula?

The table alone doesn't answer these questions. (Incomplete information is common in practice, e.g., for experimental data.) More information would help. If, say, the table records the outside temperature x hours after midnight on January 1, 1995, then (as in the preceding example) we'd expect $n(x)$ to be continuous and to accept noninteger inputs. To "model" the situation, we might look for an explicit function that agrees with the data in the table. The complicated-looking polynomial function

$$p(x) = -\frac{152}{105}x + \frac{5}{72}x^2 + \frac{683}{240}x^3 - \frac{11}{144}x^4 - \frac{33}{80}x^5 + \frac{1}{144}x^6 + \frac{1}{70}x^7$$

is one possibility; it happens to produce the values in the table. There's no guarantee, however, that $p(x)$ gives correct temperature information for any other values of x. (Experimentation quickly shows that $p(x)$ can't possibly tell temperature for all time. For instance, $p(-10) = -98,251$—an unlikely temperature even in Minnesota winter.) ■

EXAMPLE 4   Let m be the function shown graphically as follows.

Some values of m, such as $m(-1) = -2$, are clear even without a formula. Other values of m are less clear; $m(-2)$, for example, is *about* -1.7, but the exact value cannot be determined from the graph.

As it happens, m can be represented algebraically, by a **piecewise-defined formula**. Because its graph consists of three line segments and a semicircle, the pieces are as follows: ◄

Check our work.

$$m(x) = \begin{cases} -\dfrac{3x}{2} - \dfrac{9}{2} & \text{if } -5 \leq x \leq -3; \\ -\sqrt{4 - (x+1)^2} & \text{if } -3 < x \leq 1; \\ x - 1 & \text{if } 1 < x \leq 3; \\ -x + 5 & \text{if } 3 < x \leq 5. \end{cases}$$

With the formula we can now find $m(-2)$ exactly:

$$m(-2) = -\sqrt{4 - (-2 + 1)^2} = -\sqrt{3} \approx -1.732050808. \qquad ■$$

Patched-together Functions.   Functions like m, defined "in pieces," may seem strange, but they are very useful in practice. Computer-drawn graphs, for example, are usually made up of many short line segments, or *arcs*—each of which can be described algebraically. The process of **splining** (joining properly chosen curves or line segments to form a single smooth curve) is based on elementary calculus ideas. Splines are used in computer-assisted design, engineering, and other applications.

The horizontal axis indicates time, not distance, so the graph is not a "picture" of the balloon's flight. Or could it be?

EXAMPLE 5   Hot days on the Kansas prairie generate strong updrafts and downdrafts. In the gondola of a hot air balloon, the altimeter needle swings. The function A (its graph follows) describes the altimeter readings over a certain period of time. Note the units: Time t is measured in minutes and altitude $A(t)$ in *hundreds* of feet above sea level. (East-central Kansas lies about 960 feet above sea level.) ◄

A balloon over the prairie: How high?

Reading Velocity from the Altitude Graph. The graph of A tells a lot, directly, about altitude—when the balloon was highest, when it rose, when it fell, and so on. Consider another, equally interesting (for the passengers, probably more interesting) question:

How fast was the balloon rising or falling at time t?

In other words, how does the balloon's **vertical velocity** $V(t)$ vary with time? What can the altitude graph tell us about the vertical velocity? A lot:

Up or Down? The sign of $V(t)$ tells whether the balloon is rising or falling at time t. Altitude increases as the balloon rises, so $V(t) > 0$ when the balloon is rising (e.g., in the interval $4 < t < 5$), and $V(t) < 0$ when the balloon is falling (e.g., if $2.5 < t < 3.5$).

How Fast? At $t = 1.5$, altitude increases quickly relative to time; the balloon is soaring upward. By $t = 2.4$, the balloon has leveled off. Some ups and downs follow, but at $t = 7.8$ things look bad. Altitude is plunging, and so is the balloon.

Estimating Velocity The vertical velocity $V(t)$ clearly depends on how steeply the A-graph rises or falls at time t. At $t = 6$, for instance, the A-graph is dropping, but how fast? How, in other words, might we estimate a value for $V(6)$? One approach is to focus on a *small* interval containing $t = 6$, such as $5.5 \le t \le 6.5$. A close look at the graph shows that

$$A(5.5) \approx 27 \qquad \text{and} \qquad A(6.5) \approx 17.$$

Therefore, over the 1-minute interval $5.5 \le t \le 6.5$, the balloon drops about 10 vertical units, or 1000 feet, and the **average velocity** over this interval is around -1000 feet per minute. Since the graph looks nearly straight over this short interval, the velocity at $t = 6$ should not be much different.

> *At the instant $t = 6$, the velocity is about -1000 feet per minute. In symbols:* $V(6) \approx -10.$ ▶

What Slope Means The vertical velocity $V(t)$ has an important graphical meaning: At any time t, $V(t)$ is the slope of the A-graph at time t. ▶ We just estimated that $V(6) \approx -10$. The graphical version of the same strategy runs like this: On a small interval near $t = 6$, the graph is almost straight. To estimate slope, therefore, any two nearby points should do.

The unit of velocity is hundreds of feet per minute.

Mathematical legalities arise in defining the slope of a curved graph. We'll consider them carefully later; here we can trust our intuition.

The line through $(5.5, 27)$ and $(6.5, 17)$, for instance, has slope

$$\frac{17 - 27}{6.5 - 5.5} = -10,$$

which agrees with our earlier estimate.

Plotting V We can amass enough information to plot V by estimating the velocity $V(t)$ for *many* values of t. Doing so by hand would be tedious; we used a computer.

A balloon over the prairie: How fast?

There's much to see in the new graph, but we'll limit ourselves to a final observation: At $t = 8$, things look bleak. ∎

Translation: Functions from Words

Mathematics is a language. Because real-world problems usually come phrased in "natural" languages—say French or English—solving such problems must start with translation into the special vocabulary and grammar of mathematics. From paragraphs of ordinary prose come functions, variables, equations, operations, and the like.

Translation from English to French is not mere word-for-word replacement. An excellent translation respects French grammar, syntax, and stylistic conventions as well as vocabulary. A well-translated text captures the essence of the original.

Natural languages are suited to expressing nuance, emotional tone, and even ambiguity. Mathematical language is more straightforward. It emphasizes simplicity and precision, even in matters of depth and subtlety. Nevertheless, the point just made about translation applies even more strongly to mathematical language: To translate is to express the *essence*, not merely the form, of the situation. Distilling the essence of a problem and expressing it mathematically are the first—and often the largest—steps toward a solution.

Calculus: The Language of Change. The areas of mathematics that best describe and predict distinct sorts of phenomena differ. Phenomena of change are the special province of calculus. Calculus offers both language with which to *describe* changing quantities and rules with which to *predict* their behavior. Functions are the basic objects of calculus, so in this subject translation usually means translation into functional language.

EXAMPLE 6   To make a flimsy box from an ordinary $8.5'' \times 11''$ sheet of paper, cut squares from the corners and fold up the sides, as shown. (Cut on the heavy lines and fold on the dashed lines.) Which such box has the largest volume?

Boxmaking

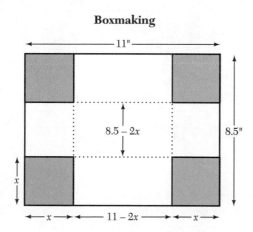

Solution   The volume V of the box depends only on the edge length—denoted by x—of the removed corners. In other words, V is a *function* of x. For any given x, V is easy to calculate algebraically:

$$V(x) = \text{length} \times \text{width} \times \text{depth} = (11 - 2x)(8.5 - 2x)x.$$

Notice that the formula makes sense in context only if all three factors are nonnegative—i.e., if $0 \le x \le 4.25$. In functional language, our problem is to find, among all feasible inputs x, the one that produces the largest possible output $V(x)$.

A graph ► shows how V varies with x:

Notice our choice of horizontal range.

Graph of $V(x) = (11 - 2x)(8.5 - 2x)x$: how volume varies

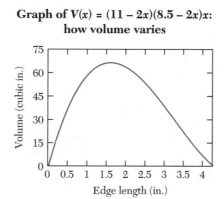

The maximum possible volume, apparently, is about 65 cubic inches; it occurs when $x \approx 1.6$ inches, i.e., for a box with dimensions $7.8'' \times 5.3'' \times 1.6''$.

By "zooming in" on the graph near $x = 1.6$, we can improve our estimates:

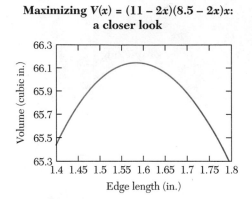

**Maximizing V(x) = (11 − 2x)(8.5 − 2x)x:
a closer look**

This sharper point of view suggests a maximum volume of about 66.1 cubic inches, achieved when $x \approx 1.59$ inches. (The dimensions then become $7.82'' \times 5.32'' \times 1.59''$.) With calculus methods, we can (and will) do better still. ◄   ■

Just a bit better; for many practical purposes the results just found would be adequate.

New Functions from Old

Starting with "old" functions and building new ones is common throughout calculus. Here's a first example.

> **EXAMPLE 7**   Start with the straight-line function $\ell(x) = x + 2$. Define a new function D, based on ℓ, as follows:
>
> *For any number x, let $D(x)$ be the distance from the origin $(0, 0)$ to the point $\bigl(x, \ell(x)\bigr)$ on the line ℓ.*
>
> Find a formula for $D(x)$; draw the graph.

Appendix A has details on the distance formula.

> **Solution**   Points on ℓ have the form $\bigl(x, \ell(x)\bigr) = (x, x+2)$. By the distance formula for points in the plane, ◄ $D(x)$ has the formula
>
> $$D(x) = \sqrt{(x - 0)^2 + (x + 2 - 0)^2} = \sqrt{x^2 + x^2 + 4x + 4} = \sqrt{2x^2 + 4x + 4}.$$
>
> Plotting $y = D(x)$ produces this graph:

Graph of $y = D(x)$

Notice the low spot on the graph of D; it seems to fall somewhere near $x = -1$. This low spot corresponds to the point on ℓ closest to the origin. ▶ ■ *See the exercises at the end of this section for more details.*

Functions Approximating Other Functions

Sometimes it's useful to approximate one function with another.

EXAMPLE 8 (**A polynomial approximation to the square root function**) The polynomial function

$$p(x) = \frac{5}{16} + \frac{15}{16}x - \frac{5}{16}x^2 + \frac{1}{16}x^3$$

turns out to approximate the square root function $f(x) = \sqrt{x}$ closely near $x = 1$. Graphs show what this means:

Approximating $\sqrt{x}$ with a polynomial

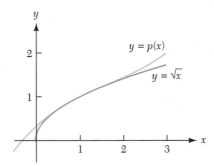

Near $x = 1$ the graphs are almost indistinguishable; elsewhere they pull apart. ▶ *For $x < 0$, $f(x)$ is not even defined, so approximation is no longer an issue.*
For example,

$$f(1.5) = \sqrt{1.5} \approx 1.22474 \quad \text{and} \quad p(1.5) \approx 1.22656$$

but

$$f(3) = \sqrt{3} \approx 1.73205 \quad \text{and} \quad p(3) = 2\,.$$

How we *chose* the polynomial p to approximate f must remain (for now) mysterious. Later in this book we'll learn to construct polynomials that approximate a wide variety of nonpolynomial functions. ■

> **Modern Times.** Years ago students learned paper-and-pencil techniques for computing square roots. Today we use calculators. Still the question remains: How does a calculator compute roots, and to what accuracy? As an experiment, try to find a systematic way of estimating square roots with a calculator without using the square root key.

BASICS

1. (a) What does the approximate population function p given in Example 1 (page 2) predict for the year 2000?
    (b) What does it "predict" for the year 1000?
    (c) What does it "predict" for 1000 B.C.E.? (B.C.E, formerly known as B.C., means "before the Common Era.")
    (d) Are these predictions reasonable? Explain.

2. Use the algebraic representation of m in Example 4 (page 4) to compute the following:
    (a) $m(-4)$ (c) $m(2.3)$
    (b) $m(0)$ (d) $m(\pi)$

3. Find a piecewise-defined formula for the function whose graph follows.

4. Let $\ell(x) = 2x + 1$.
    (a) Draw a graph of $\ell(x)$ for $0 \le x \le 5$.
    (b) Let $D(x)$ be the distance from the origin to the point $\left(x, \ell(x)\right)$. Find a formula for $D(x)$.
    (c) Sketch a graph of $D(x)$ for $0 \le x \le 5$.

5. For any $t \ge 0$, let $A(t)$ be the area of the rectangle bounded above by the line $y = 3$, below by the x-axis, on the left by the y-axis, and on the right by the line $x = t$.
    (a) Write a formula for $A(t)$.
    (b) Draw a graph of $A(t)$ for $0 \le t \le 5$.

6. Let f be the function defined by $f(x) = x^2$. Define the function m as follows: For any $x \ne 0$, $m(x)$ is the slope of the line through $(0, 0)$ and $\left(x, f(x)\right)$. For instance, $m(10) = 10$, the slope of the line through $(0, 0)$ and $(10, 100)$.
    (a) Find $m(-1)$, $m(-2)$, $m(1)$, $m(2)$, and $m(500)$.
    (b) Write a formula for $m(x)$ in terms of x.
    (c) Draw a graph of $m(x)$ for $-5 \le x \le 5$.

FURTHER EXERCISES

7. (a) At what time was the balloon in Example 5 (page 4) rising fastest? About how fast was it rising then?
    (b) At what time was the balloon in Example 5 falling fastest? About how fast was it falling then?
    (c) What happens at time $t = 8$?

8. Another possible way of depicting a balloon's flight graphically is to plot altitude vs. *horizontal position* instead of altitude vs. time. Since westerly (i.e., west-to-east) winds prevail on the American prairie, a balloon above Kansas often travels (ignoring vertical movement) due east. When this is so, it's natural to plot altitude against horizontal position, measuring the latter in miles east of a starting point. The resulting plot can be thought of as a "picture" of the balloon's flight.

    In this exercise we consider the relationship, if any, between these two types of graphs: altitude vs. time, as used in Example 5, and altitude vs. position, just discussed. In particular, can the two types of graphs possibly look the same?

    (a) Suppose that a *steady*, 60-mph westerly wind blows. Draw an altitude vs. position graph. How are the two graphs related?
    (b) Suppose that no horizontal wind blows. Draw a "picture" of the balloon's flight (i.e., an altitude-position graph). Do both types of graphs make sense?

    (c) What if an easterly wind blows? Draw a possible altitude-position graph. Does it look like an altitude-time graph?

9. When you drive an Ace Rental compact car x miles in a day, the company charges $f(x)$ dollars, where

    $$f(x) = \begin{cases} 30 & \text{if } 0 \le x \le 100; \\ 30 + 0.07(x - 100) & \text{if } x > 100. \end{cases}$$

    Describe Ace Rental's pricing policy in plain English. (Be sure to interpret the constants 30, 0.07, and 100 that appear in the pricing formula.)

Find a piecewise-defined formula for each function in Exercises 10 and 11; graphs are shown below. [**NOTE:** Each graph is made up of line segments and circular arcs.]

10.

11.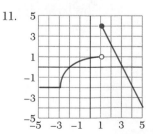

12. This problem refers to the functions ℓ and D in Example 7 (page 8).

    (a) Use a ruler to plot the straight-line function $\ell(x) = x + 2$ over the interval $-3 \le x \le 3$. Label the points $\big(3, \ell(3)\big)$ and $\big(-3, \ell(-3)\big)$ on your graph. (Use units of the same size—e.g., 1 cm—on both axes.)

    (b) For any x, $D(x)$ is the distance from the origin to the point $\big(x, \ell(x)\big)$. Use your graph and a ruler to estimate $D(3)$ and $D(-3)$. Then, for comparison, calculate $D(3)$ and $D(-3)$ using the formula.

    (c) Which point P on the line ℓ appears to be closest to the origin? Mark this point P with both its coordinates. How far is P from the origin? Relate your answer to the low point on the graph in Example 7.

    (d) We can use algebra to find point P exactly. Here's how. Completing the square under the radical sign gives $D(x) = \sqrt{2x^2 + 4x + 4} = \sqrt{2(x+1)^2 + 2}$. Use this fact to explain why $D(-1) = \sqrt{2}$ is the smallest possible value of D.

13. For any $t \ge 0$, let $A(t)$ be the area of the right triangle bounded above by the line $y = x$, below by the x-axis, and on the right by the line $x = t$.

    (a) Write a formula for $A(t)$.

    (b) Draw a graph of $A(t)$ for $0 \le t \le 5$.

14. Let $\ell(x) = 2x + 1$, and let $A(t)$ be the area of the trapezoid bounded by the x-axis, the y-axis, ℓ, and the line $x = t$.

    (a) Write a formula for $A(t)$.

    (b) Draw a graph of $A(t)$ for $0 \le t \le 5$.

15. Let f be the function defined by $f(x) = x^2$. Define the function g by this rule: $g(x)$ is the distance from $(0, 0)$ to $\big(x, f(x)\big)$. For instance, $g(10) = \sqrt{10{,}100}$, the distance from $(0, 0)$ to $(10, 100)$.

    (a) Find $g(0)$, $g(-1)$, $g(1)$, $g(2)$, and $g(500)$.

    (b) Write a formula for $g(x)$.

    (c) Draw a graph of $g(x)$ for $-5 \le x \le 5$.

16. Let f be the function defined by $f(x) = x^2$. Define the function j as follows: For any $x \ne 1$, $j(x)$ is the slope of the secant line through $(1, 1)$ and $\big(x, f(x)\big)$. For instance, $j(10) = 11$, the slope of the line through $(1, 1)$ and $(10, 100)$.

    (a) Find $j(2)$, $j(1.1)$, $j(1.01)$, $j(0.99)$, $j(0.9)$, and $j(0)$.

    (b) Write a formula for $j(x)$.

    (c) Draw a graph of $j(x)$ for $-2 \le x \le 2$.

17. Let f be the function defined by $f(x) = x^2$. Define the function k for any $x \ne 0$ by

    $$k(x) = \frac{f(2 + x) - f(2 - x)}{2x}.$$

    For instance,

    $$k(0.1) = \frac{f(2.1) - f(1.9)}{0.2} = \frac{2.1^2 - 1.9^2}{0.2} = 4.$$

    (a) Find $k(-1)$, $k(-0.1)$, $k(0.1)$, $k(1)$, and $k(1000)$.

    (b) Write a formula for $k(x)$.

    (c) Draw a graph of $k(x)$ for $-5 \le x \le 5$.

18. Let f be the function $f(x) = x^2 + 3x - 4$ and ℓ the line that intersects the graph of f at the points $x = 2$ and $x = 2 + h$.

    (a) Find an equation of the line that intersects f at $x = -1$ and $x = 2$.

    (b) Find an algebraic expression for the slope of ℓ. [**NOTE:** The answer depends on h.]

    (c) Use your answer to part (b) to find an equation of the line ℓ.

    (d) Use your answer to part (c) to find an equation of the line that intersects f at $x = 2$ and $x = 2.003$.

    (e) Use your answer to part (c) to find an equation of the line that intersects the graph of f at $x = 2$ and $x = -1$.

19. Example 8 (page 9) describes one way to approximate the square root function, using a polynomial function. This problem is about a **piecewise-linear** approximation (i.e., a piecewise-defined function built from straight lines) of the square root function.

    (a) Draw a graph of $f(x) = \sqrt{x}$ for $0 \le x \le 9$. Mark the points $(0, 0)$, $(1, 1)$, $(4, 2)$, and $(9, 3)$ on your graph, then connect them with three line segments.

    (b) Let $g(x)$ be the function whose graph consists of the three line segments in part (a). Find a piecewise-defined formula for g.

    (c) Fill in the following table (round your answers to three decimal places).

x	0	0.5	0.9	3	4	6	8	8.9	8.99
$g(x)$									
$\sqrt{x}$									

20. Here's a highly simplified version of the U.S. federal income tax system. A taxpayer pays no federal tax on the first \$15,000 of annual income, 15% on the the next \$25,000 of income, and 28% on all earnings above \$40,000. Let x denote a taxpayer's annual income (in dollars), and let $T(x)$ denote the tax owed (in dollars) at income level x.

    (a) The tax function T is piecewise-linear—its graph consists of straight lines. Use this fact to draw the graph of $T(x)$ for $0 \le x \le 60{,}000$. [**HINT:** First calculate $T(x)$ for some convenient values of x.]

    (b) Give a piecewise-defined formula for $T(x)$.

    (c) A taxpayer earning, say, \$50,123 yearly would have to pay 28 cents more tax if he or she earned *one more dollar*. A taxpayer earning \$20,123 a year would need to pay only 15 cents tax on the next dollar of income.

In tax jargon, the first taxpayer has a **marginal tax rate** of 28%, and the second has a **marginal tax rate** of 15%. Let $MR(x)$ denote the marginal tax rate (as a decimal) at income level x. Give a piecewise-defined formula for $MR(x)$.

(d) Plot MR as a function of x for $0 \le x \le 60{,}000$. Describe in words how the graphs of T and MR are related.

21. (Exercise 20 continued.) Let $F(x)$ denote the fraction (written as a decimal) of a taxpayer's total annual income paid in federal tax. For instance, a taxpayer with income $40{,}000$ pays 3750, so $F(40{,}000) = 3750/40{,}000 = 0.09375$—just over 9%.

(a) For all $x > 0$, $F(x) = T(x)/x$. Explain why.

(b) Write a piecewise-defined formula for $F(x)$. [**HINT**: Start with the piecewise-defined formula for $T(x)$.]

(c) Plot $F(x)$ for $0 < x \le 60{,}000$. [**HINT**: A calculator may help you plot the separate pieces.]

(d) Could any taxpayer pay 25% of total income in taxes? Why or why not?

(e) Could any taxpayer pay 28% of total income in taxes? Why or why not?

22. An open box is made by cutting squares w inches on a side from the four corners of a sheet of cardboard that is 24 inches by 32 inches, and then folding up the sides.

(a) Express the volume of the box as a function of w. [**HINT**: Draw a picture.]

(b) Estimate the value of w that maximizes the volume of the box.

23. Coffee mugs can be manufactured for 60 cents each. At a price of 1 dollar each, 1000 mugs can be sold. For each penny by which the price is lowered, 50 more mugs can be sold. Let N be the number of coffee mugs manufactured (and sold!), and let x be the price reduction (in pennies).

(a) Explain why $N(x) = 1000 + 50x$.

(b) Explain why the profit on each mug sold is $(1 - 0.01x) - 0.60$ dollars.

(c) Find a function that relates total profit and x.

(d) What's the *best* price? [**HINT**: Graph the function from part (c).]

1.2 Graphs

Every equation in x and y has a **graph**—the set of ordered pairs (x, y) of real numbers that satisfy the equation. For example, the graph of any equation of the form

$$(x - a)^2 + (y - b)^2 = r^2$$

is a circle (with center (a, b) and radius r). Associating an equation (an algebraic object) to a graph (a geometric object) is a powerful tool for understanding both.

Graphs can be simple, complicated, or downright bizarre. Most graphs in calculus are of the everyday, $y = f(x)$ variety; almost every page of this book contains an example. Following, for a change, are some odder graphs.

EXAMPLE 1   Following are the graphs of three equations: $x^2 + y^2 = -1$, $x^2 + y^2 = 0$, and $x^2 + y^2 = 1$.

Is there some mistake? No; *no* pair (x, y) of real numbers satisfies the first equation, so the first graph is empty. Only the origin $(0, 0)$ satisfies the second equation, so its graph is a single point. The third graph plays no tricks. ■

Graphs of Functions

The function $f(x) = x^2$ corresponds in the obvious way to the equation $y = x^2$ and hence to the familiar parabolic curve. In the same way, every function f corresponds to an equation $y = f(x)$ and therefore to a graph. ▶

> **Definition** The graph of a function f is the set of points (x, y) that satisfy the equation $y = f(x)$.

In other words:

The graph of a function is the set of all points of the form $(x, f(x))$.

The generic picture looks like this:

Not every equation (and therefore not every graph) corresponds to a function. Circles are obvious counterexamples: On a circle, most values of x correspond to two values of y.

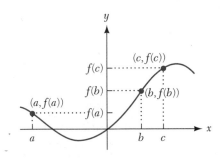

The graph of a function shows geometrically how the function varies over a range of inputs. Because calculus concerns precisely this question—how quantities vary with respect to each other—graphs play crucial roles throughout the subject.

Drawing the graph of a function given by an explicit formula, such as $p(x) = x^3 - 2x^2 + x - 2$, ▶ is an old, standard, and still important problem in elementary calculus. But other problems are just as important, especially now that machines draw graphs easily and accurately. ▶ Most important is to understand clearly what graphs say about functions and how they say it.

Drawing graphs is sometimes useful, but more interesting is the connection between analytic properties of a function (e.g., whether it's increasing or decreasing) and geometric properties of its graph (e.g., whether it's rising or falling). Calculus tools unlock many such secrets.

A graph of this function follows.

Not every machine-drawn graph is accurate. More on this subject in the next section.

Graphs—With and Without Formulas

Too much time and effort spent drawing graphs of "nice" functions with "nice" formulas can obscure an important fact:

Not every function has a nice formula—or any formula at all.

In real life, convenient formulas are the exception. In a sense, having a formula amounts to knowing *everything* about a function. In this imperfect world, little is that certain.

EXAMPLE 2   (**Function, formula, and graph**)   In the best possible circumstances we are given a function, its graph, and an explicit algebraic formula. Here, for instance, is a view of the graph of $p(x) = x^3 - 2x^2 + x - 2$. ◄

Technically speaking, we see only part of the graph. The full graph is an infinite object.

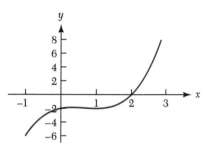

Graph of $y = p(x)$

Observe:

A y-intercept The graph intercepts the y-axis at height -2. In symbols, $p(0) = -2$.

No breaks The graph shown is free of "jumps" or "breaks"; it could be drawn manually, without lifting the pencil. In mathematical language, the function p is **continuous** on the domain shown. ◄

Later, we'll define continuity carefully.

An x-intercept The graph's x-intercept is at $x = 2$, so $p(2) = 0$. Factoring the formula shows the same thing:

$$p(x) = x^3 - 2x^2 + x - 2 = (x - 2)(x^2 + 1).$$

(The factored form tells us something more: $p(x) = 0$ *only* if $x = 2$.)

The graph rises as we trace it from left to right.

Where Is p Increasing? The graph of p *rises* from $x = 1$ to $x = 3$. ◄   In calculus language, the function p is said to be **increasing** for $1 < x < 3$. The picture ends at $x = 3$, so we can't tell from the graph alone what happens for larger values of x. It can be seen from the factored algebraic form of p, however, that $p(x)$ increases for *all* $x > 1$. (Soon we'll develop calculus methods to show the same thing more conveniently.)   ■

EXAMPLE 3   (**A bare graph: no formula, no context**)   Following is a portion of the graph of a garden-variety calculus function f. No formula for f is given. Two interesting points are labeled.

Graph of $y = f(x)$

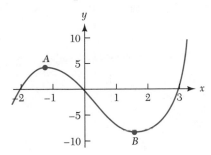

What can the graph tell us about the function f? What *can't* it say? Some important samples follow. Notice especially the words that appear in boldface; they'll reappear often.

- Because the point $(0, 0)$ lies on the graph, $f(0) = 0$. (Strictly speaking, we should say $f(0) \approx 0$, since graphical data are always approximate. Here we'll trust our eyes.)

- Judging from the graph, $f(1) \approx -6$. To know whether $f(1) = -6$ *exactly*, we would need more information.

- What's $f(10)$? The graph doesn't say. We might *guess* from the graph's general shape that $f(10)$ is some large positive number. Yet, from the graph alone, we can't be certain even that $f(10)$ exists.

- The points A and B correspond to **local maximum** and **local minimum values** ▶ of the function f. The coordinates of A—approximately $(-1.2, 4)$—imply that $f(-1.2) \approx 4$. Similarly, B's coordinates imply that $f(1.8) \approx -8$. Putting these data together, we would say (allowing for approximation errors) that f is *increasing* for $-2.5 < x < -1.2$, *decreasing* for $-1.2 < x < 1.8$, and *increasing* again for $1.8 < x < 3.5$.

  Informally, these are high and low spots on the graph. We'll define these terms formally in the next chapter.

- For which values of x is $f(x) < -5$? One approach is to draw the horizontal line $y = -5$. ▶ This line hits the graph of f at three different points—near $x = -2.4$, $x = 0.7$, and $x = 2.5$. Therefore, we would estimate that $f(x) < -5$ if either $x < -2.4$ or $0.7 < x < 2.5$.

  Do so now, with a ruler.

  Two cautions are necessary. First, every number mentioned is only approximate. Second, we have no idea how $f(x)$ behaves for x outside the interval shown.

- Later we'll define and study the slightly subtler idea of a graph's **concavity**. Concavity is a geometric property. The bowl of a spoon (in its usual position) is **concave up**; an umbrella (on a rainy day) is **concave down**. ▶ Clearly, the f-graph is concave down at point A and concave up at B. Somewhere between A and B, therefore, must be an **inflection point**, ▶ where the direction of concavity switches. Where does this happen?

  As rough synonyms, try concave up for "holds water" and concave down for "spills water."

  A close look shows that concavity changes somewhere between $x = 0$ and $x = 1$, but it's hard to be more precise. Soon, using calculus ideas, we'll see how to use the formula for p to locate inflection points precisely. ■

  Another term we'll define more precisely later.

EXAMPLE 4   **(Graphs in context)**   [This example was drawn from *Calculus in Context*, a project of the Five Colleges consortium.]   In certain disease epidemics, the affected population divides naturally into three groups: those *susceptible*, those *infected*, and those *recovered*. The epidemic's progress can be charted by observing how these populations vary over time. The following three graphs, labeled *S*, *I*, and *R*, show how the susceptible, infected, and recovered populations might vary over a 25-day period during a measles epidemic in a large school district.

The course of an epidemic

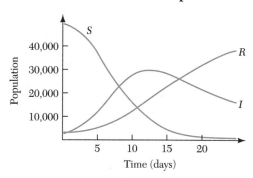

Try to answer them for yourself before reading the answers.

With these graphs, we can answer many questions about the epidemic. ◄

1. *How many children were in each population on day 10?*

   Reading the height of each graph on day 10 shows (allowing for estimation errors):

   $$\text{Susceptible} \approx 13,000;$$
   $$\text{Infected} \approx 26,000;$$
   $$\text{Recovered} \approx 11,000.$$

2. *On which day is the infected population largest?*

   The *I*-graph, and hence the infected population, peaks around day 12. (In calculus language, *I* assumes a local maximum value at day 12.)

3. *When is the epidemic at its worst?*

   The answer depends on how we measure "badness." The number *I* of ill people is one reasonable measure. By this criterion the worst day is around day 12.

   Another possible measure of "badness" is the number of people who fall ill on a given day. In the situation at hand, to fall ill means to leave the susceptible population. From this point of view, the worst day occurs when the susceptible population drops most rapidly. A careful look at the *S*-graph suggests that this happens around day 7, where the graph points most steeply downward. (In calculus language, the day in question corresponds to an inflection point of the *S*-graph.) ■

EXAMPLE 5 (**Graphs from tabular data**) Suppose we know only what this table tells about a function n: ▶

The entries should look familiar—see Section 1.1.

x	-3	-2	-1	0	1	2	3	4
$n(x)$	-4	-9	-1	0	1	8	3	-2

How should we plot n? One problem is that many different functions agree with n (i.e., have the same values) for $x = -3, -2, \ldots, 4$. In graphical language, many different possible graphs pass through the eight data points.

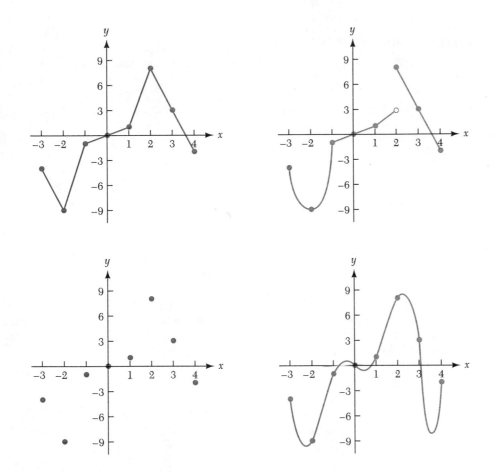

Is one of these graphs "right"? Without more information, we can't say. If we knew, say, that the function n represented outdoor temperatures, hour by hour, on a cold winter night, we might choose one as being more likely than the others. ▶

Notice, in particular, the **discontinuity** in the graph at upper right. (The hollow dot indicates a "jump" at $x = 2$.) Physical experience says that air temperature varies continuously over time, so this graph seems unlikely to tell temperature. ▪

Which one would you choose?

New Functions from Old: Operations with Constants

Creating new functions—hence new graphs—from old is a recurring theme in calculus. There are many ways to "operate on" old functions to form new ones. The simplest such operations involve only **constant parameters** attached in various ways to an original function. From the ordinary sine function, for instance, we can produce mutants such as

$$m(x) = 42 + 31 \sin\left(\frac{2\pi}{52}x\right).$$

In Section 1.6 we'll discuss a real-life application of this function. Our present goal is more general: to understand how constant parameters affect *all* functions and their graphs.

Constant Parameters

Given an "old" function f and a constant a, we can create such new functions as

$$f(x) + a, \qquad f(x + a), \qquad f(ax), \qquad \text{and} \qquad af(x).$$

How do these new functions differ from plain old f? How is the particular value of a involved?

Additive Constants: Vertical and Horizontal Translations

Given a function f and a constant a, consider the new functions g and h defined by

$$g(x) = f(x) + a; \qquad h(x) = f(x + a).$$

If $a = 0$, then $f = g = h$.

Read this sentence again; it's important.

In each definition, a is an **additive constant**. ◄   Though similar in appearance, g and h are quite different: We form g from f by adding a to *outputs* of f; we form h by adding a to *inputs* of f. ◄

Here are graphs of g for several values of a:

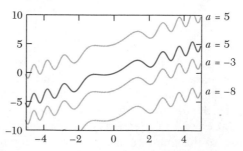

Vertical translations:
$y = f(x) + a$ for various values of a

The graphs show how the value of a affects g:

The graph of $y = f(x) + a$ is the result of shifting the f-graph a units upward. ▶

If $a < 0$, the shift is downward.

In mathematics, moving an object without changing its shape is called **translation**; thus, a's effect is a **vertical translation**.

The relationship between f and h is different. Compare these graphs:

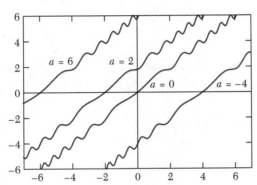

Horizontal translations:
$y = f(x + a)$ **for various values of** a

This time a causes a **horizontal translation**:

The graph of $y = f(x + a)$ is the result of shifting the f-graph a units to the left.

Left, Not Right? Yes, Left. It might be surprising that the a in $f(x + a)$ moves the graph of f left, not right. To see why, consider the function $h(x) = f(x + 2)$. Then $h(-2) = f(0)$, $h(-1) = f(1)$, $h(0) = f(2)$, and so on. In effect, h has a two-unit head start on f: Whatever h does, f does two units later.

Multiplicative Constants: Stretching, Compressing, and Reflecting

Multiplicative parameters also make new functions from old. As before, we start with a function f and a constant a, and form two new functions:

$$g(x) = af(x); \qquad h(x) = f(ax).$$

Notice that g results from multiplying f's *outputs* by a; h comes from multiplying f's *inputs* by a. ▶

If $a = 1$, then $f = g = h$.

Vertical Stretching and Reflection. Consider these graphs carefully:

Vertical stretching:
$y = af(x)$ for various values of a

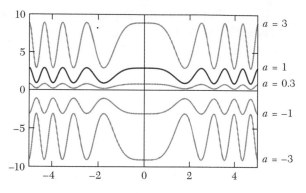

The graphs show the effect of a on $g(x) = af(x)$: Multiplying outputs from f by a causes **vertical stretching** by a factor of $|a|$ and, if $a < 0$, **reflection** about the x-axis.

Stretching the Language? Our use of such terms as "stretch," "compress," and "reflect" deserves some comment.

- Stretching by a factor $a > 1$ is simplest. The graph of $y = 3f(x)$ (labeled $a = 3$ in the preceding figure), for instance, results from *tripling* the y-coordinate of each point on the f-graph.
- If $0 \le a < 1$ (the case $a = 0.3$ is shown) stretching by a factor of a actually *decreases* vertical distances, flattening the f-graph toward the x-axis. In such situations we'll sometimes speak of **compressing** graphs.
- A negative multiplicative factor adds a reflection (about the x-axis) to the picture. The picture shows what happens for $a = -1$ and $a = -3$. In the simpler case, the graphs of f and $-f$ are mirror images. The graph of $-3f$, by comparison, requires *both* a reflection and a vertical stretch, with factor 3, starting from f.

Horizontal Stretching and Reflection. The function $h(x) = f(ax)$ is formed by multiplying f's inputs by a. We just saw that multiplying *outputs* by a stretches the f-graph vertically. For similar reasons, multiplying *inputs* by a stretches or compresses the graph horizontally. The following graphs show h for several values of a.

$y = f(x)$; $a = 1$

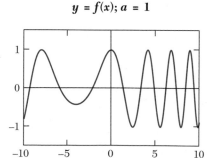

$y = f(0.5x)$; $a = 0.5$

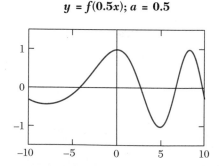

$y = f(2x); a = 2$ $y = f(-x); a = -1$

          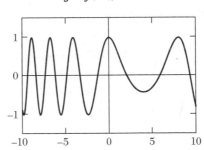

Notice the similarities to—but also the differences from—vertical stretching:

- The graphs of $f(x)$, $f(2x)$, and $f(0.5x)$ show that if $a > 0$, the graph of $f(ax)$ is the result of compressing the f-graph horizontally by a factor of a (or, alternatively, stretching it by a factor of $1/a$). ▶

- A negative value of a causes a reflection—this time, about the y-axis, The graph of $f(-2x)$, for instance, results from both compressing by a factor of 2 and reflecting about the y-axis.

Did you expect things to go the other way? If so, try this: If x measures time, then $f(x)$ is a quantity that varies with time. From this point of view, the parameter $a = 2$ in $h(x) = f(2x)$ makes $h(x)$ vary twice as fast. Anything f does, h does in half the time.

BASICS

1. The graph of a function f follows. Use the graph to complete the following table. (Estimate as necessary.)

	x	$f(x)$
(a)	-2	
(b)	0	
(c)	1	
(d)		-1
(e)		2

2. Use the following sketch to answer each question.

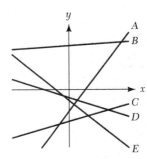

(a) Which lines have positive slope? Which have negative slope?

(b) Let m_A be the slope of line A, m_B be the slope of line B, and so on. Rank the numbers $m_A, \ldots, m_F$ in increasing order.

3. The graph of a function f follows. Several points of interest are labeled.

Graph of f

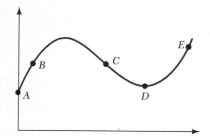

(a) Between which pairs of adjacent labeled points is f increasing? Decreasing?
(b) At which labeled point does f have an inflection point?
(c) Between which pairs of adjacent labeled points is f concave up?
(d) At which labeled point does f have a local minimum value?
(e) Between which pair of adjacent labeled points does f have a local maximum value?

Use the following graph of the function g to sketch a graph of each of the functions in Exercises 4–11.

Graph of g

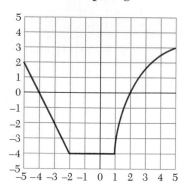

4. $f(x) = g(x) + 1$ 8. $f(x) = -g(x)$
5. $f(x) = g(x + 1)$ 9. $f(x) = g(-x)$
6. $f(x) = g(2x)$ 10. $f(x) = -2g(-x)$
7. $f(x) = 2g(x)$ 11. $f(x) = 3g(x - 2) + 1$

FURTHER EXERCISES

12. Let (x_1, y_1) and (x_2, y_2) be points on the line $y = -2(x - 1) + 3$. If $x_1 < x_2$, what can be said about y_1 and y_2? How do you know?

13. Let (x_1, y_1) and (x_2, y_2) be points on the line $y = 3(x + 1) - 2$. If $y_1 < y_2$, what can be said about x_1 and x_2? How do you know?

14. Let L_1 be the line through $(0, 0)$ and $(2, 1)$, and let L_2 be the line that passes through $(0, 0)$ and $(3, 7)$. For which values of m does the line $y = mx$ lie between the lines L_1 and L_2?

15. Suppose that the graph of a function g lies between a line A, with slope 2, and a line B, with slope -1, for all $x > 2$, as follows:

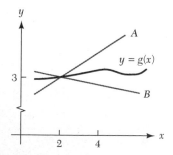

(a) Find an equation for line A.

(b) What is the height of line B at $x = 3$?
(c) Find numbers L and U such that $L < g(4) < U$.

Use the following graph of f for Exercises 16–24.   [**NOTE:** There may be more than one possible answer to each exercise.]

Graph of f

16. Find an interval over which $-10 \le f(x) \le 20$.
17. Is $f(x) \le 20$ if $-2 \le x \le -1$?
18. Find a number L such that $L \le f(x)$ if $-2 \le x \le 1$.
19. Find a number U such that $f(x) \le U$ if $-2 \le x \le 1$.
20. Find a number U such that $|f(x)| \le U$ if $-2 \le x \le 1$.
21. Find numbers L and U such that $L \le f(x) \le U$ if $2 \le x \le 3$.

22. Find a number U such that $|f(x)| \leq U$ if $2 \leq x \leq 3$.
23. Find an interval over which $|f(x)| \leq 30$.
24. Find an interval over which $|f(x) + 10| \leq 20$.
25. Let $g(x) = \dfrac{x}{x^2 + 4}$. Find an interval on which $0.1 < g(x) < 0.3$.
26. Let $h(x) = 2x^3 - 4x^2 + 5x - 6$. Find an interval on which $|h(x) - 5| < 10$.
27. Let $f(x) = 1/(1 + x^2)$.
    (a) Does f have a maximum value? If so, find it. If not, explain why not.
    (b) Does f have a minimum value? If so, find it. If not, explain why not.

Suppose that $-5 \leq f(x) < 11$ when $-3 < x \leq 8$. In Exercises 28–35, indicate whether each statement *must* be true, *might* be true, or *cannot* be true. Justify each answer with a sentence or a sketch.

28. $f(x) > -6$ if $-3 < x \leq 8$
29. $f(x) \leq 100$ if $-3 < x \leq 8$
30. $|f(x)| \leq 53$ if $|x| < 2$
31. $f(2) = 11$
32. $f(-3) = 11$
33. $|f(0)| = 5$
34. $f(13) = -6$
35. $-4 < f(x) \leq 9$ if $-3 < x \leq 8$
36. Examine this sketch of a hemispherical water tank.

— 10 ft —

Water depth

Let the letter D represent the depth of the water in the tank, and let V be the volume of the water in the tank. Regard volume as a function of water depth.
(a) What is $V(0)$?
(b) What is $V(10)$?
(c) Which of the graphs A–D could be the graph of $V(D)$? Explain your reasoning.

Graph A

Graph B

Graph C

Graph D

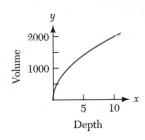

37. (a) Sketch a graph of temperature as a function of time that is consistent with the scenario described in the following sentence. "The child's temperature has been rising for the last 2 hours, but not as rapidly since we gave her the antibiotic an hour ago." (Let $t = 0$ correspond to the time when the sentence was written.)
    (b) What is the significance of the fact that the graph in part (a) is increasing on the interval $t = -2$ to $t = 0$?
    (c) What is the significance of the fact that the graph in part (a) is concave down on the interval $t = -1$ to $t = 0$?

In Exercises 38–40, sketch the graph of a function described by each of the following sentences. Briefly explain why your graph has the shape you sketched.

38. The cost of a new car continues to increase at an increasing rate.
39. The population is growing more slowly now than it was 5 years ago.
40. Upper Midwest Industries' profit growth is slowing.
41. The graph of a function f follows.

Graph of f

(a) Is f increasing on the interval $(-3, 1)$? Explain.
(b) Estimate the endpoints of the largest interval over which f is concave down.
(c) Estimate the x-value at which f has a local minimum value.

42. Sketch the graph of a function f that has *all* of these properties:
    (i) f is continuous on the interval $[-5, 5]$;
    (ii) f is concave up on $[-5, -1]$;
    (iii) f has a local minimum at $x = -2$;
    (iv) f has a local maximum at $x = 3$.
43. Can a function have more than one local maximum value? If not, why not? If so, give a graphical example.
44. The following figure shows the line $y = mx + b$.

    Sketch the following lines on a copy of the figure. (Be sure to identify each line.)
    (a) $y = mx - b$ (c) $y = -mx - b$
    (b) $y = -mx + b$
45. Let $g(x) = |3x + 2|$. Explain how to obtain the graph of g from the graph of the absolute-value function $f(x) = |x|$ by stretching, compressing, translating, and so on.
46. Let $f(x) = x^2$. For each of the following, (i) give a formula for $g(x)$, (ii) describe how to obtain a graph of g from a graph of f using translation, compression, stretching, and so on, and (iii) sketch a graph of g.
    (a) $g(x) = f(x) + 2$ (d) $g(x) = f(2x)$
    (b) $g(x) = f(x + 2)$ (e) $g(x) = f(-2x)$
    (c) $g(x) = 2f(x)$ (f) $g(x) = -2f(x)$
47. Let $f(x) = x^2$ and $g(x) = x^2 + 4x + 3$.
    (a) Complete the square to show that
        $g(x) = f(x + 2) - 1$.
    (b) Explain how to obtain the graph of g from the graph of f.
48. Let $f(x) = x^2$ and $g(x) = 2x^2 - 4x + 5$.
    (a) Complete the square to show that
        $g(x) = 2f(x - 1) + 3$.
    (b) Explain how the graphs of f and g are related.
49. Suppose that f is a function such that the slope of the secant line through the points $(2, f(2))$ and $(8, f(8))$ is -4.
    (a) Let $g(x) = f(x) + 3$. What is the slope of the secant line through the points $(2, g(2))$ and $(8, g(8))$? Justify your answer. [**HINT**: How are the graphs of f and g related?]
    (b) Let $h(x) = f(x + 3)$. What is the slope of the secant line through the points $(-1, h(-1))$ and $(5, h(5))$? Justify your answer. [**HINT**: How are the graphs of f and h related?]
    (c) Let $j(x) = f(2x)$. What is the slope of the secant

line through the points $(1, j(1))$ and $(4, j(4))$? Justify your answer.
    (d) Let $k(x) = f(-x)$. What is the slope of the secant line through the points $(-8, k(-8))$ and $(-2, k(-2))$? Justify your answer.
    (e) Let $m(x) = 5 - 3f(x)$. What is the slope of the secant line through the points $(2, m(2))$ and $(8, m(8))$? Justify your answer.
50. Suppose that g is a function such that the slope of the secant line through the points $(-3, g(-3))$ and $(5, g(5))$ is 4.
    (a) If $g(x) = f(x - 2)$, what is the slope of the secant line through the points $(-5, f(-5))$ and $(3, f(3))$? Justify your answer.
    (b) If $g(x) = f(2x)$, what is the slope of the secant line through the points $(-6, f(-6))$ and $(10, f(10))$? Justify your answer.
    (c) If $g(x) = 2f(x)$, what is the slope of the secant line through the points $(-3, f(-3))$ and $(5, f(5))$? Justify your answer.
    (d) If $g(x) = 4f(2x + 3) + 1$, what is the slope of the secant line through the points $(-3, f(-3))$ and $(1, f(1))$? Justify your answer.
51. Suppose that the minimum value of a function f is -7 and its maximum value is 3.
    (a) What is the maximum value of $g(x) = f(x) + 2$?
    (b) What is the minimum value of $g(x) = 2f(x) + 3$?
    (c) What is the maximum value of
        $g(x) = -3f(x + 2) + 5$?
    (d) What is the maximum value of $g(x) = |f(x)|$?
52. Suppose that the minimum value of f occurs at $x = 4$, the maximum value of f occurs at $x = -3$, and $-5 < f(x) < 2$ if $0 \le x \le 6$.
    (a) Where does the minimum value of $g(x) = 3f(x - 1) + 2$ occur? The maximum value?
    (b) Let g be the function defined in part (a). Find numbers a and b such that $-13 < g(x) < 8$ if $a \le x \le b$.
53. Suppose that the minimum value of f occurs at $x = 2$, the maximum value of f occurs at $x = 5$, and $-1 < f(x) < 3$ if $-4 \le x \le 6$.
    (a) Where does the minimum value of $g(x) = -3f(x) + 2$ occur? The maximum value?
    (b) Let g be the function defined in part (a). Find numbers L and U such that $L < g(x) < U$ if $-4 \le x \le 6$.
54. Suppose that the minimum value of f occurs at $x = -3$, the maximum value of f occurs at $x = 5$, and $-6 < f(x) < 4$ if $0 \le x \le 7$.
    (a) Where does the minimum value of $g(x) = 4f(2x + 1) - 3$ occur? The maximum value?
    (b) Let g be the function defined in part (a). Find a number U such that $|g(x)| < U$ if $1 \le x \le 15$.

55. Suppose that the minimum value of f occurs at $x = 4$, the maximum value of f occurs at $x = -5$, and $-1 < f(x) < 3$ if $-3 \le x \le 6$.
    (a) Where does the minimum value of $g(x) = 2f(4 - 3x) + 1$ occur? The maximum value?
    (b) Let g be the function defined in part (a). Find numbers a and b such that $-5 < g(x) < 3$ if $a \le x \le b$.

Suppose that f is increasing on the interval $(1, 10)$. In Exercises 56–69, indicate whether each statement *must* be true, *may* be true, or *cannot* be true. Justify each answer with a sentence or a sketch.

56. $f(8) < f(3)$.
57. $f(5) > 0$.
58. f is concave down on the interval $(1, 10)$.
59. $g(x) = f(x) - 25$ is increasing on the interval $(3, 7)$.
60. $g(x) = f(2x)$ is increasing on the interval $(1, 4)$.
61. $g(x) = 5f(x)$ is decreasing on the interval $(1, 2)$.
62. $g(x) = -3f(x)$ is decreasing on the interval $(1, 3)$.
63. $g(x) = f(x + 4)$ is increasing on the interval $(-2, 2)$.
64. $g(x) = -f(x)$ is decreasing on the interval $(1, 10)$.
65. $g(x) = -f(x)$ is increasing on the interval $(-10, -1)$.
66. $g(x) = |f(x)|$ is increasing on the interval $(1, 10)$.
67. $g(x) = f(-x)$ is decreasing on the interval $(1, 10)$.
68. $g(x) = f(-x)$ is increasing on the interval $(-10, -1)$.
69. $g(x) = f(|x|)$ is decreasing on the interval $(-10, -1)$.

Suppose that f is concave down on the interval $(1, 10)$. In Exercises 70–83, indicate whether each statement *must* be true, *may* be true, or *cannot* be true. Justify each answer with a sentence or a sketch.

70. $f(8) < f(3)$.
71. $f(5) > 0$.
72. f is increasing on the interval $(1, 10)$.
73. $g(x) = f(x) - 25$ is concave down on the interval $(3, 7)$.
74. $g(x) = f(2x)$ is concave up on the interval $(1, 4)$.
75. $g(x) = 5f(x)$ is concave down on the interval $(1, 2)$.
76. $g(x) = -3f(x)$ is concave up on the interval $(1, 3)$.
77. $g(x) = f(x + 4)$ is concave down on the interval $(-2, 2)$.
78. $g(x) = -f(x)$ is concave up on the interval $(1, 10)$.
79. $g(x) = -f(x)$ is concave down on the interval $(-10, -1)$.

80. $g(x) = |f(x)|$ is concave up on the interval $(1, 10)$.
81. $g(x) = f(-x)$ is concave up on the interval $(1, 10)$.
82. $g(x) = f(-x)$ is concave down on the interval $(-10, -1)$.
83. $g(x) = f(|x|)$ is concave up on the interval $(-10, -1)$.
84. If a function f is concave up on the interval $[a, b]$, then the line segment joining the points $(a, f(a))$ and $(b, f(b))$ lies above the graph of f over this interval.
    (a) Suppose that f is concave up on the interval $[a, b]$. Explain why
    $$f(a) + \frac{f(b) - f(a)}{b - a}(x - a) > f(x)$$
    for all x such that $a < x < b$.
    (b) Show that if f is concave up on the interval $[a, b]$, then
    $$\frac{f(x) - f(a)}{x - a} < \frac{f(b) - f(a)}{b - a}$$
    for all x such that $a < x < b$. [**HINT:** Use part (a).]
    (c) Suppose that g is concave down on an interval $[a, b]$. What is the relationship between the line segment joining the points $(a, g(a))$ and $(b, g(b))$ and the graph of g? [**HINT:** If g is concave down on an interval, $-g$ is concave up on that interval.]
85. Suppose that f is concave up on the interval $[0, 10]$, that $f(1) = 4$, and that $f(5) = 2$.
    (a) Explain why $f(3) < 3$. [**HINT:** Use part (a) of Exercise 84.]
    (b) Find a number U such that $f(4) < U$. Justify your answer.
86. Suppose that g is concave down on the interval $[0, 10]$, that $g(1) = 4$, and that $g(5) = 2$.
    (a) Explain why $g(3) > 3$.
    (b) Find a number L such that $g(4) > L$. Justify your answer.
87. Suppose that f is concave up on the interval $[a, b]$. Show that if $0 < t < 1$, then $f\big((1 - t)a + tb\big) < (1 - t)f(a) + tf(b)$. [**HINT:** Use part (a) of Exercise 84 and let $t = (x - a)/(b - a)$.]

1.3 Machine Graphics

Graphics—the ability to represent mathematical objects geometrically—is the most important advantage mathematical computing offers. Without computing, graphs can be tedious, if not impossible, to draw. Computing makes graphs and the insights they offer easily accessible. In this section we illustrate some of the power—and a few of the pitfalls—of computer graphics.

Graphical computing comes in many forms. Hardware ranges from handheld calculators to supercomputers. Software varies from simple plotting programs to sophisticated graphical environments. Plotting commands may be given by means of typed instructions, calculator keystrokes, mouse clicks, or in more exotic ways.

The *Maple* command

```
plot( sin(x), x=-5..5 );
```

produces a graph something like this:

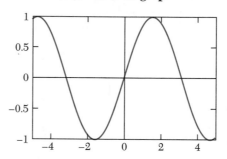

A machine graph

The *Mathematica* command

```
Plot[ Sin[x], {x,-5,5} ]
```

When plotting trigonometric functions in calculus, be sure your machine is set to radians, not degrees.

gives a similar result. (In both cases, the input variable x denotes radians, not degrees.) ◄

All such differences aside, several important ideas and terms apply across the board. Learn them well. They have to do not just with quirks of computing but with the essence of functions and graphs.

The Viewing Window

The sine curve shown earlier is incomplete. It shows what happens for $-5 \leq x \leq 5$, but the sine function undulates merrily on, in both directions, forever. Strictly speaking, we see only part of the sine graph, not the graph itself. We're not that fastidious—we'll usually say "graph," not "piece of the graph." But the point remains: Partial pictures are all we can expect. The full graph of a function may be infinitely long, so no finite picture can show all of it.

The term **viewing window** describes what *is* shown. In the preceding graph, for instance, the viewing window is the rectangle

$$\{ (x, y) \mid -5 \leq x \leq 5, \ -1 \leq y \leq 1 \}.$$

The square brackets in $[-5, 5]$ indicate a closed interval—one that contains its endpoints. See Appendix A for more on intervals and interval notation.

(In **product notation** we'd denote the same rectangle by $[-5, 5] \times [-1, 1]$.) ◄

The viewing window is the "product" of two intervals, one horizontal and one vertical. We'll call them, respectively, **xrange** and **yrange**. In computer jargon, xrange and yrange are **plotting parameters**. (The term "xdomain" might be more accurate than "xrange"; alas, it's never used.) ◄

The next section has more details on domain and range.

> **No Standard Jargon.** The names "xrange" and "yrange", although suggestive, aren't standard; there *is* no standard terminology. Other common names include "hrange," "vrange," "inputrange," and "outputrange."

Six Views of the Sine Function

The appearance of a graph depends (sometimes startlingly) on the viewing window.
Here are six views of—believe it or not—the same function:

Graph 1: $y = \sin(x)$
Window = [–30, 30] × [–5, 5]

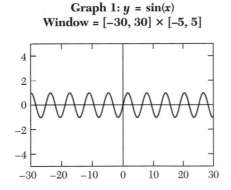

Graph 2: $y = \sin(x)$
Window = [–30, 30] × [–1, 1]

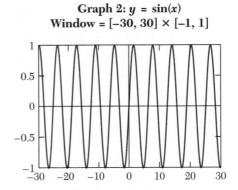

Graph 3: $y = \sin(x)$
Window = [–1, 1] × [–1, 1]

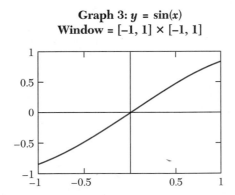

Graph 4: $y = \sin(x)$
Window = [–0.1, 0.1] × [–0.1, 0.1]

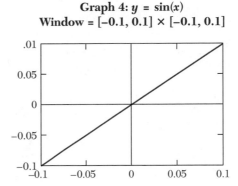

Graph 5: $y = \sin(x)$
Window = [3.13, 3.15] × [–0.1, 0.1]

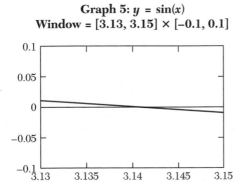

Graph 6: $y = \sin(x)$
Window = [3.13, 3.15] × [–0.01, 0.01]

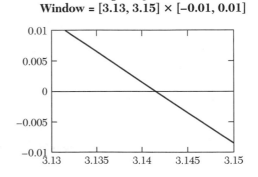

A graph's shape depends—strongly—on the viewing window. Therefore, every graphing device ▶ must somehow let the user choose the viewing window, whether with typed commands, mouse clicks, movable cross hairs, or some other device.

Calculator, computer, sharp stick, and so on.

Who Decides?   Some devices try to guess what the user wants to see. For example, in the *Maple* and *Mathematica* commands

```
plot( sin(x), x=-5..5 );
Plot[ Sin[x], {x,-5,5} ]
```

xrange is specified explicitly, but yrange is not; the programs somehow choose yrange. Leaving choices to the machine may be appropriate, especially for complicated functions. If the machine chooses poorly, the user can always intervene. Here's the bottom line: Whether specified explicitly or chosen automatically, the viewing window strongly affects the appearance of a graph.

Plotting Jargon

On paper, all six windows on the sine graph are about the same size and shape— about 1 inch by 1.75 inches. The differences reflect various choices of aspect ratio, center point, and so on. A few technical words are handy for describing such differences:

Scale: Measures the size of horizontal or vertical units on a graph. The horizontal and vertical scales on Graph 4 are ten times as large as those on Graph 3.

Zooming In: The result of *enlarging* the scale, by a **zoom factor**, in one or both axis directions. Graph 4 results from zooming in on Graph 3 with zoom factor 10 on each axis. (Horizontal and vertical zoom factors may differ.)

Zooming Out: The result of *shrinking* the scale in one or both axis directions. Graph 3 comes from zooming out from Graph 4; both zoom factors are 10.

Aspect Ratio: Measures the relative sizes of horizontal and vertical units. The viewing window $[-1, 1] \times [-3, 3]$ has aspect ratio 3:1—horizontal units are three times as large as vertical units. Aspect ratios matter especially when we plot such objects as circles, for which the relative sizes of units matter.

Sample Size; Connecting Dots

On any machine, the number of pixels limits the accuracy of a graph.

An unbroken curve has infinitely many points, but plotting devices can handle only finite data sets. ◄   Therefore, any physical graph can only approximate the "true" graph of a function. This limitation raises interesting metaphysical questions but is seldom serious for the everyday functions of elementary calculus. For most functions, 100 data points produce a passable graph. For really simple functions (lines, for example), even fewer data points suffice.

Don't try this at home. The policy applies only to software, not to hardware.

How to Lie with Computer Graphics

A good way to understand almost anything is to try to "break" it. ◄   It's fun and easy to fool even sophisticated plotting devices. Let's try.

Plotting devices use various strategies for drawing curves that look continuous. The simplest devices simply plot one data point per horizontal pixel. ► On machines with many pixels, this strategy is impractical; some way of "connecting the dots" is needed.

One simple strategy is to plot, say, 101 ► equally spaced points, join them with line segments, and hope for the best. This usually works fine, but not always. Look what happens to the sine graph in a huge viewing window:

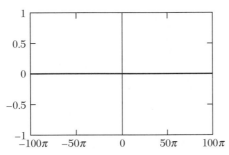

The graph is clearly wrong. ► The curve $y = \sin(x)$ actually oscillates *100 times* over the xrange shown. What went wrong?

Nothing—the plotter did as it was told. The xrange interval, $[-100\pi, 100\pi]$, was cut into 100 equal pieces, and the sine function was sampled at the end of each piece. By unlucky accident, each sample point fell at a multiple of 2π, where the sine function is zero. Our plotting sample was too small and too regularly spaced to capture the function's complicated behavior. The moral should be clear:

Distrust large windows and ill-behaved functions. ►

Luckily, problems with sample size are uncommon in calculus applications. Many "smart" plotting devices recognize sticky situations and adjust sample sizes accordingly.

Cautions

Machine graphics should be viewed skeptically. Various surprises are possible; not all are the machine's fault. Here are some pitfalls to avoid.

Too Few Data Plotting functions with irregular or ragged graphs may require many, many data points. Using smaller windows (i.e., zooming in) sometimes helps.

Awkward Windows Let $f(x) = x^{10} - x$. By simple algebra, $f(0) = 0$ and $f(1) = 0$, so the graph of f crosses the x-axis at $x = 0$ and at $x = 1$. Does the machine-drawn graph agree? Here's a look. The machine chose its own yrange. ►

One "dot" on a computer screen represents a pixel, or picture element.

Why 101? Because 100 equal subintervals have 101 endpoints.

The graph is just the x-axis. See it?

See page 352 for a truly nasty function.

The notation 1e + 07 means 1×10^7, or 10 million.

Graph of $y = x^{10} - x$
Window = $[-5, 5] \times [-10^6, 10^7]$

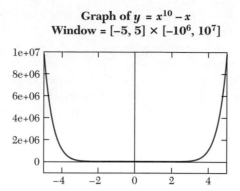

The graph shows disappointingly little near $x = 0$. The tenth power is at fault. Even for modest values of x, $f(x)$ is gigantic. To show everything, yrange must be correspondingly huge. As a result, all but the grossest behavior disappears. Here's the same function in a smaller window:

Graph of $y = x^{10} - x$
Window = $[-2, 2] \times [-10, 10]$

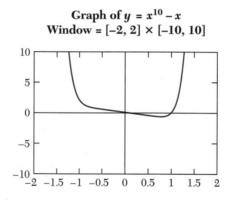

In this case, there is no really good choice of window.

Which graph is better depends—literally, in this case—on one's point of view. ◄

Scale Effects Different scales can give wildly different views of the same graph. Here are two views of the function $y = x + \sin(200x)/200$, the second a tenfold magnification of the first.

$y = x + \sin(200x)/200$

$y = x + \sin(200x)/200$

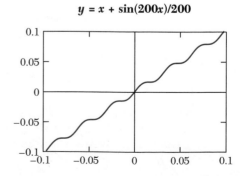

Notice the differences: The first graph looks almost straight; it's a slightly wiggly version of the line $y = x$. ▶ In the second graph, with smaller values of x, the rapid oscillations of $\sin(200x)/200$ produce a wiggly result. The moral: Viewing the same graph from different scales can show different aspects of a function's behavior.

Compared to x, the second summand, $\sin(200x)/200$, is small, so it has little effect.

Slippery Slopes Shouldn't the line $y = 10x$ look steeper than the line $y = x$? ▶ Yet the following graphs look almost identical.

The lines have slopes 10 and 1, respectively.

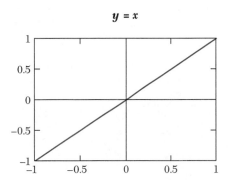

The apparent anomaly arises from different choices of yrange for the two graphs. Our plotting device chose yrange just large enough to fit the plotting data. ▶ Plotting both lines in the same window gives a very different view:

One curious effect of this policy: All lines with positive slope "look the same"; all go from lower left to upper right.

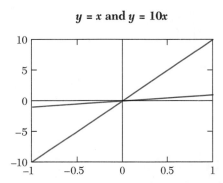

The moral: Watch vertical and horizontal scales carefully. Changing either can dramatically alter a graph's appearance.

Machine Graphs in Calculus

Graphs are crucial throughout this book. We'll draw some graphs by hand (mostly simple ones), but we'll leave most of that laborious work to machines. More important than drawing graphs is learning to read them and understand what they say about functions.

EXAMPLE 1   Find the maximum value, and where it occurs, of the function $f(x) = x \sin(x)$ on the interval $[0, 4]$.

Solution   Later we'll use a calculus concept, the derivative, to attack such problems. Here we'll simply use graphs. The ones that follow are captioned with the *Maple* commands that produce them.

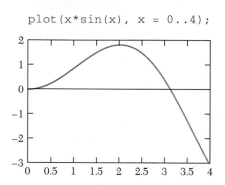

plot(x*sin(x), x = 0..4);

A maximum seems to occur around $x = 2$. Let's look more closely:

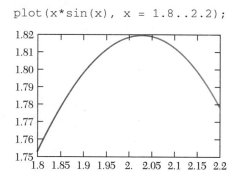

plot(x*sin(x), x = 1.8..2.2);

and still more closely:

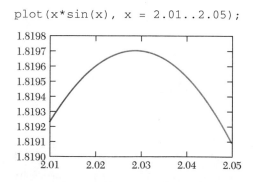

plot(x*sin(x), x = 2.01..2.05);

We conclude that the maximum value is about 1.8197; it occurs near $x = 2.03$. ■

EXAMPLE 2   For x near 0, the function $p(x) = x - x^3/6$ is a good approximation to the sine function. ▶ *How well* does p approximate the sine function on $[-1, 1]$? On $[-2, 2]$? For which values of x can we be sure that $p(x)$ differs from $\sin(x)$ by less than, say, 0.001?

Solution   First we'll compare the graphs of $p(x)$ and $\sin(x)$ ▶ for x in the interval $[-2, 2]$.

Later we'll discuss how to find simple functions, such as p, that approximate more complicated functions, such as the sine function.

Which is which? How can you tell?

Graphs of $y = \sin(x)$ and $y = x - x^3/6$

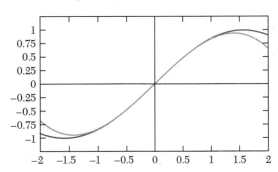

For $-1 \le x \le 1$ the approximation is good; there is little visible difference between the graphs. Things deteriorate when $|x| > 1$; by $x = \pm 2$, the graphs are far apart.

How might we *measure* how well one function approximates another on a given interval? One way is to find the largest difference, positive or negative, between the two functions on the given interval. On $[-1, 1]$, for example, we could plot the **error function** $p(x) - \sin(x)$:

Graph of $p(x) - \sin(x)$

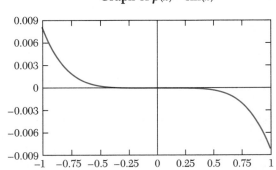

The graph shows that for x in $[-1, 1]$ the difference between the two functions is never more (in absolute value) than 0.009. In other words,

On $[-1, 1]$, $p(x)$ approximates $\sin(x)$ with error no worse than 0.009.

Not bad; for x in this interval, we could almost safely forget the $\boxed{\text{SIN}}$ button on our calculators.

The remaining question asks: *For which values of x is* $|p(x) - \sin(x)| < 0.001$? Again a graph gives the answer:

Graph of $p(x) - \sin(x)$: a closer look

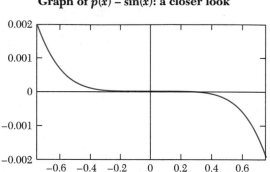

Draw the horizontal lines $y = \pm 0.001$ *to see why.*

Conclusion: $p(x)$ approximates $\sin(x)$ with error less than 0.001 if $|x| < 0.65$. ◄
■

BASICS

NOTES: *The exercises in this section require some form of plotting technology. Almost any form will do; the main requirement is some ability to choose viewing windows (e.g., by zooming). When plotting trigonometric functions, be sure to use* radians, *not degrees.*

1. In this section we discussed the fact that $\sin(x) \approx p(x) = x - x^3/6$ for x near 0. Another, simpler, approximation of the sine function says that, for x near 0, $\sin(x) \approx \ell(x) = x$. Let's pursue the latter approximation a bit further.

   (a) Complete the following table, rounding your answers to five decimal places.

x	$\sin(x)$	$\sin\lvert x - \sin(x)\rvert$	x	$\sin(x)$	$\lvert x - \sin(x)\rvert$
-1			1		
-0.5			0.5		
-0.1			0.1		
-0.01			0.01		

   (b) Earlier in this section we said, "On $[-1, 1]$, $p(x)$ approximates $\sin(x)$ with error no worse than 0.009." Does the same statement apply if $\ell(x)$ replaces $p(x)$? If not, what number could replace 0.009?

   (c) We also said: "$p(x)$ approximates $\sin(x)$ with error less than 0.001 if $-0.65 < x < 0.65$." Find an interval $(-a, a)$ on which $\ell(x)$ approximates $\sin(x)$ with error less than 0.001.

2. Use the following sketch to answer each question.

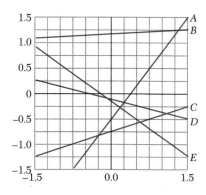

   (a) Do any lines have slope greater than 1? If so, which ones?

   (b) Where do lines A and C intersect? Justify your answer.

3. Use the following sketch to answer each question. [**HINT**: This is not the same graph as in Exercise 2.]

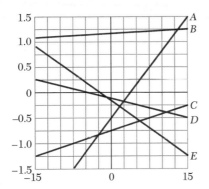

(a) Do any lines have slope greater than 1? If so, which ones?

(b) Where do lines *A* and *D* intersect? Justify your answer.

4. Use the following sketch to answer each question. [**HINT**: This is not the same graph as in Exercise 2.]

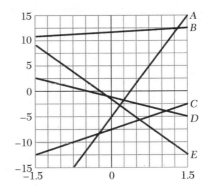

(a) Do any of the lines have slope greater than 1? If so, which ones?

(b) Where do lines *C* and *D* intersect? Justify your answer.

FURTHER EXERCISES

5. In each of the following parts, experiment to find a window in which the graph of $y = \sin(x)$ has (approximately!) the given shape or appearance. Express answers in the form $[a, b] \times [c, d]$.
   (a) A horizontal line.
   (b) A diagonal line from lower left to upper right.
   (c) A diagonal line from upper left to lower right.
   (d) A vertical line.
   (e) A rounded letter V.

6. Repeat Exercise 5, using the graph of $y = \cos x$.

7. Repeat Exercise 5, using the graph of $y = x^2$.

A number r is a **root** of a function f if $f(r) = 0$. In Exercises 8–11, use a grapher to locate the roots of each function in the interval $[-5, 5]$. (Report answers to two decimal places.)

8. $f(x) = x^2 - x$

9. $f(x) = x - \sin x$

10. $f(x) = x - \cos x$

11. $f(x) = 2^x - x^2$

12. For x near 0, the function $q(x) = 1 - x^2/2$ is a good approximation of the cosine function.
   (a) How well does q approximate the cosine function on $[-1, 1]$?
   (b) Find an interval $(-a, a)$ on which $q(x)$ approximates $\cos x$ with error less than 0.001.

Graphs of most calculus functions "look straight at small scale." (Graphs 4, 5, and 6 on page 27 illustrate what this means for the sine function.) For each of the functions in Exercises 13–18, find a window centered at the given point in which the graph of the given function looks like a straight line.

13. $y = x^2$ at $(2, 4)$

14. $y = -x^3 + 3x^2 + 2$ at $(1, 0)$

15. $y = \cos(x)$ at $(0, 1)$

16. $y = \cos(x^2)$ at $(0, 1)$

17. $y = \cos(x^2)$ at $\left(5, \cos(25)\right)$

18. $y = \cos(x^2)$ at $\left(10, \cos(100)\right)$

For each function in Exercises 19–24, use graphics to estimate the maximum and minimum values of $f(x)$ when $-10 \le x \le 10$. (Round your answers to three decimal places.)

19. $f(x) = \sin(x)$

20. $f(x) = x + \sin(x)$

21. $f(x) = \dfrac{x}{x - 11}$

22. $f(x) = x^2$

23. $f(x) = 2^x$

24. $f(x) = 2^x - x^2$

1.4 What *Is* a Function?

What exactly *is* a function? We've already seen many examples, given by formulas, by words, by graphs, and by tables. With these examples as fodder, it's time to ruminate on the *general* idea of function.

Five Examples

Any suitable definition of "function" covers a lot of ground. Functions appear in many guises; they can be described, formally or informally, in various ways. We'll refer repeatedly to five example functions at various levels of strangeness:

f $f(x) = x^2 + \dfrac{3}{x}$

g $g(t) =$ the world human population t years C.E.

h $h(u) = \begin{cases} 1 - u & \text{if } u \leq 0 \\ u^2 & \text{if } 0 < u < 2 \\ 42 & \text{if } u \geq 30 \end{cases}$

j The function given by this graph:

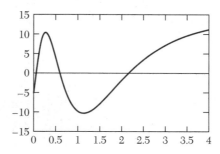

k The function given by this table:

n	−3	−2	−1	0	1	2	3	4
$k(n)$	−4	−9	−1	0	1	8	3	−2

These disparate functions have little in common except the property of *being* functions, to which we now turn.

The Definition At Last

> **Definition**   A **function** is a rule for assigning to each member of one set, called the domain, one member of another set, called the range.

The definition uses three key words: "rule," "domain," and "range." We'll discuss each separately.

Rule

A function accepts an input and—somehow—assigns an output. The **rule** describes how this assignment is made. For our example functions:

f   The rule is a simple, explicit, algebraic "recipe": *Given a real number x, compute $x^2 + 3/x$.* To 5, for example, f assigns the number $f(5) = 5^2 + 3/5 = 25.6$.

g   To a number t—say, $t = 1992.23$—g assigns the world population at exactly that time, measured in years. Thus, imaginably, $g(1992.23) = 5{,}678{,}901{,}234$.

   Admittedly there's trouble here: World population is never known exactly, and certainly no convenient formula is available. Nevertheless, g makes perfectly good sense as a function. At any time t, there is one and only one population $g(t)$—whether or not we can compute it.

   Notice, finally, that although the rule for f tells how to *calculate* outputs, the rule for g only *describes* outputs.

h   The rule for h involves "cases"; different inputs are treated differently. For example,

$$h(-5) = 1 - (-5) = 6$$

because the input -5 satisfies the first case of the definition, but

$$h(175) = 42$$

since the input 175 satisfies the third case.

   Functions defined using cases are called **piecewise-defined**, or **multiline, functions**. They arise naturally in computer graphics applications, where pieces of curves are fit together to form desired shapes. ▶

We will discuss this process, called splining, in Section 4.5. For a preview, glance at the pictures there.

j   The rule for j—like everything else about j—is graphical: $j(x)$ *is the height of the graph at horizontal position x.* Thus, $j(3) \approx 7$, $j(1) \approx -10$, and $j(10)$ is not defined.

   As with any function defined by a graph, values of j can be read only approximately. Despite this inexactness, there's plenty to learn from graphs. Approximation occurs everywhere in the real world.

k   We have only the table to use. Here is k's "rule": *For any n in the top row, k(n) is the entry below n.* Thus, $k(-2) = -9$, $k(2) = 8$, and $k(42)$ is undefined.

   Such a rule hardly seems to deserve the name. It gives no hint of *how* outputs are determined. Are we hiding some secret formula for k? No; sometimes a table contains all we know of a function. We might, with effort, find a formula or curve that "fits" tabular data, but deciding whether our formula is right would require more information. ▶

See, for instance, Example 5 in Section 1.2.

A Variable by Any Other Name.   For most of our functions, we gave names—*x, t, u, n*—to the input variables. This was for convenience; the particular names don't matter. For example, the rules

$$f(x) = x^2 + \frac{3}{x} \quad \text{and} \quad f(z) = z^2 + \frac{3}{z}$$

define exactly the same function.

Domain

A function's **domain**, denoted by $\mathcal{D}$, is its set of acceptable inputs. For our example functions:

The vertical bar inside the set brackets means "such that." For more examples of such notation, see Appendix A.

f The algebraic rule $f(x) = x^2 + 3/x$ makes good sense unless $x = 0$, so the domain of f is the set $\mathcal{D} = \{\, x \mid x \neq 0 \,\}$. In words: $\mathcal{D}$ is the set of real numbers x such that $x \neq 0$. ◄

g The rule $g(t) =$ "the world human population t years C.E." makes sense for *any* real number. Hence, the domain of g can be taken as $(-\infty, \infty)$, the full set of real numbers.

    The inputs t for which $g(t)$ can be reasonably known or estimated are more restricted. Conceding something to reality, we might arbitrarily restrict our domain to the set $[-5000, 2100]$.

h By its multiline definition, h accepts as input any real number u except those for which $2 \leq u < 30$. Therefore, the domain of h is the set $\{\, x \mid x < 2 \text{ or } x \geq 30 \,\}$ or, in interval form, $(-\infty, 2) \cup [30, \infty)$. ◄

The ∪ symbol means "union." See Appendix A for more details on set notation.

j Acceptable inputs to j lie in the interval $[0, 4]$. (In "computerese," the domain is the xrange.)

A table has only finitely many entries.

k Any function described by a table has a finite domain. ◄ The domain of k has just eight members: $\mathcal{D} = \{\, -3, -2, -1, 0, 1, 2, 3, 4 \,\}$.

"Natural" Domains. In practice, function domains are seldom stated explicitly. In this case ◄ the default is the **natural domain**—the set of inputs for which the rule makes sense. For a function such as f, given by an algebraic formula in x, the natural domain is the set of values of x for which the formula makes algebraic sense.

The most common case in calculus courses.

Range

A function's domain is its set of allowable inputs. The **range** is the corresponding set of outputs. In symbolic language,

If f has domain $\mathcal{D}$, then the range of f, denoted by $\mathcal{R}$, is the set $\{\, f(x) \mid x \in \mathcal{D} \,\}$. ◄

The expression $x \in \mathcal{D}$ means x is a member of the domain.

The number 42 (picked at random) lies in the range of a function F ◄ if some input to F produces the output 42 (i.e., if for some number x, $F(x) = 42$). For example, 42 is in the range of the function $F(x) = \sqrt{x}$ but is not in the range of the function $G(x) = -x^2$. ◄

Why F and not f? Because we already used f.

Which input to F gives output 42? Why isn't 42 in the range of G?

Reading a Function's Range from Its Graph. Defining a function's range is simple; *calculating* it—especially by hand—can be difficult. Graphs can help if viewing windows are chosen properly.

    The number 42 lies in the range of F, for instance, if and only if the horizontal line $y = 42$ hits the graph of F *somewhere*. For any function F,

The range of F is the set of y values at which a horizontal line hits the graph of F.

Here's a more illuminating way of saying the same thing:

Let lights shine horizontally, from both left and right, on the graph of F.
The range of F is the shadow thrown by the graph on the vertical axis.

The following picture shows the shadow (or **horizontal projection**) cast by the sine graph on the vertical axis:

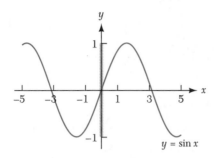

The shadow knows:
the range of $y = \sin x$ is $[-1, 1]$

Finding Ranges of Our Example Functions.   We'll use both graphical and algebraic methods to calculate the ranges of our example functions.

f The domain of f is $\mathcal{D} = \{\, x \mid x \neq 0 \,\}$, so the range of f is $\mathcal{R} = \{\, x^2 + 3/x \mid x \in \mathcal{D} \,\}$. That's easy to write, but what, exactly, does $\mathcal{R}$ contain? Does $\mathcal{R}$ contain the number -2345.678 (chosen at random)?

Here's a graph:

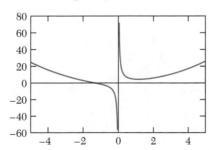

Graph of $f(x) = x^2 + 3/x$

It suggests that *all* reals lie in the range: f seems to "hit" every real value at least once. ▶   Our random number -2345.678, in particular, seems to be the output for some input just to the left of zero. ▶   To estimate which

To see why, imagine the graph's shadow on the vertical axis.

Do you agree that inputs just to the left of zero produce large negative outputs?

x gives $f(x) = -2345.678$, let's look in another window.

Graph of $f(x)$: a closer look

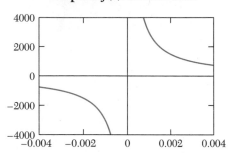

This graph suggests that $f(x) = -2345.678$ if $x \approx -0.0015$.

g By definition, g's range is the set of all world populations, for all time. This is clearly some set of nonnegative integers; saying much more is almost impossible.

h The domain of h has three pieces. To find the range, we'll decide what each domain piece contributes separately, and then assemble our answers.

Convince yourself of this; it's easy but important. A graph should help.

1. The output formula $1 - u$ holds if $u \leq 0$. As u varies from 0 through negative values, $1 - u$ covers the interval $[1, \infty)$. ◄   Thus, the range of h contains the interval $[1, \infty)$.

2. For $0 < u < 2$, $h(u) = u^2$. As u varies from 0 to 2, u^2 ranges from 0 to 4. Thus, the range of h contains the interval $(0, 4)$.

3. For $u \geq 30$, $h(u) = 42$. Thus, the third domain piece contributes the one-member set $\{42\}$ to the range.

Putting the pieces together, we find that the range of h is

$$\mathcal{R} = [1, \infty) \cup (0, 4) \cup \{42\} = (0, \infty).$$

Notice that this union is redundant—the first two intervals overlap, and the "singleton" $\{42\}$ adds nothing new. The redundancy reflects the fact that some members of the range are hit more than once.

j The j graph shows that the set of values assumed by j is the interval $[-11, 11]$. ◄

From graphs alone we can only estimate.

k The output set of k is the second row of k's table:

$$\{-4, -9, -1, 0, 1, 8, 3, -2\}.$$

The range of k, like its domain, is a finite set.

Domain and Range—a Machine's View. Computers and calculators cannot escape issues of domain and range when plotting functions. Here's a machine-drawn plot of the function $p(x) = \sqrt{x^2 + 2x}$:

Graph of $p(x) = \sqrt{x^2 + 2x}$

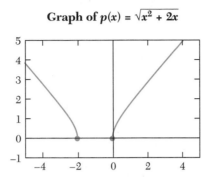

The gap from $x = -2$ to $x = 0$ is there for good reason: The natural domain of p omits this interval. The picture also suggests (but, being finite, can't prove) that the range is $[0, \infty)$. With algebra, we could verify both of these observations.

The Definition of Function: Reprise

A function is a package with three chief parts: rule, domain, and range.

The rule can be given in various ways—as an algebraic formula, as a graph, as a table, as an English description, and so on. However it's given, the rule must not be ambiguous. To each element of the domain the rule associates exactly *one* element of the range.

The domain and range of a function f are, respectively, its input and output sets. The natural domain of f is the set of inputs for which f's rule makes sense. In calculus, both range and domain are usually sets of real numbers. ▶

Elsewhere in mathematics, domains and ranges may be sets of vectors, matrices, functions, complex numbers, or more exotic objects.

Odd Functions and Even Functions: Graphical Symmetry

Certain functions, like certain numbers. are said to be either odd or even. These properties, though defined symbolically, are easily recognized graphically. First, the symbolic definitions:

> **Definition** A function f is **even** if $f(-x) = f(x)$ for all x in its domain; f is **odd** if $f(-x) = -f(x)$.

EXAMPLE 1 If $f(x) = x^2$, then

$$f(-x) = (-x)^2 = x^2 = f(x),$$

so f is even. If $g(x) = x^3$, then

$$g(-x) = (-x)^3 = -x^3 = -g(x),$$

so g is odd. The function $h(x) = x + x^2$ is neither even nor odd, because

$$h(-x) = -x + (-x)^2 = -x + x^2 \neq \pm h(x).$$ ∎

Graphs and Symmetry

For many functions we can easily check evenness or oddness using algebraic manip-ulation. But what about the function $k(x) = \sin x$, which has no algebraic formula? To decide whether k is even, odd, or neither, let's look at the graph. To see how $k(x)$ and $k(-x)$ are related, we choose an xrange centered at $x = 0$:

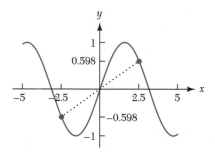

**Symmetry about the origin:
graph of $y = \sin x$**

The graph's **symmetry about the origin** means that for any input x, $k(-x) = -k(x)$. This is just what it means for k to be odd. If, say, $x = 2.5$, the bulleted points (points marked with dots) show that

$$k(2.5) = \sin(2.5) \approx 0.598 \qquad \text{and} \qquad k(-2.5) = \sin(-2.5) \approx -0.598.$$

In general,

See this section's exercises for more on odd and even functions.

> *A function f is odd if and only if the graph of f is symmetric about the origin. A function is even if its graph is symmetric about the y-axis.* ◄

Periodic Functions: Repeating Patterns

See the preceding graph.

The sine function ◄ is called 2π-periodic because it repeats itself on every interval of length 2π. (So do the other basic trigonometric functions; we'll review them soon.) Any function is called **periodic** if it repeats itself on intervals of any fixed length.

Periodic Functions: The Graphical View

Here are graphs of three periodic functions, f, g, and h:

Graphs of three periodic functions

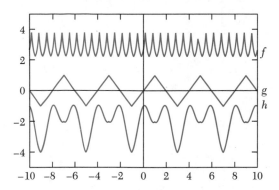

Each function's graph repeats some basic shape. How *often* a periodic function repeats itself depends on the **period**, i.e., the (horizontal) length of the basic repeating shape. It's easy to see from their graphs that g and h have period 4. Reading the period of the wigglier f is trickier. One way to avoid seasickness is to count that f completes 30 "cycles" from $x = -10$ to $x = 10$, an interval of length 20. Thus, f has period $20/30 = 2/3$. ▶

This is a graphical approximation, but a good one.

Periodic Functions: The Symbolic View

Periodicity can be defined symbolically:

> **Definition** A function f is periodic if the equation
>
> $$f(x + P) = f(x)$$
>
> holds for all x in the domain of f and a positive number P. The smallest such number P is the period of f.

On the Graph If, say, $P = 3$, then the functional equation says that $f(x + 3) = f(x)$. What does this mean about the graph of f? The equation says that $f(1) = f(4)$, $f(1.5) = f(4.5)$, $f(2) = f(5)$, $f(2.5) = f(5.5)$, and so on. Thus, for any x the height of the graph at x is the same as the height of the graph at $x + 3$.

A Shift to the Left For any function f, the graph of $y = f(x + P)$ is the result of shifting the graph of $y = f(x)$ a distance of P units to the *left*. ▶ From this point of view, the periodicity equation $f(x + P) = f(x)$ means:

The graph of f doesn't change if it's shifted P units to the left. ▶

We discussed graphical translation in Section 1.2.

Convince yourself of this for each preceding graph.

Repeats on Other Intervals Because the sine graph repeats itself on intervals of length 2π, it also repeats itself on intervals of length 4π, 6π, 8π, and so on. Remember, the period is the length of *the shortest* interval on which a function repeats.

EXAMPLE 2   Following are graphs of

$$f(x) = \frac{\sin x}{3} + 3 \quad \text{and} \quad g(x) = (\sin x)^2 + \sin(x).$$

Which is which? Why is each function 2π-periodic?

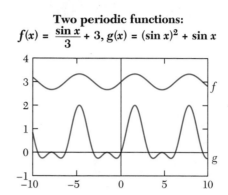

Two periodic functions:
$f(x) = \dfrac{\sin x}{3} + 3, g(x) = (\sin x)^2 + \sin x$

Solution   The upper graph is f—it's formed by compressing the sine curve vertically and translating it 3 units upward. The lower graph, therefore, is g.

Both functions appear to be 2π-periodic. So they are: They "inherit" the sine function's periodicity. A symbolic calculation shows how:

$$f(x + 2\pi) = \frac{\sin(x + 2\pi)}{3} + 3 = \frac{\sin x}{3} + 3 = f(x);$$

$$g(x + 2\pi) = \big(\sin(x + 2\pi)\big)^2 + \sin(x + 2\pi) = (\sin x)^2 + \sin x = g(x). \qquad \blacksquare$$

Functional Language: Names, Nicknames, Uses, and Abuses

Everyday speech is a rich mixture of formal and informal language, names and nicknames, careful descriptions and vague allusions. The common things of daily life—friends, places, college courses, and so on—are likeliest to attract nicknames, verbal shortcuts, and aliases.

In calculus, functions are the commonest objects of all. Mathematical language, true to form, offers various ways of talking about functions, some clearer and more precise than others. Some linguistic tips follow.

Function vs. Value.   It's sometimes important to distinguish between a function and the value of a function. The statement

Consider the function $f(x) = x^2$

may seem clear, but it blurs the distinction between f—a function—and $f(x)$—a value of f. It would be technically clearer (though less euphonious) to say ◀

Consider the function f defined by $f(x) = x^2$.

We won't always be this fastidious.

Composite Functional Expressions. If f is defined by $f(x) = x^2$, then it's clear what is meant by such expressions as $f(3)$, $f(z)$, and $f(-13/2)$. **Composite expressions**, such as

$$f(x+3), \qquad f(-x), \qquad \text{and} \qquad f\big(f(x)\big)$$

need more care.

To make sense of them, it helps to think of f as an **operator** or **procedure**: f accepts "anything" and returns its square. Thus,

$$f(x+3) = (x+3)^2 = x^2 + 6x + 9;$$
$$f(-x) = (-x)^2 = x^2;$$
$$f\big(f(x)\big) = \big(x^2\big)^2 = x^4.$$

Functions vs. Expressions.   The statement

The range of $x^2 - 6x$ is $[-9, \infty)$

is intelligible, but the *expression $x^2 - 6x$* is used—not quite correctly—to describe a function. A scrupulously correct version of the statement preserves this distinction:

Let f be the function defined by $f(x) = x^2 - 6x$. The range of f is $[-9, \infty)$.

Functions differ from expressions in subtle but important ways. Here are three ways:

- Not all functions are defined by symbolic expressions. Of the five functions that started this section, only two were defined symbolically. ▶   *Which two?*

- Several different symbolic expressions may define the same function. The following three expressions are equivalent:

$$x^2 - 6x; \qquad \frac{x^2 - 6x}{\sin^2 x + \cos^2 x}; \qquad \frac{x^4 - 6x^3 + x^2 - 6x}{x^2 + 1}.$$

  Any one of them could define f.

- Functions don't care about variable names; expressions do. For example,

$$x^2 - 6x, \qquad t^2 - 6t, \qquad \text{and} \qquad P^2 - 6P$$

  are different expressions, but all could define the same function.

Computers Know the Difference. Computer programs insist on the distinction between functions and expressions. In *Maple*, for example, the command

```
f := x^2 - 6*x ;
```

defines f only as an expression. To define *f* as a *function*, we could type

```
f := x -> x^2 - 6*x;
```

The latter form isn't pretty, but it's explicit: f is defined as an *operator* that assigns to any input *x* the output $x^2 - 6x$.

BASICS

1. Indicate which of the following curves are graphs of functions.

   (a)

   (b)

   (c)

   (d)
   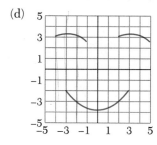

2. Let $f(x) = \begin{cases} 0 & \text{if } 0 \leq x < 1 \\ -2 & \text{if } 1 \leq x < 2 \\ x & \text{if } 2 \leq x \end{cases}$

   (a) Sketch a graph of f.
   (b) What is the domain of f?
   (c) What is the range of f?

Find the natural domain and the range of the functions in Exercises 3–6.

3. $g(x) = x^2 + x - 2$

4. $g(x) = \dfrac{x+2}{x-1}$

5. $f(x) = \dfrac{1}{(x-1)(x+2)}$

6. $h(x) = \sqrt{x+2}$

7. Indicate whether each of the following functions is even, odd, or neither.

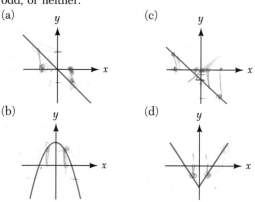

   (a)

   (b)

   (c)

   (d)

8. Suppose that f is an even function. Is the graph of f symmetric with respect to the y-axis? Justify your answer.

FURTHER EXERCISES

9. The **vertical line test** states, "A curve is the graph of a function only if each vertical line $x = k$ intersects the graph at most once." Explain why this test works.

10. (a) Sketch the solution curve of the equation $y^2 = x + 4$.
    (b) Is the curve in part (a) the graph of a function? Justify your answer.

11. (a) Can the graph of a function have more than one x-intercept? Explain.
    (b) Can the graph of a function have more than one y-intercept? Explain.

In Exercises 12 and 13, sketch the graph of each piecewise-defined function; then find the domain and range of each.

12. $f(x) = \begin{cases} 0 & \text{if } x \leq -1 \\ 3x - 2 & \text{if } 1 < x < 1.5 \\ x^2 & \text{if } 1.5 \leq x < 2 \end{cases}$

13. $f(x) = \begin{cases} \dfrac{1}{2-x} & \text{if } x < 0 \\ \\ \sqrt{1 + 4x^5} & \text{if } x > 1 \end{cases}$

Find the domain and range of each of the functions shown graphically in Exercises 14–17.

14.

$D = (-6, 6)$
$(4, 5) \cup \{5\}$

15.

16.

17.

Find the natural domain and the range of each function in Exercises 18–25.

18. $f(x) = \sqrt{x^2 - 4}$ $D = (-\infty, 2] \cup [2, \infty)$
    $R = [0, \infty)$

19. $r(x) = \dfrac{1}{\sqrt{(x+2)(1-x)}}$

20. $s(x) = \dfrac{1}{\sqrt{(x+2)(x-1)}}$

21. $t(x) = \sqrt{\dfrac{1-x}{x+2}}$

22. $h(x) = (x-2)(3-x)$

23. $j(x) = \sqrt{4 - x^2}$

24. $k(x) = (2x+5)^{1/3}$

25. $k(x) = (4 - 5x)^{2/3}$

26. Let $f(x) = \sqrt{x^4 + x^2} = \sqrt{x^2(x^2 + 1)}$ and $g(x) = x\sqrt{x^2 + 1}$.
    (a) Find the domain and range of f.
    (b) Find the domain and range of g.
    (c) Explain why $f \neq g$.

27. For each set that follows, find a function defined by a single algebraic expression that has the set as its natural domain. [**EXAMPLE:** The function $f(x) = \sqrt{(x-1)(x+3)}$ has the set $(-\infty, -3] \cup [1, \infty)$ as its domain.]
    (a) $\{x \mid x \neq -1 \text{ and } x \neq 3\}$
    (b) $[-1, 3]$
    (c) $[-1, \infty)$
    (d) $(-1, 3)$
    (e) $(-\infty, -1) \cup (3, \infty)$
    (f) $[-1, 3)$

28. Suppose that water is flowing into a cubical tank 10 feet on an edge.
    (a) Write a formula for V, the volume of water in the tank, as a function of d, the depth of water in the tank.
    (b) What is the domain of V? What is the range of V?
    (c) Write a formula for d as a function of V.

29. Let $f(x) = |x|$ and g be defined by the rule: "$g(x)$ is the slope of the graph of f at x."
    (a) Find the domain and range of f.
    (b) Explain why 0 is not in the domain of g.
    (c) Find the domain and range of g.
    (d) Find a symbolic expression for g.   [**HINT:** Use a piecewise-defined function.]

30. Let $f(x) = x + |x|$ and g be defined by the rule: "$g(x)$ is the slope of the graph of f at x."
    (a) Find the domain and range of f.
    (b) Write f as a piecewise-defined function.
    (c) Find the domain and range of g.
    (d) Find a symbolic expression for g.

31. Let $f(x) = \dfrac{x}{|x|}$ and g be defined by the rule: "$g(x)$ is the slope of the graph of f at x."
    (a) Find the domain and range of f.
    (b) Write f as a piecewise-defined function.
    (c) Find the domain and range of g.
    (d) Find a symbolic expression for g.

In Exercises 32–37, indicate whether each function is even, odd, or neither. Justify your answers with algebra. [**HINT:** Start by looking at graphs of these functions.]

32. $f(x) = 3x^4 - 5x^2 + 17$

33. $f(x) = 12x^7 + 6x^3 - 2x$

34. $f(x) = |x|$

35. $f(x) = |x + 4|$

36. $f(x) = x^4 + |x^3| - 2$

37. $f(x) = 4x^3 - 7$

38. Suppose that f is an odd function whose domain includes zero. Explain why $f(0) = 0$ must be true. [**HINT:** $-0 = 0$.]

39. Can a function be both even and odd? [**HINT:** Consider constant functions (i.e., functions of the form $f(x) = c$).]

40. Each of the following graphs is a portion of the graph of an even function over the interval $[-5, 5]$. Complete these graphs.

(a)

(b)

(c)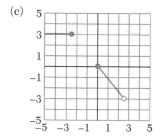

41. Which of the graphs in Exercise 40 could be a portion of the graph of an *odd* function over the interval $[-5, 5]$? Complete these graphs and explain why the others cannot be portions of the graph of an odd function.

42. Even functions are characterized by the fact that their graphs are symmetric about the y-axis. Are there functions whose graphs are symmetric about the x-axis? Explain.

43. Let $q(x) = ax^2 + bx + c$ where a, b, and c are constants.
    (a) Suppose that q is an even function. Show that $b = 0$.
    (b) Suppose that q is an odd function. What can be said about the values of the constants a, b, and c?

44. Suppose that the graph of a function f is symmetric with respect to the y-axis and that $f(3) = 1$. Evaluate $f(-3)$.

45. Suppose that f is an even function and that the point $(-3, 5)$ is on the graph of f. Find another point on the graph of f.

46. Suppose that f is an odd function and that $-12 \le f(x) \le 9$ when $x \ge 0$.
    (a) Is it possible that $f(-5) = 10$? Explain.
    (b) Is it possible that $f(-5) = -10$? Explain.

47. Suppose that f is an even function.
    (a) How are the graphs of f and $g(x) = f(|x|)$ related?
    (b) Is $h(x) = |f(x)|$ even, odd, or neither?
    (c) Is $k(x) = f(x) + |f(x)|$ even, odd, or neither?

48. Suppose that f is an odd function.
    (a) Is $g(x) = -f(x)$ even, odd, or neither? Justify your answer.
    (b) Is $h(x) = f(|x|)$ even, odd, or neither? Justify your answer.
    (c) Is $j(x) = |f(x)|$ even, odd, or neither?
    (d) Is $k(x) = f(x) + |f(x)|$ even, odd, or neither?

49. Let f be a function with domain R. Show that $g(x) = f(|x|)$ is an even function.

50. Suppose that f is a function with the following properties:
    (i) f is an even function;
    (ii) f is increasing over the interval $(2, 8)$;
    (iii) f is concave down over the interval $(3, 7)$; and,
    (iv) $4 \le f(x) \le 5$ when $1 \le x \le 6$.
    Indicate whether each of the following statements *must* be true, *may* be true, or *cannot* be true. Justify your answers. [**HINT:** Sketch a graph of f.]
    (a) f is increasing over the interval $(-8, -2)$.
    (b) f is concave down over the interval $(-7, -5)$.
    (c) $|f(-2)| < 9$
    (d) $-5 \le f(-3) \le -4$

51. Suppose that f is an odd function and that f is decreasing on the interval (a, b).
    (a) Is f increasing or decreasing over the interval $(-b, -a)$? Explain.
    (b) Is it possible that $a < 0 < b$? Explain.

52. Suppose that f is an odd function and that f is concave up on the interval (a, b).
    (a) Is f concave up or concave down over the interval $(-b, -a)$? Explain.
    (b) Is it possible that $a < 0 < b$? Explain.

53. Suppose that f is a periodic function with period $1/2$ and that
    $$f(2) = 5, \qquad f(9/4) = 2, \qquad \text{and} \qquad f(11/8) = 3.$$
    Evaluate each of the following:
    (a) $f(0)$ (d) $f(-3)$
    (b) $f(1/4)$ (e) $f(1000)$
    (c) $f(3/8)$ (f) $f(x + 5) - f(x)$

54. Suppose that f is a periodic function with period P and that g is defined by the rule: "$g(x)$ is the slope of the secant line through $(x, f(x))$ and $(x + 1, f(x + 1))$". Show that g is a periodic function with period P.

55. Suppose that f is a periodic function with period 2 and that f is increasing and concave down on the interval $[0, 1]$.
    (a) Is f increasing on the interval $[6, 7]$? Justify your answer.
    (b) Is it possible that f is concave down on the interval $[-3, -2]$? Justify your answer.
    (c) Suppose that f has a local maximum at $x = 13.5$. Where in the interval $[0, 2]$ does f have a local maximum?
    (d) Suppose that $-2 \le f(x) \le 5$ when $-4 \le x \le 3$. Explain why $-2 \le f(x) \le 5$ for all x.

56. Let f be a periodic function with period P.
    (a) Show that $f(x + 2P) = f(x)$.
    (b) What does part (a) mean graphically?

57. Suppose that f is a periodic function.
    (a) Is it possible that f is an even function? If so, sketch the graph of an even periodic function. If not, explain why not.
    (b) Is it possible that f is an odd function? If so, sketch the graph of an odd periodic function. If not, explain why not.
    (c) Is it possible that f is neither even nor odd? If so, sketch the graph of a periodic function that is neither even nor odd. If not, explain why not.

58. Suppose that f is a periodic function with period P and that $A \ne 0$ is a constant. Show that each of the following functions is periodic and find its period.
    (a) $g(x) = f(x) + A$ (c) $g(x) = Af(x)$
    (b) $g(x) = f(x + A)$ (d) $g(x) = f(Ax)$

1.5 A Field Guide to Elementary Functions

An **elementary function** is one built from certain legal basic elements, using certain legal operations. For example, the function

$$f(x) = \ln\left(\frac{\sin(2x)}{1 + 3x^4}\right)$$

is elementary because it is constructed (by addition, division, composition, and so on) ▶ from certain basic function elements: the sine and natural logarithm functions, powers of x, and constants.

We'll review these operations on functions in the next section.

Despite their name, elementary functions need not be trivial or simple-minded. Many different buildings can be built from standard sizes of brick. For the same reasons, elementary functions come in enormous variety.

What "standard" means for bricks is partly a practical matter ▶ but mainly a matter of choice. Our choice of functional building blocks is just as practical but much less arbitrary. The basic elementary functions of calculus have proved themselves useful throughout mathematics and its applications.

Bricks are twice as long as they are wide, not too heavy to lift, and so on.

A Pictorial Field Guide

This section contains brief descriptions, with pictures, of various elementary functions. Like any field guide, this one emphasizes identification, classification, and basic behavior rather than detailed study of its subjects. More details on each species of function met here can be found in later chapters or in the appendices.

Function Families

Like birds or plants, the building-block functions of calculus fall naturally into "families" (of which there are far fewer, fortunately, than there are for birds or plants!). We'll name just three.

Algebraic Functions Their rules involve only algebraic operations, such as these:

$$f(x) = x^2 + \frac{3}{x}; \qquad g(x) = \frac{x^2 + 3}{1 + \sqrt{x}}.$$

Exponential and Logarithm Functions Their rules involve exponentials or logarithms with various bases, such as these:

$$h(x) = 2^x; \qquad k(x) = e^x; \qquad m(x) = \log_2 x; \qquad n(x) = \ln x.$$

Trigonometric Functions The sine, cosine, and tangent functions, together with their descendants the cosecant, secant, and cotangent functions.

Hybrids

Like some plants, the basic calculus functions hybridize freely, combining in various ways to produce new elementary functions. How functions combine, and how their descendants' traits reflect their own, is an important theme throughout this book. In this section, our main concern is with the basic building-block functions themselves.

Algebraic Functions

An **algebraic function** is one defined using only the ordinary algebraic operations: addition, multiplication, division, raising to powers, and taking roots. Roughly speaking, algebraic functions are those that even the simplest calculator can handle. ◄ If, say, the function f is defined by the algebraic rule $f(x) = x - 3\sqrt{x+1}$, then checking with a calculator that

$2.98 buys such a machine; trigonometric, exponential, and logarithm functions start a little higher.

$$f(1.7) = 1.7 - 3\sqrt{2.7} \approx -3.229$$

requires nothing fancier than the arithmetic and square root keys.

Polynomials

Polynomials are the simplest algebraic functions; their rules require only multiplication and addition. ◄ (To review basic facts about linear, polynomial, and rational functions, see Appendix C.)

Positive integer powers of x can be found by repeated multiplication.

> **Definition** A **polynomial** in x is an expression that can be written in the form
>
> $$a_0 + a_1 x + a_2 x^2 + a_3 x^3 + \cdots + a_n x^n,$$
>
> that is, a sum of constant multiples of nonnegative integer powers of x.

EXAMPLE 1 The expressions

$$5x + 2, \qquad -x + 5x^{17}, \qquad \text{and} \qquad (x+3)(\pi x^{17} + 7)$$

Use algebra to write the last example in standard form.

are all polynomials in x. ◄ None of the expressions

$$3 - \frac{1}{x}, \qquad \sqrt{x}, \qquad \text{and} \qquad \frac{x+3}{\pi x^{17} + 7},$$

is a polynomial.

Polynomials as Functions.   The polynomial expression $2x^3 + 5x$ corresponds naturally to the function $p(x) = 2x^3 + 5x$. Any function defined by a polynomial expression is called a **polynomial function**. In practice, the technical distinction between *polynomial* and *polynomial function* seldom matters; we'll use the terms interchangeably.

Domain and Range.   Because polynomials involve only multiplication and addition, they accept all real numbers as inputs. In other words, every polynomial function has the same natural domain: all real numbers.

Polynomials' *ranges* vary drastically. The range of $p(x) = 10$ is $\{10\}$, a set with just one member. By contrast, the range of $q(x) = x^2 + 3$ is $[3, \infty)$, an infinite interval. The range of $r(x) = x^3 - 2x^2$ is not obvious from the formula, so we'll look at the graph. Here, together, are graphs of p, q, and r:

Polynomial graphs: $p(x) = 10$,
$q(x) = x^2 + 3$, $r(x) = x^3 - 2x^2$

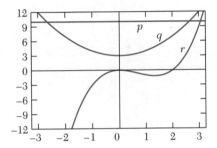

Blowing Up and Down   The picture suggests that $r(x)$ "blows up" (tends to ∞) as $x \to +\infty$ and that $r(x)$ "blows down" (tends to $-\infty$) as $x \to -\infty$. ▶ This impression is correct; for any polynomial, the term with the *largest power of x* determines what happens as x tends to plus or minus infinity. ▶ (For $r(x)$, the x^3 term "dominates.") Thus, $r(x)$ assumes *all* real values, and its range is $(-\infty, \infty)$.

▶ *Plotting r in a larger window would show this even more clearly.*

We'll show this soon, using limits.

Point of View   The appearance of *any* graph depends strongly on how the plotting window is chosen. Polynomials are no exception. Here are two more views of the same three polynomials:

Graphs of p, q, and r: a closer look

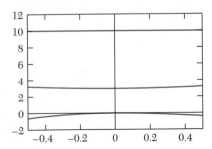

Graphs of p, q, and r: a farther view

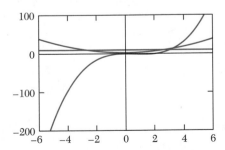

In the "farther view," the large vertical scale makes p, q, and r hard to distinguish near $x = 0$.

No Breaks Polynomial graphs are continuous and smooth everywhere—they have neither breaks or kinks. As we'll soon see, this reflects the fact that polynomials are among the simplest, best-behaved functions from the calculus point of view.

Lines and Linear Functions. The very simplest polynomials are constants, such as $p(x) = 10$. The simplest *interesting* polynomials are **linear functions**—polynomials that involve nothing higher than the first power of x. Each of the following expressions defines a linear function:

$$\ell_1(x) = 2x + 3; \qquad \ell_2(x) = 2(x - 3) + 3;$$

$$\ell_3(x) = 2x + 1; \qquad \ell_4(x) = -2x + 2.$$

Graphs of linear functions are, of course, lines:

Lines: $\ell_1(x) = 2x + 3$, $\ell_2(x) = 2(x - 3) + 3$,
$\ell_3(x) = 2x + 1$, $\ell_4(x) = -2x + 2$

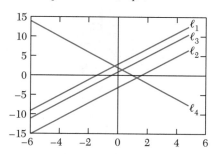

Each line's slope, y-intercept, and other important properties can be read from its equation.

The highest power of x with a nonzero coefficient.

Which is which?

Quadratics and Cubics: Characteristic Shapes. Graphs of **quadratic** and **cubic** polynomials—those with degree ◄ 2 and 3—have standard shapes, called **parabolas** and **cubic curves**. Here are three quadratics: ◄

Three quadratics: $q_1(x) = x^2$,
$q_2(x) = -4x^2 + 4$, $q_3(x) = 2x(x + 2)$

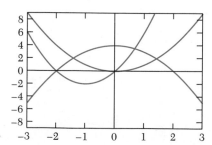

Here are three cubics: ▶

Which is which?

Three cubics: $c_1(x) = x^3,$
$c_2(x) = -3x^3 + 3,$ $c_3(x) = x(x-2)(x+2)$

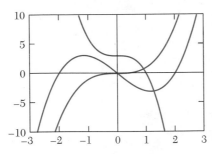

Polynomials in General: Many Shapes.   Linear, quadratic, and cubic polynomials have characteristic shapes, but polynomial graphs in general can take almost any smooth, unbroken shape.   Some shape restrictions do exist.   For instance, a polynomial of degree n can have at most n **roots**; this means, graphically, that a polynomial graph can have at most n x-**intercepts**.   (In particular, a polynomial function cannot be periodic.)

Rational Functions

Adding or multiplying two polynomials creates a new polynomial. ▶   *Dividing* two polynomials doesn't (usually) give another polynomial; it gives a rational function. Formally:

Is this really obvious? Think about it; try some examples.

> **D e f i n i t i o n** A **rational function** is a function whose rule can be written as the quotient of two polynomials.

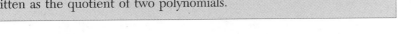

EXAMPLE 2   Each of the following functions is rational:

$$f(x) = \frac{1}{x}; \qquad g(x) = \frac{x^2}{x^2 + 3}; \qquad h(x) = \frac{2}{x} + \frac{3}{x+1}.$$

The function $m(x) = \sqrt{x} = x^{1/2}$ is not rational, because it involves a fractional power of x.

   The function h may look suspicious.   It's certainly the sum of two rational functions, but is h itself rational?   The answer is yes, because we can rewrite $h(x)$ in the form

$$h(x) = \frac{2}{x} + \frac{3}{x+1} = \frac{2(x+1)}{x(x+1)} + \frac{3x}{x(x+1)} = \frac{5x+2}{x(x+1)}.$$ ∎

Graphs of Rational Functions

Rational functions, unlike polynomials, have an important new feature: the possibility of **horizontal** and **vertical asymptotes**.

EXAMPLE 3 Consider the rational function

$$r(x) = \frac{x^2}{x^2 - 1} = \frac{x^2}{(x - 1)(x + 1)}$$

and its graph:

Graph of $r(x) = x^2/(x^2 - 1)$

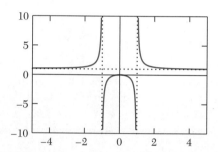

Asymptotes are straight lines toward which, "in the limit," a graph tends. We'll study asymptotes and their relations to limits in Chapter 2. Technicalities temporarily aside, it's already evident that the graph of r has three asymptotes: the vertical lines $x = -1$ and $x = 1$, and the horizontal line $y = 1$.

It's natural to expect trouble where a denominator vanishes. It may be less obvious from the formula how the function misbehaves.

Where the *vertical* asymptotes come from should be clear: They correspond to real roots of the denominator. ◄ The *horizontal* asymptote (the line $y = 1$) reflects r's "long run" behavior. It shows how r behaves for large positive and large negative inputs. Specifically, the asymptote shows that $r(x)$ approaches 1 as x approaches plus or minus infinity. Manipulating r's formula shows why:

$$r(x) = \frac{x^2}{x^2 - 1} = \frac{(x^2 - 1) + 1}{x^2 - 1} = 1 + \frac{1}{x^2 - 1}.$$

The graph of r falls toward the asymptote.

As x tends to plus or minus infinity, $1/(x^2 - 1)$ remains positive but tends to 0; hence, $r(x)$ approaches 1 from above. ◄ Tabulating values of r for inputs with large absolute values gives similar information:

Long-Run Values of $r(x) = x^2/(x^2 - 1)$					
x	10	100	1000	10,000	...
$r(x)$	1.01010101	1.00010001	1.00000100	1.00000001	...
x	−10	−100	−1000	−10,000	...
$r(x)$	1.01010101	1.00010001	1.00000100	1.00000001	...

Other Algebraic Functions

Polynomials and rational functions involve only the standard arithmetic operations: addition, subtraction, multiplication, division, and raising to integer powers. "Other" algebraic functions include ones such as

$$f(x) = \sqrt{x}, \qquad g(x) = \sqrt{x + \sqrt{x}}, \qquad \text{and} \qquad h(x) = \sqrt{\sqrt[3]{x^2 + 1} + x^{6/7}}$$

that involve nth roots. As we'll soon see, calculus methods and ideas apply just as readily to these functions as to polynomials and rational functions.

These "other" algebraic functions do pose one new problem: finding output values. To illustrate, compare the functions

$$m(x) = x + \frac{5x + 2}{x(x + 1)} \qquad \text{and} \qquad f(x) = \sqrt{x}.$$

It's easy to compute, even by hand, that $m(2) = 4$. Finding $f(2) = \sqrt{2}$ to any respectable numerical accuracy is harder, at least by hand. ▶ Even here, calculus can help. As we said in Section 1.1, ▶ calculus can help us choose polynomial functions that closely approximate the square root function.

A calculator has no trouble, of course.

See the graph on page 9.

Exponential and Logarithm Functions

Exponential and logarithm functions are among the most important and most practically useful functions in calculus. ▶ Understanding these functions requires, first of all, a familiarity with the algebra of logarithms and exponentials. Appendices E and F contain quick reviews of the necessary algebraic basics. Most basic of all are the following key facts. ▶

How Logs and Exponentials Are Related For a positive base b and real numbers x and y,

$$y = \log_b x \iff x = b^y.$$

This relation between logarithms and exponentials is crucial to understanding of either type of function. In particular, we'll use this relation to *define* logarithm functions using exponential functions.

Important Properties of Logs and Exponentials We'll use these key properties often:

$$b^x b^y = b^{x+y}; \qquad (b^x)^r = b^{xr};$$
$$\log_b xy = \log_b x + \log_b y; \qquad \log_b(x^r) = r \log_b x.$$

(**NOTE:** In each line, b must be positive. In the top two equations, x and y can take any real values. In the second two, x and y must be positive. For many more details, see the appendices.)

Section 1.7 contains a thoroughly practical application: using exponential functions to study how inflation affects college tuition.

If any are unfamiliar, review the appendices now!

Exponential Functions

> **Definition** An **exponential function** is defined by an expression of the form
>
> $$f(x) = b^x,$$
>
> where b, the **base**, is a fixed positive number.

Each of the expressions

$$2^x, \qquad 4^t, \qquad \left(\tfrac{1}{2}\right)^z, \qquad \text{and} \qquad e^x$$

defines an exponential function. (The base is fixed; the exponent varies—hence the name.) *None* of the expressions

$$x^2, \qquad z^3, \qquad (w+1)^{15}, \qquad x^\pi$$

defines an exponential function, because each involves a fixed power of a variable base.

Base e. Among all exponential functions, the one with base e turns out to be especially useful and convenient. (The fact that e is usually the default base on scientific calculators attests to its special importance.) The number e is irrational; its decimal expansion begins $e = 2.71828182\ldots$. What makes base e special—for both exponential and logarithm functions—will become clearer soon, when we study derivatives.

The function $f(x) = e^x$, with base e, is often called the **exponential function** ◄ or, sometimes, the **natural exponential function**. Alternative notations ◄ include

$$\exp(x), \qquad \exp x, \qquad \text{and} \qquad e^x.$$

The name is a bit unfortunate, because other functions with other bases are also exponential.

Parentheses are often omitted with such named functions as sin, cos, ln, and exp.

Graphs of Exponential Functions. Graphs of several exponential functions follow. The rule for each is of the form b^x—only the base b varies. Points of special interest are bulleted.

Exponential functions with various bases

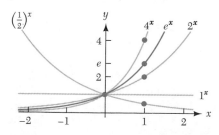

Observe:

Legitimate Bases   If $b > 0$, then the expression b^x makes sense for any real number x. ▶ Thus, any positive base b defines a sensible exponential function. (In practice, $b = 2$, $b = e$, and $b = 10$ are by far the most useful.)

Why not $b \leq 0$?
Appendix E gives details.

Domain and Range   Exponential functions accept all real inputs, so all have domain $(-\infty, \infty)$. The graphs show that unless $b = 1$, the range of the exponential function $f(x) = b^x$ is $(0, \infty)$.

Shapes   The shape of an exponential graph depends on the base b:

The greater the value of b, the faster b^x increases.

(If $b < 1$, b^x decreases. If $b = 1$, then b^x is a constant function.)

A Common Point   Every graph of the form $y = b^x$ passes through the point $(0, 1)$. In the language of functions:

If $f(x) = b^x$,   then $f(0) = b^0 = 1$.

The input $x = 1$ is also of interest. For any function $f(x) = b^x$, $f(1) = b^1 = b$. ▶

How do the graphs reflect these facts?

Exponential Functions Are Monotonic   All exponential functions $f(x) = b^x$ are **monotonic**, ▶ i.e., either nonincreasing everywhere or nondecreasing everywhere. In graphical language,

Learn the word "monotonic"; it's convenient shorthand for a longer-winded phrase.

If $b > 1$, the graph of $y = b^x$ is everywhere rising. If $0 < b < 1$, the graph of $y = b^x$ is everywhere falling. ▶

What happens if $b = 1$?

Faster and Faster   Perhaps the most important calculus property of exponential functions concerns *rates* of growth:

If $y = b^x$, then the rate of change of y is proportional to y itself.

This property of exponential functions is subtler than the others we've discussed. It can be seen only qualitatively in the graphs. The graph $y = 2^x$, for example, rises *faster* as it rises *higher*. We'll return to this important idea later, after studying rates of growth more carefully.

Exponential and Power Functions: They're Not the Same

The expressions 2^x and x^2 are typographically similar, but the resemblance ends there. They define entirely different functions, and their graphs show it:

2^x vs. x^2: which is which?

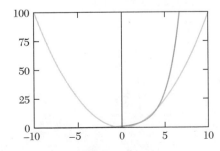

Note some differences:

Asymptotes The graph of $y = x^2$ has no horizontal (or vertical) asymptotes. The graph of $y = 2^x$ has the line $y = 0$ as a horizontal asymptote, corresponding to the fact that $2^x \to 0$ as $x \to -\infty$. In fact, *every* exponential function $y = b^x$ (with $b > 0$ and $b \ne 1$) has the line $y = 0$ as a horizontal asymptote. By contrast, *no* power function x^b (with $b > 0$) has a horizontal asymptote. ◄

If $b < 0$, then $y = 0$ is a horizontal asymptote.

Comparing Rates of Growth As $x \to \infty$, both

$$x^2 \to \infty \qquad \text{and} \qquad 2^x \to \infty.$$

However, 2^x blows up much, much faster than x^2. Graphs can only begin to suggest how quickly 2^x outruns x^2. To put the difference more dramatically, let $f(x) = x^2$ and $g(x) = 2^x$. Then $f(1000)$ dollars might buy a starter home in Hollywood, but $g(1000)$ is (more or less) the number of molecules in the universe. For even larger inputs, and larger values of b, the differences are even greater.

Logarithm Functions

This works for any positive base b except $b = 1$. Why not $b = 1$?

Notice the "mirror symmetry" in the table.

Exponential and logarithm functions come in pairs. An *exponential* function with base b corresponds to a *logarithm* function with the same base. ◄ The following table collects values of one such "matched pair" of functions, here with base $b = 2$. The values illustrate the correspondence between exponential and logarithm functions. ◄

Exponential and Logarithm Function Values								
Values of 2^x								
x	-2	-1	0	1	2	3	10	13.28771
2^x	1/4	1/2	1	2	4	8	1024	10,000
Values of $\log_2 x$								
x	1/4	1/2	1	2	4	8	1024	10,000
$\log_2 x$	-2	-1	0	1	2	3	10	13.28771

The formal definition makes the correspondence between logarithm and exponential precise:

> **Definition** The **logarithm function with base b**, denoted by
> $$f(x) = \log_b x,$$
> is defined by the condition
> $$y = \log_b x \iff b^y = x.$$

Each of the expressions

$$\log_2 x, \qquad \log_4 x, \qquad \log_{1/2} x, \qquad \text{and} \qquad \log_e x$$

defines a logarithm function with some base b. The logarithm function with base e ▶ is called the **natural logarithm function**; in print and on calculator buttons, it's usually denoted by

Recall that $e \approx 2.718$.

$$\ln(x) \qquad \text{or} \qquad \ln x.$$

Graphs of Logarithm Functions.   Graphs of logarithm functions with various bases follow. (Exponential functions with the same bases were plotted earlier; compare the pictures.) Interesting points are shown bulleted.

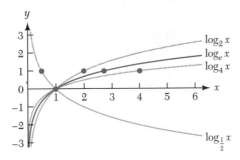

Logarithm functions with various bases

Not surprisingly, properties of logarithm function graphs reflect ▶ analogous properties of exponential function graphs.

Pun intended.

Legitimate Bases   Any positive number b except $b = 1$ can be used as the base for a logarithm function. ▶ (In practice, as for exponentials, $b = 2$, $b = e$, and $b = 10$ are most useful.)

Why not $b = 1$? Because $1^y = 1$ for all $y > 0$. Appendix F has more details.

Domain and Range   The graphs illustrate that for any suitable base b, the logarithm function accepts *positive* inputs; outputs take on *all* real values. Thus, every logarithm function has domain $(0, \infty)$ and range $(-\infty, \infty)$.

Growth Rates   As for exponential functions, each logarithmic graph's shape depends on b. This time, the *greater* the value of b, the more *slowly* $\log_b x$ increases. (If $b < 1$, $\log_b x$ decreases.) Graphs of exponential functions rise faster and faster as x increases. Graphs of logarithm functions have the opposite property: They rise more and more slowly as x increases.

A Common Point   The graph of every logarithm function passes through the point $(1, 0)$. ▶ In functional language,

Which point was common to exponentials?

$$\text{If } f(x) = \log_b x \qquad \text{then} \qquad f(1) = \log_b 1 = 0.$$

Similarly, for any $b > 0$, $f(b) = \log_b b = 1$.

That is, it is either strictly increasing everywhere or strictly decreasing everywhere.

There is no logarithm function for $b = 1$.

Logarithm Functions Are Monotonic   Every logarithm function is strictly monotonic. ◄   If $b > 1$, the graph of $y = \log_b(x)$ *rises* everywhere. If $0 < b < 1$, the graph *falls* everywhere. ◄

Exponential and Logarithm Functions as Inverses

The condition

$$y = \log_b x \iff x = b^y$$

means that a logarithm function and an exponential function with the same base are inverses of each other. ◄

We'll review the idea of inverse functions in the next section.

   EXAMPLE 4   For any base b, $f(x) = b^x$ and $g(x) = \log_b x$ are inverse functions. What does this mean if $b = 10$ and $x = 100$? What if $x = 5$?

   Solution   That the base 10 exponential and logarithm functions are inverses means that $y = \log_{10} x \iff x = 10^y$. For $x = 100$, this says that

$$y = \log_{10} 100 \iff 100 = 10^y.$$

Since the second equation holds only when $y = 2$, it follows that $\log_{10} 100 = 2$.

   What if $x = 5$? As any calculator knows, $\log_{10} 5 \approx 0.6989700043$. In this context, the inverse property says that $5 = 10^{0.6989700043}$. Your calculator should agree. ◄   ■

Does it?

Also known as e^x and $\log_e x$, respectively.

Inverse Functions: The Graphical View.   The base e exponential and logarithm functions exp and ln ◄   are inverses. The same holds for base 2. The following graphs show what this means geometrically.

Logarithm and exponential functions are inverses

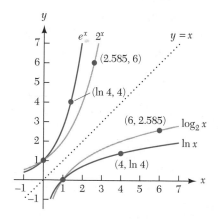

To make the picture look right, we used horizontal and vertical units of equal size.

The graphs of inverse functions are symmetric with respect to the line $y = x$. ◄   The picture suggests, too, why the base e is special. Look carefully:

At the points $(0, 1)$ and $(1, 0)$ the graphs of $y = e^x$ and $y = \ln x$ are parallel to each other and to the line $y = x$.

The other two graphs don't have this property. At the moment, this property of e may seem unimportant, but it will soon prove surprisingly useful.

Logs or Exponentials: Which Are Simpler?   The equation

$$y = \log_b x \iff x = b^y$$

can be taken to define the function $\log_b(x)$ in terms of the intuitively simpler exponential function b^x. To see that $\log_4 1.38703096913 \approx 0.236$, for example, we could check that $4^{0.236} \approx 1.38703096913$.

Does this approach to logarithms make sense? Are exponential functions really simpler than logarithm functions? In a strict logical sense, no: Knowing either implies knowing the other. Intuitively, though, exponentiation is probably simpler than finding logarithms, just as squaring is simpler than finding square roots.

Defining logarithm functions via exponential functions is not the only possibility. The reverse strategy (to start with logarithms—defined without any reference to exponentials—and proceed to exponentials) is sometimes used. Although perhaps less intuitive, this approach can skirt troublesome technicalities.

Trigonometric Functions

In geometry, sines, cosines, tangents, and other trigonometric quantities are defined as certain ratios of the sides of right triangles. Here our point of view is different: We study sines, cosines, and other trigonometric objects as functions in their own right, connecting them only occasionally to right triangles. (For a review of trigonometric basics, see Appendix G.)

Why Trigonometric Functions Matter

The most important property of the trigonometric functions—sine, cosine, tangent, and so on—is their repetitive, or periodic, behavior. The famous undulating graphs of the sine and cosine functions reflect this wavelike character:

Graphs of $y = \sin x$ and $y = \cos x$

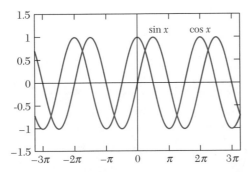

As always, the viewing window shows only part of the graph. In this case, what's shown is enough: The full graphs of the sine and cosine functions repeat the same pattern infinitely often.

Trigonometric Functions Defined: Circular Functions

Calling it the unit circle is a bit sloppy (many different circles have radius 1), but everyone does it.

Using u and v here saves x and y for later.

For calculus purposes, it's best to think of trigonometric functions in terms of circles, not triangles. The simplest circle, the **unit circle**, ◀ has radius 1 and center $(0, 0)$. In the uv plane, the unit circle's equation is $u^2 + v^2 = 1$. ◀

Imagine a point P moving *counterclockwise* around the unit circle, starting from the **east pole** $(1, 0)$. As P moves endlessly around the unit circle, many related quantities vary as well: the u-coordinate of P, the v-coordinate of P, the slope of the segment from P to the origin, and others. All the trigonometric functions can be derived from these varying quantities. (As a result, the sine and cosine functions are sometimes called **circular functions**.)

> **Definition** For any real number x, let $P(x)$ be the point reached by moving x units of distance counterclockwise around the unit circle, starting from $(1, 0)$. (If $x < 0$, go clockwise.) Then
>
> $$\cos(x) = u\text{-coordinate of } P(x);$$
>
> $$\sin(x) = v\text{-coordinate of } P(x).$$

It's worth many words; look carefully.

Here's the important picture: ◀

Defining sin x and cos x

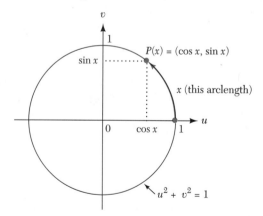

Simplest Properties of the Sine and Cosine Functions

Combining the definitions, the picture, the sine and cosine graphs, and a little reflection makes many properties of the sine and cosine functions evident.

> **Evident? Really?** Mathematicians use words like "evident," "clear," and "trivial" in peculiar ways. "Evident" does not mean "immediately obvious" or "This stuff is transparent to everyone with the slightest aptitude for mathematics; if it isn't clear to you, go away and never come back." "Evident" really means "not difficult to convince oneself of, with sufficient time and effort."

Domain and Range   For any real number x, $\cos(x)$ and $\sin(x)$ are the coordinates of $P(x)$, the point x units of distance, measured counterclockwise, from $(1, 0)$. ► Thus, every real number x is a legal input; the domain of both the sine and cosine functions is $(-\infty, \infty)$. ►

Recall: If $x < 0$, measure clockwise.

How do their graphs reflect this fact?

As our peripatetic point $P(x)$ circumnavigates the unit circle, its u- and v-coordinates fluctuate, but only between -1 and 1. Therefore, both functions share the same range: the closed interval $[-1, 1]$. In other words, for all x,

$$-1 \le \cos(x) \le 1 \qquad \text{and} \qquad -1 \le \sin(x) \le 1.$$

Famous Values   The following table collects some important values of the sine and cosine functions. ►

Check some!

Famous Sines and Cosines						
x	0	$\pi/2$	π	2π	$-\pi/2$	$\pi/4$
$P(x)$	$(1, 0)$	$(0, 1)$	$(-1, 0)$	$(1, 0)$	$(0, -1)$	$(\sqrt{2}/2, \sqrt{2}/2)$
$\cos(x)$	1	0	-1	1	0	$\sqrt{2}/2$
$\sin(x)$	0	1	0	0	-1	$\sqrt{2}/2$

Odd and Even   It follows from the definition and the symmetry of the unit circle that the sine function is odd and the cosine function is even. ► In symbols,

How do their graphs show this?

$$\sin(-x) = -\sin(x) \qquad \text{and} \qquad \cos(-x) = \cos(x).$$

Trigonometric Functions Are 2π-Periodic   For any real number x, the points $P(x)$ and $P(x + 2\pi)$ are identical. ► Therefore, both the sine and cosine functions repeat themselves, as the graphs show, on each interval of length 2π. In "math-speak," the sine and cosine functions are 2π-**periodic**. In symbols,

*Do you see why? [**HINT**: What's the circumference of the unit circle?]*

$$\sin(x + 2\pi) = \sin(x) \qquad \text{and} \qquad \cos(x + 2\pi) = \cos(x).$$

A Famous Trigonometric Identity   For any real number x, the point $P(x) = \big(\cos(x), \sin(x)\big)$ lies on the unit circle, so the coordinates of P must satisfy the defining equation $u^2 + v^2 = 1$. Therefore,

$$\big(\sin x\big)^2 + \big(\cos x\big)^2 = 1.$$

Equations like this one, that relate various trigonometric functions to each other, are called **trigonometric identities**. Many trigonometric identities exist; for examples, see Appendix G.

No Parentheses?   Trigonometric, exponential, and logarithm expressions are often typeset without parentheses. For example,

$$\sin(x), \qquad \ln(x), \qquad \sin x, \quad \text{and} \quad \ln x$$

are all syntactically correct, but the last two are typographically standard. When confusion seems likely, we'll use parentheses.

No Algebraic Formulas   Might the sine and cosine functions have simple algebraic formulas? Alas, no: Trigonometric functions are known to be **transcendental**; i.e., they *can't* be written as algebraic combinations of powers, roots, sums, products, and so on. On the other hand, the trigonometric functions are far from mysterious; they've been studied, tabulated, and used for centuries. Even cheap calculators produce highly accurate values of these functions. How do they do it? Stay tuned.

Other Trigonometric Functions

Four additional trigonometric functions are defined in terms of the sine and cosine functions.

> **Definition** For real numbers x,
>
> $$\tan x = \frac{\sin x}{\cos x}; \qquad \cot x = \frac{\cos x}{\sin x};$$
>
> $$\sec x = \frac{1}{\cos x}; \qquad \csc x = \frac{1}{\sin x}.$$

Pieces of the graphs of these functions follow. Pieces are enough: Like the sine and cosine functions, they repeat on every interval of length 2π. (In fact, the tangent and cotangent functions turn out to repeat even on intervals of length π. Appendix G explains why.)

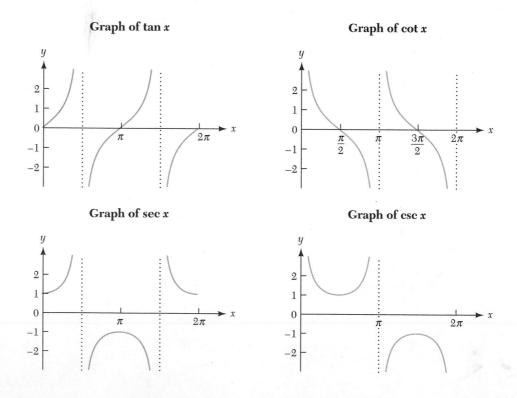

Graph of tan x   **Graph of cot x**

Graph of sec x   **Graph of csc x**

Properties of the "Other" Trigonometric Functions

As legitimate children of the sine and cosine functions, the tangent, cotangent, secant, and cosecant functions inherit certain properties from their parents, but have their own personalities too.

> **Domains and Ranges** Because the sine and cosine functions are defined for *all* real numbers, the other trigonometric functions are also defined for all real numbers except those at which a denominator is 0. The domain of the cosecant function, for instance, consists of all real numbers that are not integer multiples of π.
>
> As their graphs show, the tangent and cotangent functions have range $(-\infty, \infty)$—all real numbers are possible outputs. Outputs from the secant and cosecant functions, by contrast, must have absolute value at least 1. In interval language, the secant and cosecant functions have range $(-\infty, -1] \cup [1, \infty)$.

Asymptotes The tangent, cotangent, secant, and cosecant functions all have vertical asymptotes. Appendix G discusses the asymptotes of the tangent function in detail.

Functional Properties The following properties of the "other" trigonometric functions follow directly from their definitions. ▶ Each holds for all inputs x in the given function's domain.

$$|\sec x| \geq 1; \qquad\qquad |\csc x| \geq 1;$$
$$\tan(-x) = -\tan x; \qquad \sec(-x) = \sec x.$$

Convince yourself carefully of each one, using only the definitions and the properties of the sine and cosine functions. Then observe how the graphs reflect each property.

The Tangent Function, the Unit Circle, and Slope. Like the sine and cosine functions, the tangent function has an important and pleasing geometric interpretation in terms of the unit circle:

$$\tan x = \frac{\sin x}{\cos x} = \text{slope of the line from the origin to } P(x).$$

Interpreting the tangent function as slope shows, for instance, why $\tan(\pi/4) = 1$, and why $\tan x \to \pm\infty$ as $x \to \pi/2$.

BASICS

In Exercises 1–8, find the domain and range of each function.

1. $f(x) = e^{x^2}$
2. $f(x) = \ln(x^2)$
3. $f(x) = \ln(x^3 + 1)$
4. $f(x) = \exp((x-4)^2 - 1)$
5. $f(x) = \sin(3x - 1)$
6. $f(x) = 1 + 2\cos(3x)$
7. $f(x) = \sqrt{1 + \cos x}$
8. $f(x) = (1 + \sin x)^{-1/3}$

In Exercises 9–16, indicate whether the function is even, odd, or neither. If a function is even or odd, show this using the facts that sine is an odd function and cosine is an even function. [**HINT:** Start by examining a graph of the function.]

9. $f(x) = e^{\cos x}$
10. $f(x) = 1 + \cos x$
11. $f(x) = x + \sin x$
12. $f(x) = x \sin x$
13. $f(x) = x \cos x$
14. $f(x) = |\cos x|$
15. $f(x) = |\sin x|$
16. $f(x) = \sin^2 x$

In Exercises 17–22, find the period of each of the following functions.

17. $f(x) = \cos x - 2$
18. $f(x) = e^{\sin x}$
19. $f(x) = \sin(2x)$
20. $f(x) = \sin(x/2)$
21. $f(x) = \sin(2\pi x)$
22. $f(x) = \cot x$

23. Do the points shown in the following table lie on a line? If so, find an equation of the line. Justify your answers.

x	5.2	5.3	5.4	5.6	5.9	6.5
y	27.8	29.2	30.6	33.4	37.6	46.0

24. Do the points shown in the following table lie on a line? If so, find an equation of the line. Justify your answers.

x	0	1	3	6	10	15
y	0.000	0.785	1.249	1.406	1.471	1.504

FURTHER EXERCISES

Each of the functions graphed in Exercises 25–28 is described by a symbolic expression having *one* of the following forms, where A and B are real numbers.

(i) $f(x) = Ax + B$ (iv) $f(x) = \cos(Ax) + B$
(ii) $f(x) = Ax^2 + B$ (v) $f(x) = A\sin x + B$
(iii) $f(x) = A\cos x + B$ (vi) $f(x) = \sin(Ax) + B$
(vii) $f(x) = 3^{(x+A)} + B$
(viii) $f(x) = \log_2(x+A) + B$

Decide which expression corresponds to each function and find appropriate numerical values for the parameters A and B. Explain each choice in a sentence or two.

25.

26.

27.

28.

29. Each of the following graphs is of a function of the form $f(x) = A\sin(Bx) + D$. For each graph, identify values of the parameters A, B, and D.

(a)

(b)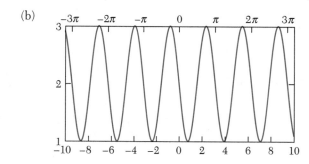

30. Each of the following graphs is of a function of the form $g(x) = A\cos(x+C) + D$. For each graph, identify values of the parameters A, C, and D.

(a)

(b)

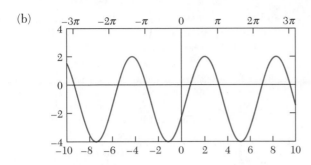

31. (a) Give an example of a polynomial of even degree that defines an even function.
    (b) Does *every* polynomial of even degree define an even function? Why or why not?

32. Let $f(x) = e^x$ and $g(x) = 10^x$.
    (a) What is the domain of g?
    (b) Show that there is a number k such that $g(x) = f(kx)$.
    (c) Explain how to obtain the graph of g from the graph of f.

33. Let $f(x) = \ln x$ and $g(x) = \log_{10} x$.
    (a) What is the domain of g?
    (b) Show that there is a number k such that $g(x) = kf(x)$.
    (c) Explain how to obtain the graph of g from the graph of f.

34. Let $f(x) = \ln(x^3)$ and $g(x) = \ln(x^2)$.
    (a) What is the domain of f?
    (b) Explain how to obtain the graph of f from the graph of $y = \ln x$.
    (c) What is the domain of g?
    (d) Explain how to obtain the graph of g from the graph of $y = \ln x$.

35. Let $f(x) = \ln(2 + \sin x)$.
    (a) Evaluate $f(0)$, $f(\pi/6)$, and $f(\pi/2)$ using a calculator.
    (b) What are the domain and range of f?
    (c) Is f even, odd, or neither? Justify your answer.
    (d) Is f periodic? If so, find its period. If not, explain why not.
    (e) Find the maximum and minimum values of f.

36. Let $f(x) = Ae^{kx}$, where $A > 0$ and k are constants. Show that the graph of $y = \ln(f(x))$ is a straight line.

37. Let $g(x) = Ax^k$, where $A > 0$ and k are constants. Show that the graph of $y = \ln(g(e^x))$ is a straight line.

38. Suppose that $y = 20x + 1$ is an equation of the line that intersects the curve $y = b^x$ at $x = 0$ and $x = 4$.
    (a) Find the number b.
    (b) What is the slope of the line that intersects the curve $y = \log_b x$ at $x = 1$ and $x = 81$?

39. Use properties of the sine and cosine functions to show that $\sec x$ is an even function.

40. Use properties of the sine and cosine functions to show that $\tan x$ is an odd function.

In Exercises 41–46, use properties of the trigonometric functions to find the period of each function.   [**HINT:** $\cos(x+\pi) = -\cos x$; $\sin(x + \pi) = -\sin x$.]

41. $f(x) = |\sin x|$
42. $f(x) = 2|\cos(3x)|$
43. $f(x) = \cos^2 x$
44. $f(x) = \cos x \sin x$
45. $f(x) = \cos(|x|)$
46. $f(x) = \sin(|x|)$

47. Use properties of the cosine function to explain why $|\sec x| \geq 1$ for every x in the domain of the secant function.

48. Use properties of the sine function to explain why $|\csc x| \geq 1$ for every x in the domain of the cosecant function.

49. Explain why $\dfrac{1}{3} \leq \dfrac{1}{2 + \cos x} \leq 1$ for all x.

50. Show that $0.3 < \dfrac{1}{1 + \tan x} \leq 1$ if $0 \leq x \leq 1$.
    [**HINT:** $\tan x$ is increasing on the interval $[0, 1]$ and $\tan(\pi/3) = \sqrt{3} < 2$.]

51. Find a number U such that $|1 + \sec x| \leq U$ if $2 \leq x \leq 4$.

52. Each of the following statement shows that the sine function is *not* a polynomial. Explain why.
    (a) The equation $\sin x = 0$ has infinitely many solutions ($x = n\pi$ is a solution for every $n \in \mathbb{Z}$).
    (b) The sine function is bounded below by -1 and above by 1 (i.e., for every x, $-1 \leq \sin x \leq 1$).

53. Suppose that f is a linear function (i.e., $f(x) = ax + b$, where a and b are constants).
    (a) Show that $f(5) - f(3) = f(10) - f(8)$.
    (b) Find a number c such that $f(4) - f(1) = f(c) - f(-9)$.
    (c) Show that $3(f(2) - f(-1)) = f(10) - f(1)$.
    (d) Assume that $h \neq 0$ is a constant. Show that the value of the expression

$$\boxed{\dfrac{f(x + h) - f(x)}{h}}$$

does not depend on either x or h. What is the geometric significance of this expression?

54. Suppose that f is an exponential function (i.e., $f(x) = Ab^x$, where A and $b > 0$ are constants).
    (a) Explain why $f(5)/f(3) = f(10)/f(8)$.
    (b) Find a number a such that $f(4)/f(1) = f(a)/f(-9)$.
    (c) Explain why $(f(2)/f(-1))^3 = f(10)/f(1)$.
    (d) Show that the value of the expression $f(x + h)/f(x)$ does not depend on x.
    (e) The value of the expression in part (d) does depend on h. Show that the value of the expression

$$\dfrac{\ln(f(x + h)/f(x))}{h}$$

does not depend on h. (Assume $h \neq 0$.)

55. The points listed in the following table lie on the graph of an exponential function f.

x	1.5	2.0	3.0	5.0	8.0
y	8.075	8.470	9.317	11.273	15.005

  (a) Evaluate $f(12)$. [**HINT**: See the previous exercise.]
  (b) Evaluate $f(0)$.
  (c) Evaluate $f(1)$.
  (d) Find a symbolic expression of the form Ab^x for f.
  (e) Find a symbolic expression of the form Ae^{kx} for f.

56. Suppose that the line $y = 2$ is a horizontal asymptote of the function f.
  (a) Explain why the line $y = 5$ is a horizontal asymptote of the function $g(x) = f(x) + 3$.
  (b) Find the equation of a horizontal asymptote of the function $g(x) = f(x + 1)$.
  (c) Find the equation of a horizontal asymptote of the function $g(x) = 3f(x + 4) + 1$.

57. Suppose that the line $x = -3$ is a vertical asymptote of the function f.
  (a) Explain why the line $x = -3$ is a vertical asymptote of the function $g(x) = f(x) + 3$.
  (b) Find the equation of a vertical asymptote of the function $g(x) = f(x + 1)$.
  (c) Find the equation of a vertical asymptote of the function $g(x) = 3f(x + 4) + 1$.

58. Let $f(x) = \dfrac{\sqrt{x} - 3}{x - 9}$ and $g(x) = \dfrac{1}{\sqrt{x} + 3}$.

  (a) Show that $f(x) = g(x)$ for all $x \geq 0$ such that $x \neq 9$.
  (b) Find the domain and range of f.
  (c) Find the domain and range of g.
  (d) How are the graphs of f and g related?
  (e) Does the graph of f have a horizontal asymptote? If so, find an equation for this asymptote.
  (f) Does the graph of f have a vertical asymptote? If so, find an equation for this asymptote.

59. Let $f(x) = \dfrac{x - 3}{x^2 - 9}$ and $g(x) = \dfrac{1}{x + 3}$.
  (a) Show that $f(x) = g(x)$ for all x such that $x \neq \pm 3$.
  (b) Find the domain and range of f.
  (c) Find the domain and range of g.
  (d) How are the graphs of f and g related?
  (e) Does the graph of f have a horizontal asymptote? If so, find an equation for this asymptote.
  (f) Does the graph of f have a vertical asymptote? If so, find an equation for this asymptote.

60. Find a formula for a rational function that has a vertical asymptote at $x = 1$ but has no horizontal asymptote. [**HINT**: There are many possible answers. Check your answer by graphing.]

61. Find a formula for a rational function that has the line $y = 1$ as a horizontal asymptote but has no vertical asymptotes. [**HINT**: There are many possible answers. Check your answer by graphing.]

62. Find a formula for a rational function that has vertical asymptotes at $x = -1$ and $x = 2$ and that has the line $y = 3$ as a horizontal asymptote. [**HINT**: There are many possible answers. Check your answer by graphing.]

1.6 New Functions from Old

Operations

Multiplication by -1.

Ordinary addition, as in $2 + 3 = 5$, is an **operation on numbers**: Two old numbers are operated on to produce a "new" number. Subtraction, multiplication, division, and negation ◄ are also operations on numbers. Each begins with one or more old numbers and—somehow or other—produces a new number.

Operations on functions are similar to operations on numbers. Addition, for instance, makes sense for functions: The function $h(x) = x + \cos x$ is, in a natural way, the *sum* of the functions $f(x) = \cos x$ and $g(x) = x$.

A mathematical operation starts with one or more objects of some type and produces a new object *of the same type*. In particular:

> *An operation on functions accepts one or more functions as input and produces a new function as output.*

Operations on Functions: The Machine View

Schematic diagrams like these

represent functions as input–output machines; they accept inputs and produce outputs. ▶ From this "mechanical" point of view, function operations hook simple machines together to produce fancier machines.

The first function name is generic; the second refers to the squaring function.

> **A Literal Function Machine.** The boxes that represent functions in our machine diagrams resemble the function keys on a scientific calculator. This is a happy coincidence, because calculator function keys, such as $\boxed{\text{SIN}}$, nicely illustrate the idea of a function machine. Pressing $\boxed{\text{SIN}}$ takes an input, does something to it, and returns an output.

Adding Machines.   The following diagram gives a machine view of function addition: Two function machines, f and g, are linked in parallel to form a new function machine, h:

Building a new function machine: addition

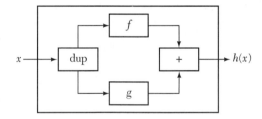

The first box duplicates an input x, because one copy is needed for f, and one for g. The last box adds its two inputs. The large outer box represents the new function, h. Its presence emphasizes that the sum of two functions is another function.

The definition of function addition makes precise what the picture depicts. ▶

The other algebraic operations have similar definitions and similar machine diagrams. We'll handle them less formally.

> **Definition** Let f and g be two functions. Their sum, denoted by $f + g$, is the new function whose rule is
> $$(f + g)(x) = f(x) + g(x).$$

Algebraic Operations on Functions

Algebraic operations—addition, subtraction, multiplication, and division—on symbolically defined functions are simple and familiar. The following table illustrates various operations applied to the functions f and g defined by $f(x) = \cos x$ and $g(x) = x$.

Algebraic Operations on Functions			
Operation	**Output**	**Rule**	**Domain**
Addition	$f + g$	$\cos(x) + x$	$(-\infty, \infty)$
Subtraction	$f - g$	$\cos(x) - x$	$(-\infty, \infty)$
Multiplication	$f \cdot g$	$x \cos(x)$	$(-\infty, \infty)$
Division	f/g	$\cos(x)/x$	$\{x \mid x \neq 0\}$

The table contains few surprises, but note the following.

Outputs Are Functions Each output is a new function; its rule appears in the third column.

Domains The natural domain of f/g is smaller than the domains of f and g, because the quotient makes sense only if $g(x) \neq 0$. Sometimes, as the next example shows, domains need closer attention.

Algebraic Operations on "Other" Functions. Not all functions have symbolic formulas. With due care taken, however, algebraic operations still make sense.

EXAMPLE 1   Let f be defined by the formula $f(x) = x^2 + 3/x$, and let g be defined by the table

x	-3	-2	-1	0	1	2	3	4
$g(x)$	-4	-9	-1	2	0	8	3	-2

Find the natural domain and some output values of f/g.

Solution   The expression $f(x)/g(x)$ makes sense if and only if both numerator and denominator are defined and $g(x) \neq 0$. If, say, $x = 3$, then

$$\frac{f}{g}(3) = \frac{f(3)}{g(3)} = \frac{10}{3}.$$

However,

$$\frac{f}{g}(2.3) = \frac{f(2.3)}{g(2.3)}, \qquad \frac{f}{g}(0) = \frac{f(0)}{g(0)}, \qquad \text{and} \qquad \frac{f}{g}(1) = \frac{f(1)}{g(1)}$$

Why? Convince yourself.   are all undefined—each for a different reason. ◀   In fact, $f(x)/g(x)$ is defined only for x in the finite set $\{-3, -2, -1, 2, 3, 4\}$. Thus, the domain of f/g is smaller than either the domain of f or the domain of g.   ■

Composition

Let f and g be the functions defined by $f(x) = x^2$ and $g(x) = \cos x$. The new functions h and k defined by

$$h(x) = \cos(x^2) \qquad \text{and} \qquad k(x) = (\cos x)^2$$

are constructed from f and g by **composition**, i.e., by following one function with the other. More precisely:

> **Definition** Let f and g be functions. The **composition** of f and g, denoted $f \circ g$, is the function defined by the rule
> $$(f \circ g)(x) = f\big(g(x)\big).$$

For the preceding functions f, g, h, and k,

$$(f \circ g)(x) = f\big(g(x)\big) = f(\cos x) = (\cos x)^2 = k(x);$$

$$(g \circ f)(x) = g\big(f(x)\big) = g(x^2) = \cos(x^2) = h(x).$$

Notice the following properties of composition:

Order Matters   If $f(x) = \sqrt{x}$ and $g(x) = x + 5$, then

$$(f \circ g)(x) = f(x+5) = \sqrt{x+5};$$

$$(g \circ f)(x) = g\big(\sqrt{x}\big) = \sqrt{x} + 5.$$

The two functions $f \circ g$ and $g \circ f$ are different. For instance, $(f \circ g)(4) = \sqrt{9} = 3$, while $(g \circ f)(4) = \sqrt{4} + 5 = 7$.

Domains   A legal input x to $f \circ g$ must first be a legal input to g. To keep things moving, the output $g(x)$ must also be a legal input to f. ▶   Thus, the domain of $f \circ g$ is a subset of the domain of g. For the preceding functions, the domains of $f \circ g$ and $g \circ f$ are, respectively, the intervals $[-5, \infty)$ and $[0, \infty)$.

The following machine diagram illustrates the idea: A legal input to $f \circ g$ must pass successfully through the whole contraption.

Composition: Function Machines in Series

The machine picture of composition is especially simple:

Mechanical composition

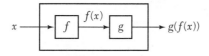

The machines f and g are linked "in series"; the output of f is "piped" through g.

Composing Functions Defined by Formulas: Substitution

Composing functions defined by explicit symbolic formulas, such as $f(x) = x + 2$ and $g(x) = \sin x$, is a straightforward matter:

$$(f \circ g)(x) = \sin(x) + 2; \qquad (g \circ f)(x) = \sin(x+2).$$

Like f and g themselves, the composite functions $f \circ g$ and $g \circ f$ have explicit formulas. Finding these formulas amounts to substituting one symbolic expression into another.

Composing "Other" Functions

The following examples illustrate various ways of composing functions that are not given by formulas. The common thread is the idea of composition itself.

EXAMPLE 2   **(Composing with a tabular function)**   Let $f(x) = x + 2$; let g be defined by the table

x	-3	-2	-1	0	1	2	3	4
$g(x)$	-4	-9	-1	2	0	8	3	-2

Describe the functions $f \circ g$ and $g \circ f$; describe their domains.

Solution   Because $f(x) = x + 2$,

$$(f \circ g)(x) = f(g(x)) = g(x) + 2 \quad \text{and} \quad (g \circ f)(x) = g(f(x)) = g(x + 2).$$

The formulas permit calculations such as these:

$$(f \circ g)(2) = g(2) + 2 = 8 + 2 = 10;$$
$$(g \circ f)(2) = g(2 + 2) = g(4) = -2.$$

But not just *any* input works;

$$(f \circ g)(10) = g(10) + 2$$

is undefined. The problem is with domains. Although f accepts any real number as input, g has a tiny domain—the finite set $\{-3, -2, \ldots, 3, 4\}$. The effect is to restrict sharply the domains of both $f \circ g$ and $g \circ f$.

Which inputs pass safely through the composite functions $f \circ g$ and $g \circ f$? Because only g presents an obstruction, the domain of $f \circ g$ is the full domain of g.

The domain of $g \circ f$ is different. Because $(g \circ f)(x) = g(x + 2)$, an input x survives the trip through f and g only if $x + 2$ is in the domain of g, i.e., only if $x \in \{-5, -4, \ldots, 2\}$.

The following table gives full details on both $g \circ f$ and $f \circ g$; gaps in the table reflect gaps in the respective domains.

Composition with a Tabular Function										
x	-5	-4	-3	-2	-1	0	1	2	3	4
$g(x)$			-4	-9	-1	2	0	8	3	-2
$(f \circ g)(x)$			-2	-7	1	4	2	10	5	0
$(g \circ f)(x)$	-4	-9	-1	2	0	8	3	-2		

Both f and g actually have fairly simple formulas. Try to guess them if you like, but the graphs are the point here.

EXAMPLE 3   **(Graphical composition)**   Let f and g be the following functions: ◄

Composing graphical functions

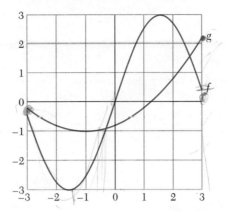

Find approximate values of the composite function $f \circ g$ for integer inputs in $[-3, 3]$; use them to plot $f \circ g$. Is there an x for which $(f \circ g)(x) = 0$?

S o l u t i o n   By definition, $(f \circ g)(3) = f(g(3))$. Reading approximate values ▶ from the graphs gives

Check these estimates for yourself.

$$(f \circ g)(3) = f(g(3)) \approx f(2.2) \approx 2.4.$$

Similar graphical estimates produced this table of values:

Composing Graphical Functions							
x	-3	-2	-1	0	1	2	3
$(f \circ g)(x)$	-0.6	-2.2	-2.5	-2.2	-0.6	2.2	2.4

With enough such data, we can plot $f \circ g$ as follows (bold curve). The tabulated data points are bulleted.

Composing graphical functions

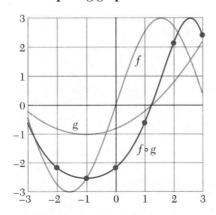

EXAMPLE 4   **(Composing functions described in words)** Motoring across Montana from Billings to Missoula, one's *altitude* (measured in feet above sea level) varies with *distance* traveled (in miles from Missoula). Distance, in turn, varies with *time* (in hours since departure), so altitude also varies with time. Interpret the situation in the language of functions and composition.

Solution   Let A, d, and t denote altitude, distance, and time. Altitude is a function—say f—of distance; distance is another function—say, g—of time. In function notation,

$$A = f(d) \quad \text{and} \quad d = g(t).$$

The composite function

$$A(t) = (f \circ g)(t) = f(g(t))$$

represents altitude (in feet) as a function of time (in hours).   ∎

The Identity Function

The numbers 0 and 1 are **identities** for (ordinary) addition and multiplication, respectively, because

$$a + 0 = 0 + a = a \quad \text{and} \quad a \cdot 1 = 1 \cdot a = a$$

for any real number a.

A similar notion of identity applies for composition. Composition is an operation on functions, ◄   so an identity for composition must also be a function. The identity function, denoted by I, has a simple definition: ◄

We'll assume that all functions here have sets of real numbers as their domain and range. This does no harm and skirts distracting technicalities.

As a function machine, I is just a hollow tube; inputs pass through unchanged.

> **Definition** The **identity function** for composition is defined by the rule
>
> $$I(x) = x.$$

To deserve the name "identity function," I should satisfy the equations $f \circ I = I \circ f = f$. It does; for any input x,

$$(f \circ I)(x) = f(I(x)) = f(x) \quad \text{and} \quad (I \circ f)(x) = I(f(x)) = f(x).$$

In words:

Composing any function f with I, in either order, gives f.

Inverse Functions

But all the meanings are related—in each case, one object is the inverse of another with respect to a certain operation.

Beware: $f^{-1}(x) \neq 1/f(x)$.

The word "inverse" has various meanings in mathematics. ◄   The numbers 3 and -3 are additive inverses because their sum is 0, the additive identity. Similarly, 3 and 1/3 are multiplicative inverses because their product is 1, the multiplicative identity.

The term "inverse function" has yet another meaning. Two functions f and g are inverses if *composing* them—in either order—gives the identity function. Thus the inverse of a function f is another function, denoted by f^{-1}, ◄   that undoes the effect of the original function f.

The following examples illustrate the idea of inverse function in the best and worst possible cases.

EXAMPLE 5 If $f(x) = x + 3$, then $f^{-1}(x) = x - 3$. ▶ For any input x,

f adds three; f^{-1} takes it away.

$$\left(f^{-1} \circ f\right)(x) = f^{-1}(x + 3) = (x + 3) - 3 = x = I(x);$$

$$\left(f \circ f^{-1}\right)(x) = f(x - 3) = (x - 3) + 3 = x = I(x).$$

In other words, $f^{-1} \circ f = I = f \circ f^{-1}$. As required, both compositions give the identity function. ■

EXAMPLE 6 Let f be the constant function $f(x) = 3$. This function has no inverse. The problem is that for every function g, $f \circ g$ is still a constant function; ▶ thus $f \circ g$ cannot be the identity function I. Specifically, if g is any function and x any input, then

Composition with a constant function always produces a constant function. Do you see why?

$$(f \circ g)(x) = f(g(x)) = 3.$$ ■

Here is the scrupulously correct definition:

> **Definition** Let f and g be functions. If the equations
>
> $$(f \circ g)(x) = x \qquad \text{and} \qquad (g \circ f)(x) = x$$
>
> hold for all x in the domains of g and f, respectively, then f and g are **inverse functions**. In this case, we write $g = f^{-1}$ (and $f = g^{-1}$).

Machine Views Viewing f and g as function machines, the definition says this:

Inverses

In effect, the machine g is really the machine f running in reverse.

Famous Families: Logs and Exponentials Logarithm and exponential functions with the same base are the most important and useful examples of inverse functions. ▶ Indeed, we *defined* logarithms as inverses of exponentials. In base e, for instance, the natural logarithm function is defined by the condition

We reviewed these functions in Section 1.5. See the graphs on page 60. See also Appendix F.

Notice how log and exponential keys appear in "inverse" positions on a calculator keypad.

$$y = \ln x \iff x = e^y.$$

Similarly, in base 10, $y = \log_{10} x \iff x = 10^y$. ▶

Equivalent Conditions   If f and g are inverse functions, then

$$f(a) = b \iff g(b) = a,$$

for every a in the domain of f.

Domains and Ranges   If f and g are inverses, then inputs to g are outputs from f, and vice versa. Therefore, the domain of f is the range of g, and the range of f is the domain of g.

Which Functions Have Inverses?

Does every number have an additive inverse? A multiplicative inverse?

We saw earlier that the constant function $f(x) = 3$ has no inverse function. ◄ Which functions *do* have inverses?

One-to-One Functions.   Every function f, invertible or not, sets up a correspondence between its domain and its range; to an input x from the domain, f assigns $f(x)$, one member of the range. An inverse function—if one exists—reverses this correspondence. To be invertible, therefore, f must be **one-to-one**, i.e., each output must correspond to only one input. A constant function fails this test miserably: it sends *every* input to the same output.

Recognizing Invertibility Graphically.   Graphically speaking, one-to-one functions are those that pass the **horizontal line test**:

No horizontal line crosses the graph more than once.

The test guarantees that every possible output y corresponds to no more than one input x.

EXAMPLE 7   The functions $f(x) = x^2$ and $h(x) = 2x + 3$ are shown graphically. Is either function invertible? If so, find its inverse.

  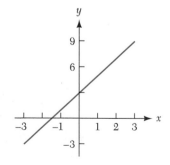

The graph of f fails the horizontal line test at $y = 4$, for example.

Solution   The function f is not one-to-one on the interval $[-3, 3]$, ◄ so it's not invertible on the domain $[-3, 3]$. The function h is one-to-one and therefore invertible.

With algebra we can find an explicit formula for h^{-1}: ▶

Check each step.

$$y = h^{-1}(x) \iff x = h(y) \iff x = 2y + 3 \iff y = h^{-1}(x) = \frac{x-3}{2}.$$

∎

Inversion and Symmetry

The definition of inverse function, interpreted graphically, says this:

A point (a, b) lies on the graph of f if and only if (b, a)—the reflection of (a, b) across the line $y = x$—lies on the graph of f^{-1}.

Graphs of h and h^{-1} from the preceding example illustrate what this means:

Graphs of h and h^{-1}

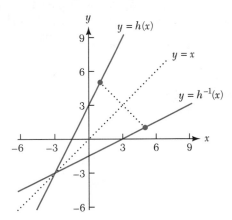

The two graphs are symmetric about the line $y = x$: Every point on either graph is the mirror image of a point on the other. One symmetric pair of points, $(1, 5)$ and $(5, 1)$, is bulleted. ▶

The graphs look symmetric because equal scales are used on the two axes.

Restricting Domains to Make Functions Invertible

Is the function $f(x) = x^2$ invertible? Earlier we said no, judging from the graph over the interval $[-3, 3]$. But isn't $g(x) = \sqrt{x}$ an inverse? Resolving this quandary requires a closer look at domains.

EXAMPLE 8   Make careful sense of the functions $f(x) = x^2$ and $g(x) = \sqrt{x}$ as inverses.

It's "two-to-one." Do you understand why?

Solution   The function $f(x) = x^2$ is not one-to-one ▶ on its natural domain—the full set of real numbers. However, f *becomes* one-to-one if its domain is restricted to nonnegative inputs. With this restriction, f *is* invertible and has the expected inverse $g(x) = \sqrt{x}$, as the following graphs show. ▶

Check algebraically, too, that $(f \circ g)(x) = x = (g \circ f)(x)$ for all appropriate values of x.

Inverses: $f(x) = x^2$ and $g(x) = \sqrt{x}$

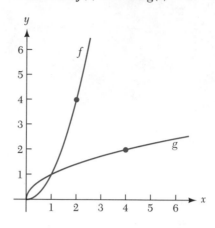

This is expected, because f and g are inverses.

The graphs of f and g are, as expected, mirror images through the line $y = x$. The bulleted points, for instance, are such mirror images; they show that $f(2) = 4$ and $g(4) = 2$. ◄   The conclusion: $f(x) = x^2$ is invertible on the domain $[0, \infty)$; its inverse there is $g(x) = \sqrt{x}$.   ■

A Moral about Domains.   For most calculus purposes, it's enough to treat domains informally. For most functions the **natural domain** (the set of all inputs for which the function's rule makes sense) is clear from context and seldom requires explicit attention.

The question of finding inverses is an exception to this rule. As the preceding example showed, a function may well be invertible on one domain but not on another. The practical moral is to take care with domains when working with functions and their inverses.

BASICS

1. (This problem is related to Example 1 on page 70.) Let f be defined by the formula $f(x) = x^2 + 3/x$, and let g be defined by the following table.
   (a) Fill in as many entries in the following table as possible. (A few gaps are inevitable.)

x	-3	-2	-1	0	1	2	3	4
$f(x)$	8				4		10	
$g(x)$	-4	-9	-1	2	0	8	3	-2
$(f+g)(x)$								
$(f \cdot g)(x)$								
$(f/g)(x)$								
$(f \circ g)(x)$								

   (b) Briefly explain each gap in the table.
   (c) State the domains and ranges of $f + g$, $f \cdot g$, f/g, and $f \circ g$.

2. The functions f and g shown in Example 3, page 72, are defined by the formulas $f(x) = a \sin x$ and $g(x) = (x + b)^2/5 + c$ for certain constants a, b, and c.
   (a) Use the graphs in Example 3 to find numerical values of a, b, and c.
   (b) Use the formulas for f and g to find values for $(f \circ g)(-3)$, $(f \circ g)(-2)$, $(f \circ g)(-1)$, $(f \circ g)(0)$, $(f \circ g)(1)$, $(f \circ g)(2)$, and $(f \circ g)(3)$. (Approximate values are tabulated in Example 3.)
   (c) Use the formulas you found for f and g in part (a) to find a formula for $f \circ g$. Use a calculator or computer to plot $f \circ g$ in the window $[-5, 5] \times [-3, 3]$. (Compare your plot with the one in Example 3.)

In Exercises 3–14, let $f(x) = \sin x$, $g(x) = x^2$, and $h(x) = x + 3$. Write each of the following functions as a composition of one or more of the functions f, g, and h.

3. $j(x) = \sin(x + 3)$
4. $j(x) = (x + 3)^2$
5. $j(x) = \sin^2 x$
6. $j(x) = \sin(x^2)$
7. $j(x) = x^2 + 3$
8. $j(x) = 3 + \sin x$
9. $j(x) = \sin^2(x + 3)$
10. $j(x) = \sin\big((x + 3)^2\big)$
11. $j(x) = \sin(x^2) + 3$
12. $j(x) = \big(\sin(x + 3)\big)^2$
13. $j(x) = x^4$
14. $j(x) = x + 6$

In Exercises 15–22, write each function as the composition of two (or more) simpler functions. For example, the function $h(x) = (x + 1)^2$ can be written as $h(x) = (f \circ g)(x)$ where $f(x) = x^2$ and $g(x) = x + 1$.

15. $h(x) = \big(x^3 - 4x + 5\big)^6$
16. $h(x) = \sqrt{(x + 1)^3}$
17. $h(x) = \ln(2 + \sin x)$
18. $h(x) = e^{x^2}$
19. $h(x) = \sqrt{1 + e^x}$
20. $h(x) = \big|x^2 - 4\big|$

21. $h(x) = \sin\big(x^2 - 4x + 4\big)$
22. $h(x) = \cos^2 x + 4\cos x + 4$
23. Consider the function $f(x) = x^2$ restricted to the domain $(-\infty, 0]$. Since f is one-to-one on this interval there is an inverse function f^{-1}.
    (a) Sketch graphs of both f and f^{-1} on the same axes. [**HINT:** Reflect the graph of f across the line $y = x$.]
    (b) What's the formula for f^{-1}?
    (c) What are the domain and range of f? What are the domain and range of f^{-1}? How do the graphs of f and f^{-1} reflect this?
24. Suppose that the point $(-3, 5)$ is on the graph of f. Find a point on the graph of f^{-1}.
25. Suppose that $f^{-1}(2) = 3$. Find a point on the graph of f.
26. Suppose that f is a nonconstant linear function (i.e., $f(x) = ax + b$ and $a \neq 0$).
    (a) Explain why f^{-1} exists.
    (b) Show that f^{-1} is a linear function.
    (c) How are the slopes of f and f^{-1} related?
    (d) Why is the assumption $a \neq 0$ needed in part (a)?

FURTHER EXERCISES

27. Let $f(x) = x^3$ and $g(x) = \sqrt{x + 1}$.
    (a) What is the domain of $f + g$?
    (b) What is the domain of $f \cdot g$?
    (c) What is the domain of $f \circ g$?
    (d) What is the domain of $g \circ f$?
    (e) What is the domain of $g \circ g$?
    (f) What is the domain of f/g?
    (g) What is the domain of g/f?
28. Let $f(x) = \sqrt{4 - x}$ and $g(x) = x^2$. What are the domain and range of $f \circ g$?
29. Let $f(x) = \sqrt{x - 2}$ and $g(x) = -x$.
    (a) What are the domain and range of $f \circ g$?
    (b) What are the domain and range of $g \circ f$?
30. Let $h(x)$ be the concentration of oxygen (in atoms/cm^3) in the air x meters above sea level, and let $g(x)$ be the number of breaths per minute an average person has to take when the concentration of oxygen in the air is x atoms/cm^3. Which of the following tells something useful: $g \circ h$ or $h \circ g$? What does it tell?
31. Suppose that f is an odd function, g is an even function, and h is a function that is neither even nor odd, all with domain $\mathbb{R}$.
    (a) Let $k = g \circ f$. Is k even, odd, or neither? Explain.
    (b) Let $j = h \circ g$. Is j even, odd, or neither? Explain.
    (c) Let $m = f \circ f$. Is m even, odd, or neither? Explain.
32. Let $f(x) = ax + b$ and $g(x) = cx + d$. What values of the parameters a, b, c, and d make each of the following statements true?
    (a) $f \circ g = g \circ f$
    (b) $f \circ g = f$
    (c) $f \circ g = g$
33. Let f, g, and h be functions defined for all real numbers.
    (a) Find functions f, g, and h for which $f \circ (g + h) \neq f \circ g + f \circ h$.
    (b) Find functions f, g, and h for which $f \circ (g + h) = f \circ g + f \circ h$.
34. Let $f(x) = 1$ when x is irrational, and $f(x) = 0$ when x is rational. For which values of x, if any, is $(f \circ f)(x) = 0$?
35. This problem is about inverting the function $f(x) = \sin x$.
    (a) Explain why f is *not* invertible on its full domain, $(-\infty, \infty)$.
    (b) Explain why f *does* have an inverse function f^{-1} when f is restricted to the domain $[-\pi/2, \pi/2]$.
    (c) On one set of axes, carefully draw graphs of f (restricted to the domain $[-\pi/2, \pi/2]$) and f^{-1}. [**HINT:** Exploit symmetry to draw f^{-1}.]
    (d) The function f^{-1} you drew in part (c) is known as the **arcsine function**. It's usually written either $\arcsin(x)$ or $\sin^{-1}(x)$. Use a calculator or computer to plot the sine and arcsine functions together in the viewing window $[-\pi/2, \pi/2] \times [-1, 1]$. (We'll study the arcsine and other "inverse trigonometric functions" carefully later in this book.)

36. Let f be the function shown graphically as follows.

Graph of f

(a) Explain why f has an inverse.
(b) What are the domain and range of f^{-1}?
(c) Fill in the following table. (Because f is given graphically, your answers will be estimates.)

x	-3	0	2
$f^{-1}(x)$			
$f^{-1}(x+2)$			
$f^{-1}(x)+2$			
$2f^{-1}(x)$			
$f^{-1}(2x)$			

37. Let f be the function graphed below. (The graph is made up of a line segment and a quarter-circle.)

Graph of f

(a) What is the domain of f^{-1}?
(b) Evaluate $f^{-1}(-4)$, $f^{-1}(0)$, $f^{-1}(2)$, and $f^{-1}(3)$.
(c) Find a formula for $f^{-1}(x)$.

In Exercises 38–49, find the domain, the range, and a symbolic expression for the inverse of each function. Also, sketch graphs of f and f^{-1} on the same axes.   [**HINT:** In some cases you will need to restrict the domain of the given function to make it invertible.]

38. $f(x) = x^{1/3}$
39. $f(x) = -\sqrt{x}$
40. $f(x) = x^3 + 1$
41. $f(x) = 1/x$
42. $f(x) = 1/x^2$
43. $f(x) = 2/(3 + 4x)$
44. $f(x) = (x^2 + 1)^2$
45. $f(x) = x/(x^2 - 4)$
46. $f(x) = x/(x - 1)$
47. $f(x) = \sqrt{1 - x^2}$
48. $f(x) = e^{x^2}$
49. $f(x) = \ln\big(x/(x + 1)\big)$

50. Suppose that f is a periodic function with domain R. Explain why f does not have an inverse.
51. Sketch the graph of a function that is concave down and whose inverse is concave up.
52. If the domain of $f(x) = \sin(1/x)$ is taken to be an interval of the form $(0, r)$, where r is a positive real number, then f^{-1} does not exist. Explain why.
53. Let g be an odd function. Some values of g are shown in the following table. Fill in as many of the missing entries as possible.

x	-5	-2	1	3	4
$g(x)$	6	3	2	-1	-5
$g^{-1}(x)$					

54. Suppose that f and f^{-1} both have domain $(-\infty, \infty)$ and that f is an odd function. Show that f^{-1} is also an odd function.
55. Suppose that f is an even function, that the domain of f is $(-\infty, \infty)$, and that $f(0) = 1$. Explain why the domain of f^{-1} cannot be $(-\infty, \infty)$.
56. Suppose that $1 \le f(x) \le 5$ if $-2 \le x \le 3$. Is it possible that $f^{-1}(0) = 0$? Explain.
57. The graph of a function g is the reflection about the line $y = x$ of the function $f(x) = x^3 - 2$. Find a formula for $g(x)$.
58. Let $f(x) = e^{2x}$ and $g(x) = f^{-1}(x)$. Evaluate $f\big(g(\ln 2)\big)$.
59. (a) Suppose that the function f has an inverse and that $g(x) = f(x + a)$. Show that $g^{-1}(x) = f^{-1}(x) - a$.
    (b) Use the result in part (a) to find an expression for f^{-1} if $f(x) = \ln(x + 1)$.
60. (a) Suppose that the function f has an inverse and that $h(x) = f(ax)$. Show that if $a \ne 0$, then $h^{-1}(x) = f^{-1}(x)/a$.
    (b) Use the result in part (a) to find an expression for f^{-1} if $f(x) = e^{2x}$.
61. Suppose that $h = f \circ g$ where f and g are invertible functions.
    (a) Show that $h^{-1} = g^{-1} \circ f^{-1}$.
    (b) Use the result in part (a) to obtain the result stated in part (a) of Exercise 59.
    (c) Use the result in part (a) to obtain the result stated in part (a) of Exercise 60.
62. Find all values of the parameter k for which the function $f(x) = kx + |x|$ with domain $(-\infty, \infty)$ is invertible.

1.7 Modeling with Elementary Functions

Why do the basic elementary function families—algebraic, exponential, and trigono-metric—deserve their starring roles in calculus? What makes these particular functions uniquely valuable and important?

One answer is entirely practical: Elementary functions can be used to model real-world phenomena. We'll illustrate with several examples.

What Is Mathematical Modeling?

To *model* a phenomenon—the U.S. Consumer Price Index, snowshoe hare and fox populations in the Arctic, gravitational forces on spacecraft, the shape of the Golden Gate Bridge, heating requirements in Minnesota, demand for hard red Montana wheat, risks and benefits of birth control drugs, varying weights of 1-pound loaves, the progress of AIDS, Big Mac prices over the years—mathematically is to *describe* or *represent* the phenomenon quantitatively, in mathematical language. All mathematical models have two important features:

Simplification No model, mathematical or not, reproduces every property or aspect of the thing modeled. Every model necessarily *simplifies*, to some extent, the reality it describes. A useful model is simple enough to permit calculations but detailed enough to capture the essence of the phenomenon modeled.

Prediction A useful mathematical model must do more than describe already observed behavior. It should also help us predict or understand the behavior.

The process of mathematical modeling involves three steps:

Description Describing a phenomenon in mathematical language, using mathematical ingredients: functions, variables, equations, etc.

Deduction Deducing mathematical consequences of the mathematical description.

Interpretation Interpreting mathematical results in the language of the phenomenon under study.

Polynomials, Rational Functions, and the Physical World

Polynomials and rational functions are among the simplest functions, because they're built using only the familiar arithmetic operations. ▶ Despite their simplicity, these functions nicely model a surprising number and variety of real-world physical phenomena.

Addition, multiplication, subtraction, and division.

Parabolic Trajectories. What sort of path (or **trajectory**) does a projectile (a baseball, say, or a high jumper) describe during its flight from the ground into the air and back again? The general shape—some sort of arch—is clear enough, but how can we describe the trajectory mathematically? Is it a parabola? Part of a circle? One arch of a sine curve? Here are curves of all three types:

**A semicircle, a parabola, and a sine curve:
is one a trajectory?**

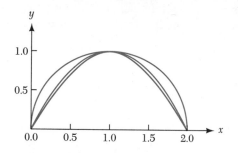

Actually, one is less convincing than the other two. For more information, see this section's exercises.

For example, ignoring wind resistance, glider effects, and other stray forces.

All three curves look possible at first glance, ◄ but only one is correct. It's a striking fact of mathematical physics that under ideal conditions ◄ a projectile influenced only by the force of gravity follows a *parabolic* trajectory. Later, using calculus tools, we'll be able to explain exactly why this fact holds. For now we'll assume it, in the following form.

> **Fact** Suppose that a projectile leaves the origin with initial velocity v_0, aimed along the line $y = mx$, and influenced only by gravity. The projectile's trajectory is a parabola, given by the quadratic equation
>
> $$y = mx - \frac{g}{2v_0^2}(1 + m^2)x^2$$
>
> (g is the [constant] acceleration due to gravity).

We'll use the formula in several ways. First note several remarks:

Initial Angle The projectile's line of aim can be described either by its slope m (as above) or by the angle θ the line makes with the positive x-axis. In this case, $m = \tan\theta$, ◄ and the trajectory equation becomes

$$y = x\tan\theta - \frac{g}{2v_0^2}(1 + \tan^2\theta)x^2.$$

Is it clear that $m = \tan\theta$? For more details see the exercises at the end of this section.

Units The preceding formulas work only with consistently chosen units of measurement. If x and y (which measure horizontal and vertical distance) are measured in feet and v_0 is measured in feet per second, then $g = 32$ feet per second per second. If, instead, x and y are measured in meters and v_0 in meters per second, then the *downward* acceleration g is (approximately) 9.8 meters per second per second.

Up or Down? Gravity causes *downward* acceleration—that's why the downward acceleration g is *positive* ($g \approx 9.8$ in metric units). We might have said, equivalently, that the *upward* acceleration is −9.8 meters per second per second.

EXAMPLE 1   A pitcher standing at the origin throws baseballs along the positive x-axis, always with initial velocity 128 feet per second ▶ but aiming along various lines $y = mx$. Describe the possible trajectories.

About 87 miles per hour—typical in the major leagues.

S o l u t i o n   In present units, $v_0 = 128$ and $g = 32$, so a trajectory with initial slope m has the form ▶

Check to make sure we've worked out the constants correctly.

$$y = mx - \frac{1 + m^2}{1024}x^2 = -\frac{1 + m^2}{1024}x\left(x - \frac{1024m}{1 + m^2}\right).$$

Here are trajectory curves for many values of m: ▶

The equations make sense in context only when $y \geq 0$, of course.

Baseball trajectories

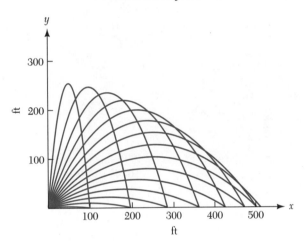

Observe:

Various Shapes Every trajectory parabola begins at the origin. Depending on its initial slope m, a trajectory may be either tall and narrow or short and squat. As the picture shows, balls that fly farthest have trajectories somewhere between these two extremes. In the next example we'll discuss how to maximize the ball's horizontal range.

Back to Earth When a ball hits the ground, its trajectory function is zero. As the preceding formulas show, ▶ each trajectory parabola has two roots: $x = 0$, where the ball starts, and

Check our work.

$$x_{\max} = 1024\frac{m}{1 + m^2}$$

where the ball lands. (If, say, $m = 1$, then $x_{\max} = 1024/2 = 512$ feet.)

At the Peak Knowing $x_{\max}$ lets us find $y_{\max}$, the ball's height at its peak. The symmetry of parabolas ▶ means that every trajectory peaks exactly halfway from $x = 0$ to $x = x_{\max}$, i.e., at $x_{\text{middle}} = 512m/(1 + m^2)$. This value of x gives the ball's maximum height for any given m:

A parabola is symmetric about the vertical line through its vertex.

$$y_{\max} = 256\frac{m^2}{1 + m^2}.$$ ■

EXAMPLE 2   Consider again the situation of Example 1. With what initial slope m should the ball be thrown to rise as high as possible? What initial slope gives the maximum possible horizontal range?

And we'll hope that our calculations agree …

Solution   We'll start with common sense. ◄   Intuition says that, for a ball to go as high as possible, it should be thrown straight up. To throw as *far* as possible, we want a trajectory that's neither too tall nor too flat: setting $m = 1$ (or, equivalently, $\theta = 45$ degrees) seems a reasonable compromise.

To check these guesses algebraically, we consider two functions of the initial slope m. For any given slope $m \geq 0$, let $f(m)$ be the peak *height* attained, and $g(m)$ the horizontal *distance* traveled. As we showed above, f and g can be written

Both f and g are rational functions of m.

as functions of m: ◄

$$f(m) = 256 \frac{m^2}{1 + m^2}; \qquad g(m) = 1024 \frac{m}{1 + m^2}.$$

Plotting f and g suggests answers to our questions:

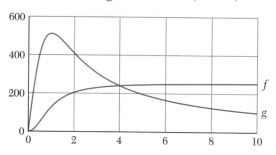

**Graphs of $f(m) = 256m^2/(1 + m^2)$
and $g(m) = 1024m/(1 + m^2)$**

Note the horizontal asymptote.

As its graph suggests, ◄   the height function f increases with m. Specifically, as $m \to \infty$,

$$256 \frac{m^2}{1 + m^2} \to 256.$$

See this section's exercises for an algebraic approach to the same problem.

Do real baseball players obey this rule? Do shot-putters?

In baseball terms, as the slope m gets larger and larger (and the line of aim approaches the vertical), the ball's maximum height approaches 256 feet.

The g-graph suggests that (as we guessed) the range function peaks at or near $m = 1$, where $g(1) = 512$ feet. Using ideas from calculus, we'll soon be able to show this rigorously by calculation. Until then, graphical evidence will do. ◄   In baseball terms: *For maximum range, aim the ball 45 degrees above the horizon.* ◄

■

Exponential Functions, College Tuition, and Growth with Interest

Exponential functions model phenomena that grow, like undisturbed bank accounts, with "interest." A fable will illustrate.

Ingrid and Eric.   Ingrid was born on Labor Day, 1991. Her family is poor, but her wealthy Uncle Eric is determined that Ingrid will someday enroll at his alma mater, St. Lena College. To this end, Eric resolves to pay Ingrid's first-year comprehensive fee at St. Lena. (After a year, Eric thinks, college students should fend for themselves.) If all goes well, Ingrid will enter St. Lena on Labor Day, 2010, her 19th birthday.

Eric's Problem.   Uncle Eric, ever prudent, wants to deposit the money now, at Ingrid's birth. Tax-sheltered 19-year bonds that pay 8% annually are available. Eric wants to cover Ingrid's first-year fee, but not trips to Fort Lauderdale or other foolishness. How much should he deposit now? Would $14,300—the 1991 fee at St. Lena—be enough?

The crucial question is St. Lena's fee in the year 2010. Uncle Eric can't hope to know it 19 years ahead, but perhaps he can *estimate* it on the basis of historical data. He starts by collecting information on St. Lena's annual fee from 1961, when the fee was $1575, through 1992, when it will be $15,400. Looking for a pattern, he plots the data. (For convenience, he lets $t = 0$ correspond to the year 1960.)

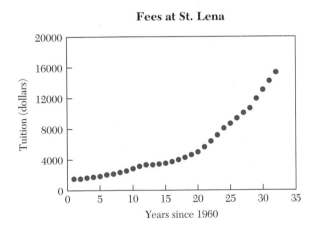

Fees at St. Lena

"Hmmm," thinks Eric. "Looks exponential to me. Might some exponential-type formula fit these data? If so, I could use the formula to estimate St. Lena's fee in 2010."

Uncle Eric Fits a Curve.   Curious, Uncle Eric plots functions of the form

$$f(t) = ae^{bt}$$

for various constants a and b, along with the fee data. Rusty on theory, he relies on trial and error. Eventually he settles on the function

$$f(t) = 1350e^{0.07606t},$$

based on this graph:

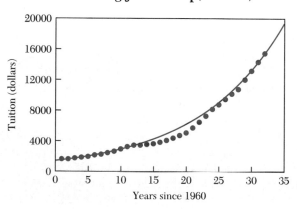

Fees at St. Lena:
fitting $y = 1350 \exp(0.07606t)$

The year 2010 corresponds to $t = 50$.

"A nice fit," thinks Eric. "If this keeps up, in 2010 ◄   I'd expect St. Lena to charge about

$$f(50) = 1350\, e^{0.07606\cdot 50} \approx 60{,}527.91.$$

The cost of books in 2010 we leave to the reader's imagination.

What the heck—I'll make it a round $60,550; the kid will need something for books." ◄

Uncle Eric Plans Ahead.   How much should Eric deposit for Ingrid *now*?

Eight percent interest compounded annually means that on each Labor Day Ingrid's old balance will be multiplied by 1.08. In functional language, let $P(t)$ denote the value of the deposit after t years; then the initial deposit—still to be determined—is $P(0)$. Here's what happens over time:

$$P(1) = P(0) \cdot 1.08;$$

$$P(2) = P(1) \cdot 1.08 = P(0) \cdot 1.08^2;$$

$$P(3) = P(2) \cdot 1.08 = P(0) \cdot 1.08^3;$$

$$\vdots$$

$$P(t) = P(t-1) \cdot 1.08 = P(0) \cdot 1.08^t.$$

Now we can solve for $P(0)$, the initial deposit. We want $P(19) = P(0) \cdot 1.08^{19} = 60{,}550$, so

$$P(0) = \frac{60{,}550}{1.08^{19}} \approx 14{,}030.17.$$

How It Ended, and the Moral.   Uncle Eric put $14,030.17—not quite this year's fee—into 19-year bonds. Baby Ingrid, although still an infant, already looks like college material.

The fable hints at an important—the most important—use of exponential functions: modeling the behavior of quantities that grow (or shrink) under the influence of "interest." Financial quantities most familiarly exhibit such behavior, but so do others: populations, radioactive masses, and many more. ▶

We'll revisit exponential growth in Chapter 4.

Do Models Describe or Predict?   With a little mathematics and lots of hindsight, we could fit the *past* behavior of the St. Lena fee as well as we like. For example, it's possible (but messy) to find a polynomial function, of degree 30 or so, whose graph hits every one of the 31 points Uncle Eric plotted. Such a function might be said to describe the past, but must it also accurately predict the future?

No! How accurately *any* mathematical model can be expected to predict behavior is a difficult and subtle question. The answer depends on many factors, some nonmathematical: the nature and complexity of the phenomenon being modeled, the reliability and validity of the observed data, and other factors.

Trigonometric Functions: Modeling Weather

Minnesota weather is famous for its wild swings. Averaged over many years, however, even the weather becomes surprisingly predictable. Weather phenomena are by nature periodic, so it's natural to use trigonometric functions in weather models.

Temperature in Minneapolis.   Plotting average weekly temperatures in Minneapolis ▶ produces a familiar shape:

From U.S. Weather Service data. Notice the horizontal scale; the "year" starts around March 21.

Average weekly temperature in Minneapolis

Although the plot resembles a sine curve, the unembellished formula $y = \sin x$ clearly doesn't fit the data:

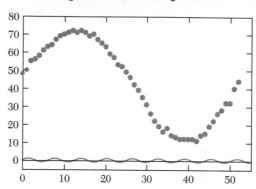

Fitting the data: start with sin x
Temperature data and $y = \sin x$

Unpromising as the preceding graph may seem, let's persevere. Weather varies periodically (in an average sense, at least), so it's reasonable to use some **sinusoidal function**

$$a \sin(bx + c) + d$$

to model our data. Our problem, then, is to choose appropriate values for a, b, c, and d.

A close look at the temperature scatter plot tells a lot.

A Vertical Center   The temperature data are vertically centered around 42 degrees Fahrenheit. Therefore, an appropriate sinusoid should be lifted 42 units up. We'll try $d = 42$.

There's no shame in the word "about"— mathematical modeling is an inexact science!

High and Low Temperatures   The maximum and minimum temperatures— 72 and 11 degrees, respectively—suggest a vertical variation of about ◄ 31 degrees Fahrenheit in each direction. This argues for stretching the sine function vertically with factor $a = 31$.

Stretching the Period   The period for temperature data is 52 weeks; the sine function's period is 2π. To match things up, we must stretch the sine curve horizontally, with factor $52/(2\pi)$. Setting $b = 2\pi/52 \approx 0.12083$ has this effect.

We arranged this by starting our 1-year period in spring. This does no harm. Had we insisted on an ordinary calendar year we would have needed a nonzero horizontal shift.

No Horizontal Shift   The scatter plot suggests that no horizontal shift is needed: the sinusoid seems to start at the left. Thus, we set $c = 0$. ◄

Combining these observations argues for selecting the sinusoid

$$y = 31 \sin\left(\frac{2\pi}{52}x\right) + 42$$

as our best guess to fit the data. The remaining plots show the effects of adding these parameters one by one to the basic sine function.

Setting $d = 42$ produces a vertical shift:

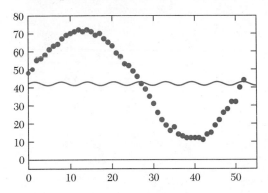

Fitting the data: a vertical translation
Temperature data and $y = \sin x + 42$

The graph is far too flat. Setting $a = 31$ turns up the "amplitude":

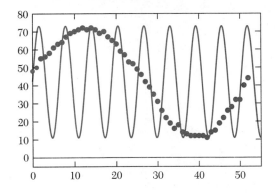

Fitting the data: increase amplitude
Temperature data and $y = 31 \sin x + 42$

Now it's wiggling too fast. Including $b = 2\pi/52$ stretches the period:

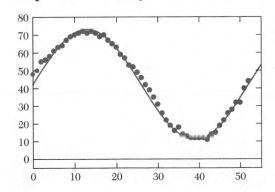

Fitting the data: make the period 52
Temperature data and $y = 31 \sin (2\pi x/52) + 42$

This graph is much better.

EXAMPLE 3   Predict the mean temperature in Minneapolis on December 31.

Solution   December 31 falls 40 weeks and 5 days (i.e., 40.71 weeks) after the spring equinox. The formula predicts a mean temperature of

$$42 + 31 \sin\left(\frac{2\pi}{52} \cdot 40.71\right) \approx 11.7°\text{F}. \qquad \blacksquare$$

BASICS

1. Consider again the three curves in the picture on page 82. (The curves are candidates for trajectories of an ideal projectile.)

    (a) The three curves have three points in common. Assign these points the coordinates $(0,0)$, $(1,1)$, and $(2,0)$, respectively. In this system, the three curves have equations $y = \sqrt{1-(x-1)^2}$, $y = 1 - (x - 1)^2$, and $y = \sin(\pi x/2)$. Which curve has which equation? How do you know?

    (b) The angle a trajectory curve makes with the ground (i.e., the horizontal axis) is the angle at which the projectile is "fired." Estimate these angles (in radians) for the three curves shown. Using the result, explain why the semicircle cannot be a trajectory curve.

    (c) The Fact box on page 82 says that among the three candidate curves, only the quadratic curve $y = 1 - (x - 1)^2 = 2x - x^2$ is a trajectory. Suppose that this curve *is* a trajectory and that the x and y units represent feet (so the projectile travels 2 feet horizontally and has a maximum height of 1 foot). Find the initial slope m and the initial velocity v_0. (Use appropriate units throughout; time is measured in seconds.)

    (d) Repeat part (c), but assume now that the x and y units represent meters (so that the projectile travels 2 meters horizontally and has a maximum height of 1 meter; time is still measured in seconds). How do your answers compare with those in the previous part?

2. We claimed in this section that if θ is the angle from the positive x-axis to the line $y = mx$, then $m = \tan\theta$. Explain why, using the right triangle with vertices $(0,0)$, $(1,0)$, and $(1,m)$.

3. All parts of this exercise refer to the situation of Example 1 (page 83). In particular, assume throughout that a ball thrown from the origin with initial slope m and initial velocity 128 feet per second follows the trajectory curve

$$y = mx - \frac{1+m^2}{1024}x^2.$$

    (a) Show that the ball lands at $x = \dfrac{1024m}{1+m^2}$ feet. (The ball lands where $y = 0$.)

    (b) Let m be given. At what value of x is the ball highest? (Since the trajectory is a parabola, the vertex occurs halfway between the x-intercepts.)

    (c) Let m be given. How high is the ball at its peak? (Use the preceding result.)

    (d) Use results of part (c) to complete the following table:

m	0.1	0.5	1.0	2.0	10.0	
Peak height						255.9

    (e) A ball thrown straight up with initial velocity 128 feet per second rises 256 feet. Give an informal reason why, using results from part (d).

4. In Example 2 (page 84) we claimed that if $g(m) = 1024m/(1 + m^2)$, then $g(1) = 512$ is the largest possible value of g. Equivalently, $1/2$ is the largest possible value of $m/(1 + m^2)$. Prove the latter claim by justifying each of the following steps.

$$0 \le (1-m)^2 \implies 0 \le 1 - 2m + m^2$$
$$\implies 2m \le 1 + m^2$$
$$\implies \frac{m}{1+m^2} \le \frac{1}{2}.$$

5. **(The danger zone)**   A curious bird flies over the baseball field of Example 1 (page 83). The bird wants to see what's happening but also wants to avoid collisions with thrown balls. How high must the bird fly to be safe? (If you find this scenario unconvincing, then try this: What

shape should the stadium's roof have to avoid being hit by thrown balls?) The answer depends, of course, on x: The bird's danger zone is tall for x near zero but gets shorter as x takes progressively larger positive values; for sufficiently large x, the bird is safe even on the ground. The bird's danger zone is defined by still another curve. Here's the picture:

Baseball trajectories and their upper envelope

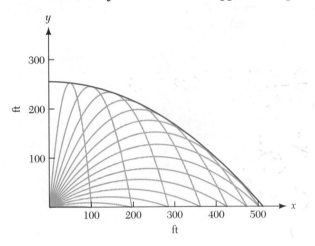

ft

The curve on top is called the **upper envelope** of our family of baseball trajectories. In this exercise we'll study the upper envelope curve.

(a) We said in Example 2 (page 84) that (i) a ball thrown straight upward at 128 feet per second rises to 256 feet and (ii) the maximum horizontal range for any trajectory is 512 feet. Use these facts to label the x- and y-intercepts of your upper-envelope curve.

(b) The upper-envelope curve looks like half a parabola, but is it really parabolic? If so, its equation must be of the form $y = a - bx^2$ for some constants a and b. Use this fact and results from part (a) to find numerical values for a and b.

(c) Using your results from the previous part, plot the parabola $y = a - bx^2$ and the trajectory curves for $m = 0.5$, $m = 1.0$, $m = 2.0$, and $m = 4.0$—all on the same axes. Use the window $[0, 600] \times [0, 300]$. Does your picture suggest that you've chosen the upper envelope curve correctly? Why or why not?

6. Exercise 5 gives convincing graphical evidence that the upper-envelope curve for the baseball trajectories is (like the trajectories themselves) a parabola. A rigorous proof of this fact requires some careful algebraic manipulation. Provide such a proof, using the following outline.

(a) Let $q(t) = at^2 + bt + c$ be a quadratic polynomial in t; a, b, and c are constants, with $a \neq 0$. Show that $q(t)$ can be written in the form

$$q(t) = a\left(t + \frac{b}{2a}\right)^2 + \left(c - \frac{b^2}{4a}\right).$$

Conclude from this (explain your reasoning) that the parabola $y = q(t)$ has its vertex at $t = -b/2a$.

(b) Let u be the function whose graph is the upper envelope for the baseball trajectories. We want to show that for any fixed x_0 between 0 feet (the minimum horizontal range) and 512 feet (the maximum range), $u(x_0)$ is given by a quadratic expression in x_0. (We'll use x_0 rather than x to emphasize that, for now, x_0 is a fixed number.) Let $h(m)$ denote the height at x_0 of the trajectory curve with initial slope m. Explain the statement:

$u(x_0)$ is the maximum possible value of $h(m)$.

(c) Show that, for $m > 0$,

$$h(m) = mx_0 - (1 + m^2)\frac{x_0^2}{1024}$$
$$= -\frac{x_0^2}{1024}m^2 + x_0 m - \frac{x_0^2}{1024}.$$

(d) Part (c) shows that $h(m)$ is a quadratic function of m. Since the coefficient of m^2 is negative, h takes a maximum value at its vertex. Show that this maximum occurs at $m = 512/x_0$ and that the maximum value is $256 - x_0^2/1024$.

(e) Conclude that $u(x_0) = 256 - x_0^2/1024$, so the upper envelope is indeed a parabola.

7. According to Uncle Eric's model, St. Lena's tuition doubles every X years. What is X? Why?

1.8 Chapter Summary

This chapter is a general introduction and fly-by overview of some of the objects, ideas, uses, and techniques of calculus.

Functions: More than Formulas.   The most important idea in this chapter is that of **function**. We expect most readers are familiar, from precalculus experience, with functions such as

$$f(x) = x^2 + \frac{3}{x} + \sqrt{x}$$

that are defined by formulas. Such functions will indeed play important roles in this book. More characteristic of the chapter, however, are functions defined in other ways: by graphs, by tables, and in words. An important goal of this book is to present calculus ideas from graphical and numerical (as well as algebraic) points of view. In real-world practice, functions seldom appear ready-made as formulas.

Graphs: Pictures of Functions.   Formulas, even simple ones, give little sense of how a function behaves for various inputs. Graphs, by contrast, are ideally suited for this purpose. A graph gives a function a geometric, not just symbolic, meaning. Best of all, graphs show how functions vary—when they rise, when they fall, when they're concave up or concave down, and so on. All of these properties are crucially important in calculus itself, but even more so in applications of the subject.

Machine Graphics.   With computers and calculators, drawing accurate graphs becomes easy. What remains is the good stuff: interpreting what graphs say. Doing so is an unending refrain throughout this book.

Using machines effectively to plot functions requires some care (things can go wrong) and some familiarity with newfangled jargon: **xrange**, **yrange**, **plotting windows**, **aspect ratio**, etc. Section 1.3 discussed such niceties.

Functions, Formally.   Truly understanding the idea of function means understanding its three main parts: **rule**, **domain**, and **range**. Section 1.4 treated such things, with many examples. The variety is important: Seeing function ideas and terminology in various settings helps clarify what they mean in *any* setting.

A Field Guide to Elementary Functions.   Section 1.5 introduced the cast of leading functional characters in calculus. Many of the functions we'll treat in calculus are **elementary**—that is, built by combining members of the leading function families: algebraic, trigonometric, exponential, and logarithm functions. We met these function families briefly in Section 1.5. More details about each family appear in the appendices and in later sections.

New Functions from Old.   Section 1.6 described various ways of combining old functions to form new ones. Algebraic combinations—sums, differences, products, and quotients—are the simplest; **composition** and **inversion**, perhaps less familiar, are equally important in calculus.

Modeling with Elementary Functions.   The basic elementary function families—algebraic, exponential, and trigonometric—deserve their starring roles in calculus because they're practically useful, both mathematically and in applications. Section 1.7 described several uses of elementary functions in modeling real-world phenomena.

2

The Derivative

2.1 Amount Functions and Rate Functions: The Idea of the Derivative

Introduction

Calculus is the mathematics of change. The immense practical power of calculus derives from its ability to describe—and predict—the behavior of changing quantities. Falling apples, orbiting spacecraft, growing populations, decaying radioactive materials, rising consumer prices—all can be modeled through the calculus of change.

The calculus tool that makes all this possible is called the derivative function. ▶ In coming chapters we'll spend a lot of time and effort studying how to define derivative functions formally and how to calculate them symbolically. In the process, various subtleties and puzzles will inevitably arise. Nevertheless, the basic idea behind derivatives is quite simple: *For a given function f, the derivative function, denoted by f',* ▶ *tells the rate of change of f.* This principle, suitably interpreted, lies behind all our calculations and applications of the derivative.

The name isn't very descriptive, unfortunately. There are many ways to derive one function from another.

Read: "eff prime."

Whose Idea? No one person invented the derivative. Related ideas and methods appear throughout mathematical history, spanning at least 2000 years. The modern European development of derivative happened largely in the 17th and 18th centuries, ranging from Pierre de Fermat (1601–1665; French) through Isaac Newton (1642–1727; English) and Gottfried Leibniz (1646–1716; German)—generally considered the cofounders of modern calculus—to Leonhard Euler (1707–1783; Swiss) and Joseph-Louis Lagrange (1736–1813; French). Lagrange, building on Euler's idea of function, may have been first to use both the phrase "derivative function" ▶ and the prime symbol to denote it.

In French, fonction dérivée.

About This Chapter. This chapter is about the concept of the derivative function. The relationship between f and f' is the main theme; we'll study what it means for f' to be the rate function derived from f, what each function says about the other and especially how graphs of f and f' are related. Formal definitions appear toward the end of the chapter, and symbolic computational techniques after that.

Well-Behaved Functions.   It must be admitted that functions f exist for which the derivative function f' is either somehow ill-behaved or, worse, doesn't exist at all. ◄   In this chapter, though, we assume that (unless otherwise noted) our functions are well-behaved—i.e., that a given function f has a derivative f' and that both f and f' have smooth, unbroken graphs. Later, as we develop new language and tools, we'll study more carefully some of what can go wrong.

We'll see some examples soon.

The Derivative as a Rate Function

Any function f can be used to build new functions, derived in one way or another from f. Starting with $f(x) = \sin(x)$, for instance, we can concoct all sorts of new functions. The rules

$$g(x) = \sin(2x), \qquad h(x) = \sin(x + \pi/2), \qquad \text{and}$$
$$k(x) = \frac{\sin(x + 0.1) - \sin x}{0.1}$$

all define "relatives" of f; the possibilities are endless.

Among all the possible functions one might derive from a given function f, the derivative function f' is unquestionably most important. Here's an informal description:

> **Definition** (The Derivative as Rate Function) Let f be any function. The new function f', called the **derivative** (or **rate function**) of f, is defined by the rule
>
> $$f'(x) = \text{instantaneous rate of change of } f \text{ at } x.$$

(We'll use "derivative" and "rate function" interchangeably to describe f'; when we do, we'll call f an **original function** or an **amount function**.)

We'll have to say, for instance, exactly what "instantaneous" means, mathematically.

We'll revisit and refine our definition of derivative in later sections. ◄   First, however, we'll try out the ideas of rate function and amount function "on the road."

Rates, Amounts, and Cars: The Prime Example

Moving cars provide the most familiar setting for comparing rates and amounts; cars nicely illustrate most of the basic ideas and even some of the subtleties. Here's a typical situation, stripped to essentials:

At noon on October 1, 1995, a car enters Interstate 94, heading east, from the State Capitol on-ramp in Bismarck, North Dakota. For several hours, the car moves—sometimes east, sometimes west—at varying speeds.

To put this scenario in functional language, let's define some symbols:

t   the elapsed time, measured in hours, since a reference time $t = 0$

P(t) the car's position at time t, measured in miles *east* of a reference point at time t

V(t) the car's *eastward* velocity at time t, measured in miles per hour (mph)

Here are possible graphs of the position and velocity functions:

Graph of P(t)

Graph of V(t)

There's a lot to see: ▶

Watch for several uses of the prime notation.

Points of Reference.   Choices of a reference time and a reference point were entirely up to us. We could have used, say, the moment in the year 800 when Charlemagne was crowned Emperor of the West as our reference time, and downtown Glendive, Montana, as our reference point. But it seems natural to choose noon on October 1, 1995, and the State Capitol on-ramp in Bismarck.

Position and Velocity as Functions of Time.   Once we've specified the necessary reference points and units of measurement, P and V become usable functions of t; at any sensible input time t, the car has a well-defined position $P(t)$ and velocity $V(t)$. For example, the statements

$$P(2) = 100 \qquad \text{and} \qquad V(2) = 70$$

mean that at 2 P.M. the car is 100 miles *east* of Bismarck, moving *east* at 70 mph.

Amount vs. Rate.   It's natural to think of P as an amount function and V as the associated rate function. ▶   For any time t, $V(t)$ is the *instantaneous rate of change of position with respect to time*. In symbols, $V(t) = P'(t)$, or simply $V = P'$.

The units—miles and miles per hour, respectively—support this view.

Positive and Negative Values.   Both P and V can take either positive or negative values. The statements

$$P(6.5) = -40 \qquad \text{and} \qquad V(6.5) = -60$$

Maybe the driver missed his exit and still hasn't noticed it.

Equivalently, the P-graph is rising.

Speedometer dials, for instance, contain no negative numbers.

In fact, $V(4) = -70$. What does this mean about direction?

Do the graphs support these claims?

mean that at 6:30 P.M. the car is 40 miles *west* of Bismarck, traveling *west* at 60 mph. ◄ When $P(t) > 0$, the car is *east* of the State Capitol. When $V(t) > 0$, the car is headed *eastward*. ◄

Position vs. Distance, Velocity vs. Speed.

Our functions describe the car's position and velocity. Why not use the more familiar terms "distance" and "speed"? The difference is technical but nevertheless important. In physical language, distance and speed connote *nonnegative* quantities. ◄ Because position and velocity can take either positive or negative values, they can convey information about direction (in this case, east or west).

What the Speedometer Tells; What It Doesn't.

Speedometers can't tell direction, so a speedometer reading of 70 mph at $t = 4$ means that $V(4) = \pm 70$. ◄ The speedometer reading, moreover, is an instantaneous speed. It means that if the car maintains the same speed for an hour, then it covers 70 miles.

Acceleration—Another Rate.

Velocity, like position, has its own rate function. In physical language, this new rate function—the rate of change of velocity with respect to time—is called acceleration. Like position and velocity, the acceleration varies with time; in symbols: $A(t) = V'(t)$. Acceleration, too, can take either positive or negative values. The statement $A(2) > 0$, for instance, means that at $t = 2$ the car's eastward velocity is *increasing*. Similarly, $A(1) < 0$ implies that the car's eastward velocity is *decreasing*. ◄

Velocity—Rate or Amount? It Depends.

Earlier we viewed V as the *rate* function derived from P. Now our point of view has changed: When we consider acceleration, velocity becomes the associated *amount*.

Here's the moral and some advice: *Any function can be thought of either as a rate or as an amount, depending on the situation.* To avoid confusion, think of rate functions and amount functions together, as inseparable pairs.

Units of Velocity and Acceleration.

Every rate, properly understood, is the rate of change of something with respect to something else. The proper units for measuring rates depend both on the quantities that vary and on the units used to measure those quantities.

In the current setting, velocity is the rate of change of position (measured in miles) with respect to time (measured in hours), so the unit of velocity is miles per hour. Acceleration is the rate of change of velocity with respect to time, so its appropriate unit of acceleration is miles per hour per hour—the ratio of the unit of velocity to the unit of time. ◄

Acceleration is sometimes expressed in other units, such as miles per hour per second.

The variable name t suggests variation in time, but this is only an aid to intuition; mathematically, rates need not involve time.

Rate Functions in Other Contexts

The ideas of amount function and rate function are intuitively natural when applied to physical quantities, such as distance and velocity, but the ideas of rate and amount make good sense in any context. *Any* function f can be thought of as an amount function: To an input t, f assigns the amount $f(t)$. ◄ How one interprets f' as a rate depends on context. ▶   If $f(t)$ is the position of a moving object at time t, then $f'(t)$ is the velocity at time t, i.e., the rate of change of position with respect to time. If $f(t)$ represents altitude, then $f'(t)$ represents *rate of ascent*, i.e., the rate of change of altitude with respect to time. If $f(t)$ represents the varying value of a bank deposit, then $f'(t)$ represents the rate, at time t, at which money flows into the account. If $f(x)$ tells how many gallons of gasoline are consumed in covering x miles, then $f'(x)$ represents the varying rate of consumption, in gallons per mile. ▶

The units of f' depend on the situation, too.

Here we use gallons per mile—not the more familiar miles per gallon.

Even if f is a "pure" mathematical function—unattached to cars, balloons, money, or anything else—the idea of a rate function still makes good sense. To see how, notice that in each instance above, f' measured how fast the *output* from f increased with respect to the *input* to f. The same idea applies to "pure" mathematical functions: If f is any function, and $y = f(x)$, then $f'(x)$ represents the rate, at x, at which y increases with respect to x. For instance, the statement $f'(3) = 5$ means, that if $x \approx 3$, increasing x by a small amount produces about five times as much increase in $f(x)$.

The Derivative Graphically: f' as a Slope Function

What does the derivative mean *graphically*? How do the graphs of f and f' reflect the amount–rate relationship?

Slope is the key idea that links graphs and rates. By definition, slope is a rate of change, i.e., the *ratio* ▶   of two changes, Δy and Δx. Specifically, the slope of the f-graph at any point $\big(x, f(x)\big)$ is the instantaneous rate of change of f with respect to x. In a nutshell:

"Rate" and "ratio" have common roots.

> **Definition** (**The Derivative as a Slope Function**) Let f be any function. The derivative function f' is given by the rule
> $$f'(x) = \text{slope of the } f\text{-graph at } x.$$

Like our first, informal definition of derivative, this one needs—and will get—plenty of refinement and discussion. ▶   First let's see what the definition means.

We'll have to say, for instance, precisely how to calculate a graph's "slope at a point."

Derivatives of Linear Functions

It's especially easy to find derivatives of linear functions.

EXAMPLE 1   Let f be the linear function

$$f(x) = 3x + 1.$$

Explain, graphically and in terms of rates, why the derivative is the *constant* function

$$f'(x) = 3.$$

Solution   The graph of f is a straight line, so its slope—at *any* value of x—is 3. In symbols, $f'(x) = 3$ for all x. Here are graphs of f and f':

Graph of $f(x) = 3x + 1$

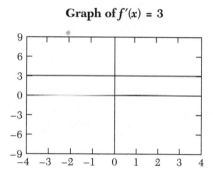

Graph of $f'(x) = 3$

That f is linear with constant slope 3 means that the f-graph rises 3 units for every unit it runs. In rate language, f increases everywhere at the rate of 3 y-units per x-unit. ■

The example illustrates a general fact:

> **Fact** For any linear function $f(x) = ax + b$, the derivative (rate) function is the constant function $f'(x) = a$.

The Slope of a Graph at a Point: Tangent Lines

For straight lines, the idea of slope is simple: A line has only one slope, and any two points suffice to compute it. But what about a *curved* graph—how can we measure *its* slope at a point? Indeed, does a curve even *have* slope at a point?

Defining and finding the slope of a curve at a point are key ideas (perhaps *the* key ideas) of calculus. In this section we take a first graphical look; in the next section we'll return to the problem in more detail.

The slope of a curve at a point P can be thought of as the slope of a certain line through P (called the **tangent line at P**) that "fits" the curve at P. Equivalently, the tangent line at P "points in the direction of the curve." ◄   In car talk:

The words in quotation marks are an intuitive description. Formal definitions come later in this chapter.

The curve is a highway, seen from above. A car, headlights blazing through fog, follows the road. When the car is at point P, its headlights point in the direction of the tangent line at P. ▶

If the driver lost control at P, the car would (literally) go off on a tangent.

Using these intuitive ideas, one can usually draw tangent lines with reasonable accuracy and estimate their slopes. Here are short segments of tangent lines at several points along a curve:

Examples of tangent lines

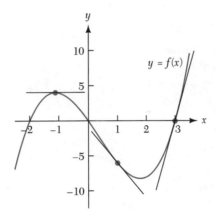

The leftmost tangent segment, at about $x = -1.1$, looks approximately horizontal. In function notation: $f'(-1.1) \approx 0$. Estimating slopes of the other two tangent segments suggests that $f'(1) \approx -5$ and $f'(3) \approx 15$. ▶

Do these estimates look reasonable to you?

Relating Amount and Rate Functions

How are an amount function f and its rate function f' related to each other? What can either function say about the other? These questions are at the heart of calculus. We'll revisit them often, in graphical, numerical, and symbolic forms. The rest of this section begins the discussion.

The Racetrack Principle: Fast Dogs Win

Here's a safe bet:

If dogs f and g start together and dog g always runs faster than dog f, then dog g always leads and therefore wins the race.

Credit for this terminology goes to H. Porta and J. J. Uhl, University of Illinois.

This so-called **racetrack principle**, ▶   appropriately rephrased, is surprisingly useful. The first step is to interpret it as a property of rate and amount functions. ▶

f' tells the velocity of dog f.

Here's one way to do so:

F a c t (**The Racetrack Principle**) Let f and g be functions defined for all x in $[a, b]$, and suppose that $f(a) = g(a)$. If

$$f'(x) \leq g'(x)$$

for all x in $[a, b]$, then

$$f(x) \leq g(x)$$

for all x in $[a, b]$.

Observe these fine points.

Fair Start The requirement that $f(a) = g(a)$ means that f and g "start together" at $x = a$.

Rates of Growth The condition that $f'(x) \leq g'(x)$ means that f "grows no faster" than g for $a \leq x \leq b$.

Never Ahead The conclusion means that f "never outruns" g for $a \leq x \leq b$.

In the strict version, $<$ replaces $\leq$.

Strict Inequalities A strict version ◄ of the racetrack principle also holds: If $f(a) = g(a)$ and $f'(x) < g'(x)$ for x in $[a, b]$, then $f(x) < g(x)$ for every x in (a, b).

The racetrack principle is easiest to use when either f' or g' is a constant function. In that case, we can compare f or g to a linear function.

EXAMPLE 2   Let f be a function such that (i) $f(0) = 0$; (ii) $f'(x) \leq 10$ if $x \geq 0$. Show that $f(1) \leq 10$.

A dog with a top speed of 10 meters per second can't run more than 10 meters in one second.

Solution   Informally speaking, (ii) means that, for $x \geq 0$, $y = f(x)$ increases at a rate no greater than 10 y-units per x-unit. Equivalently, the slope of the f-graph doesn't exceed 10. Thus, as x runs 1 unit (from $x = 0$ to $x = 1$), y rises no more than ten units (from $y = 0$ to $y = 10$), so $f(1) \leq 10$. ◄

The racetrack principle restates these ideas a little more formally. To apply it, let g be the linear function $g(x) = 10x$. (We chose this particular g because it has constant slope 10, so $g'(x) = 10$, and because $g(0) = 0 = f(0)$.) Now all the ingredients are ready. Since $f(0) = g(0)$ and $f'(x) \leq 10 = g'(x)$, the racetrack principle guarantees that for $x \geq 0$, $f(x) \leq g(x) = 10x$. Setting $x = 1$ gives $f(1) \leq g(1) = 10$.   ■

More Safe Bets: Racetrack Variants

The basic racetrack principle—that faster dogs pull ahead—can be restated in several other useful forms by changing the conditions slightly. In the three variants that follow, we no longer assume that dogs f and g start together. Each principle appears first in words, then in symbolic notation. ◄

Does each word version seem reasonable? Do the symbols really say the same thing?

C represents the constant distance between the dogs.

Equal Speeds, Equal Distances In words: If two dogs run equally fast from time a to time b, then the distance between them remains constant. In symbols: If $f'(x) = g'(x)$ for $a \leq x \leq b$, then for some constant C, $f(x) - g(x) = C$ for $a \leq x \leq b$. ◄

Faster Means Farther   In words: If dog f runs no faster than dog g, then dog f runs no farther than dog g from time a to time x—regardless of where each dog starts. In symbols: If $f'(x) \leq g'(x)$ for $a \leq x \leq b$, then

$$f(x) - f(a) \leq g(x) - g(a).$$

for $a \leq x \leq b$. (Note that $f(x) - f(a)$ and $g(x) - g(a)$ represent, respectively, the net changes in f and g from a to x. In racing terms, $f(x) - f(a)$ and $g(x) - g(a)$ are the net distances covered by dogs f and g from time a to time x.)

Photo Finish   In words: If dogs f and g finish together, then the faster dog cannot have been ahead at any time. In symbols: Suppose that $f'(x) \leq g'(x)$ for $a \leq x \leq b$ and $f(b) = g(b)$. Then, for $a \leq x \leq b$,

$$f(x) \geq g(x).$$

EXAMPLE 3   Suppose that $f(0) = 3$ and $f'(x) \leq 2$ for all x. What can be said about the values $f(2)$ and $f(-4)$? Interpret the result graphically.

Solution   We apply the racetrack principle with $g(x) = 2x + 3$. We chose this g because (i) $g(0) = f(0) = 3$; and (ii) $g'(x) = 2 \geq f'(x)$. ▶ To estimate $f(2)$ we use the ordinary racetrack principle on the interval $[0, 2]$. It says that

Why is $g'(x) = 2$? Because g defines a line with slope 2.

$$f(0) = g(0) \qquad \text{and} \qquad f'(x) \leq g'(x) \implies f(2) \leq g(2) = 7.$$

To estimate $f(-4)$, we'll use the photo-finish version on the interval $[-4, 0]$. It says that

$$f(0) = g(0) \qquad \text{and} \qquad f'(x) \leq g'(x) \implies f(-4) \geq g(-4) = -5.$$

All of this can be understood geometrically. Here are possible graphs of f and g:

Racing functions: graphs of f and g

As the picture shows, f and g agree at $x = 0$, and f never rises faster than g. As a result, says the racetrack principle, f lies above g before $x = 0$, and below g after $x = 0$. ▶

Compare, for example, the points at $x = -4$ and $x = 2$.

Derivatives in Action: Using Rate and Amount Functions

Many real-world phenomena naturally involve rate and amount functions; derivatives are just what's needed to describe and predict their behavior. We end this section with some examples of derivatives in action. As before, we defer calculations until later sections.

Free Fall Without Air Resistance

A moving object is said to be in **free fall** if it is affected only by gravity. Other possible forces—air resistance, engine thrust, etc.—are assumed to play no role.

Gravity exerts the same downward acceleration on all falling objects regardless of their size, weight, initial velocity, or other characteristics. (For objects near the surface of the earth, this acceleration is essentially constant.) Galileo Galilei (1564–1642) is supposed to have discovered this fact by tossing a variety of objects from the Leaning Tower of Pisa and noticing that all landed at about the same time.

> **Can Air Resistance Be Ignored?**   In practice, air resistance matters. Parachutists depend on it for survival; cannonballs fall faster than, say, live chickens. *What* Galileo dropped isn't known; indeed, the whole Pisa story seems unlikely, with all those tourists milling around below. For physical reasons, air resistance is usually negligible for small, dense objects moving at moderate speeds, as when a coin falls to the floor.

That is, v is the rate function of h; a is the rate function of v.

A falling object has, at any time t, a certain height $h(t)$, an upward velocity $v(t)$, and an upward acceleration $a(t)$. Given appropriate measurement scales for time and distance, h, v, and a are all functions of t, and the derivative relations $v = h'$ and $a = v'$ hold. ◄   Galileo's discovery that an object in free fall has constant downward acceleration says symbolically that $a(t) = -g$ or, equivalently, that

$$v'(t) = -g$$

The g stands for gravity.

for some positive constant g. ◄   (The negative sign is there because the object accelerates downward.) In English units, $g \approx 32$ feet per second per second; in metric units, $g \approx 9.8$ meters per second per second.

We reasoned similarly earlier, when we said that if a function is linear, then its derivative is constant.

Formulas for Distance and Velocity.   In the late 1600s Isaac Newton used Galileo's discovery to derive his laws of motion. Newton reasoned that if (as in free fall) the acceleration function $a(t)$ is *constant*, then the velocity function must be linear. ◄

$v_0 > 0$ means that the object rises initially, perhaps by having been thrown upward.

Suppose that at time $t = 0$ seconds an object is h_0 feet above ground level, and has *upward* velocity v_0. ◄   Newton's laws of motion imply rather simple formulas for $h(t)$ and $v(t)$, the functions for altitude (in feet) and (vertical) velocity (in feet per second):

$$h(t) = h_0 + v_0 t - 16t^2; \qquad v(t) = v_0 - 32t.$$

With these formulas ▶ we can answer many questions related to free fall.

Consider: Why are the minus signs present?

EXAMPLE 4 Could a major league pitcher throw a ball to the 195-foot ceiling of the Hubert H. Humphrey Metrodome, home of the Minnesota Twins?

Solution Not likely. A pitcher throws a fast ball with an initial horizontal velocity of around 95 mph, or 140 ft/sec. Throwing straight up is awkward, so let's assume a smaller initial *vertical* velocity, say 100 ft/sec. If the ball starts from an initial height of 6 feet, then the preceding formulas say that

$$h(t) = 6 + 100t - 16t^2; \qquad v(t) = 100 - 32t.$$

Height is greatest at the instant when upward velocity is zero, ▶ that is, when

Why?

$$v(t) = 100 - 32t = 0 \iff t = \frac{100}{32}.$$

At that time, ▶

Check the calculation.

$$h\left(\frac{100}{32}\right) = 162.25 \text{ ft},$$

lower than the Metrodome's ceiling. Graphs of h and v follow.

Graph of $h(t) = 6 + 100t - 16t^2$

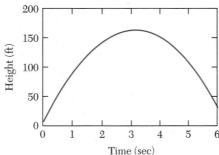

Graph of $v(t) = 100 - 32t$

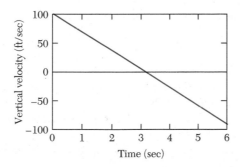

Differential Equations

Any equation that involves derivatives is called a **differential equation (DE).** The equations

$$f'(x) = 3; \qquad v'(t) = -g + kv(t)^2; \qquad f'(x) = f(x)$$

are all DEs. A **solution** to a differential equation is a *function* that satisfies the differential equation. ◄ (As we saw, $f(x) = 3x + 1$ is one solution to the first differential equation. We'll see solutions to the others later.)

NOTE: *A solution to a DE is a function, not a number.*

Many of the most important applications of the calculus—especially physical applications—involve posing problems as differential equations and then finding and using solutions.

Falling with Air Drag

It's a fact of physics that when small, dense objects (such as baseballs) fall at moderate speeds, the deceleration due to air drag is (approximately) proportional to the *square* of the velocity. In derivative notation, this fact says that the total acceleration is

$$v'(t) = -g + kv(t)^2$$

for some positive proportionality constant k. (The first term, $-g$, represents the downward acceleration due to gravity.) The numerical value of k depends on physical factors—smoothness of the object, air density, and so on. For a falling baseball, the value $k \approx 0.003$ is reasonable.

E X A M P L E 5   On an outing with Dad, Bud and Sis play catch on the observation deck of the IDS Tower in Minneapolis. To Dad's horror their baseball slips over the edge, squarely beaning a lawyer who is on a lunchtime stroll along the Nicollet Mall, 600 feet below. What happens next?

S o l u t i o n   Thanks to air drag, Dad is sued for personal injury, not negligent homicide. Here's why.

Note that $h(0) = 600$; what does this mean in the present context?

Without air drag, Newton's free-fall model applies, so ◄

$$h(t) = 600 - 16t^2,$$
$$h'(t) = v(t) = -32t, \qquad \text{and}$$
$$v'(t) = -32.$$

Taking air drag into account leads to another differential equation:

$$v'(t) = -32 + 0.003v(t)^2.$$

Solving this differential equation symbolically is hard. We won't try, but we'll show graphs of the functions v and h that apply if air drag is taken into account. In the following pictures, the solid curves show the effect of air drag; the dotted curves ignore it. Notice, for example, the difference between the two velocity graphs. Ignoring air drag, the ball accelerates at a constant rate. Taking account of air drag gives a more realistic result: as the ball gathers speed, its acceleration tapers toward zero.

Effects of air drag on acceleration

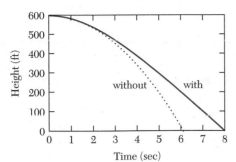

Effects of air drag on velocity

The two velocity curves show the advantage (to both Dad and the lawyer) of air drag. *With* air drag, the ball hits the lawyer at about 100 feet per second—an unpleasant blow, certainly, but a slow pitch by major league standards. *Without* air drag, the ball would be traveling more than twice as fast when it landed. ▶

■ *Of course, it might also then miss the strolling lawyer.*

Understanding the Universe. The idea of derivative is among the most important and powerful in all of mathematics—indeed, in all of human history. This concept, more than any other, distinguishes calculus from other branches of mathematics and accounts for the great power of the calculus to describe, and therefore predict, physical phenomena.

Planetary motion is one such phenomenon. In the early 1600s Johannes Kepler discovered (by direct observation) that the orbit of Mars is an ellipse; he suggested the same about other planets. Isaac Newton, using his early theory of calculus, demonstrated all this (and more) mathematically toward the end of the 1600s. As George Simmons put it,[a]

Newton's enormous success revived and greatly intensified the almost-forgotten Greek belief that it is possible to understand the universe in a rational way. This new confidence in its own intellectual powers permanently altered humanity's perception of itself, and over the past 300 years almost every department of human life has felt its consequences.

[a] *Calculus Gems: Brief Lives and Memorable Mathematics*, McGraw-Hill, 1992, pp. 334–335.

BASICS

1. An airplane takes off from one airport and, 45 minutes later, lands at another airport 300 miles away. Let t represent the time since the airplane took off, $x(t)$ the *horizontal* distance traveled, and $y(t)$ the *altitude* of the airplane.
   (a) What is the physical meaning of $x'(t)$? Sketch plausible graphs of x and x'.
   (b) What is the physical meaning of $y'(t)$? Sketch plausible graphs of y and y'.

2. The graph of a function f is shown on the right. Rank the values of $f'(-3)$, $f'(-2)$, $f'(0)$, $f'(1)$, and $f'(4)$ in increasing order.

Graph of f

3. Suppose that $f(x) = x^3 - 5x^2 + x + 1$ and that $g(x) = x^3 - 5x^2 + x + 4$. Explain why $f'(x) = g'(x)$ for every x. [**HINT**: How are the graphs of f and g related?]

4. Let $f(x) = 5$, $g(x) = x$, and $h(x) = -2x$. Evaluate $f'(11)$, $g'(-53)$, and $h'(37)$.

5. Let $f(x) = 3x + 2$ and $g(x) = 3x - 4$.
   (a) Evaluate $f'(0)$ and $g'(0)$.
   (b) Give a graphical explanation of the fact that $f'(x) = g'(x)$ for all values of x.
   (c) Find a function h such that $h(1) = 2$ and $h'(x) = f'(x)$ for all values of x.

6. Suppose that $f(x) = x - 2$ and $g(x) = 4x + 1$.
   (a) Plot graphs of f and g on the same axes.
   (b) Show that $f'(x) \le g'(x)$ for all values of x.
   (c) Show how the racetrack principle implies that $f(x) \le g(x)$ for all $x \ge -1$.
   (d) What does the racetrack principle say about the relationship between the numbers $f(x)$ and $g(x)$ when $x \le -1$?

FURTHER EXERCISES

7. Interpret each sentence below as a statement about a function and its derivative. In each case, indicate clearly what the function is, what each variable means, and appropriate units.
   (a) The price of a product decreases as more of it is produced.
   (b) The increase in demand for a new product decreases over time.
   (c) The work force is growing more slowly now than it was five years ago.
   (d) Health care costs continue to rise, but at a slower rate than three years ago.
   (e) During the past two years, the United States has continued to cut its consumption of imported oil.

8. Let $G(v)$ be the number of miles per gallon that a vehicle gets as a function of its speed, v, in miles per hour. Interpret the statement $G'(35) = 0.4$ in plain English using terms such as gas mileage and speed.

9. A company must budget for research and development of a new product. Let x represent the amount of money invested in R&D and T be the time until the product is ready to market.
   (a) Give reasonable units for T and x. What is T' in these units?
   (b) What economic interpretation can be given to T'?
   (c) Would you expect T' to be positive or negative? Explain.

10. The height of a ball (in feet) t seconds after it is thrown is given by $h(t) = -16t^2 + 38t + 74$; its (upward) velocity at time t is $v(t) = -32t + 38$.
    (a) From what height was the ball thrown?
    (b) What was the ball's initial velocity? Was it thrown up or down? How can you tell?
    (c) Was the ball's height increasing or decreasing at time $t = 2$?
    (d) At what time did the ball reach its maximum height? How high was it then?

    (e) How long was the ball in the air?

11. Suppose that $g(x) = f(x) + 3$ and that $f'(x)$ exists for all x.
    (a) Explain how the graphs of f and g are related.
    (b) How is the graph of g' related to the graph of f'? Explain.
    (c) If $f'(1) = 5$, what is $g'(1)$?
    (d) Let $h(x) = f(x) + k$ where k is a constant. How is the graph of h' related to the graph of f'?

12. Let $f(x) = \sin x$ and $g(x) = \cos x$. Explain why the graph of g' is the graph of f' shifted $\pi/2$ units to the left. [**HINT**: $\cos x = \sin(x + \pi/2)$.]

13. Let $g(x) = f(x + 3)$ and suppose that $f'(x)$ exists for all x.
    (a) Explain how the graphs of f and g are related.
    (b) How is the graph of g' related to the graph of f'? Explain.
    (c) If $f'(2) = 4$, what is $g'(-1)$?
    (d) Let $h(x) = f(x + k)$ where k is a nonzero constant. How is the graph of h' related to the graph of f'?

14. Suppose that $g(x) = 3f(x)$ and that $f'(x)$ exists for all x.
    (a) Explain how the graphs of f and g are related.
    (b) How is the graph of g' related to the graph of f'? Explain.
    (c) If $f'(-1) = 2$, explain why $g'(-1) = 6$.
    (d) Let $h(x) = kf(x)$ where $k > 0$ is a constant. How is the graph of h' related to the graph of f'?
    (e) Let $j(x) = kf(x)$ where $k < 0$ is a constant. How is the graph of j' related to the graph of f'?

15. Suppose that f is periodic with period 5 and that $f'(x)$ exists for all x. Explain why f' is also a periodic function with period 5.

16. Suppose that $f'(3) = -2$.
    (a) Assume that f is an *even* function. Explain why $f'(-3) = 2$. [**HINT**: Give a graphical argument.]
    (b) Now assume that f is an *odd* function. What is $f'(-3)$?

17. Suppose that $f'(x) = 1$ for all x. Find an algebraic expression for $f(x)$ assuming that
    (a) $f(0) = 2$. (c) $f(0) = -2$.
    (b) $f(1) = 2$. (d) $f(-1) = -2$.

18. Suppose that $f(a) = b$ and $f'(x) = c$ for all x. Find an algebraic expression for $f(x)$.

19. Suppose that f and g are defined on the interval $[a, b]$, that $f(a) = g(a)$, and that $f'(x) \le g'(x)$ for all x in $[a, b]$.
    (a) Use car talk to explain why $f(b) \le g(b)$.
    (b) Suppose that $f(b) = g(b)$. What does this imply about f' and g'?

20. Suppose that f and g are defined on the interval $[a, b]$, that $f(a) = g(a)$, and that $f'(x) \le g'(x)$ for all x in $[a, b]$. Use the basic racetrack principle to show that $f(x) - f(a) \le g(x) - g(a)$ for all x in $[a, b]$.

21. Suppose that $f(1) = -2$ and $f'(x) \le 3$ for all x in $[-10, 10]$.
    (a) Show that $f(2) \le 1$.
    (b) Show that $f(5) \le 10$.
    (c) Show that $f(0) \ge -5$.
    (d) Is $f(9) < 31$? Justify your answer.
    (e) Is $f(-5) > -23$? Justify your answer.
    (f) Could $f(4) = 8$? Justify your answer.
    (g) Could $f(-7) = -25$? Justify your answer.
    (h) Could $f(8) = -25$? Justify your answer.
    (i) Let $g(x) = 4x + 3$. Compare the numbers $f(7)$ and $g(7)$.

22. Suppose that $f(1) = 3$ and $f'(x) \ge 2$ for all x in $[0, 5]$.
    (a) Show that $f(4) \ge 9$.
    (b) Show that $f(0) \le 1$.
    (c) Let $g(x) = x + 1$. Compare the numbers $f(3)$ and $g(3)$.

23. Suppose that $f(2) = 1$ and $-4 \le f'(x) \le 3$ for all x in $[-10, 10]$.
    (a) Find upper and lower bounds on the value of $f(6)$.
    (b) Find upper and lower bounds on the value of $f(-5)$.

24. Suppose that $f(1) = 2$ and $f'(x) \le -3$ for all x in $[-5, 5]$.
    (a) What does the racetrack principle imply about the value of $f(4)$?
    (b) What does the racetrack principle imply about the value of $f(-2)$?

25. Suppose that $f(0) = 2$ and $|f'(x)| \le 1$ for all x in $[-5, 5]$.
    (a) Find upper and lower bounds on the value of $f(1)$.
    (b) Find upper and lower bounds on the value of $f(-3)$.

26. Suppose that $f'(x) \le 0$ for all x in $(2, 7)$. Explain why $f(3) \ge f(6)$.

27. Suppose that $f'(x) > 0$ for all x in $(2, 7)$. Explain why $f(4) < f(5)$.

28. The graph of the derivative of a function f is shown below. Use the graph of f' and the racetrack principle to answer the following questions about f. [**NOTE:** The graph of f is not shown.]

Graph of f'

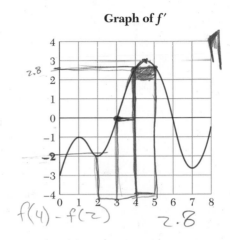

(a) Explain why $f(5) - f(4) > 2$.
(b) Is $-6 < f(2) - f(0) < -2$? Justify your answer.
(c) Show that $-5 < f(2) - f(0)$.
    [**HINT:** $f(2) - f(0) = \big(f(2) - f(1)\big) + \big(f(1) - f(0)\big)$.]
(d) Which is larger: $f(3)$ or $f(6)$? Justify your answer.
    [**HINT:** Use part (a).]
(e) Which is larger: $f(0)$ or $f(3)$? Justify your answer.

29. The graph of the derivative of a function g is shown below. Use the graph of g' and the racetrack principle to answer the following questions about g. [**NOTE:** The graph of g is not shown.]

Graph of g'

(a) Is $g(2) < g(5)$? Justify your answer.
(b) Is $g(6) > g(8)$? Justify your answer.
(c) Rank the five numbers 0, 1, $g(2) - g(1)$, $g(8) - g(7)$, and $g(9) - g(8)$ in increasing order.
(d) Rank the four numbers $g(0)$, $g(1)$, $g(3)$, and $g(5)$ in increasing order.

30. Suppose that $f(-1) = -2$ and that f' is the function shown below. Use the racetrack principle to answer the following questions about f. [**NOTE**: The graph of f is not shown.]

Graph of f'

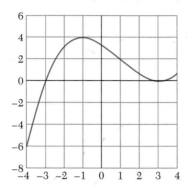

(a) Could $f(3) = -6$? Justify your answer.
(b) Could $f(3) = 2$? Justify your answer.
(c) Show that f achieves its largest value in the interval $[-4, 4]$ at $x = 4$. [**HINT**: Show that $0 < f(-4) - f(-3) < 6$ and $f(1) - f(-2) > 6$.]
(d) Where in the interval $[-4, 4]$ does f achieve its smallest value?

31. Suppose that k is a real number and that f is a function such that $f'(x) \le k$ for all x in an interval $[a, b]$. Show that $f(x) \le k(x - a) + f(a)$ for all x in $[a, b]$.

32. The converse of the racetrack principle says that if $f(a) = g(a)$ and $f(x) \le g(x)$ for all x in $[a, b]$, then $f'(x) \le g'(x)$ for all x in $[a, b]$. Show that this statement is not true in general by giving graphical examples of functions f and g that satisfy the hypotheses of the statment but not its conclusion.

33. (a) Suppose that $f'(x) \le 0$ for all x in $[a, b]$ and $f(a) = c$. Use the racetrack principle to show that $f(x) \le c$ for all x in $[a, b]$.
    (b) Suppose that $f'(x) \ge 0$ for all x in $[a, b]$ and $f(a) = c$. Use the racetrack principle to show that $f(x) \ge c$ for all x in $[a, b]$.
    (c) Use parts (a) and (b) to prove that if $f'(x) = 0$ for all x in $[a, b]$, then f is a constant function on the interval $[a, b]$.

34. Suppose that $f'(x) = g'(x)$ for all x in $[a, b]$. Use the previous problem to show that there is a constant C such that $g(x) = f(x) + C$ for all x in $[a, b]$. [**HINT**: If $h(x) = g(x) - f(x)$, then $h'(x) = 0$ for all x in $[a, b]$.]

2.2 Estimating Derivatives: A Closer Look

In the preceding section we described the derivative function in terms of rates and slopes. Given a function f and an input value $x = a$, ◀

Writing $x = a$ helps "fix" our attention on a single point of the domain.

> $f'(a)$ is the instantaneous rate of change of f with respect to x at $x = a$.

In graphical language,

> $f'(a)$ is the slope of the f-graph at $x = a$.

This brief section is—literally—a closer look at the derivative at a point. By "zooming in" on a function or its graph near an input value $x = a$, we can estimate numerical values of f'; we can also see more clearly what the derivative means, both graphically and numerically.

Estimating $f'(a)$: The Slope of a Graph at a Point

For a line, finding slope is a routine matter. For a curve, the idea of slope at a point is a little subtler, but a lot more interesting.

Tangent Lines

Remember the headlights-through-fog analogy? See page 99.

The slope of a curve at a point P can be thought of as the slope of the curve's tangent line at P, i.e., the straight line through P that, roughly speaking, "points in the direction of the curve" at P. ◀ This description, although vague by strict mathematical standards, is perfectly adequate to illustrate the idea. For nearly all calculus functions, moreover, it leads reliably to good derivative estimates.

EXAMPLE 1   Let $f(x) = x^2$. Use tangent lines to estimate $f'(1)$ and $f'(-1.5)$. Interpret the results as rates of change.

Solution   Here's the graph of f, together with reasonable candidates ▶ for tangent lines at $x = 1$ and $x = -1.5$:

Do you agree that they're reasonable?

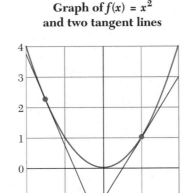

**Graph of $f(x) = x^2$
and two tangent lines**

We can now estimate the slope of each tangent line directly from its graph. The line at $x = 1$, for instance, seems to rise 2 units as it runs 1, so its slope is 2. Similarly, the tangent line at $x = -1.5$ seems to have a slope around -3. ▶ We conclude that

Check our estimates.

$$f'(1) \approx 2; \qquad f'(-1.5) \approx -3.$$

In the language of rates of change, the results mean that (i) at $x = 1$, y is increasing twice as fast as x, and (ii) at $x = -1.5$, y *decreases* three times as fast as x *increases*.   ■

Zooming In to Estimate Slope

Drawing a tangent line is one way to estimate the graph's slope at a point. The strategy works because, near the point in question, the graph and the tangent line are almost indistinguishable.

Another strategy—often more convenient, given plotting technology—is to "zoom in" on the point in question, perhaps repeatedly, until the graph *itself* looks like a straight line. Once the graph looks straight, it's easy to estimate the slope.

EXAMPLE 2   Let $f(x) = x^2$, as before. Estimate $f'(1)$ by zooming. Use the result to find an equation for the tangent line at $x = 1$.

Note the plotting windows.

S o l u t i o n   Zooming in successively on the f-graph near $(1, 1)$ gives these results: ◄

Graph of $f(x) = x^2$:
a closer look

Graph of $f(x) = x^2$:
an even closer look

 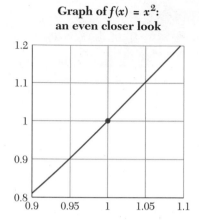

Check the estimate; use the grid.

Sketch this line into the two preceding pictures.

In the last picture, the graph looks virtually straight; its slope seems to be about 2. ◄ The tangent line in question has slope 2 and passes through $(1, 1)$; its equation, therefore, is $y = 2(x - 1) + 1$. ◄   ∎

When Zooming Works: Local Linearity.   Remarkably, the just-illustrated strategy of zooming in to estimate slope almost always works. Zooming in on the graph of almost any calculus function f, at almost any point $(a, f(a))$, eventually produces what looks like a straight line with slope $f'(a)$. A function with this property is sometimes called **locally linear** (or **locally straight**) at $x = a$. ◄   Local linearity says, in effect, that f "looks like a line" near $x = a$ and therefore has a well-defined slope at $x = a$.

These aren't formal definitions, just descriptive phrases.

Example 2 showed that $f(x) = x^2$ is locally linear at $x = 1$. Now let's try zooming in on a trigonometric function.

EXAMPLE 3   Is $f(x) = \sin x$ locally linear at $x = 1.4$? What does the answer say about f'?

S o l u t i o n   Here are two views of the sine graph near the target point $P = (1.4, \sin 1.4) \approx (1.4, 0.9854)$:

The sine graph near $P = (1.4, \sin 1.4)$

Zooming in near $P = (1.4, \sin 1.4)$

 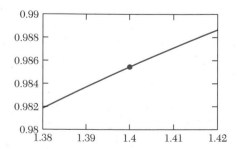

In the second window, the sine graph is almost straight. To estimate the slope at P we can use any two points from the graph. Using $x = 1.38$ and $x = 1.42$, we find (approximately, of course) a slope of 0.17. ▶ In function notation, $f'(1.4) \approx 0.17$.

Check for yourself. Your estimate may differ in the second decimal place.

When Zooming Fails: Local Nonlinearity.   Not every function is locally linear everywhere. A kink in a graph can't be smoothed out by repeated zooming in. At such a point the derivative doesn't exist, and our model of slope breaks down.

EXAMPLE 4   Let $f(x) = |x|$. Does $f'(0)$ exist? Does $f'(a)$ exist if $a \neq 0$?

Solution   Here's the result of zooming near the origin:

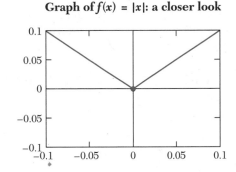

Graph of $f(x) = |x|$ **Graph of $f(x) = |x|$: a closer look**

There's no mistake in the pictures. Zooming in had no effect on the kink at the origin. Zooming again would make no difference; that kink is there to stay, and $f'(0)$ doesn't exist.

For $a \neq 0$, on the other hand, the graph is already straight (with slope ± 1), so there's no problem with derivatives:

$$f'(a) = 1 \quad \text{if } a > 0; \qquad f'(a) = -1 \quad \text{if } a < 0.$$

In words: $f(x) = |x|$ is not locally linear at $x = 0$, but is locally linear everywhere.

Zoom Factors and Distortion.   Throughout this section, "zooming in" means "zooming in with equal zoom factors in both directions." Requiring equal zoom factors is more than a technicality. Zooming repeatedly with *different* zoom factors can so distort the picture that all information is lost.

Consider, for example, the kink at $x = 0$ on the graph $y = |x|$, as in Example 4. Zooming in repeatedly—but *only* in the x-direction (e.g., with zoom factors 4 in x and 1 in y)—gradually flattens the vee. Eventually, the graph looks horizontal. Try it.

Plotting the Rate Function

Because our slope results are estimates, the plot is only approximate. But so is every plot.

Zooming in or drawing tangent lines to the graph of f lets us estimate $f'(a)$ for various values of a. By pooling such information we can *plot* the derivative function. ◄ The results are sometimes surprising.

EXAMPLE 5   For $f(x) = x^2$, plot $f'(x)$. Then guess a formula for $f'(x)$.

About the y-axis.

Solution   In Example 1 we estimated that $f'(1) = 2$ and $f'(-1.5) = -3$. From the symmetry ◄   of the graph $y = x^2$, a few more values come easily:

x	−1.5	−1	0	1	1.5
$f'(x)$	−3	−2	0	2	3

Our guess is correct. Soon we'll see exactly why, by symbolic calculation.

Plots of these points suggest that the graph of f' is a straight line with slope 2, i.e., that $f'(x) = 2x$. ◄

Graph of f'

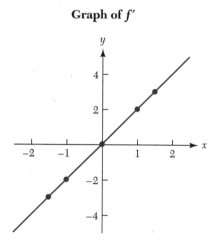

What's Natural about e^x?   Applying the same technique to the natural exponential function produces a real surprise—one that helps explain why the adjective "natural" fits.

EXAMPLE 6   Let $f(x) = e^x$. Use tangent lines to plot f'; then guess a formula for f'.

Solution   Here's a graph of f (shown dotted); at several points, small segments of tangent lines (shown solid) are drawn:

Tangent segments to the graph of
$$f(x) = e^x$$

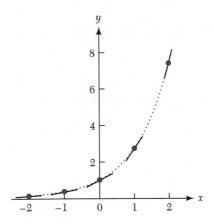

Using a ruler or any convenient plotting technology, one can estimate the slope of each tangent segment. Several such results (rounded to one decimal place) are tabulated below. ▶

Check some table entries yourself. Are they plausible?

The Derivative of e^x—Graphical Estimates					
x	-2	-1	0	1	2
y	0.135	0.368	1.000	2.718	7.389
Slope at (x, y)	0.1	0.4	1.0	2.7	7.4

Plotting the slope data (with a smooth curve through the points) gives a striking result:

The derivative of $f(x) = e^x$

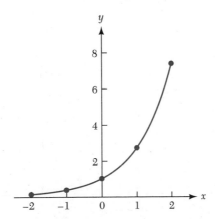

We'll prove this symbolically later.

Judging from the graphs, f and f' are identical! They *are* identical: ◄

If $f(x) = e^x$, then $f'(x) = e^x$. *The natural exponential function is its own derivative.*

In other words: $f(x) = e^x$ solves the **differential equation** $f'(x) = f(x)$. This remarkable property makes the natural exponential function one of the most important and useful in all of mathematics. ■

BASICS

1. The graph of a function f is shown below.

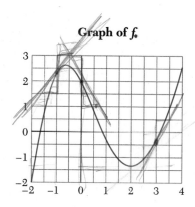

Graph of f.

(a) Fill in the missing entries in the table below.
[**NOTE:** Because f is given graphically, your answers will be estimates. However, your answers should approximate the exact values within ± 0.5.]

x	-2	-1	0	1
$f'(x)$	6.0			-2.0
x	2	2.5	3	4
$f'(x)$				3.5

(b) Use your answers to part (a) to sketch a graph of f'.

2. The graph of a function f is shown below.

Graph of f

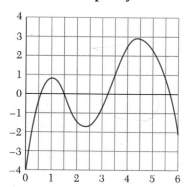

(a) Check the entries in the table below and replace any that are wrong with reasonable estimates.

x	1	2	3	4	5	6
$f'(x)$	0	2	6	3	-2	42

(b) Sketch a graph of f'.

3. The graph of the derivative of a function f appears below. [**NOTE:** The graph of f is not shown.]

Graph of f'

look @ graph, then go t is ax's, is then the slope

(a) Suppose that $f(1) = 5$. Find an equation of the line tangent to the graph of f at $(1, 5)$.

$f(1) =$ on graph, ② - slope!

(b) Suppose that $f(-3) = -6$. Find an equation of the line tangent to the graph of f at $(-3, -6)$.

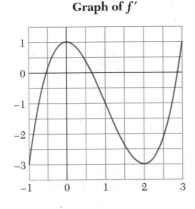

FURTHER EXERCISES

4. Graphs of a function f and its derivative f' are shown below.

Graph of f

Graph of f'

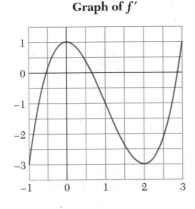

(a) Draw short segments of the lines tangent to f at $x = -1$, $x = 0$, $x = 1$, $x = 2$, $x = 3$, and $x = 3.5$ on the graph of f.

(b) Estimate the slope of each of the tangent lines you drew in part (a).

(c) Are your answers in part (b) consistent with the graph of f' shown above? Explain.

(d) Find equations for the lines tangent to f at $x = -1$, $x = 0$, $x = 1$, $x = 2$, and $x = 3$.

5. The graph of a function f is shown below.

Graph of f

(a) Fill in the missing entries in the table below.

x	-3	-2	-1	0	1
$f'(x)$	-4.9			0.8	
x	2	3	4	5	6
$f'(x)$		1.1			-4.0

(b) Sketch a graph of f'.

6. The graph of a function f is shown below.

Graph of f

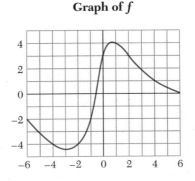

(a) Estimate $f'(-5)$, $f'(-3)$, $f'(-1)$, $f'(2)$, and $f'(4)$.

(b) Sketch a graph of f'.

7. Let f be the function whose graph is shown below.

Graph of f

(a) Estimate $f'(0)$, $f'(1)$, $f'(2)$, $f'(3)$, $f'(4)$, and $f'(5)$.
(b) Use your results from part (a) to sketch a graph of the function f'.

8. Let $f(x) = x^2$.
   (a) Use zooming to estimate $f'(0)$, $f'(1/2)$, $f'(1)$, $f'(2)$, and $f'(3)$.
   (b) Use your results from part (a) to sketch a graph of f' over the interval $[-3, 3]$. [**HINT**: Why is $f'(-1) = -f'(1)$?]
   (c) Use your results from parts (a) and (b) to guess an algebraic formula for $f'(x)$.

9. Let $f(x) = \sqrt{x}$.
   (a) Use zooming to estimate $f'(1/4)$, $f'(1)$, $f'(9/4)$, $f'(4)$, $f'(25/4)$, and $f'(9)$.
   (b) Use your results from part (a) to sketch a graph of f' over the interval $[0, 9]$.
   (c) Why is $f'(-1)$ undefined? Why is $f'(0)$ undefined? What is the domain of f'? [**HINT**: Look at a graph of f.]
   (d) Use your results from part (a) to guess an algebraic formula for $f'(x)$. [**HINT**: Evaluate $1/f(x)$ at $x = 1/4$, $x = 1$, $x = 9/4$, $x = 4$, $x = 25/4$, and $x = 9$.]

10. Let $f(x) = x^3$ and $g(x) = x^{1/3}$.
    (a) Use zooming to estimate $f'(0)$, $f'(1/2)$, $f'(1)$, $f'(3/2)$, and $f'(2)$.
    (b) Use your results from part (a) to sketch a graph of f' over the interval $[-2, 2]$. [**HINT**: Why is $f'(-1) = f'(1)$?]
    (c) Use your results from parts (a) and (b) to guess an algebraic formula for $f'(x)$.
    (d) Estimate $g'(-1)$ and $g'(1)$.
    (e) Why is $g'(0)$ undefined? What is the domain of g'? What is the domain of g?
    (f) It can be shown that $f'(4) = 48$. Use this fact to find $g'(64)$. [**HINT**: How are the graphs of f and g related? The point $(4, 64)$ is on the graph of f and the point $(64, 4)$ is on the graph of g.]

11. Let $f(x) = 1/x$.
    (a) Use zooming to estimate $f'(1/4)$, $f'(1/2)$, $f'(1)$, $f'(2)$, and $f'(3)$.
    (b) Use your results from part (a) to sketch a graph of f' over the intervals $[-3, 0)$ and $(0, 3]$. [**HINT**: Why is $f'(-1) = f'(1)$?]
    (c) Use your results from parts (a) and (b) to guess an algebraic formula for $f'(x)$.

12. Let $f(x) = e^x$.
    (a) Use zooming to estimate $f'(-1)$, $f'(0)$, $f'(1)$, $f'(1.5)$, and $f'(2)$.
    (b) Use your results from part (a) to sketch a graph of f' over the interval $[-1, 2]$.
    (c) Use your results from part (a) to guess a formula for $f'(x)$. [**HINT**: Evaluate $f(x)$ at $x = -1$, $x = 0$, $x = 1.5$, and $x = 2$.]

13. Let $f(x) = \ln x$.
    (a) Use zooming to estimate $f'(1/10)$, $f'(1/5)$, $f'(1/2)$, $f'(1)$, $f'(2)$, and $f'(5)$.
    (b) Use your results from part (a) to sketch a graph of f' over the interval $[0, 5]$.
    (c) Use your results from part (a) to guess a formula for $f'(x)$.

14. Let $f(x) = \sin x$.
    (a) Use zooming to estimate $f'(0)$, $f'(\pi/6)$, $f'(\pi/4)$, $f'(\pi/3)$, $f'(\pi/2)$, and $f'(\pi)$.
    (b) Use the symmetry of the sine function and your answers from part (a) to estimate $f'(2\pi/3)$, $f'(3\pi/4)$, and $f'(5\pi/6)$. Check your answers using zooming.
    (c) Use your results from parts (a) and (b) to guess an expression for $f'(x)$.

15. Let $f(x) = \cos x$.
    (a) Use zooming to estimate $f'(0)$, $f'(\pi/6)$, $f'(\pi/4)$, $f'(\pi/3)$, $f'(\pi/2)$, and $f'(\pi)$.
    (b) Use your results from part (a) to guess an expression for $f'(x)$.

16. (a) Zoom in on the graph of $y = e^x$, near the point $(0, 1)$, until the graph looks like a straight line. What's the slope of this line?
    (b) Zoom in on the graph of $y = \ln x$, near the point $(1, 0)$, until the graph looks like a straight line. What's the slope of this line?

17. The previous exercise explains (partly) why base e is special. The italicized statement on p. 60 holds *only* for exponential and logarithm functions with base e. For other bases, a slightly different relation holds between the slope of $y = b^x$ at $(0, 1)$ and the slope of $y = \log_b x$ at $(1, 0)$. Let's "discover" this relationship for $b = 10$.
    (a) Zoom in on the graph of $y = 10^x$ near the point $(0, 1)$ until the graph looks like a straight line. (This is the "tangent line" to the graph of $y = 10^x$ at $(0, 1)$.) Find the slope as accurately as possible.

(b) Zoom in on the graph of $y = \log_{10} x$ near the point $(1, 0)$ until the graph looks like a straight line. (This is the "tangent line" to the graph of $y = \log_{10} x$ at $(1, 0)$.) Find the slope as accurately as possible.

(c) How are the slopes in parts (a) and (b) related? What "variant" of the italicized statement referred to above holds in this situation?

18. In Section 1.5 we claimed:

> If $y = b^x$, then the rate of change of y is proportional to y itself.

In this exercise we explore this claim.

(a) By zooming in, estimate the slope of the tangent line to the graph of $y = e^x$ at $x = 0$, $x = 1$, and $x = 2$. Are your results consistent with the above statement? With what constant of proportionality?

(b) By zooming in, estimate the slope of the tangent line to the graph of $y = 10^x$ at $x = 0$, $x = 1$, and $x = 2$. Are your results consistent with the above statement? With what constant of proportionality?

19. (a) Sketch the graph of a function f that is consistent with these data:

x	-2	-1	1	2
$f(x)$	1	-1	-1	2
$f'(x)$	-3	0	-1	-2

(b) Sketch the graph of a function g that is consistent with these data:

x	-2	-1	1	2
$g(x)$	1	-1	-1	2
$g'(x)$	1	0	-1	-2

20. The line tangent to f at $(5, 2)$ passes through the point $(0, 1)$. Find $f(5)$ and $f'(5)$. Justify your answers.

21. Suppose that the line tangent to the curve $y = f(x)$ at $x = 3$ passes through the points $(-2, 3)$ and $(4, -1)$. Find $f(3)$ and $f'(3)$.

22. The line tangent to the curve $y = f(x)$ at $x = 2$ has slope -1 and crosses the x axis at $x = 5$. Find $f(2)$ and $f'(2)$.

23. Suppose that the line tangent to the curve $y = f(x)$ at $x = -2$ is described by the equation $y = 4x + 3$. Find $f(-2)$ and $f'(-2)$.

Exercises 24–32 concern a function f whose derivative, f', is the function shown below. [**NOTE:** The graph of f is not shown.]

Graph of f'

24. Explain why each of the following *cannot* be the line tangent to the curve $y = f(x)$ at $x = 1$: (i) $y = x + 1$; (ii) $y = 3$; (iii) $y = -2x - 1$; (iv) $y = 2x$.

25. Which of the following could be the line tangent to the curve $y = f(x)$ at $x = 6$: (i) $y = -5$; (ii) $y = 6(x + 1)$; (iii) $6y = 2 - 3x$; (iv) $y = -6x$; (v) $y = 6x + 1$?

26. Find the equation of a line parallel to the line tangent to the curve $y = f(x)$ at $x = 4.5$.

27. Suppose that $f(0) = 1$. Explain why $y = -2x + 3$ is *not* the line tangent to the curve $y = f(x)$ at $x = 2$.

28. Suppose that $f(0) = 1$. Explain why $y = -2x - 1$ is *not* the line tangent to the curve $y = f(x)$ at $x = 2$.

29. Let $\ell(x) = f'(5)(x - 5) + f(5)$.
   (a) Is $\ell(6)$ greater than, less than, or equal to $f(6)$? Justify your answer.
   (b) Is $\ell(8)$ greater than, less than, or equal to $f(8)$? Justify your answer.

30. Let $g(x) = -3x + f(0)$.
   (a) Explain why $g(1) < f(1)$.
   (b) Is $g(2) < f(2)$? Justify your answer.
   (c) Is $g(8) < f(8)$? Justify your answer.

31. Let ℓ be the line tangent to f at $x = 1$.
   (a) Is $\ell(2) \leq f(2)$? Justify your answer.
   (b) Is $\ell(0) \leq f(0)$? Justify your answer.

32. Let ℓ be the line tangent to f at $x = 7$.
   (a) Is $\ell(8) \leq f(8)$? Justify your answer.
   (b) Is $\ell(6) \leq f(6)$? Justify your answer.
   (c) Is $\ell(2) \leq f(2)$? Justify your answer.

33. The graph of the derivative of a function g is shown below. [**NOTE:** The graph of g is not shown.]

Graph of g'

(a) Suppose that $g(3) = 11$. Find an equation of the line tangent to g at the point $(3, 11)$.

(b) Let ℓ be the line tangent to g at $x = 2$. Is $g(5) \leq \ell(5)$? Is $g(1) \leq \ell(1)$? Justify your answers.

(c) Let h be the line tangent to g at $x = 6$. Explain why $g(x) \leq h(x)$ if $6 \leq x \leq 8$.

(d) Let h be the line tangent to g at $x = 6$. How do the values of $g(x)$ and $h(x)$ compare if $0 \leq x \leq 6$? Justify your answer.

34. Suppose that $f(2) = 4$ and $f'(x) = \sqrt{x^3 + 1}$ for $x \geq -1$.
   (a) Find an equation for the line tangent to f at $x = 2$.
   (b) Let ℓ be the tangent line from part (a). Does $\ell(0)$ overestimate or underestimate $f(0)$? Justify your answer.
   (c) Does $\ell(3)$ overestimate or underestimate $f(3)$? Justify your answer.

For each pair of functions in Exercises 35–40, the constant A can be chosen so that $f'(x) = Ag(x)$ for every x in the domain of g. Find this value of A. [**HINT:** Examine graphs of f and g.]

35. $f(x) = x^3$, $g(x) = x^2$
36. $f(x) = \sqrt{x}$, $g(x) = x^{-1/2}$
37. $f(x) = \sin x$, $g(x) = \cos x$
38. $f(x) = \cos x$, $g(x) = \sin x$
39. $f(x) = e^x$, $g(x) = e^x$
40. $f(x) = \ln x$, $g(x) = 1/x$

2.3 The Geometry of Derivatives

The geometric relationship between a function f and its derivative f' is easy to state:

For any input a, $f'(a)$ is the slope of the line tangent to the f-graph at $x = a$.

This statement—innocuous as it appears—is one of the most important in this book. The next two sections (and much of this book) explore its meaning and implications.

Graphs of f and f': An Extended Example

But don't peek.

Corresponding points (e.g., A and A') have the same x-coordinate.

Graphs of a function f and its derivative f' follow. (For the moment, no formulas are given—or needed. We'll return to these functions symbolically at the end of the next section.) ◄   Three interesting points are labeled on each. ◄

Graph of f

Graph of f'

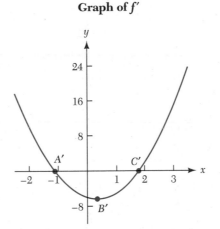

A First Look

We've only asserted, not proved, that f' has the graph shown. Tools to be developed in later sections will let us *prove* this; here we'll just assume it.

The f'-graph tells how slopes of tangent lines to the f-graph behave. At B, for instance, the f-graph seems ▶ to have a slope around −6; for this reason, the f'-graph has height −6 at B'.

Check carefully for yourself.

First let's observe several straightforward geometric relationships between the two graphs, introducing some useful new terminology in the process.

> **The Sign of f'**   The graph of f rises to the left of A, falls between A and C, and rises again to the right of C. The sign of $f'(x)$ tells whether the line tangent to the f-graph at x points up or down. At A' and C', ▶ f' changes sign. Thus, at A and C, f itself changes direction.

A' and C' correspond to roots of f'.

> **In the Long Run**   The graph suggests that f blows up to the right of C and blows down to the left of A. In geometric terms, as $x \to \infty$, the graph of $f(x)$ rises more and more steeply. What happens over the same intervals on the f'-graph? To the right of C', f' is positive and increasing. To the left of A', f' is positive and decreasing.

> **Stationary, Maximum, and Minimum Points**   The points A and C, with approximate coordinates $(-1.1, 4)$ and $(1.8, -8)$, ▶ where the f-graph is horizontal, are obviously of interest. They mark, respectively, high and low points of the graph. The situation looks clear, but to avoid later confusion, it will pay to be extremely picky with language here—especially about the distinction between inputs to f (called points) and outputs from f (called values).
>
> Here, in full detail, is the situation at $A \approx (-1.1, 4)$. The domain point $x = -1.1$ is called a local maximum point of f; the corresponding output—$f(-1.1) \approx 4$—is called a local maximum value of f. At C ▶ the situation is similar: $x = 1.8$ is a **local minimum point** of f, and $f(1.8) \approx -8$ is the corresponding **local minimum value** of f. The

We'll improve these guesses later by symbolic methods.

$C \approx (1.8, -8)$.

Careful definitions of all these terms will follow.

x-coordinates of both A and C are called **stationary points** of f. ◄ (We say local rather than global because elsewhere in its domain f may assume larger or smaller values.) The corresponding points A' and C' occur where the f'-graph crosses the x-axis (i.e., at roots of f').

Any More Stationary Points? Are A and C the only stationary points of f? The picture doesn't say. It shows only the interval $-2.5 \le x \le 3.5$; additional stationary points could lie outside this interval.

Concavity and Inflection The point B, near $x = 0.3$, is an **inflection point** of f: At B, the f-graph's **direction of concavity** changes, from concave down to concave up. ◄ The point B has another special geometric property: At B, the graph of f points most steeply downward.

Recall: "Holds water" and "spills water" are rough synonyms for "concave up" and "concave down."

Locating B or B' precisely by magnifying either graph doesn't work very well. Viewed closely, both graphs look like straight lines.

The corresponding point on the f'-graph, B', is easier to see; it's a local minimum point. Later we'll use this property and some algebra to find the *exact* location of B. ◄

What f' Says about f

Interpreting the derivative function f' in terms of the slopes of tangent lines on the f-graph has many important geometric implications. We summarize several below.

> **Nicely Behaved Functions.** Many ideas in this section depend on both f and f' being reasonably, "nice," or "well behaved." "Nice enough" means, roughly that neither f nor f' displays sudden jumps or breaks. Any function f whose derivative function f' is continuous (i.e., without breaks) is nice enough for us. Fortunately, most calculus functions *are* nice in this sense. As we proceed we'll see more of what nice means and why it matters.

Increasing or Decreasing?

From left to right!

A function f **increases** where its graph rises ◄ and **decreases** where its graph falls. The following definition captures these natural ideas in analytic language.

> **Definition** Let I denote the interval (a, b). A function f is **increasing** on I if
>
> $$f(x_1) < f(x_2) \qquad \text{whenever} \qquad a < x_1 < x_2 < b.$$
>
> f is **decreasing** on I if
>
> $$f(x_1) > f(x_2) \qquad \text{whenever} \qquad a < x_1 < x_2 < b.$$

In these definitions, $\le$ and $\ge$ replace $<$ and $>$.

(If $f(x_1) \le f(x_2)$ for $x_1 < x_2$, then f is **nondecreasing** on I; if $f(x_1) \ge f(x_2)$ for $x_1 < x_2$, then f is **nonincreasing** on I. ◄ In either case, f is **monotonic** on I.)

The derivative detects whether a function increases or decreases on an interval I.

> **Fact** If $f'(x) > 0$ for all x in I, then f increases on I. If $f'(x) < 0$ for all x in I, then f decreases on I.

This fact certainly *sounds* reasonable. To say that $f'(x) > 0$ means that the tangent line to the f-graph at x points upward. With any luck, so should the f-graph itself. Racetrack ▶ language makes the fact seem even simpler: If dog f has positive velocity, then dog f always runs *forward*.

See page 99 for more on the racetrack principle.

Plausible as the fact sounds, proving it rigorously requires one of calculus's "big theorems": the mean value theorem (we'll see it later).

Behavior at a Point.   Functions are often said to increase or decrease *at a point*. To say, for example, that f increases *at* $x = 3$ means that f increases on some interval—perhaps a small one—containing $x = 3$. Now we can restate the preceding fact ▶ as follows:

Here, too, we're assuming that both f and f' are free of discontinuities and other naughty behavior.

> **Fact** If $f'(a) > 0$, then f is increasing at $x = a$. If $f'(a) < 0$, then f is decreasing at $x = a$.

At $x = 1$, for instance, $f'(1) < 0$. The Fact says—and the picture agrees—that at $x = 1$ the f-graph is decreasing.

Derive Carefully.   The previous fact's **converse** reads like this: *If f is increasing at $x = a$, then $f'(a) > 0$.* Tempting as it is, the converse is false. The next example shows why.

EXAMPLE 1   Where is the function $f(x) = x^3$ increasing?   Where is $f'(x) > 0$? What happens at $x = 0$?

Solution   As its graph suggests,

Graph of $f(x) = x^3$

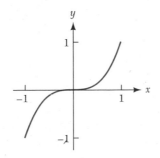

Imagine tangent lines.

the function $f(x) = x^3$ increases *everywhere*. The graph suggests, too, ◄ that $f'(x) > 0$ if $x \neq 0$, but that $f'(0) = 0$.   ■

The Derivative of an Increasing Function.   What *can* be said about f' where f increases? This:

Fact If f increases at $x = a$, then $f'(a) \geq 0$; if f decreases at $x = a$, then $f'(a) \leq 0$.

The derivative of an increasing function must be *nonnegative*, but not necessarily *strictly positive*.

Maximum Points, Minimum Points, and Stationary Points: Formal Definitions

See page 119.

Look carefully to see why.

The point A on the first graph in this section ◄ corresponds to a local, but not global, maximum value of f. ◄ Successful use of such technical language requires clear definitions.

> **Definition** Let f be a function and x_0 a point of its domain.
>
> - x_0 is a **stationary point** of f if $f'(x_0) = 0$.
> - x_0 is a **local maximum point** of f if $f(x_0) \geq f(x)$ for all x in some open interval containing x_0. The number $f(x_0)$ is a **local maximum value** of f.
> - x_0 is a **global maximum point** of f if $f(x_0) \geq f(x)$ for all x in the domain of f. The number $f(x_0)$ is the **global maximum value** of f.
>
> (Local and global **minimum points** are defined similarly.)

Several linguistic subtleties deserve mention.

Points versus Values Local and global maximum *points* are inputs to a function; local and global maximum *values* are the corresponding outputs. The difference, although important, is sometimes blurred in informal speech. (The first sentence of the next paragraph is impeccably correct on this score.)

Local versus Global At a *local* maximum point, a function f may or may not assume a *global* maximum value. Minimum points, too, may be local, global, or both.

Extreme Values The word **extremum** (plural extrema), which means "either maximum or minimum value," is a convenient shorthand. We'd say, for instance, that for the function f at the beginning of this section, both points A and C correspond to extrema of f.

Finding Maximum and Minimum Points

Geometric intuition says that at a local maximum point or local minimum point, a smooth graph must be "flat." ▶   More succinctly:

See A and C on the preceding graphs.

> **Fact** On a smooth graph every local maximum or local minimum point x_0 is a stationary point—i.e., a root of f'.

This fact has immense practical value. To find maximum and minimum values of f, the fact says, we can limit our search to roots of f'. Each such root is a stationary point and therefore, possibly, a maximum or minimum point (but not a sure thing—a stationary point may be only a "flat spot" in the graph). ▶

The next example—an important one—shows how to sort out all the possibilities.

See the previous example: $f(x) = x^3$ has a "flat spot"—but not a maximum or minimum—at $x = 0$.

EXAMPLE 2   The graph of a function f' appears as follows; the f-graph is not shown (for now). Three points of interest are bulleted. Where, if anywhere, does f have local maximum or local minimum points? Why?

Graph of f'

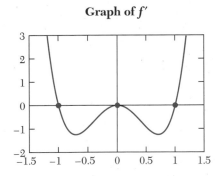

Solution   The three bullets on the graph—at $x = -1$, $x = 0$, and $x = 1$—represent roots of f' and therefore correspond to stationary points of f. What type of stationary point is each one: a local maximum, a local minimum, or just a flat spot? The key to deciding is to check the sign of f' just before and just after each stationary point. We take each root of f' in turn.

At $x = -1$ Just before (i.e., to the left of) $x = -1$, $f'(x) > 0$. Therefore (by an earlier Fact), f increases until $x = -1$. Just after $x = -1$, $f'(x) < 0$, so f decreases immediately after $x = -1$. This means that f has a local maximum at $x = -1$.

At $x = 1$ Consider values of x near $x = 1$. The graph shows that if $x < 1$, $f'(x) < 0$; if $x > 1$, $f'(x) > 0$. Thus, reasoning as above, f decreases before $x = 1$ and increases after $x = 1$. This means that f has a local minimum at $x = 1$.

At $x = 0$ This time the graph shows that $f'(x) < 0$ on both sides of $x = 0$. Thus, f must decrease before *and* after $x = 0$, so $x = 0$ is neither a maximum nor a minimum point, but just a flat spot in the f-graph.

Here, at last, is a possible f-graph. It agrees with everything we said.

Graph of f

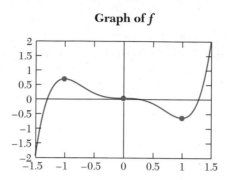

Let's summarize what we know about stationary points.

F a c t (**First Derivative Test**) Suppose that $f'(x_0) = 0$.

- If $f'(x) < 0$ for $x < x_0$ and $f'(x) > 0$ for $x > x_0$, then x_0 is a local minimum point.
- If $f'(x) > 0$ for $x < x_0$ and $f'(x) < 0$ for $x > x_0$, then x_0 is a local maximum point.

The next example shows what happens on a crooked graph.

At $x = 0$ this function is not nice in the sense discussed earlier.

EXAMPLE 3   The function $f(x) = |x|$ has a local (and global) minimum point at $x = 0$, but not a stationary point. ◄   The graph shows why:

Graph of $f(x) = |x|$:
a local minimum at $x = 0$

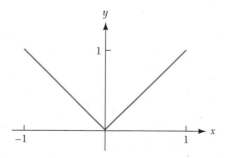

At $x = 0$ the graph has a kink and therefore *no* tangent line and no well-defined slope. In functional language, $f'(0)$ does not exist. ∎

Concave Up or Concave Down?

So far we've described concavity and inflection points informally, in graphical language. Here's a more formal, analytic definition:

> **Definition** The graph of f is **concave up** at $x = a$ if the slope function f' is increasing at $x = a$. The graph of f is **concave down** at $x = a$ if f' is decreasing at $x - a$. Any point at which a graph's direction of concavity *changes* is an **inflection point** of the graph.

The graphs of f and f' on page 119 ▶ illustrate all of this.

Look again.

Concave Down To the left of B, the f-graph is concave down; to the left of B', the f'-graph is decreasing.

Concave Up To the right of B, the f-graph is concave up; to the right of B' the f'-graph is increasing.

An Inflection Point B is an inflection point of f; f' has a minimum point at B'.

Finding Inflection Points from the Graph of f′

The direction of concavity of the graph of f depends, as the definitions show, on whether f' increases or decreases. An inflection point occurs wherever f' *changes direction*—i.e., wherever f' has a local minimum or a local maximum.

EXAMPLE 4 Discuss concavity of the sine function. Find all inflection points; explain the results in derivative language.

Solution It's a fact—we'll show it carefully later—that if $f(x) = \sin x$, then $f'(x) = \cos x$. The graphs certainly make the claim plausible: ▶

Do you agree? Are the roots of f' where they should be?

Graph of $f(x) = \sin x$

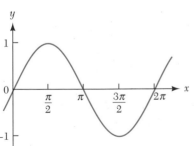

Graph of $f'(x) = \cos x$

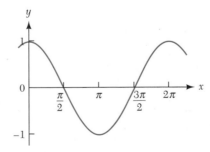

Notice:

Stationary Points f has stationary points (a local ▶ maximum point and a local minimum point) $x = \pi/2$ and at $x = 3\pi/2$—exactly the roots of f'.

Global, too, in this case.

Increasing or Decreasing? f increases on the intervals $(0, \pi/2)$ and $(3\pi/2, 2\pi)$; on the same intervals, f' is positive.

Concavity f is concave down on $(0, \pi)$—where f' decreases—and concave up on $(\pi, 2\pi)$—where f' increases. In fact, f illustrates every possible combination of increasing/decreasing and concavity behavior.

Inflection Points f has an inflection point at each multiple of π—precisely where f' assumes a local maximum or local minimum. ∎

A Sample Problem

Here's a problem in the spirit of this section.

EXAMPLE 5   The graph of the derivative of a function f is shown.

Graph of f'

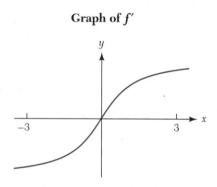

a. The equation $f(x) = 0$ can have no more than two solutions in the interval $[-3, 3]$. Explain why.

b. Explain why f cannot have two roots in the interval $[0, 3]$.

c. Suppose that $f(0) = 1$. How many solutions does the equation $f(x) = 0$ have in the interval $[-3, 3]$? Explain.

Solution   Because f' is negative on the interval $(-3, 0)$ and positive on the interval $(0, 3)$, f itself must decrease on $(-3, 0)$ and increase on $(0, 3)$. Thus f has a local minimum at $x = 0$ and no other stationary points. Therefore:

a. The graph of f is U-shaped, so it can intersect the interval $[-3, 3]$ in at most two places.

b. The graph of f increases on the interval $(0, 3)$, so it can cross the x-axis no more than once on that interval.

c. If $f(0) = 1$, then f has no roots in the interval $[-3, 3]$, because f has its minimum value at $x = 0$. ∎

BASICS

1. The graph of a function f is shown below. Several points are labeled.

Graph of f

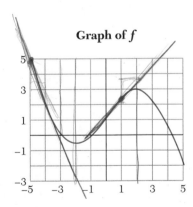

(a) At which labeled points is f' positive? Explain.
(b) Between which pairs of labeled points does f have a stationary point?
(c) Between which pair of labeled points does f have an inflection point?
(d) Between which pair of labeled points is f' increasing?
(e) Between which pair of labeled points does f' achieve its minimum value?

2. The graph of a function f is shown below.

Graph of f

(a) On which intervals is f' negative? Positive?
(b) On which intervals is f' increasing? Decreasing?
(c) Where does f' achieve its maximum value? Estimate this value of f'.
(d) Where does f' achieve its minimum value? Estimate this value of f'.
(e) Sketch a graph of f'.   [**NOTE:** Your sketch should be consistent with your answers to parts (a)–(d).]

3. The graph of a function f is shown below

Graph of f

Which of the following figures could be the graph of f'? For each graph that you did *not* select, explain why it can't be the graph of f'.

Graph I   **Graph II**

Graph III   **Graph IV**

4. The graph of the derivative of a function g is shown below. Use the graph of g' to answer the following questions about g.   [**NOTE:** The graph of g is not shown.]

Graph of g'

(a) Where does g have stationary points?
(b) Where does g have local maxima? Local minima?
(c) The graph of g' has a local maximum at $x = 3.8$ and a local minimum at $x = 7.4$. What do these facts say about the graph of g?

(d) Is g concave up or concave down at $x = 5$? At $x = 8$? Justify your answers.

(e) Suppose that $g(0) = 0$. Sketch a graph of g.

5. The graph of the derivative of a function f appears below.   [**NOTE**: The graph of f is not shown.]

Graph of f'

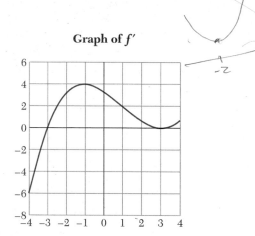

(a) Where does f have stationary points?

(b) Where does f have local maxima? Local minima? Points of inflection?

(c) Is f concave up or concave down at $x = 1$? Justify your answer.

(d) Sketch the graphs of *two* functions that could be f.

6. Suppose that the derivative of a function f is $f'(x) = x + 2$.

(a) On which intervals, if any, is f increasing?

(b) At which values of x, if any, does f have a local maximum? A local minimum?

(c) On which intervals, if any, is f concave down?

(d) Sketch a graph of f.

Repeat the previous exercise for the derivative functions in Exercises 7–18. (Report approximate answers to two decimal places.)

7. $f'(x) = (x - 2)(x + 4)$

8. $f'(x) = (x - 2)^2(x + 4)$

9. $f'(x) = (x + 1)^2(x - 2)$

10. $f'(x) = (x^2 + 9)(x + 6)$

11. $f'(x) = (x^2 - 9)(x + 3)$

12. $f'(x) = \dfrac{1}{x^2 + 1}$

13. $f'(x) = \dfrac{x - 2}{x^2 + 5}$

14. $f'(x) = \dfrac{x + 1}{\sqrt{x^2 + 1}}$

15. $f'(x) = \dfrac{x^2 + 2}{\sqrt{x^4 + 8}}$

16. $f'(x) = e^{-x^2}$

17. $f'(x) = (9 - x^2)e^{-x/4}$

18. $f'(x) = \ln(1 + x^2)$

FURTHER EXERCISES

19. In the left-hand column below are graphs of several functions. In the right-hand column—in a different order—are graphs of the associated *derivative* functions. Match each function with its derivative.   [**NOTE**: The scales on the graphs are not all the same.]

20. Sketch a graph of the derivative of each function labeled (i)–(v) in the right column of the preceding problem.

21. For each function labeled (a)–(e) in the left column of Exercise 19, sketch the graph of a function whose *derivative* is the function shown.

22. Graphs of f, f', and g (not the derivative of f) are shown below. Which is which? Explain how you can tell.

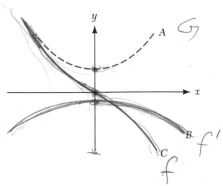

23. Graphs of f, f', and g (not the derivative of f) are shown below. Which is which? Explain how you can tell.

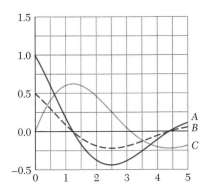

24. (a) Sketch the graph of a function f for which $f(x) > 0$ and $f'(x) > 0$ for all x.
    (b) Sketch the graph of a function g for which $g(x) > 0$ and $g'(x) < 0$ for all x.
    (c) Give formulas for functions f and g with these properties.

25. (a) Sketch the graph of a function f for which $f(x) < 0$ and $f'(x) > 0$ for all x.
    (b) Sketch the graph of a function g for which $g(x) < 0$ and $g'(x) < 0$ for all x.
    (c) Give formulas for functions f and g with these properties.

26. The graph of the derivative of a function f is shown below. Use the graph of f' to answer the following questions about f. [**NOTE:** The graph of f is not shown.]

Graph of f'

(a) On which intervals is f increasing? Decreasing?
(b) Where does f have a stationary point?
(c) Where does f have a local maximum? A local minimum?
(d) On which intervals is f concave up? Concave down?
(e) Where does f have a point of inflection?
(f) Where does f achieve its maximum value on the interval $[0, 2]$? Its minimum value?
(g) Where does f achieve its maximum value on the interval $[3, 6]$? Its minimum value?
(h) Assume that $f(0) = 0$. Sketch a graph of f. (Your graph need only have the right general shape.)
(i) How does your answer to part (h) change if $f(0) = 5$?

27. Suppose that $f(3) = -1$ and that f' is the function graphed in the previous exercise.
(a) Let ℓ be the line tangent to f at the point $(3, -1)$. Find an equation for ℓ.
(b) Explain why ℓ lies below the curve $y = f(x)$ near $x = 4$.
(c) Does ℓ lie above or below the curve $y = f(x)$ near $x = 2$? Justify your answer.

28. The graph of the *derivative* of a function f is shown below.

Graph of f'

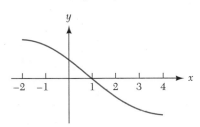

(a) Where in the interval $[-2, 4]$ is f increasing? Where is f decreasing? Why?

(b) Where in $[-2, 4]$ does f have a stationary point? Is it a local maximum, a local minimum, or neither?

(c) Describe in words the general shape of the f-graph over the interval $[-2, 4]$.

29. Let $f(x) = e^{\sin x}$. How many roots does f' have in the interval $[0, 2\pi]$? Where are they located? Justify your answers.

30. Suppose that the following information is known about a function f:
    (i) f is positive on $[-7, 8]$
    (ii) f is increasing on $[-2, 3]$
    (iii) f is decreasing on $(4, 10)$

(iv) f' is decreasing on $[-3, 6]$

(v) $f'(4) = 0$

(vi) f has an inflection point at $x = 7$

(a) Explain why f has a local extremum at $x = 4$. Is it a local maximum or a local minimum?

(b) Where is f concave down?

(c) Where is f' nonnegative?

(d) Rank the four values $f'(4)$, $f(5)$, $f'(6)$, and $f(7)$ in increasing order.

(e) Sketch the graph of a function f with these properties.

2.4 The Geometry of Higher-Order Derivatives

As before, "nice" means that all the derivatives in question exist and have continuous graphs.

"Differentiate" means "find the derivative of."

The derivative of a nice function f is another nice function, f'. ◄   Given the many uses of f', the next step is natural: Differentiate ◄   again. The resulting function, denoted by f'', is called the **second derivative** of f. Like f', the second derivative f'' carries useful information about f. What information? How do we read it?

Prime Advice.   With several functions $(f, f', f'',$ etc.) in the picture, it's easy to lose one's bearings. One organizing principle helps:

> *The derivative of any function tells whether, and how fast, that function increases or decreases.*

We defined concavity in terms of f' in the last section.

If, say, $f''(a) > 0$, then f' is *increasing* at $x = a$. Therefore, it follows that f must be concave up at $x = a$. ◄

What f'' Says about f—Concavity Revisited

The graph of f is concave up or down at $x = a$ depending on whether f' is increasing or decreasing at $x = a$. The preceding principle, applied to f', reads like this:

> *If $f''(a) > 0$, then f' increases at $x = a$. If $f''(a) < 0$, then f' decreases at $x = a$.*

This leads to a useful test for concavity:

Fact The graph of f is concave up where $f'' > 0$ and concave down where $f'' < 0$.

(The fact gives no information if $f''(a) = 0$. In that case, as we'll see, f may be concave up, concave down, or neither.)

E X A M P L E 1   Interpret the concavity of the sine function using the second derivative.

S o l u t i o n   If $f(x) = \sin x$, then $f'(x) = \cos x$ and $f''(x) = -\sin x$. (We'll derive both of these formulas carefully later in the chapter. For the moment, we'll assume that they hold.) Following are graphs of f and f'' on the same axes.

Graphs of f and f''

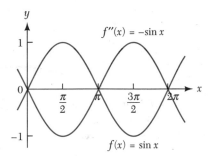

It's easy to see ▶ that the advertised relationship holds: f is concave up precisely where $f'' > 0$ and is concave down where $f'' < 0$. ■   *See for yourself.*

Wave Claims.   We claimed earlier, on graphical evidence alone, that if $f(x) = \sin x$, then $f'(x) = \cos x$. We also claimed that if $g(x) = \cos x$, then $g'(x) = -\sin x$. Both claims require careful proof.

Notice, though, that the two claims are essentially the same. Since the graphs of $f(x)$ and $g(x)$ differ only by a horizontal translation, so must the graphs of their derivatives.

Finding Inflection Points

A function f has an inflection point wherever the *sign* of f'' changes. It follows, therefore, that for nice functions,

> *Every inflection point of f occurs at a root of f''.*

On the sine graph, for instance, an inflection point occurs at every root of f''.

Derive Carefully, Again.   The converse of the preceding statement is false:

> *$f''(a)$ may be zero even if no inflection occurs at $x = a$.*

The next example illustrates why: If $f''(a) = 0$, then a is a stationary point of f' but not necessarily a local maximum or local minimum.

The functions shown have simple formulas: $f(x) = x^4$, $f'(x) = 4x^3$, *and* $f''(x) = 12x^2$. *We "hid" the formulas here in the margin because they're not necessary to this problem.*

EXAMPLE 2   Graphs of a function f and its derivatives f' and f'' follow. ◄ Discuss concavity and inflection points of f.

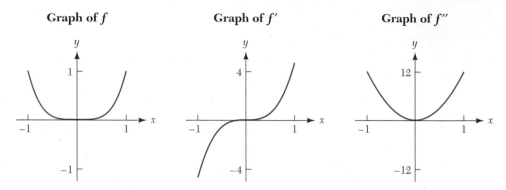

Graph of f Graph of f' Graph of f''

S o l u t i o n   Although $f''(0) = 0$, $(0, 0)$ is *not* an inflection point. On the contrary, the graph of f is concave up everywhere, because f' is everywhere increasing.

Notice too how the shape of the f-graph reflects the fact that $f(0) = f'(0) = f''(0) = 0$. If *several* derivatives of a function are zero at a point $x = a$, the graph of f looks flat near $x = a$. ◄   ■

The greater the number of derivatives that vanish at $x = a$, the flatter the graph.

All Together Now: Maximum Points, Minimum Points, Concavity, and the Second Derivative

A stationary point on the graph of a function f may be a local maximum point, a local minimum point, or neither. If an accurate graph of f is available, it's usually obvious which case applies. If not, concavity often helps us distinguish among these alternatives. The idea is simple:

> *At a local minimum point, the graph of f is concave up.*
> *At a local maximum point, the graph of f is concave down.*

The following claims are all immediate consequences. Together, they're known as the second derivative test.

Fact (Second Derivative Test) Suppose that $f'(a) = 0$.

If $f''(a) > 0$, then f has a local minimum at $x = a$.

If $f''(a) < 0$, then f has a local maximum at $x = a$.

If $f''(a) = 0$, anything can happen: f may have a local maximum, a local minimum, or neither at $x = a$, and the f-graph may be concave up, be concave down, have an inflection point, or have none of these properties.

If—but only if—appropriate care is taken, the second derivative test can be a convenient tool for finding maxima and minima of functions.

Stirring in the Formulas

We return, finally, to the functions f, f', and f'' we first met at the beginning of the preceding section—this time with algebraic formulas added. Here are all three formulas:

$$f(x) = x^3 - x^2 - 6x; \qquad f'(x) = 3x^2 - 2x - 6; \qquad f''(x) = 6x - 2.$$

Here are all three graphs:

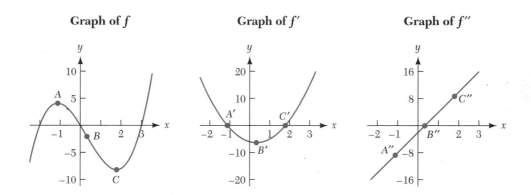

Combining graphs *and* formulas, we can sharpen and reinterpret our earlier observations:

What the Derivative Says Both the graph and the algebraic definition of f' suggest that as $x \to \infty$, $f'(x) \to \infty$. Therefore, as $x \to \infty$, the slope of the f-graph is (a) positive and (b) increasing. Equivalently, the f-graph itself is (a) rising and (b) rising faster and faster.

Stationary Points The f-graph has horizontal tangent lines at the stationary points A and C. According to the derivative formula, stationary points occur at values of x for which

$$f'(x) = 3x^2 - 2x - 6 = 0$$

By great (and, in some respects, atypical) good luck, $f'(x)$ happens to be a quadratic polynomial. By the quadratic formula, the roots of f' fall at $x = (1 \pm \sqrt{19})/3$. In decimal form, the roots are $x_1 \approx -1.1196$ and $x_2 \approx 1.7863$.

How Many Stationary Points? The preceding calculation shows, too, that f' has *only* two roots. Hence, A and C are the only stationary points of f.

Finding the Inflection Point The inflection point B occurs where f' takes its minimum value, somewhere near $x = 0.3$. Where *exactly* is B? The formula for f'' answers the question: A root of f'' occurs at $x = 1/3$, where

$$f''(x) = 6x - 2 = 0.$$

As its graph shows, f'' changes sign at $x = 1/3$, so f changes its direction of concavity there. ▶

The formula for f'' shows a little more: f'' changes sign only at $x = 1/3$. Therefore, f changes concavity only once.

Morals

We have mentioned a large, even bewildering, variety of connections among f, f', and f''. The reader may be tempted to memorize complicated lists of cases and subcases, particularly as they apply to concavity, inflection points, and the second derivative. That would be a mistake. A better strategy is to learn and understand the single organizing principle behind this welter of information:

> *The derivative function f' tells whether, and how fast, the original function f increases or decreases.*

A Sample Problem

The following example problem gives some idea of the variety of uses to which derivatives and higher derivatives can be put.

EXAMPLE 3   The graph of f', the derivative of a certain function f, follows. (The f-graph is not shown!)

Graph of f'

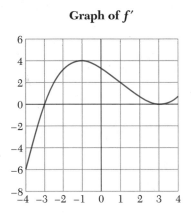

a. Estimate $f''(-3)$.

b. Where is f concave up? Concave down?

c. Where does f have inflection points?

Solution   Looked at properly, the graph answers all our questions. It's crucial to remember always that the picture shows f', not f or f''.

a. By definition, $f''(-3) = (f')'(-3)$, i.e., the slope of the graph above at $x = -3$. A careful look reveals that this slope is about 5.

b. Since $f''(-3) > 0$, f is concave up at $x = -3$. More generally, f is concave up wherever $f'' > 0$, i.e., where f' is increasing. The picture shows that this occurs for $-4 \leq x \leq -1$ and for $3 \leq x \leq 4$. For the same reason, f is concave down for $-1 \leq x \leq 3$.

c. The function f has an inflection point wherever its direction of concavity changes. Part (b) shows that this happens at $x = -1$ and at $x = 3$. ■

BASICS

1. The graph of a function f appears below.

Graph of f

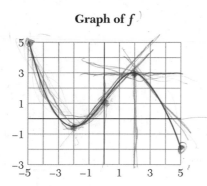

(a) For which values of x, if any, is $f''(x)$ negative? Positive? Zero?

(b) Rank the four numbers $f'(0)$, $f(2)$, $f'(2)$, and $f''(2)$ in increasing order.

2. The graph of a function f appears on the right. [**NOTE:** f has points of inflection near $x = 0.82$ and near $x = 1.93$; f has stationary points near $x = 0.52$, $x = 1.19$, and $x = 2.42$.]

Graph of f

(a) Where is f'' positive? Negative? Zero?

(b) Rank the four values 0, $f''(0.5)$, $f''(1.2)$, and $f''(2.4)$ in increasing order.

3. Water flows at a constant rate into a large tank that is a cylinder with a circular base. Let $V(t)$ be the volume of water in the tank at time t and $H(t)$ be the height of the water in the tank at time t. Give a physical interpretation of each of the following expressions and indicate whether each is positive, negative, or zero at the time t^* when the tank is one-quarter full:

(a) $V'(t)$
(b) $H'(t)$
(c) $V''(t)$
(d) $H''(t)$

FURTHER EXERCISES

4. The graph of the second derivative of a function g is shown below. Use this graph to answer the following questions about g and g'.

Graph of g''

(a) Where is g concave up?
(b) Where does g have points of inflection?
(c) Rank the four numbers $g'(0)$, $g'(1)$, $g'(2)$, and $g'(3)$ in increasing order.
(d) Suppose that $g'(0) = 0$. Is g increasing or decreasing at $x = 2$? Justify your answer.
(e) Suppose that $g'(0) = -4$. Is g increasing or decreasing at $x = 2$? Justify your answer.
(f) Suppose that $g'(1) = -1$. Is g increasing or decreasing at $x = 2$? Justify your answer.

5. Suppose that $g' = f$ where f is the function defined in Exercise 2 above.

(a) Estimate $g''(0)$, $g''(0.5)$, $g''(1)$, $g''(1.5)$, $g''(2)$, and $g''(3)$.

(b) Sketch a graph of g'' over the interval $[0, 3]$.

6. In the left-hand column below are graphs of several functions. In the right-hand column—in a different order—are graphs of the *second* derivatives of these functions. Match each function with its second derivative function. [**NOTE:** The scales on the graphs are not all the same.]

7. Graphs of f, f', and f'' appear below. Which is which? How can you tell?

8. Graphs of f, f', and f'' appear below. Which is which? How can you tell?

9. Graphs of f, f', f'', and g (not a derivative of f) are shown below. Which is which? How can you tell?

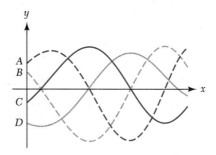

10. Suppose that the functions f, f', and f'' are continuous on $(-\infty, \infty)$ and that f has the following properties:
    (i) f is negative on $(-10, 1)$ and positive on $(3, 10)$.
    (ii) f is increasing on $(-\infty, 5)$ and decreasing on $(5, \infty)$.
    (iii) f is concave up on $(-\infty, 2)$ and concave down on $(2, \infty)$.
    Rank the five values $f'(0)$, $f'(1)$, $f'(2)$, $f''(2)$, and $f'(8)$ in increasing order.

11. Repeat Problem 3 assuming that the tank is a sphere.

12. Repeat Problem 3 assuming that the tank is an inverted cone (i.e., a cone with its vertex pointing down).

13. Interpret each sentence below as a statement about a function and its derivatives. In each case, indicate clearly what the function is and what each symbol means.
    (a) The child's temperature is still rising, but the penicillin seems to be taking effect.
    (b) The cost of a new car is increasing at an increasing rate.
    (c) Upper Midwest Industries' profit growth slows.

14. Let $U(t)$ be the number of people unemployed in a country t months after the election of an economically conservative president. Translate each of the following facts about the graph of $U(t)$ into statements about the unemployment situation. (Don't forget to specify units in your answers.)
    (a) The y-intercept of $U(t)$ is 2,000,000.
    (b) $U(2) = 3,000,000$.
    (c) $U'(20) = 10,000$.
    (d) $U''(36) = 800$ and $U'(36) = 0$.

15. (a) Sketch the graph of a function f for which $f(x) > 0$, $f'(x) > 0$, and $f''(x) > 0$ for all x.
    (b) Sketch the graph of a function f for which $f(x) > 0$, $f'(x) < 0$, and $f''(x) > 0$ for all x.

16. (a) Sketch the graph of a function f for which $f(x) < 0$, $f'(x) < 0$, and $f''(x) < 0$ for all x.
    (b) Sketch the graph of a function f for which $f(x) < 0$, $f'(x) > 0$, and $f''(x) < 0$ for all x.

17. Is there a function f for which $f(x) > 0$, $f'(x) < 0$, and $f''(x) < 0$ for all x? If so, sketch the graph of such a function. If not, explain why no such function exists.

18. Is there a function f for which $f(x) < 0$ and $f''(x) > 0$ for all x? If so, sketch the graph of such a function. If not, explain why there is no function with these properties.

19. Assume that f is a continuous function defined on the closed interval $[-3, 3]$ such that $f(-3) = 4$ and $f(3) = 1$. Furthermore, assume that f' and f'' are continuous on $(-3, 3)$ and that the information in the table below is known about these functions.

x	$-3 \le x < -1$	-1	$-1 < x < 0$	0
$f'(x)$	$+$	0	$-$	$-$
$f''(x)$	$-$	$-$	$-$	0

x	$0 < x < 1$	1	$1 < x \le 3$
$f'(x)$	0	$-$	$-$
$f''(x)$	$+$	0	$-$

    (a) Sketch the graph of a function that has the given properties.
    (b) What are the x-coordinates of the points where f

achieves its maximum and minimum values on the interval $[-3, 3]$? Justify your answer.

20. Sketch the graph of a continuous function g that has *all* of the following properties:
    (i) The domain of g is the interval $[-3, 5]$.
    (ii) $g''(x) < 0$ for all $x \in (-2, 1)$
    (iii) $g''(x) > 0$ for all $x \in (1, 2)$
    (iv) $g'(-2) < 0$
    (v) $g'(4) > 0$
    (vi) $g'(1) = 0$ but $g(1)$ is neither the maximum nor the minimum value of g over the interval $[-2, 4]$

21. Sketch the graph of a continuous function that satisfies the following conditions:
    (i) $f'(x) < 0$ for all $x \ne 4$
    (ii) $f'(4)$ does not exist
    (iii) $f''(x) < 0$ for all $x < 4$
    (iv) $f''(x) > 0$ for all $x > 4$

22. It can be shown that if f is an even function, then f' is an odd function, and that if f is an odd function, then f' is an even function. Use these facts to show that if f is an even function, then f'' is an even function.

23. Suppose that $1 \le f''(x) \le 3$ if $2 \le x \le 4$, that $f'(2) = 5$, and that $f(2) = -6$.
    (a) Explain why $f'(3) \le 8$.
    (b) Use part (a) to show that $f(3) \le 2$.
    (c) Explain why $f'(3) \ge 6$.
    (d) Find upper and lower bounds on $f'(4)$.
    (e) Find upper and lower bounds on $f(4)$.

24. Suppose that the functions f, f', and f'' are continuous on $(-\infty, \infty)$, that $f'(2) = -1$, and that $0.1 < f''(x) < 0.5$ if $2 < x < 4$. Find upper and lower bounds on the value of $f'(3)$. [**HINT**: Apply the racetrack principle.]

25. Suppose that the line $y = 2x - 1$ is tangent to the graph of f at $x = 3$ and that $f''(x) > 0$, if $0 < x < 5$.
    (a) Find $f(3)$ and $f'(3)$.
    (b) Explain why $7 < f(4)$ must be true.
    (c) Find a lower bound on the value of $f(2)$.

26. Suppose that $f(2) = 5$, $f'(2) = 2$, and that f is concave up everywhere.
    (a) How many roots could f have? What can you say about the location of these roots?
    (b) Find a lower bound on the value of $f(0)$.
    (c) Suppose that $f(-2) = -1$. Show that $f(0) < 3$.

27. Is the following statement true or false? Justify your answer.

    If $f(2) = 3$, $f'(2) = 1$, *and* $f''(2) = 0$, *then the graph of* $f(x)$ *has an inflection point at* $(2, 3)$.

28. Suppose that $f'(c) = f''(c) = 0$, but $f'''(c) > 0$. Does f have a local minimum, a local maximum, or an inflection point at c? Explain.

2.5 Average and Instantaneous Rates: Defining the Derivative

In earlier sections we introduced and studied derivatives informally, interpreting them as rates or slopes. In this section we take a more formal point of view.

Given an original function f and an input $x = a$, what exactly do we mean, mathematically, by $f'(a)$? We've thought of $f'(a)$ as the slope of f at $x = a$ or as the instantaneous rate of change of y with respect to x at $x = a$, but important questions remain: What, exactly, is the *instantaneous* rate of change of a function at a point? Exactly how is the tangent line to a graph at a point defined? Most important of all:

How can we use a formula for f to calculate—not just estimate—values of f'?

The Rate-of-Change Problem

In this no-frills model, drivers calculate speed from odometer readings.

Exactly at midnight, Professor X drives away in her brand-new Göttingen automobile. The odometer reads zero. There is no speedometer. ◄ Consulting her odometer and wristwatch, Professor X can find, for any time t (in hours), her total distance $D(t)$ (in miles) traveled since midnight. She might even plot her readings, as we did:

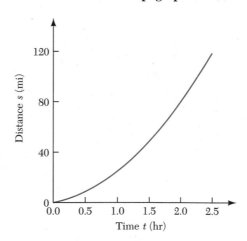

Professor X's trip: graph of $D(t)$

How can Professor X compute her speed at time $t = 2$, when she expects to reach a notorious speed trap? Reading the odometer at $t = 2$ won't help. It may say that she's traveled, say, 80 miles in 2 hours and hence that her *average* speed was 40 mph for the trip; but what about her speed at the *instant* $t = 2$?

She's right: The inexpensive Göttingen responds slowly both to throttle and to brake.

Professor X reasons that her *average* speed over a short interval—say $t = 1.99$ to $t = 2.00$—should be about equal to her *instantaneous* speed at any time during the same interval. ◄ The odometer reports 0.6985 miles of travel during the interval.

"Hence," ▶ thinks Professor X, "my average speed over the short period must have been

Professors say "hence," even to themselves.

$$\frac{0.6985 \text{ miles}}{0.01 \text{ hours}} = 69.85 \text{ mph}.$$

I'd better slow down."

The Tangent-Line Problem

The tangent-line problem asks:

Given a curve C and a point P on C, describe the line tangent to C at P.

EXAMPLE 1   Which line is tangent to the graph of $f(x) = x^2 + x$ at $x = 2$?

Solution   Here's part of the graph, with a plausible tangent line at the point $P = (2, 6)$.

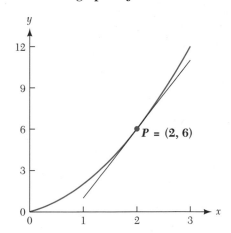

A tangent line to the graph of $y = x^2 + x$

From the graph alone we can estimate that the tangent line L has slope 5. ▶ If we're right, then L has equation $y = 5(x - 2) + 6$.

Check this for yourself.

Finding the slope of L *exactly* is complicated by the fact that we know for certain only one point on L. To start, we might look at a line L' through *two* nearby points on the graph, such as those at $x = 2$ and $x = 2.01$. We expect that L and L' will have similar slopes. Knowing *two* points on L' means that its slope is easy to find:

$$\frac{\Delta f}{\Delta x} = \frac{f(2.01) - f(2)}{2.01 - 2} = \frac{6.0501 - 6}{0.01} = 5.01.$$

This numerical evidence further supports our guess that $f'(2) = 5$. ▶   ■

Because of the shape of the graph near $x = 2$, the slope of L' slightly overestimates the slope of L.

Sharpening the Tangent-Line Problem

We stated the tangent-line problem in intuitive but slightly vague terms. What does it mean for a curve to be "given," and how should we "describe" the line tangent to C at P?

Given by Equations   Curves are often given as equations (e.g., $x^2 + y^2 = 4$) or as functions (e.g., $f(x) = x^2 + x$). Some curves are given *only* graphically or numerically, without any algebraic formula attached. In that case tangent lines can be found only approximately. If a curve C *has* an exact formula, we'd like to use it to find an exact formula for the line tangent to C at P.

Wanted: Slope   A line L is completely determined by its slope and a point P on L. In the tangent-line problem, a point P is given from the start, so "finding" L means, in effect, finding the *slope* of L.

Estimating Slope   If a curve C has a tangent line at P, we can estimate its slope either by hand—i.e., by drawing a line that fits C at P—or by zooming in—i.e., by viewing C near P in a plotting window small enough that C appears straight.

Draw a picture to convince yourself.

No Tangent Line   Not every curve C has a tangent line at a given point P. For example, the graph of $y = |x|$ has a sharp corner ◀   at $P = (0, 0)$, so no single straight line can be said to fit the graph at P.

EXAMPLE 2   We guessed earlier that the line L with equation $y = 5(x - 2) + 6$ is tangent to C, the graph of $f(x) = x^2 + x$, at $x = 2$. Does zooming support this guess? How?

Solution   Here are graphs of C and L at four scales:

C and L: window $[-2, 6] \times [-18, 30]$

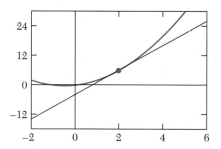

C and L: window $[1, 3] \times [0, 12]$

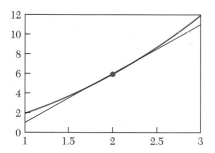

C and L: window $[1.5, 2.5] \times [3, 9]$

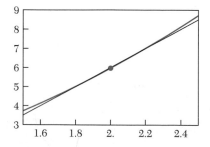

C and L: window $[1.9, 2.1] \times [5.4, 6.6]$

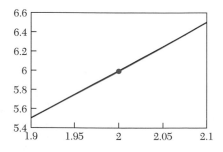

The graphs show that C, viewed close up, does resemble L, as claimed. In particular, because L has slope 5, so does the curve C at the point $(2, 6)$. ■

The Problems Solved: Secant Lines and Tangent Lines, Average Rates and Instantaneous Rates

Graphical methods (zooming until straight, drawing tangent lines by hand, and so on) give simple and useful *estimates* of slopes and rates. For more precision, and to get deeper under the skin of the derivative, we need clear, concise definitions, in mathematical language, for instantaneous rates and for slopes of tangent lines. The concept of limit turns out to be just the tool we need. ▶

We'll see much more on limits in the next two sections.

We'll solve our two problems in parallel. First we pose them explicitly.

The Tangent-Line Problem Let $y = f(x)$ be a function. Find the slope of the line tangent to the graph of f at $x = 2$.

The Instantaneous-Speed Problem Let the function $s = D(t)$ give the distance, in miles, traveled by a car up to time t, in hours. Find the instantaneous speed at time $t = 2$.

Secant Lines, Tangent Lines, and Limits

Given two points P and Q on a curve C, we can draw the **secant line** through P and Q. If P and Q are close together, it's reasonable to expect the secant line they determine to resemble the line tangent to C at P. ▶

Finding slopes of tangent lines poses a challenge, but finding slopes of secant lines is easy. If C is the graph of a function $y = f(x)$, and $P = \big(a, f(a)\big)$ and $Q = \big(b, f(b)\big)$ are points on the graph, then, as the following picture shows,

Before reading on, draw some curves and secant lines to convince yourself of this easy but important idea.

The secant line from $x = a$ to $x = b$

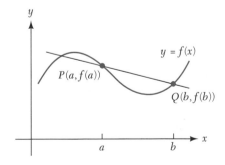

the slope of the secant line from P to Q (i.e., from $x = a$ to $x = b$) is

$$\frac{\Delta f}{\Delta x} = \frac{f(b) - f(a)}{b - a}.$$

Earlier we calculated that the secant line to $f(x) = x^2 + x$ from $x = 2$ to $x = 2.01$ has slope 5.01. This result suggests our strategy: We'll take the slope of the tangent line to be the limit of slopes of secant lines over smaller and smaller intervals.

What Is a Limit?   The **limit** is a mathematical tool used to describe how a function $f(x)$ behaves for x *near* a specific value $x = a$. We'll discuss limits carefully in the next two sections. In this section we treat limits informally, in context.

EXAMPLE 3   Compute slopes of secant lines to the graph of $f(x) = x^2 + x$ over small intervals near $x = 2$, and interpret the results in limit language. Draw (and defend) a conclusion about $f'(2)$.

Solution   Results for various small intervals appear below:

Slopes of Secant Lines over Small Intervals				
Interval	[1.99, 2]	[1.999, 2]	[2, 2.001]	[2, 2.01]	[2, 2.1]
Slope	4.99	4.999	5.001	5.01	5.1

Only a mathematician could doubt that these slopes approach 5.

To put the situation in limit language, consider the interval from $x = 2$ to $x = 2 + h$. ◄   Then the secant line from $x = 2$ to $x = 2 + h$ has slope

We think of h as a small (positive or negative) increment.

$$\frac{\Delta f}{\Delta x} = \frac{f(2 + h) - f(2)}{(2 + h) - 2} = \frac{f(2 + h) - f(2)}{h}.$$

The slope of the tangent line at $x = 2$ is the limit as h tends to zero of this ratio. In symbols,

$$f'(2) = \lim_{h \to 0} \frac{f(2 + h) - f(2)}{h}.$$

At last we can put this problem out of its misery, by calculating the slope of the tangent line at $x = 2$:

$$\lim_{h \to 0} \frac{f(2 + h) - f(2)}{h} = \lim_{h \to 0} \frac{(2 + h)^2 + (2 + h) - 6}{h}$$

$$= \lim_{h \to 0} \frac{5h + h^2}{h}$$

$$= \lim_{h \to 0} (5 + h)$$

$$= 5. \qquad ■$$

(The preceding limit calculation should be, for the moment, intuitively reasonable. We'll study the algebra of limits carefully in the next two sections.)

Average Speeds, Instantaneous Speeds, and Limits

The rate-of-change problem, for Professor X, is to convert information about distance traveled to information about instantaneous speed. Its solution closely mimics that for the tangent-line problem.

Average speed over an interval of time is easy to compute. As Professor X knows,

$$\text{average speed} = \frac{\text{distance traveled}}{\text{time elapsed}}.$$

The remaining problem is to connect average speeds (over intervals) with instantaneous speeds. Again, limits provide the missing link: The instantaneous speed at time t_0 is the limit of average speeds over small time intervals that contain t_0. More succinctly put:

Consider the interval from $t = 2$ to $t = 2 + h$ hours. ▶ *The average speed from $t = 2$ to $t = 2 + h$ is*

Again, h represents a small positive or negative increment.

$$\frac{\Delta D}{\Delta t} = \frac{D(2 + h) - D(2)}{(2 + h) - 2} = \frac{D(2 + h) - D(2)}{h}.$$

Therefore, the instantaneous speed at $t = 2$ is

$$\lim_{h \to 0} \frac{D(2 + h) - D(2)}{h}.$$

We confess that, all along, we've had the function $D(t) = 10t + 15t^2$ secretly in mind for Professor X's trip. Finding her speed at $t = 2$ hours is now a simple calculation: ▶

But check each step!

$$\lim_{h \to 0} \frac{D(2 + h) - D(2)}{h} = \lim_{h \to 0} \frac{10(2 + h) + 15(2 + h)^2 - 80}{h}$$

$$= \lim_{h \to 0} \frac{70h + 15h^2}{h}$$

$$= \lim_{h \to 0} (70 + 15h)$$

$$= 70.$$

At *exactly* 2:00 A.M., Professor X's Göttingen was doing *exactly* 70 mph.

Derivative: The Formal Definition

The tangent-line and rate-of-change problems turned out to be identical. Both problems—and many others—lead to limits of a common form. This form appears in the following definition.

> **Definition** Let f be a function defined near and at $x = a$. The **derivative of f at $x = a$**, denoted by $f'(a)$, is defined by the limit
>
> $$f'(a) = \lim_{h \to 0} \frac{f(a + h) - f(a)}{h}.$$

The most important in all of calculus.

This important definition ◄ deserves some amplification:

Must $f'(a)$ Exist? In a word, no. The limit in the definition may or may not exist. If it *does* exist, we say that f is differentiable at $x = a$ or that f has a derivative at $x = a$. Most of the functions we treat in calculus are differentiable wherever they make sense. The absolute value function $g(x) = |x|$ is one exception. The kink at $x = 0$ means that g has no tangent line—and hence no derivative—at $a = 0$. (We'll see in the next section why the limit that defines $g'(0)$ does not exist.)

But only if the algebra is done carefully.

Dividing by Zero? The limit in the derivative definition *always* involves a denominator tending to zero; evaluating it always involves some trickery or other. In practice, things often work out as in the preceding examples: After some algebra in the numerator, the troublesome h in the denominator goes away. ◄

Other Forms The limit in the derivative definition can be written in other equivalent forms. Here are several popular ones:

$$\lim_{h \to 0} \frac{f(a + h) - f(a)}{h} \qquad\qquad \lim_{\Delta x \to 0} \frac{f(a + \Delta x) - f(a)}{\Delta x}$$

$$\lim_{x \to a} \frac{f(x) - f(a)}{x - a} \qquad\qquad \lim_{\Delta x \to 0} \frac{\Delta f}{\Delta x}$$

Depending on f, one form may be simpler or easier to manipulate than another, but all convey the same information.

Difference Quotients, Average Rates of Change, and Their Limits

The quotients in the preceding limits are all instances of a general idea:

> **Fact** For any function f, the ratio
>
> $$\frac{f(a + h) - f(a)}{h}$$
>
> is called a **difference quotient**. It measures the *average rate of change* of the function f from $x = a$ to $x = a + h$.

To understand what it tells, let's consider the various parts of the difference quotient separately

The Denominator The difference

$$h = (a + h) - a = \Delta x$$

represents a change, or increment, in the input variable, from $x = a$ to $x = a + h$. (The change may be either an increase or a decrease.)

The Numerator As x moves from $x = a$ to $x = a + h$, the output $f(x)$ changes too. The numerator

$$\Delta f = f(a + h) - f(a)$$

measures this change. Like Δx, Δf may have either sign.

The Quotient The difference quotient is the *ratio* of two changes Δf and Δx, so it represents a *rate* of change of $f(x)$ with respect to x. Because the changes occur over an interval in x, the quotient represents an *average* rate. The sign of $\Delta f/\Delta x$ tells whether x and $f(x)$ change in the same direction. If, say, $\Delta f/\Delta x < 0$, then an increase in x produces a decrease in $f(x)$.

From Average to Instantaneous The difference quotient is an average rate of change over the interval $[a, a + h]$. Therefore, the limit

$$\lim_{h \to 0} \frac{f(a + h) - f(a)}{h}$$

is the *instantaneous* rate of change at the instant $x = a$.

At Any Rate. In everyday speech, the words "rate" and "instantaneous" suggest quantities varying in time. "Rate of change" should be understood more generally in calculus. If f is *any* function—perhaps having nothing to do with time—the difference quotient $\Delta f/\Delta x$ and its limit are still thought of as rates; they measure how fast outputs from f change relative to inputs x.

A case could be made for replacing the customary phrase "rate of change"—which seems to apply equally well to increasing and decreasing quantities—with "rate of increase." "Rate of increase" makes the bookkeeping a little simpler: If $\Delta f/\Delta x > 0$, the rate of increase of $f(x)$ with respect to x is positive; this means that $f(x)$ increases as x does. If $\Delta f/\Delta x < 0$, the rate of increase is negative; $f(x)$ decreases as x increases.

Difference Quotients and Derivatives: What They Mean in Various Settings

Any function f has many possible interpretations. Our old favorite $f(x) = x^2 + x$ could represent the weight in ounces of a kitten x days after birth, the area of a rectangle with variable sides, or almost anything else. Any particular interpretation of f leads to a corresponding interpretation of the difference quotient. Several examples follow. In every case the difference quotient and derivative are rates of change. Notice especially the various units of measurement: Choices of units for x and y determine appropriate units for the difference quotient and the derivative.

Slopes On any graph $y = f(x)$, the difference quotient represents the *slope of the secant line* from $x = a$ to $x = a + h$. The derivative $f'(a)$, as the limit of slopes of secant lines, represents the *slope of the tangent line* at $x = a$.

Velocities If $y = f(x)$ represents the position of a moving object at time x, then the difference quotient $\Delta f / \Delta x$ represents *average* velocity over the interval from $x = a$ to $x = a+h$. The derivative $f'(a)$, as a limit of average velocities, represents *instantaneous* velocity at time a. ◄

Recall that speed is the absolute value of velocity.

Gas Mileage Let y be the total distance (in miles) traveled by Professor X on x gallons of fuel. Then y is a function—say, g—of x. (Even the Göttingen has a gas gauge.) The professor takes these readings:

x (gallons)	0	1.5	1.9	1.99	2
y (miles)	0	43.2725	55.4993	58.3969	58.7206

A difference quotient, such as

$$\frac{g(2) - g(0)}{2 - 0} = \frac{58.7206 \text{ miles}}{2 \text{ gallons}} = 29.3603 \text{ miles per gallon,}$$

represents the *average* rate of gas consumption over the interval in which the first two gallons of gas were consumed. The *instantaneous* rate of gas consumption varies from moment to moment. To find the instantaneous rate at $x = 2$, we need to find a limit. Without an explicit formula for g, we can only estimate. The following table shows average rates for various intervals in x.

Interval	[0, 2]	[1.5, 2]	[1.9, 2]	[1.99, 2]
Average Miles per Gallon	29.3603	30.8962	32.213	32.37

A sensible guess, then, is that at the moment when $x = 2$, gas mileage was about 32.4 mpg.

Units What units make sense for difference quotients and derivatives? It depends on the units chosen for x and y. For Professor X's speed we had $s = D(t)$, with s in miles and t in hours; the natural rate unit, therefore, is miles per hour (i.e., the quotient miles/hours). For fuel consumption we used miles and gallons as units, so rates were measured in miles per gallon. Had Professor X chosen another vehicle and other units of distance and fuel (e.g., horse, leagues, and bales) the proper units for difference quotient and derivative would have been different (e.g., leagues per bale).

The Tangent-Line Problem in Historical Perspective

The problem of finding a straight line that fits a curve at a given point is very old. For the Greeks the problem had less to do with description—how to *define* the tangent line rigorously—than with construction—how, given a specific curve, to use ruler, compass, and other tools to *draw* tangent lines. For a circle, constructing the tangent line L in the Greek sense is easy: L is the line perpendicular to the radius at P. The Greeks could construct tangent lines for circles, parabolas, and a few other curves of classical geometry. In his treatise *On Spirals* (ca. 250 B.C.E.) Archimedes constructed tangent lines to certain spirals. The following figure illustrates the situation for circles and spirals.

A tangent to a circle   **A tangent to a spiral**

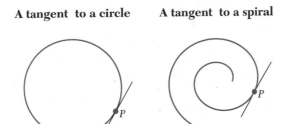

If C is a circle, the tangent line L at P happens to be the only line through P that touches ▶ the circle only at P. This touches-only-at-P criterion works (i.e., identifies a line which deserves to be called "tangent") for some curves but not for many others we meet in modern calculus courses. For the curve $y = \sin x$, for example, most tangent lines cross the curve at several points.

Tangere is Latin for "to touch."

The Greeks seem to have taken the touching criterion as their *definition* of tangent line. That they did so is a bit surprising, because even for the Archimedean spiral the criterion fails: Every tangent line eventually crosses the spiral. Perhaps the Greeks accepted such an inadequate definition because they were less interested in defining tangent lines than in constructing them.

Finding the "right" definition of tangent line is slightly subtle. In one sense, the tangent line is anything we say it is; we can define it any way we like. We might, for example, foolishly define the line tangent to C at P to be the horizontal line through P. This definition is not so much logically incorrect as useless: A tangent line by this definition would tell us nothing about the direction of C near P. A good definition must be both simple enough to work with and precise enough to capture the essence of the idea in question.

A good mathematical definition starts with a general intuition and makes it precise. For tangent lines we have plenty of geometric intuition to build on. Given a curve and a point on it, it is usually easy to draw a tangent line and agree that it's correct (allowing duly for imperfect pencils, rulers, eyesight, and so on). The challenge is to describe in *mathematical language* the line we would draw anyway.

The right definition of tangent line involves a limit. In retrospect, this conclusion seems obvious, but it was arrived at only after a long historical process. Although the question goes back to ancient times, not until the work of Cauchy in the 1820s—100 years after Newton's death—were limits in the present-day sense known. That is not to say that nothing happened in the meantime. By 1630, Fermat had essentially the modern idea of tangent line. He used his idea to solve problems like those that appear in calculus courses today.

BASICS

1. Amy traveled from Chicago to Milwaukee. Due to road construction, she drove the first 10 miles at a constant 20 mph. For the next 30 miles she drove at a constant 60 mph. She stopped at a rest area for a 10-minute snack, then drove the last 45 miles at a constant 45 mph.
   (a) Plot Amy's distance along the road from her starting point as a function of time.
   (b) Plot Amy's velocity as a function of time.
   (c) What was Amy's average speed for the trip (including the stop at the rest area)?

2. If $f(1) = 2$ and the average rate of change of f from $x = 1$ to $x = 5$ is 3, what's $f(5)$?

3. If $f(2) = 1$ and the average rate of change of f from $x = -3$ to $x = 2$ is 4, what's $f(-3)$?

4. True or false? Justify your answers.
   (a) Velocity is the rate of change of position with respect to time.
   (b) If the average rate of change of a function over an interval is zero, then the function must be constant.
   (c) If the average rate of change of a function over every interval is 17, then the graph of the function is a straight line.

5. The graph below shows how the price of a certain stock varied over a recent trading day.

(a) For each time interval below, find the total change in price and the average rate of change of price. (Be sure to indicate the units used to measure these quantities.)
   (i) 8:00 to 11:00
   (ii) 9:00 to 1:00
   (iii) 9:30 to 2:00
   (iv) 11:00 to 1:00
(b) Estimate the instantaneous rate of change of the stock's price at each of the following times. (Be sure to indicate units with your answers.)
   (i) 9:15 A.M.
   (ii) 10:30 A.M.
   (iii) 12:15 P.M.
   (iv) 1:45 P.M.

6. The position $p(t)$ of an object at time t (in seconds) along a line with units marked in meters is given by $p(t) = 3t^2 + 1$.
   (a) Find the change in position of the object between time $t = 1$ and $t = 3$.
   (b) Find the average velocity of the object between time $t = 1$ and $t = 3$.
   (c) On one set of axes, sketch a graph of $p(t)$ and the secant line from $t = 1$ to $t = 3$. Find an equation of the secant line.
   (d) Find the average velocity of the object between time $t = 1$ and time $t = 1 + h$. What are the units of your answer?
   (e) Find the instantaneous velocity of the object at time $t = 1$.
   (f) Find an equation of the line tangent to the graph of $p(t)$ at $t = 1$. Add this line to the sketch you made in part (c) above.

7. For each function f below, use a difference quotient with $h = 0.001$ to estimate the value of f' at the given point.
   (a) $f(x) = x^2 - 3x$ at $x = 3$
   (b) $f(x) = x^{1/3}$ at $x = 2$
   (c) $f(x) = x^{-2}$ at $x = 1$
   (d) $f(x) = \sin x$ at $x = 0$
   (e) $f(x) = \cos x$ at $x = 0$
   (f) $f(x) = \cos(3x)$ at $x = \pi/2$
   (g) $f(x) = 2^x$ at $x = 0$
   (h) $f(x) = \ln x$ at $x = 1$

8. For each of the following functions, use a graph of
   $$g(\Delta x) = \frac{f(1 + \Delta x) - f(1)}{\Delta x}$$
   over a small interval centered at $\Delta x = 0$ to estimate $f'(1)$.
   (a) $f(x) = x^3$ (c) $f(x) = \sin(\pi x)$
   (b) $f(x) = x + (1/x)$ (d) $f(x) = e^x$

9. Suppose that f is a function for which $f'(2)$ exists. Use the values of f given below to estimate $f'(1.99)$, $f'(2)$, $f'(2.01)$, and $f'(2.1)$. Explain how you obtained your estimates.

x	1.9	1.99	1.999	2.0
$f(x)$	25.34	33.97	34.896	35
x	2.001	2.01	2.1	
$f(x)$	35.104	36.05	46.18	

10. Suppose that f is a function for which $f'(2)$ exists. Use the values of f given at right to estimate $f'(1.9)$, $f'(2)$, and $f'(2.02)$.

x	1.9	1.97	2.0	2.02	2.2
$f(x)$	6.6	6.905	7	7.059	7.5

FURTHER EXERCISES

11. On a 50-minute trip, a car travels for 20 minutes with an average velocity of 30 mph and then for 30 minutes with an average velocity of 50 mph. Find the total distance traveled and the average velocity over the entire trip.

12. (a) A car travels for 30 miles with an average velocity of 40 mph and then for another 30 miles with an average velocity of 60 mph. What is the average velocity of the car for the entire trip?

    (b) Another car travels for 30 minutes at 40 mph and then for 30 minutes at 60 mph. Find the average velocity over the 1-hour time period.

    (c) A car is to travel 2 miles. It went the first mile at an average velocity of 30 mph. The driver wishes to average 60 mph for the entire 2-mile trip. Is this possible? Explain.

13. The number of deer in a forest at time t years after the beginning of a conservation study is shown on the graph below.

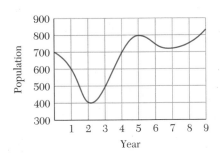

    (a) Over which two of the following periods did the population of deer decline at an average rate of 50 deer per year?
        (i) 1–2   (ii) 1–3   (iii) 1–4   (iv) 2–3   (v) 5–6

    (b) Approximately how fast was the deer population increasing (or decreasing) at time $t = 1.5$? At time $t = 3$? (Include appropriate units.)

14. Let $f(x) = \sqrt{x}$. Tabulate $f(x)$ at $x = 1.998, 1.999, 2.000, 2.001$, and 2.002 and use your results to answer the following questions. (Round numerical values of $f(x)$ to seven decimal places.)

    (a) Explain how these tabulated values show that f is "locally linear" near $x = 2$.

    (b) Estimate $f'(1.999)$, $f'(2)$, and $f'(2.001)$ using the tabulated values of f.

    (c) Does $\big(f(2.001) - f(2)\big)/0.001$ overestimate or underestimate the exact value of $f'(2)$? Why? [**HINT:** Look at a graph of $f(x)$ near $x = 2$.]

    (d) Estimate $f''(2)$ using the results from part (b).

15. Repeat the previous problem for each of the following functions.
    (a) $f(x) = x^2$ (c) $f(x) = e^x$
    (b) $f(x) = \sin x$ (d) $f(x) = \ln x$

16. Let $f(x) = |x|$.
    (a) Sketch a graph of f over the interval $[-5, 5]$.
    (b) Explain why $f'(x) = \begin{cases} -1, & \text{if } x < 0 \\ 1, & \text{if } x > 0. \end{cases}$
    (c) Explain why $f'(0)$ doesn't exist.

17. Let $f(x) = 2^x$.
    (a) Find the average rate of change of f between $x = 0$ and $x = 0.01$. Use this result to estimate $f'(0)$. Explain why this estimate is too large.
    (b) Find the average rate of change of f between $x = -0.01$ and $x = 0$. Use this result to estimate $f'(0)$. Explain why this estimate is too small.
    (c) Find the average rate of change of f between -0.01 and 0.01. Use this result to estimate $f'(0)$.
    (d) Why is the estimate of $f'(0)$ found in part (c) more accurate than the estimates computed in parts (a) and (b)?
    (e) Describe a method based on part (c) for estimating $f'(a)$ from tabulated data.

18. Each of the following limits is $f'(a)$ for some function f and some number a. In each case, identify f and a.
    (a) $\lim\limits_{x \to 4} \dfrac{\sqrt{x} - 2}{x - 4}$ (c) $\lim\limits_{u \to \pi/4} \dfrac{\cos u - \sqrt{2}/2}{u - \pi/4}$
    (b) $\lim\limits_{h \to 0} \dfrac{5^{2+h} - 25}{h}$ (d) $\lim\limits_{s \to 8} \dfrac{\log_2 s - 3}{s - 8}$

2.6 Limits and Continuity

We broached the subject of limits in the preceding section, in defining the derivative. In this section we study limits in their own right. Then we use limits to define precisely another important concept: continuity of a function at a point of its domain.

The Basic Idea; Simplest Examples

The general idea of limit is simple. Suppose that f is a function defined near, but not necessarily *at*, $x = a$. To say that

$$\lim_{x \to a} f(x) = L$$

means, informally, that $f(x)$ approaches L as x approaches a. In other *symbols*,

$$f(x) \to L \qquad \text{as} \qquad x \to a.$$

In other *words*,

> As inputs to f approach a, outputs from f approach L.

In still other words,

$$f(x) \approx L \qquad \text{whenever} \qquad x \approx a.$$

Intuitive ideas aren't formal definitions. We'll give a definition soon. First, here are some examples.

EXAMPLE 1   What is $\lim_{x \to 3} x^2$? Why?

Solution   The answer is no surprise:

$$\lim_{x \to 3} x^2 = 9.$$

The squaring function $f(x) = x^2$ is well behaved near $x = 3$. In particular, $f(x) \approx 9$ whenever $x \approx 3$. (In fact, $f(x)$ *equals* 9 when $x = 3$, but that's irrelevant to the limit.)   ■

EXAMPLE 2   The statement

$$\lim_{x \to 3} \frac{x^2 - 9}{x - 3} = 6$$

is about the limit of a certain function at a certain point. Which function? Which point? What does the statement say? Is it true?

Solution   The function—we'll call it g—is defined by the rule $g(x) = (x^2 - 9)/(x - 3)$. The point in question is $x = 3$.

Why not?   Notice that $g(x)$ is not defined ◄   at the interesting point $x = 3$. (Everywhere else, the rule for g makes good sense.) The limit statement says that, although g has no ordinary *value* at $x = 3$, it does the next best thing: $g(x)$ approaches 6 as x approaches 3. In symbols,

$$g(x) \to 6 \qquad \text{as} \qquad x \to 3.$$

That's what the statement says. Is it true? The formula for g isn't immediately helpful, precisely because $g(3)$ isn't defined. The graph of g helps:

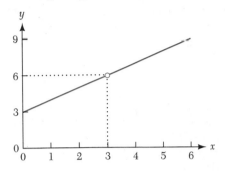

Graph of $g(x) = (x^2 - 9)/(x - 3)$

It shows convincingly that the limit statement is true. Despite the gap ▶ at $x = 3$, $g(x)$ does approach 6 as x approaches 3, from either right or left.
Numerical evidence points to the same conclusion:

How does the gap appear on the graph?

x	2.9	2.99	2.999	2.9999	...	3.0001	3.001	3.01	3.1
$g(x)$	5.9	5.99	5.999	5.9999	...	6.0001	6.001	6.01	6.1

Observe, finally, that the graph of g (except at $x = 3$) looks suspiciously like the line $y = x + 3$. A little algebra shows why: If $x \neq 3$,

$$g(x) = \frac{x^2 - 9}{x - 3} = \frac{(x - 3)(x + 3)}{x - 3} = x + 3.$$

(We repeat: The last equation holds only for $x \neq 3$—but that's good enough, because we're interested in $g(x)$ only for values *near* $x = 3$.) Now it's clear that

$$\lim_{x \to 3} g(x) = \lim_{x \to 3} (x + 3) = 6.$$ ■

Why Limits Matter: Functions with Gaps

The simplest limit expressions (e.g., $\lim_{x \to 3} x^2 = 9$ and $\lim_{x \to 3}(x, +3) = 6$) illustrate what limits are, but not why they matter. Judging only from such examples, limits may seem to be much ado about little. That impression is wrong. Limits do matter in calculus. One reason is that functions with gaps in their domains (such as g in Example 2) arise often in calculus. Finding derivatives is the most common and most important way in which such functions arise.

EXAMPLE 3 Let $f(x) = x^2$. Use the limit definition to find the derivative $f'(3)$. What does the result mean graphically?

Difference quotients can be written in various forms. See page 144.

S o l u t i o n   By definition, the derivative is a limit of difference quotients. We write ours like this: ◀

$$f'(3) = \lim_{x \to 3} \frac{f(x) - f(3)}{x - 3} = \lim_{x \to 3} \frac{x^2 - 9}{x - 3}.$$

That's the limit we found in Example 2, so $f'(3) = 6$. In graphical terms: *At $x = 3$ the curve $y = x^2$ has slope 6.*   ■

Limits in Theory

Functions with gaps or other anomalies in their domains arise in connection with instantaneous phenomena, including tangent lines and rates of change. Example 3 showed how the concept of limit permits us to handle such functions.

With a satisfactory theory of limits, based on precise definitions, mathematicians can make sense of—and calculate reliably with—mysterious-seeming concepts such as "infinitely small quantities." Without such theory as footing, we would remain mired (as mathematicians did for hundreds of years) in the vague, metaphorical—sometimes even mystical—language of "fluxions," "fluents," and "indivisibles."

Seeing Ghosts.   A precise idea of limit is surprisingly subtle. Developing a rigorous but usable definition of limit was an important historical step toward putting calculus on a sound logical footing.

The modern idea of limit stems from the work of the French mathematician Augustin-Louis Cauchy, around 1820—over 150 years after Isaac Newton and Gottfried Leibniz developed calculus. Newton's and Leibniz's own understanding of limit could not have been, by modern standards, rigorous. Sensing this, Bishop Berkeley satirized the intellectual pretensions of mathematicians in his 1734 philosophical treatise *The Analyst: A Discourse Addressed to an Infidel Mathematician*, which ridiculed Newton's limits as "ghosts of departed quantities."

Limits in Practice

Most of the limits encountered in calculus are relatively simple and straightforward. Combining algebraic, graphical, and numerical tools makes them even more so. The rest of this section aims to support this claim.

E X A M P L E 4   Let $f(x) = \sin x$. What limit defines $f'(0)$? What does the limit mean graphically? Find the limit.

S o l u t i o n   The derivative $f'(0)$ is the slope of the tangent line to the curve $y = \sin x$ at $x = 0$. As we saw in the last section, this slope is the limit of slopes of secant lines over small intervals starting (or ending) at $x = 0$. To investigate that limit, we'll define a new function, g, by the rule

$$g(t) = \text{slope of secant line joining } (0, 0) \text{ and } (t, \sin t).$$

The rule makes sense for any $t \neq 0$, as the next picture shows.

Defining the function g

Even though $g(t)$ is not defined at $t = 0$, we'll find the limit $\lim_{t \to 0} g(t)$.

Although we defined g in words, finding an explicit formula isn't hard. For $t \neq 0$,

$$g(t) = \frac{\Delta y}{\Delta x} = \frac{\sin t}{t}.$$

We want to find

$$\lim_{t \to 0} g(t) = \lim_{t \to 0} \frac{\sin t}{t}.$$

Because $g(0)$ isn't defined, we examine values of $g(t)$ for t near 0. Here are some numerical results, rounded to five decimals: ▶

Check some entries with a calculator; use radians.

t	-0.1	-0.01	-0.0001	...	0.0001	0.01	0.1
$g(t)$	0.99833	0.99998	1.00000	...	1.00000	0.99998	0.99833

The numbers suggest, rightly, that the limit is 1. ▶ An airtight proof would take more work. We won't give one here, but see Appendix I.

The answer means that the line $y = x$, with slope 1, is tangent to the sine curve at $x = 0$. ▶ ■

For graphical evidence, plot $g(t)$ for t near 0.

Draw this line on the preceding graph to understand what tangency means.

The Importance of Being Algebraic. We calculated that

$$\lim_{x \to 3} \frac{x^2 - 9}{x - 3} = \lim_{x \to 3} (x + 3) = 6.$$

Won't something similar work with $\lim_{x \to 0} \sin(x)/x$?

No. The difference is that $x^2 - 9$ is an *algebraic* expression and so permits factoring, division, and other algebraic operations. The numerator $\sin x$, by contrast, is not algebraic, so ordinary factoring and division aren't possible.

No Limit. A function f may not *have* a derivative at a point of its domain. The absolute value function $f(x) = |x|$, for example, has a kink in its graph at $x = 0$. By rights, therefore, $f'(0)$ shouldn't exist. The next example shows exactly why it doesn't.

*Here, again, is our favorite
"not-nice" function.*

EXAMPLE 5   Let $f(x) = |x|$. ◀   Show, using limits, that $f'(0)$ doesn't
exist.

Solution   By definition,

$$f'(0) = \lim_{x \to 0} \frac{f(x) - f(0)}{x - 0} = \lim_{x \to 0} \frac{|x|}{x},$$

if the limit exists.

To see that the limit doesn't exist, consider the function h defined by $h(x) = |x|/x$. Does $\lim_{x \to 0} h(x)$ make sense? A moment's thought reveals that $h(x) = 1$ if $x > 0$, $h(x) = -1$ if $x < 0$, but $h(0)$ is undefined. The graph agrees:

Graph of $h(x) = |x|/x$

The limit $\lim_{x \to 0} h(x)$ does not exist; there is no *single* number to which $h(x)$ tends as x tends to 0. The problem is that, as $x \to 0$ from the *left*, $h(x) \to -1$, but as $x \to 0$ from the *right*, $h(x) \to 1$. Because the limit doesn't exist, neither does $f'(0)$.   ■

Definitions of Limit—Informal and Precise

Let L be any finite number. The limit expression

$$\lim_{x \to a} f(x) = L \tag{2.1}$$

means, informally, that $f(x)$ approaches L as x approaches a. To be precise, we'll say exactly what we mean by Equation 2.1, first informally and then rigorously.
Equation 2.1 has two types of ingredients:

- real numbers a and L;
- a function f defined for all x near $x = a$, but not necessarily *at* $x = a$.

Definition (**Limit, Informally**)   If $f(x)$ tends to a single number L as x approaches a, then $\lim_{x \to a} f(x) = L$.

> **Definition** (Limit, Formally)   Suppose that for every positive number ϵ (epsilon), no matter how small, there is a positive number δ (delta) so that
>
> $$|f(x) - L| < \epsilon \quad \text{whenever} \quad 0 < |x - a| < \delta.$$
>
> Then
>
> $$\lim_{x \to a} f(x) = L.$$

Observe:

Estimating Limits   If $\lim_{x \to a} f(x) = L$, then $f(x) \approx L$ if $x \approx a$. This means that we can investigate limits numerically (by evaluating $f(x)$ for x near a) and graphically (by graphing $f(x)$ for x near a). ▶

We took both of these approaches above.

A Gap at a? No Problem   It's an important fact that $\lim_{x \to a} f(x)$ may exist *even if $f(a)$ is undefined*. Indeed, $f(a)$ is *usually* undefined in interesting cases, such as these:

$$\lim_{x \to 4} \frac{\sqrt{x} - 2}{x - 4}; \qquad \lim_{x \to 4} \frac{x^2 - 16}{x - 4}; \qquad \lim_{x \to 0} \frac{\cos x - 1}{x}.$$

None of these functions is "defined at a." In a sense, the problem of limits is to *find* a suitable value for $f(a)$—if one exists.

No Guarantees   Even if $f(a)$ is defined, there is no guarantee that $\lim_{x \to a} f(x) = f(a)$. The limit definition "doesn't see" $f(a)$. Whether or not $f(a)$ exists, the definition refers only to values of x for which $|x - a| > 0$—i.e., $x \neq a$. (See the next example.)

From Greek to English   In words, the formal definition of limit says:

We can ensure that $f(x)$ differs from L by less than any prearranged amount ϵ by requiring that x be within δ of a.

The full idea is subtle, but the following "generic" picture gives some idea of how ϵ and δ are related.

**Epsilon and delta:
the generic picture**

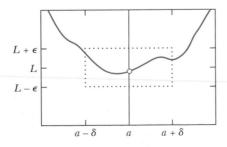

Graphically, the formal definition means this:

> *For any $\epsilon > 0$ we can choose a viewing window centered at (a, L), of width 2δ and height 2ϵ, from which the f-graph never escapes out the top or bottom, except possibly at $x = a$.*

We'll Be Informal   In elementary calculus—and in this book—informal views of the limit are almost always adequate. In more advanced work (e.g., in "real analysis"), a rigorous understanding of limits is essential.

Airtight Proofs.   Careful proofs of limit statements are essential at higher levels of mathematical analysis; inevitably, such proofs involve careful balancing of epsilons and deltas. We are more interested in the *concept* of limit, and how to use it, than in formal proofs. We'll discuss some theoretical properties of limits, but mainly as aids to concrete computation.

Variations on a Theme: One-Sided Limits

We'll see the latter in the next section.

Notice the unusual use of "+" and "−" as superscripts.

The most basic and most important limit idea is the one just studied: the (two-sided) limit of a function at a point. However, several variants on the same idea arise often enough to deserve mention: "one-sided" limits and limits that involve the ∞ symbol. ◄   Slightly different definitions apply to these variants. Informal versions will be enough: ◄

Definition (Right-Hand Limit) $\lim\limits_{x \to a^+} f(x) = L$ means that $f(x) \to L$ as $x \to a$ from the right.

Definition (Left-Hand Limit) $\lim\limits_{x \to a^-} f(x) = L$ means that $f(x) \to L$ as $x \to a$ from the left.

EXAMPLE 6   In Example 5 we saw (e.g., from the graph) that if $h(x) = |x|/x$, then $h(x)$ tends either to 1 or to -1 depending on whether x tends to 0 from the right or from the left. In one-sided limit notation,

$$\lim_{x \to 0^-} h(x) = -1 \qquad \text{and} \qquad \lim_{x \to 0^+} h(x) = 1. \qquad \blacksquare$$

Because these one-sided limits aren't equal, we concluded that the ordinary limit $\lim_{x \to 0} h(x)$ does not exist. There's a useful principle lurking here:

Fact The ordinary limit $\lim\limits_{x \to a} f(x)$ exists if and only if both corresponding one-sided limits exist, and are equal. In symbols,

$$\lim_{x \to a} f(x) = L \iff \lim_{x \to a^+} f(x) = L = \lim_{x \to a^-} f(x).$$

Continuity

To say that f is **continuous** on an interval I means, roughly speaking, that the graph of f is "unbroken" for x in I. This intuitive idea, although not a formal definition, works reliably enough for most calculus functions.

EXAMPLE 7   Each of the following functions is continuous on the full set $(-\infty, \infty)$ of real numbers:

$$f(x) = x^2; \qquad g(x) = \sin x; \qquad h(x) = \frac{x^2}{x^2 + 1}; \qquad k(x) = |x|.$$

(The graph of k has a kink at $x = 0$, but that doesn't destroy continuity.)

By contrast, $\sqrt{x}$ is continuous on the interval $[0, \infty)$. The function $m(x) = 1/x$ is continuous on $(0, \infty)$ and on $(-\infty, 0)$ but not on any interval that contains 0.   ■

EXAMPLE 8   Following is a function n, shown graphically. Where is n continuous?

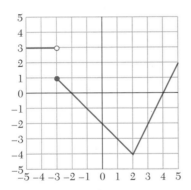

Solution   The only discontinuity occurs at $x = -3$. In particular, n is continuous on any interval that omits $x = -3$.   ■

Continuity Defined: Limits Again

Graphical intuition is helpful, but a precise definition of continuity cannot depend on pictures. ▶ The idea of limit turns out to be the right one to express, in analytic terms, the geometric idea of "unbrokenness" at a point. Here's the formal definition:

One problem is that drawing accurate pictures is difficult. In higher dimensions, things are even worse.

> **Definition**   Let f be a function defined on an interval I containing $x = a$. If
>
> $$\lim_{x \to a} f(x) = f(a),$$
>
> then f is continuous at $x = a$. (If a is the left or right endpoint of I, then a right-hand limit or a left-hand limit replaces the preceding two-sided limit.) To say that f is continuous on I means that f is continuous at each point of I (including endpoints of I, if any).

The formal definition illustrates the important theoretical role played by limits in calculus. In practice, however, the intuitive idea of continuity as unbrokenness is usually adequate. We'll content ourselves with several brief remarks.

Continuous Where?   A function f may be continuous on one set but discontinuous on another set. Thus, the common phrase "f is continuous" is technically ambiguous—it doesn't say *where* f is continuous. Here's the rule of thumb:

> *Unless otherwise qualified, "continuous" means "continuous throughout the domain of f."*

No Surprises   The definition says that f is continuous at a if, whenever $x \to a$, $f(x) \to f(a)$. In other words, inputs near a produce outputs near $f(a)$. In effect, this means that f behaves "predictably" at $x = a$.

Continuity Is the Rule   As a rule, the elementary functions of calculus are continuous wherever they're defined. For example, $f(x) = \sqrt{x-1}$ is continuous for $x \geq 1$, i.e., on the natural domain of f. Similarly, every polynomial function $p(x)$ is continuous on the full set of real numbers. (Proofs of such facts are important in higher analysis; in this book we'll deemphasize them.)

Easy Limits   Finding limits of continuous functions is as easy as can be. If f is continuous at $x = a$, then—by the very definition of continuity—

$$x \to a \Longrightarrow f(x) \to f(a).$$

In words: The *limit* of f as x approaches a is the *value* of f at a.

EXAMPLE 9   Let $k(x) = x^2 + 3x + 5$. Find $\lim_{x \to 2} k(x)$.

Solution   The problem looks easy; neither the formula nor the graph of k suggests any surprises near $x = 2$.

Graph of $k(x) = x^2 + 3x + 5$

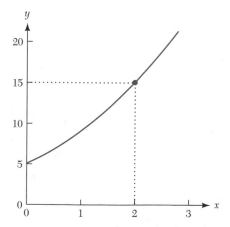

The obvious guess,

$$\lim_{x \to 2} k(x) = k(2) = 15,$$

is correct, precisely because k is continuous at $x = 2$. ▶

As we said earlier, a polynomial function is continuous everywhere.

Discontinuity: Three Possible Problems

A function f is **discontinuous** at $x = a$ if—for any reason—the limit condition in the definition doesn't hold. Three things can go wrong:

1. $f(a)$ may not be defined; this happens for $f(x) = 1/x$ at $x = 0$.

2. The limit $\lim_{x \to a} f(x)$ may not exist, this happens at $x = -3$ for the function n in Example 8.

3. The limit $\lim_{x \to a} f(x)$ may exist but differ from $f(a)$.

The next example illustrates the last form of pathology.

EXAMPLE 10   Consider the function f ▶   defined by

$$f(x) = \begin{cases} x & \text{if } x \neq 0 \\ 1 & \text{if } x = 0 \end{cases}.$$

What's $\lim_{x \to 0} f(x)$?

Plot this function by hand to understand the example.

Solution   The fact that $f(0) = 1$ makes no difference to the limit. Because $f(x) = x$ for $x \neq 0$, we have

$$\lim_{x \to 0} f(x) = \lim_{x \to 0} x = 0.$$

In a sense, f has the wrong value at $x = 0$; for this reason, f is called discontinuous at $x = 0$. ■

Limits and Computer Graphics. Most plotting devices can successfully plot a function over an interval containing one suspicious point. They will, for instance, draw reasonable pictures for $h(x) = |x|/x$ on $[-1, 1]$ even though h isn't defined at $x = 0$. If a device "complains" or produces nonsense, one can still use intervals that avoid the problem point. For $h(x) = |x|/x$ we might try graphs over $[-1, -0.01]$ and $[0.01, 1]$.

Sample Problems

Here are two problems that combine ideas from this section.

EXAMPLE 11   Let $f(x) = \begin{cases} bx^2 + 1 & \text{if } x < -2 \\ x - 2 & \text{if } x \geq -2 \end{cases}.$

What value of b makes f continuous at $x = -2$?

Solution   To make f continuous at $x = -2$, we need

$$\lim_{x \to -2} f(x) = f(-2) = -2.$$

Viewing the two lines of the definition separately, we see that

$$f(x) \to 4b + 1 \quad \text{as} \quad x \to -2^-;$$
$$f(x) \to -2 \quad \text{as} \quad x \to -2^+.$$

For the two one-sided limits to be equal, we need $4b + 1 = -2$, or $b = -3/4$. ∎

EXAMPLE 12   Interpret $\lim\limits_{h \to 0} \dfrac{e^h - 1}{h}$ as a derivative; estimate its value.

Check this equation.

Solution   If $f(x) = e^x$, then ◄

$$\lim_{h \to 0} \frac{e^h - 1}{h} = \lim_{h \to 0} \frac{f(h) - f(0)}{h}.$$

We took one in Example 6, Section 2.2. Look back or plot the curve yourself.

Thus, the given limit is (by the definition of derivative) the slope of the curve $y = e^x$ at $x = 0$. A close look at this curve ◄ suggests that this slope is 1. ∎

BASICS

1. Let $f(x) = \begin{cases} 1, & \text{if } x \text{ is an integer} \\ 2, & \text{if } x \text{ is not an integer.} \end{cases}$

(a) Draw a graph of f over the interval $[0, 5]$.
(b) Evaluate $\lim\limits_{x \to 4} f(x)$.
(c) Evaluate $\lim\limits_{x \to 5/2} f(x)$.
(d) For which values of a does $\lim\limits_{x \to a} f(x)$ exist? Why?

Let f be the function whose graph is shown below. Using the graph, evaluate each of the limits in Exercises 2–15 or explain why the limit does not exist. [**NOTE:** $f(1) = 2$.]

Graph of f

2. $\lim\limits_{x \to 0} f(x)$
3. $\lim\limits_{x \to -3} f(x)$
4. $\lim\limits_{x \to -1} f(x)$
5. $\lim\limits_{x \to 2^-} f(x)$

6. $\lim\limits_{x \to 2^+} f(x)$
7. $\lim\limits_{x \to 2} f(x)$
8. $\lim\limits_{x \to -2^-} f(x)$
9. $\lim\limits_{x \to 1^+} f(x)$

10. $\lim\limits_{x \to -4^+} f(x)$
11. $\lim\limits_{x \to 4^-} f(x)$
12. $\lim\limits_{x \to 3} f'(x)$

13. $\lim\limits_{x \to -1^+} f'(x)$
14. $\lim\limits_{x \to -1^-} f'(x)$
15. $\lim\limits_{x \to -3} f'(x)$

16. Over which intervals is f continuous?

Let g be the function whose graph is shown below. Using the graph, evaluate each of the limits in Exercises 17–26 or explain why the limit does not exist.

Graph of g

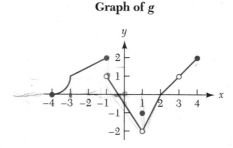

17. $\lim\limits_{x \to -4^+} g(x)$
18. $\lim\limits_{x \to -3} g(x)$
19. $\lim\limits_{x \to -2} g(x)$
20. $\lim\limits_{x \to -1^+} g(x)$
21. $\lim\limits_{x \to -1} g(x)$

22. $\lim\limits_{x \to 0^-} g(x)$
23. $\lim\limits_{x \to 1} g(x)$
24. $\lim\limits_{x \to 2} g(x)$
25. $\lim\limits_{x \to 3} g(x)$
26. $\lim\limits_{x \to 4^-} g(x)$

27. Over which intervals is g continuous?

28. Let $s(t)$ be the slope of the secant line from $x = 3$ to $x = t$ on the graph of the function $f(x) = x^2 + x$.
    (a) Find a formula for $s(t)$.
    (b) Find $\lim_{t \to 3} s(t)$.

(c) What does your answer to part (b) imply about the graph of f? Draw graphs of f and the line tangent to f at $x = 3$ over the interval $[2, 4]$.

FURTHER EXERCISES

29. Suppose that f is continuous at $x = 1$ and that $f(1) = -2$. Evaluate $\lim_{x \to 1} f(x)$.

30. Suppose that f is continuous at $x = 3$ and that $\lim_{x \to 3} f(x) = 17$. Indicate which of the following statements about f *must* be true, which *might* be true, and which *cannot* be true. Justify your answers.
    (a) 3 is in the domain of f.
    (b) $f(3) = 17$.
    (c) $\lim_{x \to 3^-} f(x) = 17$.

31. Let $f(x) = \begin{cases} \frac{\sin x}{x} & \text{if } x \neq 0 \\ 1 & \text{if } x = 0. \end{cases}$
    Explain why f is continuous at $x = 0$.

32. Let $g(x) = \begin{cases} \frac{\sin x}{|x|} & \text{if } x \neq 0 \\ 1 & \text{if } x = 0. \end{cases}$
    Is g continuous at $x = 0$? Justify your answer.

33. Let $h(x) = \begin{cases} \frac{\sin x}{\sqrt{1 - \cos^2 x}} & \text{if } x \neq 0 \\ 1 & \text{if } x = 0. \end{cases}$
    Is h continuous at $x = 0$? Justify your answer.

34. Let $f(x) = \begin{cases} ax + 1 & \text{if } x < 2 \\ x^2 & \text{if } x \geq 2. \end{cases}$
    Find the value of a for which $\lim_{x \to 2} f(x)$ exists.

35. Suppose that the function f is continuous at $x = 2$ and that f is defined by the rule

    $f(x) = \begin{cases} ax^2 + 3 & \text{if } x < 2 \\ 3x - 5 & \text{if } x \geq 2. \end{cases}$

    (a) Find the value of a.
        Evaluate each of the following limits.
    (b) $\lim_{x \to 3} f(x)$ (d) $\lim_{x \to 2} f(x)$
    (c) $\lim_{x \to 0} f(x)$ (e) $\lim_{x \to 1} f(x)$

36. Let $g(x) = \begin{cases} x^2 + a & \text{if } x < -1 \\ x + a^2 & \text{if } x = -1 \\ a - x & \text{if } x > -1. \end{cases}$
    (a) Find *all* values of a for which $\lim_{x \to -1} g(x)$ exists.
    (b) Find *all* values of a for which g is continuous at $x = -1$.

37. Let $f(x) = \begin{cases} x^2 - 2 & \text{if } x \leq b \\ x & \text{if } x > b. \end{cases}$
    Find *all* values of b for which $\lim_{x \to b} f(x)$ exists.

Each of the limits in Exercises 38–47 is $f'(a)$ for some function f and some number a. Identify f and a, then use a calculator or grapher to estimate the value of the limit.

38. $\lim_{h \to 0} \frac{\tan h}{h}$

39. $\lim_{h \to 0} \frac{\cos h - 1}{h}$

40. $\lim_{h \to 0} \frac{\sin(3h)}{h}$

41. $\lim_{x \to \pi} \frac{\sin x}{x - \pi}$

42. $\lim_{x \to \pi/2} \frac{\cos x}{x - \pi/2}$

43. $\lim_{x \to 16} \frac{\sqrt[4]{x} - 2}{x - 16}$

44. $\lim_{x \to 1} \frac{2^x - 2}{x - 1}$

45. $\lim_{h \to 0} \frac{\sqrt{9 + h} - 3}{h}$

46. $\lim_{h \to 0} \frac{\sin(\pi/6 + h) - 1/2}{h}$

47. $\lim_{h \to 0} \frac{\ln(1 + h)}{h}$

48. Suppose that $\lim_{x \to a} f(x) = L$. Explain why $\lim_{h \to 0} f(a + h) = L$.

49. Suppose that $\lim_{h \to 0} \frac{f(a+h) - f(a)}{h} = L$. Explain why $\lim_{x \to a} \frac{f(x) - f(a)}{x - a} = L$.

50. Suppose that g is the linear function $g(x) = mx + b$.
    (a) Show that $g'(a) = \lim_{h \to 0} \frac{g(a+h) - g(a)}{h} = m$ for every value of a.
    (b) Interpret the result in part (a) in terms of slopes and rates of change.

In Exercises 51–56 use the graphs of the functions f (see page 160) and g (see page 160) to evaluate each limit. If a limit does not exist, explain why.

51. $\lim_{x \to 0} \frac{f(x) - 2}{x}$

52. $\lim_{x \to 3} \frac{f(x)}{x - 3}$

53. $\lim_{x \to -1^+} \frac{f(x)}{x + 2}$

54. $\lim_{h \to 0} \frac{g(-2 + h) - 3/2}{h}$

55. $\lim_{h \to 0^-} \frac{g(-1 + h) - 2}{h}$

56. $\lim_{h \to 0} \frac{g(2 + h)}{h}$

57. Let $f(x) = x^2$. Carefully explain each step in the following evaluation of $f'(1)$:
    (a) $f'(1) = \lim_{x \to 1} \frac{x^2 - 1}{x - 1}$
    (b) $= \lim_{x \to 1} (x + 1)$
    (c) $= 2$.

58. Let $f(x) = x^2$. Carefully explain each step in the following evaluation of $f'(3)$:
    (a) $f'(3) = \lim_{h \to 0} \frac{(3 + h)^2 - 9}{h}$
    (b) $= \lim_{h \to 0} \frac{6h + h^2}{h}$
    (c) $= \lim_{h \to 0} (6 + h)$
    (d) $= 6$.

In Exercises 59–65, use the limit definition of the derivative to compute $f'(4)$ for each function.

59. $f(x) = 3$
60. $f(x) = 2x + 1$
61. $f(x) = x^2 - 3x$
62. $f(x) = \sqrt{x}$
63. $f(x) = x^{-1}$
64. $f(x) = 3x^{-2}$

65. $f(x) = x^{-1/2}$

66. Suppose that $f(x + y) = f(x) + f(y) + 3xy$ for all real numbers x, y, and that $\lim_{h \to 0} f(h)/h = 42$.
    (a) Explain why $f(0) = 0$.   [**HINT**: Continuity of f *cannot* be assumed.]
    (b) Find a symbolic expression for $f'(x)$.

67. Suppose that $f'(2) = 3$. Find $\lim\limits_{h \to 0} \dfrac{f(2-h) - f(2)}{h}$.

68. Each of the following limits has the form

$$\lim_{x \to a} f(x) = L.$$

For each, use a grapher to find a value of δ that guarantees that $|f(x) - L| < 0.001$ when $|x - a| < \delta$.

(a) $\lim\limits_{x \to 2} 3x = 6$ (c) $\lim\limits_{x \to 1} \dfrac{x+1}{x+3} = 0.5$

(b) $\lim\limits_{x \to 0} \dfrac{1 - \cos x}{x} = 0$ (d) $\lim\limits_{x \to 0} \dfrac{1 - \cos(2x)}{x^2} = 2$

In Exercises 69–73, use the value of δ given to prove the limit. (Assume $\epsilon < 1$.)

69. $\lim\limits_{x \to 5} x = 5$; $\delta = \epsilon$

70. $\lim\limits_{x \to 4} x^2 = 16$; $\delta = \epsilon/10$

71. $\lim\limits_{x \to 2} 6 = 6$; $\delta = 0.2$

72. $\lim\limits_{x \to 3} \dfrac{x^2 - 9}{x - 3} = 6$; $\delta = \epsilon$

73. $\lim\limits_{x \to 0} |x| = 0$; $\delta = \epsilon$

74. Use the formal definition of limit to show that $\lim_{x \to 1} 0.99999x \neq 1$.

75. Let $f(x) = \begin{cases} x^2 + 1 & \text{if } x \leq 0 \\ -x & \text{if } x > 0. \end{cases}$
    Use the formal definition of limit to show that f is not continuous at $x = 0$.   [**HINT**: $f(0) - f(x) > 1$ when $x > 0$.]

76. Let f be the function defined in the previous problem.
    (a) Show that for every positive number δ, no matter how small, there is a positive number ϵ such that $|f(x) - 1| < \epsilon$ whenever $0 < |x| < \delta$.
    (b) Why doesn't the result in part (a) show that f is continuous at $x = 0$?   [**HINT**: Compare part (a) with the formal definition of limit.]

2.7 Limits Involving Infinity; New Limits from Old

In the preceding section we defined limits, explored several examples and extensions of the idea, and then applied limits to define continuity. This section continues the limit theme. First we define another variant on the basic limit idea—limits that involve the ∞ symbol—and interpret results graphically in terms of asymptotes. Then we consider how to combine simple, known limits to obtain new, more challenging limits.

Limits at Infinity, Infinite Limits, and Asymptotes

Limits at infinity and infinite limits (they're different!) describe the behavior of functions when either inputs or outputs increase or decrease "without bound." Informal definitions and examples will suffice to give the idea.

> **Definition** (**Limit at Infinity**) $\lim\limits_{x \to \infty} f(x) = L$ means that $f(x) \to L$ as $x \to \infty$ (i.e., as x increases without bound).

> **Definition** (**Infinite Limit**) $\lim\limits_{x \to a} f(x) = \infty$ means that $f(x) \to \infty$ (i.e., $f(x)$ "blows up") as $x \to a$.

For instance, limits as $x \to -\infty$, and one-sided infinite limits.

Additional variants exist. ◄ Defining them all separately would be tedious. What each means is usually clear from context.

EXAMPLE 1   Discuss the meaning of each of the following limit statements:

$$\lim_{x \to \infty} \frac{1}{x} = 0; \qquad\qquad \lim_{x \to 0} \frac{1}{x^2} = \infty;$$

$$\lim_{x \to 0^-} \frac{1}{x} = -\infty; \qquad\qquad \lim_{x \to \infty} \sin x \text{ does not exist.}$$

Solution   The limit $\lim_{x \to \infty} \sin x$ does not exist because the sine function oscillates forever, never settling down near any single value. The following graphs should help the reader interpret the first three limits. ▶

What other limit statements do the graphs suggest?

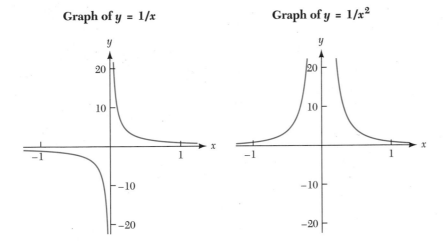

Graph of $y = 1/x$ Graph of $y = 1/x^2$

Notice especially the graphical meaning of each infinite limit and limit at infinity. Each one corresponds to an asymptote, either vertical or horizontal. We'll refine this observation in what follows.   ∎

Asymptotes and Limits

The connections among asymptotes, limits at infinity, and infinite limits are easily summarized:

> **Horizontal Asymptotes; Limits at Infinity**   The horizontal line $y = L$ is an asymptote to the graph of f if (and only if) f approaches L as $x \to \pm\infty$—that is, if
>
> $$\lim_{x \to \infty} f(x) = L \qquad \text{or} \qquad \lim_{x \to -\infty} f(x) = L.$$

> **Vertical Asymptotes; Infinite Limits**   The vertical line $x = a$ is an asymptote to the graph of f if (and only if) f blows up (or blows down) at a—that is, if
>
> $$\lim_{x \to a^-} f(x) = \pm\infty \qquad \text{or} \qquad \lim_{x \to a^+} f(x) = \pm\infty.$$

Limits and Asymptotes of Rational Functions

Asymptote information is useful for any function. Horizontal asymptotes reflect long-term behavior; vertical asymptotes reveal sudden spikes and other anomalies. For polynomials and rational functions (quotients of polynomials) the story is especially simple.

Can Polynomials Have Asymptotes?   Polynomials do not have vertical asymptotes. A vertical asymptote occurs at a *finite* value of x, near which a function's value is unbounded. Since polynomials have the pleasant form

$$p(x) = a_0 + a_1 x + \cdots + a_n x^n \, ,$$

free of troublesome denominators, polynomials cannot blow up (or down) except as x approaches $\pm \infty$.

Draw the graph of $p(x) = 3$ to convince yourself.

Only a constant polynomial can have a *horizontal* asymptote. If, say, $p(x) = 3$, then the p-graph is itself horizontal—it is its own asymptote. ◀   Nonconstant polynomials cannot have horizontal asymptotes. Moreover:

Fact If $p(x)$ is a nonconstant polynomial, then

$$\lim_{x \to \pm\infty} |p(x)| = \infty.$$

We won't prove this fact formally, but we'll illustrate why it holds.

EXAMPLE 2   Let $p(x) = 2x^3 - 53x^2 - 123x$. How does p behave as $x \to \pm\infty$?

Solution   It's clear at a glance that as $x \to \infty$, $2x^3 \to \infty$, $-53x^2 \to -\infty$, and $-123x \to -\infty$. But what about the sum?

Convince yourself of each step. Watch the last step in each operation carefully— we'll say more soon about algebra with the ∞ symbol.

The trick is to *factor out the largest power of x*: ◀

$$\lim_{x \to \infty} \left(2x^3 - 53x^2 - 123x\right) = \lim_{x \to \infty} x^3 \left(2 - \frac{53}{x} - \frac{123}{x^2}\right)$$

$$= \lim_{x \to \infty} x^3 \cdot \lim_{x \to \infty} \left(2 - \frac{53}{x} - \frac{123}{x^2}\right)$$

$$= \infty \cdot 2 = \infty \, .$$

Similarly,

$$\lim_{x \to -\infty} \left(2x^3 - 53x^2 - 123x\right) = \lim_{x \to -\infty} x^3 \cdot \lim_{x \to -\infty} \left(2 - \frac{53}{x} - \frac{123}{x^2}\right)$$

$$= -\infty \cdot 2 = -\infty \, .$$

More on this fact soon.

(In each calculation we used the fact that the limit of a product is the product of the limits. ◀   )   ∎

Algebra with the Infinity Symbol.   In Example 2 we breezed past two possibly suspicious claims: $\infty \cdot 2 = \infty$ and $-\infty \cdot 2 = -\infty$. Do they really make sense? Since ∞ and $-\infty$ are not numbers, their appearance in equations should always raise flags of caution. Here, fortunately, we're OK—neither $\infty \cdot 2$ nor $-\infty \cdot 2$ is ambiguous. The following table collects some rules of etiquette for the use of the infinity symbol.

Good and Bad Uses of ∞	
Expression	**What It Means; Remarks**
$1/\infty = 0$	Usually OK, but if we really mean, say, $\lim_{x\to\infty} 1/x = 0$, why not say so?
$3 \cdot \infty = \infty$	OK as mathematical shorthand. It means: If any quantity increases without bound, so does three times that quantity.
$\infty + \infty = \infty$	OK again. It means: If two quantities increase without bound, so does their sum.
$1/0 = \infty$	Wrong. Division by 0 is not defined for real numbers. Worse, $1/x \to \infty$ as $x \to 0$ from the right, but $1/x \to -\infty$ as $x \to 0$ from the left.
$\infty - \infty = 0$	Wrong again. For example, as $x \to \infty$, $x^3 \to \infty$ and $x^2 \to \infty$, but $x^3 - x^2 \to \infty$.
$\infty/\infty = 1$	Wrong. For example, as $x \to \infty$, $x^3 \to \infty$ and $x^2 \to \infty$, but $x^3/x^2 = x \to \infty$.

Horizontal Asymptotes of Rational Functions: Appeal to Higher Powers.
Finding horizontal asymptotes of rational functions means finding limits at infinity.
A simple algebraic strategy applies:

> *To find limits at infinity of rational functions, factor out the highest powers of x from numerator and denominator.*

Let's apply it:

EXAMPLE 3   Consider the three rational functions

$$r(x) = \frac{2x^3 - x}{x^2 - 1}, \qquad s(x) = \frac{2x^3 - x}{x^3 - 1}, \qquad \text{and} \qquad t(x) = \frac{2x^3 - x}{x^4 - 1}.$$

Find all horizontal asymptotes.

Solution   The graphs follow soon, but we won't peek. Instead we'll use the preceding factoring strategy:

$$r(x) = \frac{2x^3 - x}{x^2 - 1} = \frac{x^3(2 - 1/x^2)}{x^2(1 - 1/x^2)} = x \cdot \frac{2 - 1/x^2}{1 - 1/x^2};$$

$$s(x) = \frac{2x^3 - x}{x^3 - 1} = \frac{x^3(2 - 1/x^2)}{x^3(1 - 1/x^3)} = \frac{2 - 1/x^2}{1 - 1/x^3};$$

$$t(x) = \frac{2x^3 - x}{x^4 - 1} = \frac{x^3(2 - 1/x^2)}{x^4(1 - 1/x^4)} = \frac{1}{x} \cdot \frac{2 - 1/x^2}{1 - 1/x^4}.$$

Now we can see that as $x \to \infty$,

$$r(x) \to \infty, \qquad s(x) \to 2, \qquad \text{and} \qquad t(x) \to 0. \blacktriangleright \qquad \qquad \textit{Convince yourself.}$$

As $x \to -\infty$,

$$r(x) \to -\infty, \qquad s(x) \to 2, \qquad \text{and} \qquad t(x) \to 0.$$

Thus, s and t have *horizontal* asymptotes, at $y = 2$ and $y = 0$, respectively; r has none. $\blacksquare$

Vertical Asymptotes of Rational Functions: Look for Roots. If a rational function has vertical asymptotes, they are located at roots of the denominator.

> **EXAMPLE 4**   Let r, s, and t be as before. Find all vertical asymptotes.

> **Solution**   A rational function has a vertical asymptote at any position where the denominator is zero and the numerator isn't. (If both numerator and denominator are zero at the same place, they have a common factor; for present purposes, it can be canceled.) To locate the asymptotes, factor the denominator. ◄   For r, s,

Factoring polynomials of high degree can be hard or impossible. Authors usually choose polynomials that are easy to factor. So will we.

and t:

$$r(x) = \frac{2x^3 - x}{x^2 - 1} = \frac{2x^3 - x}{(x+1)(x-1)};$$

$$s(x) = \frac{2x^3 - x}{x^3 - 1} = \frac{2x^3 - x}{(x-1)(x^2 + x + 1)};$$

$$t(x) = \frac{2x^3 - x}{x^4 - 1} = \frac{2x^3 - x}{(x^2 - 1)(x^2 + 1)} = \frac{2x^3 - x}{(x-1)(x+1)(x^2+1)}.$$

The factored denominators show that, for each function, $x = \pm 1$ are the suspicious values, or singular points. Since the numerator $2x^3 - x$ is not zero at $x = \pm 1$, it follows that r and t have vertical asymptotes at $x = 1$ and at $x = -1$; s has a vertical asymptote at $x = 1$.

On either side of a vertical asymptote, a function may tend either to ∞ or to $-\infty$. What happens at each of the preceding asymptotes? Graphs make the answers obvious, but even without graphs, the question is easy to answer. We need only determine the *sign* of each function just to the right and just to the left of each asymptote.

For $r(x)$, just to the left of $x = -1$, we see that

$$2x^3 - x < 0, \qquad x + 1 < 0, \qquad \text{and} \quad x - 1 < 0.$$

r blows down to $-\infty$ as $x \to -1$ from the left.

Therefore, $r(x) < 0$, so $\lim_{x \to -1^-} r(x) = -\infty$. ◄   A similar sign check reveals that r blows *up* to ∞ as $x \to -1$ from the *right*. ◄

Can you mimic this analysis for the functions s and t?

We've waited long enough. Here are the graphs:

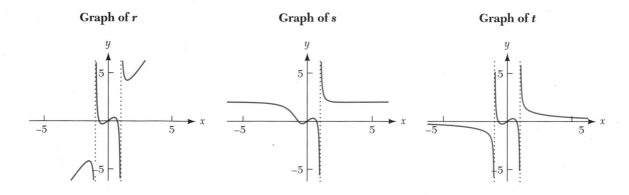

Graph of r   **Graph of s**   **Graph of t**

Asymptotes of Other Algebraic Functions

Like rational functions, other algebraic functions may have asymptotes, with exactly the same connection to infinite limits and limits at infinity. Calculating these limits algebraically may take extra ingenuity.

EXAMPLE 5 The algebraic function defined (for $x > 0$) by $f(x) = \sqrt{x^2 + x} - x$ has a horizontal asymptote. Find it.

Solution How $f(x)$ behaves as $x \to \infty$ isn't obvious from the formula, ▶ but the following table of values argues persuasively for a limit of $1/2$. ▶

Is it? Why or why not?

So would a graph; draw one.

As $x \to \infty$, $\sqrt{x^2 + x} - x \to 1/2$								
x	1	2	4	8	16	32	64	128	256
$f(x)$	0.4142	0.4495	0.4721	0.4853	0.4924	0.4962	0.4981	0.4990	0.4995

With algebra we'll clinch our case. The first step is a trick: multiplying top and bottom by the conjugate.

$$\lim_{x \to \infty}(\sqrt{x^2 + x} - x) = \lim_{x \to \infty} \frac{(\sqrt{x^2 + x} - x)(\sqrt{x^2 + x} + x)}{(\sqrt{x^2 + x} + x)}$$

$$= \lim_{x \to \infty} \frac{x}{\sqrt{x^2 + x} + x}$$

$$= \lim_{x \to \infty} \frac{x}{x\left(\sqrt{1 + 1/x} + 1\right)}$$

$$= \lim_{x \to \infty} \frac{1}{\sqrt{1 + 1/x} + 1} = \frac{1}{2}. \qquad \blacksquare$$

New Limits from Old

When calculating limits, we often take a few basic limits to be obvious and try to reduce more complicated limits to combinations of these simpler ones. The next example illustrates that process.

EXAMPLE 6 Find $\lim_{x \to \infty} \dfrac{x^2 + 3}{2x^2 - 5}$.

Solution With a graph it would be easy to guess that the limit is $1/2$. Algebra makes the case airtight: ▶

The key step is dividing numerator and denominator by x^2.

$$\lim_{x \to \infty} \frac{x^2 + 3}{2x^2 - 5} = \lim_{x \to \infty} \frac{x^2 \cdot (1 + 3/x^2)}{x^2 \cdot (2 - 5/x^2)}$$

$$= \lim_{x \to \infty} \frac{1 + 3/x^2}{2 - 5/x^2}$$

$$= \frac{\lim\limits_{x\to\infty} 1 + \lim\limits_{x\to\infty} 3/x^2}{\lim\limits_{x\to\infty} 2 - \lim\limits_{x\to\infty} 5/x^2}$$

$$= \frac{1+0}{2-0} = 1/2 \,.$$ ∎

We used several plausible facts in the preceding calculations: that $\lim_{x\to\infty} 3/x^2 = \lim_{x\to\infty} 5/x^2 = 0$, that the limit of a sum or quotient is the sum or quotient of the limits, and so on.

The following theorem justifies combining old limits to find new ones. Each part tells how limits behave with respect to some algebraic operation on functions.

Theorem 1 (Algebra with Limits)   Suppose that

$$\lim_{x\to a} f(x) = L \qquad \text{and} \qquad \lim_{x\to a} g(x) = M,$$

where L and M are finite numbers. Let k be any constant. Then

(i) $\lim\limits_{x\to a} kf(x) = kL$;

(ii) $\lim\limits_{x\to a} [f(x) + g(x)] = L + M$;

(iii) $\lim\limits_{x\to a} f(x) \cdot g(x) = L \cdot M$;

(iv) $\lim\limits_{x\to a} \dfrac{f(x)}{g(x)} = \dfrac{L}{M}$ (if $M \neq 0$).

Remarks on the Theorem.

Why They Hold   Each rule is intuitively plausible, but rigorous proofs would require the ϵ–δ definition. We'll skip the proofs here, but a typical proof of this kind appears in Appendix I.

Limits That Involve Infinity   All the rules hold for limits at infinity (i.e., when a is $\pm\infty$). They also hold for infinite limits (i.e., when $L = \pm\infty$ and/or $M = \pm\infty$), but only if appropriate care is taken with with uses of the infinity symbol, as described in the preceding table.

"Obvious" Limits   Using these algebraic rules, we can find many new limits—provided that we know some old limits. To get such a foothold, we'll take as obvious ◄   limits such as these:

To be fully rigorous we'd have to prove even these limits.

"Obvious" Limits	
$\lim\limits_{x\to 0} |x| = 0$	
$\lim\limits_{x\to a} x = a$	for any real number a
$\lim\limits_{x\to\infty} x^n = \infty$	for any $n > 0$
$\lim\limits_{x\to\infty} \dfrac{1}{x^n} = 0$	for any $n > 0$

New Continuous Functions from Old: Algebraic Combinations

If f and g are continuous functions, it sounds reasonable that such algebraic combinations as $f + g$, f/g, and $3f - 2g$ should also define continuous functions. As a rule, they do. Because continuity is defined via limits, ▶ it's not surprising that properties of limits are reflected in properties of continuous functions. The next theorem gives the fine print.

See the previous section!

Theorem 2 (Algebra with Continuous Functions) Suppose that both f and g are continuous at $x = a$. Then each of the functions

$$f + g, \qquad f - g, \qquad \text{and} \qquad fg$$

is also continuous at $x = a$. If $g(a) \neq 0$, then f/g is continuous at $x = a$.

Proof. Each claimed property of continuous functions follows directly from the analogous property of limits. The first claim, for instance, goes like this:

$$\lim_{x \to a}(f+g)(x) = \lim_{x \to a} f(x)+g(x) = \lim_{x \to a} f(x)+\lim_{x \to a} g(x) = f(a)+g(a) = (f+g)(a).$$

The result is just what's needed to show that $f + g$ is continuous at $x = a$.

Composition, Limits, and Continuity

Limits often involve functions built up by composition, so it's helpful to know how the operations of composition and taking limits interact. First let's discuss some straightforward examples.

EXAMPLE 7   Find and discuss each of the following limits. ▶

Guess all three limits before reading on. With any luck you'll guess correctly.

$$L_1 = \lim_{x \to 0} \ln(\cos x); \qquad L_2 = \lim_{x \to 0} \ln\left(\frac{\sin x}{x}\right); \qquad L_3 = \lim_{x \to \infty} \exp\left(\frac{x+1}{x+2}\right).$$

Solution   The first limit is easiest:

As $\quad x \to 0, \quad \cos x \to \cos 0, \quad$ so $\quad \ln(\cos x) \to \ln(\cos 0) = \ln 1 = 0.$

Why did things work so well? Because both functions involved are continuous at the points that matter: As $x \to 0$, $\cos x \to \cos 0$ *because the cosine function is continuous at $x = 0$; as $\cos x \to 1$, $\ln(\cos x) \to \ln 1$ because the natural logarithm function is continuous at $x = 1$.*

To find L_2, we reason similarly. Recall, first, that as $x \to 0$, $(\sin x)/x \to 1$. ▶ Therefore:

We've used this fact repeatedly. See, for example, Example 4, page 152, for graphical and numerical evidence.

As $\quad x \to 0, \quad \dfrac{\sin x}{x} \to 1, \quad$ so $\quad \ln\left(\dfrac{\sin x}{x}\right) \to \ln 1 = 0.$

Continuity helps again: Because $\ln$ is continuous at $x = 1$,

$$\frac{\sin x}{x} \to 1 \implies \ln\left(\frac{\sin x}{x}\right) \to \ln 1.$$

Convince yourself of the first implication.

The limit L_3 is "at infinity," but the same sort of reasoning applies as for L_1 and L_2: ◄

$$x \to \infty \implies \frac{x+1}{x+2} \to 1 \implies \exp\left(\frac{x+1}{x+2}\right) \to \exp(1) = e.$$

Again, continuity is key: The second implication holds because the exponential function is continuous at $x = 1$. ∎

Limits of Composite Functions: A General Principle. One basic principle applies to all three limits in Example 7. We'll state it in general terms. ◄ (The formal proof, which we omit, appeals to the ϵ–δ definition of limit.)

In the following Fact, the constant a may be $\pm\infty$.

> **Fact** (Limit of a Composite Function) Let f and g be functions, such that $\lim_{x \to a} g(x) = b$ and f is continuous at $x = b$. Then
> $$\lim_{x \to a} (f \circ g)(x) = \lim_{x \to a} f(g(x)) = f(b).$$

We used this fact repeatedly in Example 7.

The precise version says exactly where f and g must be continuous.

Composing Continuous Functions. Recall that, by definition, a function f is continuous at a point $x = a$ if and only if $\lim_{x \to a} f(x) = f(a)$. ◄ With this definition and the preceding Fact, it's easy to show that if f and g are continuous functions, then so is $f \circ g$. More precisely: ◄

> **Theorem 3** Suppose that g is continuous at $x = a$ and f is continuous at $g(a)$. Then $f \circ g$ is continuous at $x = a$.

Proof. If $x \to a$, $g(x) \to g(a)$, because g is continuous at a. As $g(x) \to g(a)$, $f(g(x)) \to f(g(a))$, because f is continuous at $g(a)$. Putting everything together,

$$x \to a \implies (f \circ g)(x) \to (f \circ g)(a).$$

That's exactly what we needed to show. ∎

Limits and Inequalities: The Squeeze Principle

Algebraic limit rules don't help with limits such as

$$\lim_{x \to 0} x \cdot \sin\left(\frac{1}{x}\right) \qquad \text{and} \qquad \lim_{x \to \infty} \frac{\sin x}{x}.$$

The product-of-limits rule doesn't apply to the first limit, for example, because the second factor, $\sin(1/x)$, has no limit as $x \to 0$. With graphs the limits are easy to guess, but let's try to compute them directly.

Squeezing Inequalities. The idea is to use carefully chosen inequalities to "squeeze" the function of interest between two simpler functions. For the first of the preceding limits, the key inequalities say that, for any x,

$$-|x| \le x \sin\left(\frac{1}{x}\right) \le |x|.$$

A graph shows convincingly that the inequality is valid:

Finding a limit by squeezing

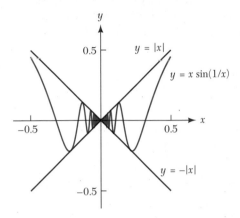

As $x \to 0$, both the left and the right sides of this three-term inequality tend to 0. Squeezed between $-|x|$ and $|x|$, $x \sin(1/x)$ has only one place to go:

$$x \sin\left(\frac{1}{x}\right) \to 0 \quad \text{as} \quad x \to 0.$$

There's a general theorem here. The preceding graph makes it reasonable; we omit the formal proof.

> **Theorem 4** (The Squeeze Principle)   Suppose that
> $$f(x) \le g(x) \le h(x)$$
> for all x near $x = a$. If $f(x) \to L$ and $h(x) \to L$ as $x \to a$, then $g(x) \to L$ as $x \to a$.

Note:

- The theorem is good for more than computation. It also guarantees that $\lim_{x \to a} g(x)$ exists—no matter how crazily g careens between f and h.

- The squeeze principle is conceptually easy; the hard part, in practice, is to choose the "bounding functions" f and h properly.

EXAMPLE 8   Explain why $\displaystyle \lim_{x \to \infty} \frac{\sin x}{x} = 0$.

Solution   The inequality

$$-\frac{1}{x} \le \frac{\sin x}{x} \le \frac{1}{x}$$

holds ▶   for all $x > 0$. Because the left and the right sides tend to 0 as $x \to \infty$, so must the middle expression. ■

Convince yourself; use the fact that for all x,
$-1 \le \sin x \le 1$.

Finding Limits Graphically and Numerically

If $\lim_{x \to a} f(x) = L$, then $f(x) \approx L$ whenever $x \approx a$. Therefore, in order to guess or estimate L, it's natural to look at values of f near $x = a$. We can do this either numerically, by plugging in numbers, or graphically, by plotting f near $x = a$. Each strategy has its advantages. Numerical outputs may be more accurate, but graphs show better how f varies. With technology, both approaches are almost painless.

Can you guess these limits using only mental gymnastics?

Check some with a calculator.

EXAMPLE 9   Let $f(x) = x \sin(1/x)$.   Discuss ◄ $\lim_{x \to 0} f(x)$ and $\lim_{x \to \infty} f(x)$.

Solution   We saw earlier, by squeezing, that the first limit is 0. Numerical values of f near $x = 0$ (rounded to 4 decimal places) agree: ◄

x	−0.1	−0.01	−0.001	...	0.001	0.01	0.1
$f(x)$	−0.0544	−0.0051	0.0008	...	0.0008	−0.0051	−0.0544

To investigate $\lim_{x \to \infty} f(x)$ we'll consider large values of x, first graphically

Graph of $y = \sin(1/x)$ for large x

and now numerically, to six decimal places:

x	10	100	1000	10000
$f(x)$	0.998334	0.999983	1.000000	1.000000

Again the conclusion is irresistible: $\lim_{x \to \infty} f(x) = 1$.   ■

EXAMPLE 10   Let $g(x) = \sin(1/x)$. Discuss $\lim_{x \to 0} g(x)$.

S o l u t i o n   This time both graph ▶ and numbers look suspiciously chaotic: *The graph shown is of poor quality. What's wrong with it? Why is a good graph hard to find? Would a smaller window help?*

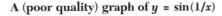

A (poor quality) graph of $y = \sin(1/x)$

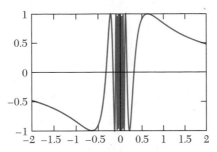

x	-0.1	-0.05	-0.025	$\cdots$	0.001	0.0025	0.0075
$g(x)$	0.544	-0.913	-0.745	$\cdots$	0.827	-0.851	0.983

In fact, the limit doesn't exist. As $x \to 0$ from the right, $1/x \to +\infty$, so $\sin(1/x)$ oscillates faster and faster between -1 and 1.   ■

BASICS

1. Let f and g be the functions whose graphs are shown below. Use the graphs to evaluate the following limits. If a limit doesn't exist, explain why. [**NOTE:** $f(1) = 2$.]

Graph of f   **Graph of g**

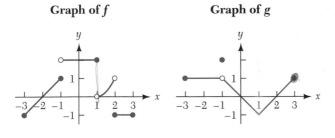

(a) $\lim\limits_{x \to 0}\big(f(x) + g(x)\big)$ (e) $\lim\limits_{x \to 0} \dfrac{f(x)}{g(x)}$

(b) $\lim\limits_{x \to 2}\big(f(x) + g(x)\big)$ (f) $\lim\limits_{x \to 0}\big(f(x) \cdot \cos x\big)$

(c) $\lim\limits_{x \to 1}\big(f(x) \cdot g(x)\big)$ (g) $\lim\limits_{x \to -2} x^2 g(x)$

(d) $\lim\limits_{x \to 2}\big(f(x) \cdot g(x)\big)$

2. Let f and g be the functions whose graphs are shown below. Use the graphs to evaluate the following limits. If a limit doesn't exist, explain why.

Graph of f   **Graph of g**

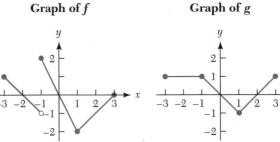

(a) $\lim\limits_{x \to -2}\left(\dfrac{f(x)}{g(x)}\right)$ (c) $\lim\limits_{x \to 0}\left(\dfrac{f(x)}{g(x)}\right)$

(b) $\lim\limits_{x \to 2}\left(\dfrac{f(x)}{g(x)}\right)$ (d) $\lim\limits_{x \to 1}\left(\dfrac{f(x)}{g(x)}\right)$

3. Use a calculator to estimate each of the following limits numerically.

(a) $\lim\limits_{x \to \infty} \dfrac{3}{x^2}$ (f) $\lim\limits_{x \to 0} \sin(\sin x)$

(b) $\lim\limits_{x \to \infty} \dfrac{5}{\sqrt{x}}$ (g) $\lim\limits_{x \to \infty} \dfrac{2^x}{x^2}$

(c) $\lim\limits_{x \to \infty} \dfrac{x^2 + 1}{x}$ (h) $\lim\limits_{x \to \infty} \dfrac{x^2}{2^x}$

(d) $\lim\limits_{x \to 0^+} \dfrac{\sqrt{x}}{\sin x}$ (i) $\lim\limits_{x \to -\infty} \dfrac{2^x}{x^2}$

(e) $\lim\limits_{x \to 0^+} \big(x \log_{10} x\big)$ (j) $\lim\limits_{x \to 0^+} x^x$

4. Use a grapher to guess whether each of the following limits exist. Estimate values of those that do exist.

(a) $\lim_{x \to \infty} \dfrac{2000x^2}{1000x^2 + \sin x}$

(b) $\lim_{x \to 0} \dfrac{1 - \cos^2(3x)}{x^2}$

(c) $\lim_{x \to 0} \cos\left(\dfrac{1 - \cos x}{x}\right)$

(d) $\lim_{x \to \infty} \dfrac{3^x - 3}{2^x - 2}$

(e) $\lim_{x \to -\infty} \dfrac{3^x - 3}{2^x - 2}$

5. Evaluate

(a) $\lim_{x \to \infty} \sin\left(\dfrac{\pi}{x}\right)$.

(b) Evaluate $\lim_{x \to 0} \sin\left(\dfrac{\pi}{x}\right)$.

(c) Evaluate $\lim_{x \to \infty} \exp\left((3x^2 + x)/2x^3\right)$.

6. Suppose that $\lim_{x \to 1} f(x) = 5$ and $\lim_{x \to 1} g(x) = -2$. Use this information to evaluate each of the following limits.

(a) $\lim_{x \to 1}\big(f(x) + g(x)\big)$

(b) $\lim_{x \to 1}\big(g(x) - f(x)\big)$

(c) $\lim_{x \to 1} \dfrac{f(x) + 2}{f(x)g(x)}$

FURTHER EXERCISES

7. Let $f(x) = \sqrt[3]{x}$. Explain why $\lim_{x \to 0} f'(x) = \infty$.

8. Suppose that f is a (nonconstant) periodic function. Explain why $\lim_{x \to \infty} f(x)$ does not exist.

9. Sketch the graph of a function f that has *all* of the following properties. (Be sure to label axes and indicate units on each axis.)

(i) f has domain $(-3, \infty)$

(ii) f has range $(-\infty, 2)$

(iii) $f(-1) = 0$; $f(0) = 0$; $f(5) = 0$

(iv) $\lim_{x \to -3^+} f(x) = -\infty$

(v) $\lim_{x \to 2} f(x) = -3$

(vi) $\lim_{x \to \infty} f(x) = 2$

10. Sketch the graph of a function f that has *all* of the following properties:

(i) f has domain $\mathbb{R}$

(ii) f has range $(-3, \infty)$

(iii) $\lim_{t \to 2} f(t) = -2$

(iv) $\lim_{t \to \infty} f(t) = -3$

(v) $\lim_{t \to -\infty} f(t) = 3$

11. Let P be the (fixed) point $(0, 1)$, let Q be a (moveable) point on the x-axis, and let ℓ be the line joining P and Q. In this exercise we'll explore how the *slope* of ℓ behaves as Q slides back and forth along the x-axis.

(a) Suppose that Q slides to the *right* along the x-axis. How does the slope of ℓ behave? Does this slope approach a limit as Q moves farther and farther to the right? What limit? Give a geometric reason for your answer—no calculation is needed. (Draw a picture to see what's going on.)

(b) Redo the previous part algebraically: First find a symbolic expression for the slope in question, then find an appropriate limit. [**HINT:** Q has coordinates $(x, 0)$.]

(c) Repeat parts (a) and (b) but assuming this time that Q slides farther and farther to the *left*.

(d) Suppose that Q slides *toward the origin* from the right. Does the slope of ℓ approach a limit? Justify your answer with an appropriate limit calculation.

12. Let O be the origin $(0, 0)$, let Q be a (moveable) point on the line $y = 2x + 3$, and let ℓ be the line joining O and Q. This exercise explores how the *slope* of ℓ behaves as Q moves back and forth along the line $y = 2x + 3$, while O remains fixed.

(a) Suppose that Q slides to the *right* along the line $y = 2x + 3$. How does the slope of ℓ behave? Does this slope approach a limit as Q moves farther and farther to the right? What limit? Give a geometric reason for your answer—no calculation is needed. (Draw a generic picture to see what's going on.)

(b) Redo the previous part algebraically: First find a symbolic expression for the slope in question, then find an appropriate limit at infinity. [**HINT:** Q has coordinates $(x, 2x + 3)$.]

(c) Does it make any difference in the two parts above if Q slides farther and farther along ℓ to the *left*? Explain your answer.

13. *This is a more general and open-ended version of Exercises 11 and 12. One or both of them should be done first.* Let P be any point in the plane, ℓ_1 any line, and Q any point on ℓ_1; assume that $P \neq Q$. Let ℓ_2 be the line through P and Q. This exercise explores what happens to the slope of ℓ_2 as Q moves to and fro along ℓ_1. [**NOTE:** No algebraic formulas or formal limit calculations are needed until the last part. In all other parts, just use pictures.]

(a) Suppose that P happens to lie on ℓ_1. How does the slope of ℓ_2 behave as Q moves along ℓ_1? (Draw a picture to guess the answer—no calculation is needed.)

(b) Suppose that ℓ_1 is a *vertical* line. (Don't assume that P lies on ℓ_1.) How does the slope of ℓ_2 behave as Q moves higher and higher along ℓ_1? Does the slope approach a limit? What limit? [**HINT:** Does it matter whether P lies to the left or right of ℓ_1?]

(c) Suppose that ℓ_1 is a *horizontal* line. (Don't assume that P lies on ℓ_1.) How does the slope of ℓ_2 behave as Q moves farther and farther to the right along ℓ_1? Does the slope approach a limit? What limit?

(d) Suppose that ℓ_1 is an *oblique* line (i.e., with slope $m \neq 0$). (Don't assume that P lies on ℓ_1.) How does the slope of ℓ_2 behave as Q moves farther and farther to the right along ℓ_1? Does the slope approach a limit? What limit?

(e) Use limits at infinity to verify as many of your answers above as possible. To do so, write P in the form (a, b) and ℓ_1 either in the form $y = mx + k$ or in the form $x = c$.

14. Show that there is only one value of $k \geq 1$ for which $\lim_{x \to 0} \dfrac{\sin(2 \sin x)}{x^k}$ is a real number. Find this value of k and evaluate the limit.

15. Let $p(x) = (x + 1)(2 - x)$ and $q(x) = x^{17} - 3x^{14} + 39$. Evaluate each of the following:

(a) $\lim_{x \to \infty} p(x)$ (c) $\lim_{x \to \infty} q(x)$

(b) $\lim_{x \to -\infty} p(x)$ (d) $\lim_{x \to -\infty} q(x)$

16. Suppose that $r(x)$ is a polynomial of degree n such that $\lim_{x \to \infty} r(x) = -\infty$ and $\lim_{x \to -\infty} r(x) = -\infty$. Can you tell if n is even or odd? Explain.

17. Let h be a function that has a vertical asymptote at $x = 3$ and the properties $\lim_{x \to -\infty} h(x) = \infty$ and $\lim_{x \to \infty} h(x) = -1$.

(a) Must h have a horizontal asymptote? If so, what is it?

(b) Could h be a rational function? Why or why not?

18. Any rational function $r(x)$ behaves like ax^n for some real number a and some integer n in the sense that

$$\lim_{x \to \infty} \frac{r(x)}{ax^n} = 1.$$

For example, $k(x) = 2x^3 / (x^2 + 1)$ behaves like $2x$ ($a = 2$ and $n = 1$) for large values of x because

$$\lim_{x \to \infty} k(x) = \lim_{x \to \infty} \frac{2x^3}{x^2 + 1}$$
$$= \lim_{x \to \infty} \left(2x \cdot \frac{x^2}{x^2 + 1} \right)$$

and $\lim_{x \to \infty} x^2 / (x^2 + 1) = 1$.

Find the appropriate values of a and n for each of the following rational functions. Then graph the function and ax^n on the same axes for large values of x (e.g., on the interval $[50, 100]$).

(a) $f(x) = \dfrac{4x^2 + 3x}{3x^4 + 2x^2 - 1}$

(b) $g(x) = \dfrac{3x^4 - 2x^3 + 4x^2 - 5x + 7}{5x^4 - 12x^2 + 100}$

(c) $h(x) = \dfrac{2x^5 - 3x^2 + 15}{3x^2 + 4x - 5}$

19. Let $f(x) = \sqrt{x^2 + 1} - x$ and $g(x) = \sqrt{x^2 + x} - x$.

(a) Use a calculator or a computer to complete the following table.

x	100	1,000	10,000	100,000	1,000,000
$f(x)$					
$g(x)$					

(b) Use the results you recorded in the table above to estimate $\lim_{x \to \infty} f(x)$ and $\lim_{x \to \infty} g(x)$.

(c) What do your answers to part (b) imply about the graphs of f and g?

20. Give examples of polynomials $p(x)$ and $q(x)$ such that $\lim_{x \to \infty} p(x) = \infty$, $\lim_{x \to \infty} q(x) = \infty$, and

(a) $\lim_{x \to \infty} \dfrac{p(x)}{q(x)} = \infty$.   (b) $\lim_{x \to \infty} \dfrac{p(x)}{q(x)} = 2$.

(c) $\lim_{x \to \infty} \dfrac{p(x)}{q(x)} = 0$.

(d) $\lim_{x \to \infty} \dfrac{p(x)}{q(x)} = 1$ and $\lim_{x \to \infty} \left(p(x) - q(x) \right) = 3$.

21. For each part of the preceding problem, give a geometric explanation of how the graphs of p and q are related. [**HINT**: Graph your polynomials p and q on the same set of axes.]

22. Let $p(x) = ax^n + q(x)$ where $q(x)$ is a polynomial of degree $m < n$.

(a) What is $\lim_{x \to \infty} \dfrac{q(x)}{x^n}$?   (b) What is $\lim_{x \to -\infty} \dfrac{q(x)}{x^n}$?

(c) If $a > 0$ and n is *even*, what is $\lim_{x \to \infty} p(x)$?

(d) If $a > 0$ and n is *even*, what is $\lim_{x \to -\infty} p(x)$?

(e) How do your answers to parts (c) and (d) change if n is *odd*?

23. Suppose that the average rate of change of a function f over the interval from $x = 3$ to $x = 3 + h$ is $4e^h - 5 \cos(2h)$. What is $f'(3)$?

24. Let $f(x) = \dfrac{2x^2 + x + a}{x^2 - 2x - 3}$.

(a) Can the number a be chosen so that $\lim_{x \to 3} f(x)$ exists? Explain your answer.

(b) Can the number a be chosen so that $\lim_{x \to \infty} f(x)$ exists? Explain your answer.

25. Explain why $\lim_{x \to \infty} \cos x$ does not exist.

26. Evaluate each of the following limits, if it exists. [**HINT**: Graphs may help.]

(a) $\lim_{x \to \infty} \dfrac{\sin x}{x}$ (b) $\lim_{x \to -\infty} \dfrac{x^3}{2 + \cos x}$

27. Suppose that neither $\lim_{x \to a} f(x)$ nor $\lim_{x \to a} g(x)$ exists. Find functions f and g so that the following limits exist or explain why it is impossible to do so.

(a) $\lim_{x \to a} \left(f(x) + g(x) \right)$ (b) $\lim_{x \to a} \left(f(x) \cdot g(x) \right)$

28. Let f be a function defined for all real numbers. Which of the following statements *must* be true about f? Which *might* be true? Which *must* be false? Justify your answers.

    (a) $\lim\limits_{x \to a} f(x) = f(a)$

    (b) If $\lim\limits_{x \to 0} \dfrac{f(x)}{x} = 2$, then $f(0) = 0$.

    (c) If $\lim\limits_{x \to 0} \dfrac{f(x)}{x} = 1$, then $\lim\limits_{x \to 0} f(x) = 0$.

    (d) If $\lim\limits_{x \to 1^-} f(x) = 1$ and $\lim\limits_{x \to 1^+} f(x) = 3$, then $\lim\limits_{x \to 1} f(x) = 2$.

    (e) If $\lim\limits_{x \to 2} f(x) = 3$, then 3 is in the range of f.

    (f) If $\lim\limits_{x \to 0} \dfrac{f(x) - f(0)}{x} = 3$, then $f'(0) = 3$.

2.8 Chapter Summary

This chapter is the heart of Calculus I. Its main idea, the **derivative**, is the most important idea in calculus.

Amount Functions, Rate Functions, Slope Functions. Section 2.1 approached the derivative informally and intuitively, interpreting it (depending on the situation) either as the **rate function** or as the **slope function** derived from an original function, or **amount function**. *Any* function can be thought of as "original"; to any function we can associate a rate function that describes how fast the original function varies.

For example, suppose that $s(t)$ describes the **position** of a moving object at time t; then its rate function $s'(t)$ (usually written $v(t)$) describes the object's **velocity**. If v is considered original, then *its* rate function $v'(t)$ (usually written $a(t)$) describes the object's **acceleration**.

The **racetrack principle** relates the behavior of rate functions to that of amount functions. In its simplest form, it says that if g grows *faster* than f over an interval, then g grows *more* than f over that interval—not a very surprising fact, but one that will prove useful later.

For a linear function the associated rate function (or slope function) is constant— its value is the slope itself. For a nonlinear function the question of slope is stickier— what does it mean for a *curve* to have slope?

The answer—which we pursued in Section 2.2—is best understood graphically: The slope of a curve at a point is the slope of the **tangent line** at that point. Zooming helps us find this slope:

To find the slope of a curve at a point, zoom in until the curve looks straight, then find the slope.

This trick really works, even for nonlinear functions. It works because almost all calculus functions are "locally straight."

Derivatives, Geometrically. Sections 2.3 and 2.4 are as important as any in this book. They described in detail how a function and its derivatives (first and second) are related graphically. The key idea is worth repeating:

The derivative f' of a function f tells whether and how fast the graph of f is rising.

Almost everything in Sections 2.3 and 2.4 derived from that fact. Where a function increases or decreases, where it's concave up or concave down, how many and what types of stationary points it has—derivatives carry all such information.

The Derivative as a Limit. The formal definition of derivative is as a **limit of difference quotients**:

$$f'(x) = \lim_{h \to 0} \frac{f(x+h) - f(x)}{h}.$$

Section 2.5 explored this important limit and what it means in various settings. For a given function f, the difference quotient has two main interpretations: It can be thought of either as the **slope of the secant line from** x **to** $x + h$ or as the **average rate of change** of f from x to $x + h$. From these interpretations it follows that the derivative itself also has two main interpretations. The derivative $f'(x)$ can be thought of either as the **slope of the tangent line** at x or as the **instantaneous rate of change** of f at x. Both viewpoints are important, and we'll use both throughout this book.

Limits. The derivative is defined as a **limit**; Sections 2.6 and 2.7 treated the idea of limit in its various forms. The basic idea is simple: $\lim_{x \to a} f(x) = L$ means that $f(x)$ approaches L as x approaches a. Section 2.6 explored this definition and some of its variants in symbolic, graphical, and numerical forms.

Once understood, the limit is applied to define **continuity** of a function f at a point $x = a$ of its domain. In graphical terms, continuity of f means that the f-graph is unbroken at $x = a$; limits offer the right analytic language for describing this phenomenon.

Section 2.7 extended the idea of limits to **infinite limits** and **limits at infinity**—limits that involve the ∞ symbol. Such limits correspond graphically to **vertical and horizontal asymptotes**, respectively. Section 2.7 also described how to combine old limits algebraically to form new ones and, in the same spirit, how to combine old continuous functions to form new ones.

A Summary Problem

Here's a problem that calls on ideas from various parts of this chapter.

EXAMPLE 1 In each case, decide whether a function with the given properties exists. If so, sketch the graph of such a function. If not, explain why there cannot be such a function.

a. $\lim_{x \to \infty} f(x) = 5$ and $\lim_{x \to \infty} f'(x) = 1$.

b. $f(x) < 0$ and $f'(x) > 0$ for all x.

c. $f(x) < 0$ and $f'(x) < 0$ for all x.

d. $f''(x) > 0$ and $f(x) < 0$ for all x.

e. $f(x) > 0, f'(x) > 0,$ and $f''(x) < 0$ for all x.

f. $f(x) > 0, f'(x) > 0,$ and $f''(x) < 0$ for all $x > 0$.

Plot it!

Solution   (Note: Read these solutions with graphs at hand. Use either some form of plotting technology or the graphs in Appendix I.)

a. No function with these properties can exist. If $\lim_{x \to \infty} f'(x) = 1$, then, for large x, $f'(x) \approx 1$. This means that, for large x, f increases at the rate of about one output unit for each input unit. Such a function cannot have a horizontal asymptote as $x \to \infty$, as the statement $\lim_{x \to \infty} f(x) = 5$ implies.

b. The conditions say that f should be *negative* and *increasing* for all x. The function $f(x) = -e^{-x}$ ◀ has these properties.

c. The conditions say that f should be *negative* and *decreasing* for all x. The function $f(x) = -e^x$ has these properties.

d. The conditions require that f be *concave up* and *negative* for all x. This is impossible. A graph that's concave up cannot stay below the x-axis forever—think about it.

e. The conditions require that f be *positive, increasing,* and *concave down* for all x. This is impossible, for the same reason as in part (d): A graph that's concave down cannot stay above the x-axis forever.

f. The conditions mean that for positive x, f is *positive, increasing,* and *concave down*. The function $f(x) = \ln(x + 1)$ has all these properties.   ■

Derivatives of Elementary Functions

3.1 Derivatives of Power Functions and Polynomials

Introduction

In Chapter 2 we introduced the idea of the derivative (or rate) function and discussed geometric relationships between the functions f and f'. Next we looked closely at $f'(a)$, the value of the derivative function for a specific input $x = a$—how it's defined formally as a limit, how slopes of secant and tangent lines arise, and how they're connected to average and instantaneous rates of change. Now it's time to *calculate* some derivatives symbolically.

Derivatives: The Symbolic View

If a function f has a simple symbolic formula, then it's reasonable to hope that the derivative f' has one, too. This hope is well founded. Indeed, we'll learn in this chapter how to find a formula for the derivative of any **elementary function**, i.e., any function "built" from algebraic, exponential, logarithmic, and trigonometric ingredients. ▶ Even for a monster such as

For much more information on elementary functions, see Section 1.5.

$$f(x) = \ln\left(\frac{\sin(2x)}{1 + 3x^4}\right),$$

we'll soon have little trouble in writing down a derivative formula.

But first things first. Before we can differentiate ▶ a built-up function such as f, we'll need to know derivatives of its most basic ingredients: the natural logarithm functions, the sine function, the fourth power function, and so on. Once we know these basic derivatives, we'll see how to combine them appropriately to differentiate *any* elementary function.

Recall: "differentiate" means "find the derivative of."

Antiderivatives

Similarly, f is an antiderivative of f′.

The relation between functions and derivatives runs both ways. If f is the derivative of F, then F is an **antiderivative** of f. ◄ (*Antidifferentiation* means finding antiderivatives.) Throughout this chapter we'll keep an eye on both derivatives and antiderivatives.

Other Notations for the Derivative

The prime notation f' for the derivative of f is typographically convenient. It also reminds us of a derivative's "ancestry": Every f' comes from some original f.

Leibniz Notation.   Alternatives to the prime notation exist. Depending on context, one may be more convenient than the others. To illustrate what's available, let $y = f(x) = x^2$. The following notations all say the same thing:

$$f'(x) = 2x; \qquad \frac{dy}{dx} = 2x; \qquad \frac{d}{dx}\left(x^2\right) = 2x; \qquad \frac{df}{dx} = 2x.$$

So do these:

$$f'(3) = 6; \qquad \left.\frac{dy}{dx}\right|_{x=3} = 6; \qquad \left.\frac{d}{dx}\left(x^2\right)\right|_{x=3} = 6; \qquad \left.\frac{df}{dx}\right|_{x=3} = 6.$$

(We'll soon see just below *why* these statements are true. Here the point is the notation.) The dy/dx form of the derivative is known as **Leibniz notation**, after Gottfried Wilhelm Leibniz (1646–1716), a contemporary of Newton. The "fractional" appearance of Leibniz notation recalls the definition of the derivative as a limit of difference *quotients*:

$$\frac{dy}{dx} = \lim_{\Delta x \to 0} \frac{\Delta y}{\Delta x}.$$

Notations for Higher Derivatives.   Both prime and Leibniz notations exist for second, third, and higher derivatives, as well. The following notations all say the same thing about the function defined by $y = f(x) = x^2$.

$$f''(x) = 2; \qquad \frac{d^2 y}{dx^2} = 2; \qquad \frac{d^2}{dx^2}\left(x^2\right) = 2; \qquad \frac{d^2 f}{dx^2} = 2.$$

We'll use both the prime and the Leibniz notations in this book, as convenience dictates.

The Derivative as a Function: The Definition Revisited

In earlier work we formally defined $f'(a)$, the derivative of f at $x = a$. The following definition is essentially the same; using the generic x in place of the specific a emphasizes that f' is a *function*.

> **Definition** Let f be a function. The derivative of f, denoted by f', is the function defined for an input x by
> $$f'(x) = \lim_{h \to 0} \frac{f(x+h) - f(x)}{h}.$$

Observe:

Domain The domain of f' is the set of inputs x for which the limit exists. In the best cases (more often than not, for our purposes), the domain of f' is the full domain of f. ▶ For some functions f the domain of f' omits a few (usually obvious) points in the domain of f. ▶

The domain of f' can't possibly be bigger than the domain of f.

For $f(x) = |x|$, the domain of f' is smaller than the domain of f.

Finding the Limit The limit in the definition can almost never be found simply by plugging in $h = 0$; some trick or algebraic manipulation is almost always needed to eradicate the troublesome denominator.

Beyond Algebra If f is given by a graph, by a table of values, in nonalgebraic form, or even as a complicated algebraic formula, the preceding limit may be hard or impossible to find exactly. All is not lost: In such cases, graphical or numerical estimates are usually possible.

Derivatives of Power Functions

Our first order of business is to find derivatives of **power functions**—i.e., functions of the form $f(x) = x^k$, where k is any constant. ▶ Each of the following expressions defines a power function: ▶

The power k need not be an integer.

What's k in each case?

$$x, \quad x^2, \quad \sqrt{x}, \quad \frac{1}{\sqrt{x}}, \quad \frac{x^5}{\sqrt{x^3}}, \quad x^\pi.$$

We start with nonnegative integer powers.

EXAMPLE 1 Find derivative formulas for the linear, quadratic, and cubic power functions $l(x) = x$, $q(x) = x^2$, and $c(x) = x^3$. Interpret results graphically.

Solution The function l is linear, with constant slope 1. Its derivative is therefore the constant function $l'(x) = 1$. Reassuringly, the limit definition says the same thing:

$$l'(x) = \lim_{h \to 0} \frac{l(x+h) - l(x)}{h} = \lim_{h \to 0} \frac{x+h-x}{h} = 1.$$

Higher powers of x require a little more algebraic work. ▶ For q:

Follow each step; watch for a crucial cancellation.

$$
\begin{aligned}
q'(x) &= \lim_{h \to 0} \frac{q(x+h) - q(x)}{h} \\
&= \lim_{h \to 0} \frac{(x+h)^2 - x^2}{h} \\
&= \lim_{h \to 0} \frac{x^2 + 2xh + h^2 - x^2}{h} \\
&= \lim_{h \to 0} (2x + h) \\
&= 2x.
\end{aligned}
$$

The cubic function c requires more of the same:

$$
\begin{aligned}
c'(x) &= \lim_{h \to 0} \frac{c(x+h) - c(x)}{h} \\
&= \lim_{h \to 0} \frac{(x+h)^3 - x^3}{h}
\end{aligned}
$$

$$= \lim_{h \to 0} \frac{x^3 + 3x^2h + 3xh^2 + h^3 - x^3}{h}$$

$$= \lim_{h \to 0} \left(3x^2 + 3xh + h^2 \right)$$

$$= 3x^2.$$

Do they look right? Do the derivative graphs plausibly represent slopes on the original graphs?

In graphical form, our results look like this: ◄

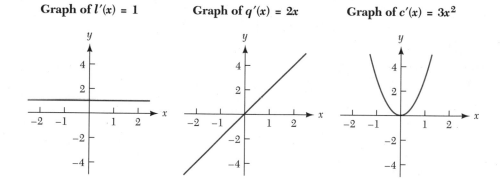

The answers aren't surprising; we've drawn such graphs before. The limit calculations assure us, for the first time, that our derivative formulas are exact.   ■

What can be said about power functions x^k for which k is *not* a nonnegative integer?

EXAMPLE 2   Find derivative formulas for the power functions $s(x) = \sqrt{x} = x^{1/2}$ and $n(x) = 1/x = x^{-1}$. What is the domain of each derivative function? Interpret the results graphically.

S o l u t i o n   This example, like the last one, calls for limit computations. Here we'll need slightly more clever algebraic methods to eliminate the troublesome denominator h. For s,

$$s'(x) = \lim_{h \to 0} \frac{s(x + h) - s(x)}{h}$$

$$= \lim_{h \to 0} \frac{\sqrt{x + h} - \sqrt{x}}{h}$$

$$= \lim_{h \to 0} \frac{\sqrt{x + h} - \sqrt{x}}{h} \cdot \frac{\sqrt{x + h} + \sqrt{x}}{\sqrt{x + h} + \sqrt{x}}$$

$$= \lim_{h \to 0} \frac{1}{\sqrt{x + h} + \sqrt{x}}$$

$$= \frac{1}{2\sqrt{x}} = \frac{1}{2}x^{-1/2}.$$

(See the trick on the third line? Multiplying top and bottom by the conjugate expression $\sqrt{x + h} + \sqrt{x}$ helped us simplify the numerator.)
For n,

$$n'(x) = \lim_{h \to 0} \frac{n(x + h) - n(x)}{h}$$

$$= \lim_{h \to 0} \frac{\frac{1}{x+h} - \frac{1}{x}}{h}$$

$$= \lim_{h \to 0} \frac{\frac{1}{x+h} - \frac{1}{x}}{h} \cdot \frac{(x + h)(x)}{(x + h)(x)}$$

$$= \lim_{h \to 0} \frac{-h}{h\, x\, (x + h)}$$

$$= \lim_{h \to 0} \frac{-1}{x(x + h)}$$

$$= \frac{-1}{x^2} = -1x^{-2} = -x^{-2}.$$

Here are all the graphs:

Graph of $s(x) = \sqrt{x}$ **Graph of $s'(x) = 1/(2\sqrt{x})$**

Observe the domains: The derivative s' is defined wherever s is defined—except at $x = 0$, where the graph of s has a vertical tangent line. The derivative n' has the same domain (all nonzero reals) as n itself. ■

A General Power Rule

Each preceding example is an instance of a more general fact:

> **T h e o r e m 1** (**Power Rule for Derivatives**) Let k be any real constant. If $f(x) = x^k$, then $f'(x) = kx^{k-1}$.

What does the rule say if
$k = 0$?

(We've just seen that the rule holds for $k = 1$, $k = 2$, $k = 3$, $k = -1$, and $k = 1/2$.) ◄ We won't prove this theorem rigorously, but at the end of this section we indicate how the **binomial theorem** can be used to show that it holds for all nonnegative integers.

Combinations of Power Functions: The Sum and Constant Rules

Knowing the derivatives of "pure" power functions, it's easy to find derivatives of *combinations* of power functions (such as polynomials), piece by piece.

EXAMPLE 3   Let $p(x) = 3x^5 + 7x^4 - x^2/3 + 11$. Find $p'(x)$.

Solution   We break p apart, differentiate each piece separately, and then reassemble the results. Here are the details, in Leibniz notation:

$$
\begin{aligned}
\frac{dy}{dx} &= \frac{d}{dx}\left(3x^5 + 7x^4 - \frac{1}{3}x^2 + 11\right) \\
&= 3\frac{d}{dx}\left(x^5\right) + 7\frac{d}{dx}\left(x^4\right) - \frac{1}{3}\frac{d}{dx}\left(x^2\right) + \frac{d}{dx}(11) \\
&= 3 \cdot 5x^4 + 7 \cdot 4x^3 - \frac{1}{3} \cdot 2x + 11 \cdot 0 \\
&= 15x^4 + 28x^3 - \frac{2}{3}x.
\end{aligned}
$$

We just made two leaps of faith, having to do with sums and constants, which we'll justify in a moment. First, observe graphically that the result *looks* right:

Graphs of p and p'

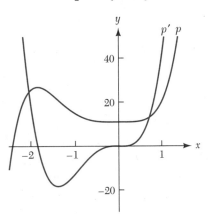

The Sum and Constant Rules for Derivatives

We made two reasonable-seeming assumptions in Example 3. Here they are explicitly, stated as theorems:

Theorem 2 (The Sum Rule for Derivatives) Let f and g be differentiable functions, and let $h = f + g$. Then

$$h'(x) = (f + g)'(x) = f'(x) + g'(x).$$

Theorem 3 (The Constant Rule for Derivatives) If k is any constant, f is any differentiable function, and $g(x) = k \cdot f(x)$, then

$$g'(x) = (k \cdot f)'(x) = k \cdot f'(x).$$

Observe these fine points.

In Words   In words, the theorems say:

> *The derivative of a sum of functions is the sum of their derivatives.*

> *The derivative of a constant times a function is the constant times the derivative of the function.*

Derivatives and Limits   The sum and constant rules for *derivatives* can be proved via the analogous rules for *limits*. The argument for the sum rule runs like this:

$$\frac{d}{dx}\big(f(x) + g(x)\big) = \lim_{h \to 0} \frac{\big(f(x+h) + g(x+h)\big) - \big(f(x) + g(x)\big)}{h}$$

$$= \lim_{h \to 0} \frac{f(x+h) - f(x)}{h} + \lim_{h \to 0} \frac{g(x+h) - g(x)}{h}$$

$$= f'(x) + g'(x).$$

Can you interpret the sum rule graphically?

Geometric Thinking   The constant rule can be interpreted geometrically. Recall that the graph of $kf(x)$ is obtained by stretching the graph of f vertically, with factor k. How does such a stretch affect a tangent line to the f-graph at $x = a$?

The constant rule gives the answer: A vertical k-stretch multiplies slopes of everything—the graph as well as the tangent line—by the same factor k. ◀

Higher Derivatives and Graphical Analysis

Higher derivatives are found symbolically exactly as first derivatives are—we just repeat the process.

As we saw in Chapter 2, geometric properties of the graph of a function f are closely linked to the first and second derivative functions f' and f''. Having formulas for all these functions makes the job of analyzing such properties much easier.

EXAMPLE 4   Following are graphs of several functions of the form $f(x) = x^3 + ax$; a takes integer values from -3 to 3.

Graphs of $f(x) = x^3 + ax$,
$a = -3, -2, -1, 0, 1, 2, 3$

Which graph is which? Which graphs have stationary points? Which have inflection points? Where, exactly, *are* these points?

Solution Since each function has the form $f(x) = x^3 + ax$, the respective first and second derivative formulas are

$$f'(x) - 3x^2 + a \qquad \text{and} \qquad f''(x) = 6x.$$

These formulas tell us everything we need to know.

Observe first that, in each case, $f'(0) = a$. Thus, the constant a represents a given graph's *slope at* $x = 0$. This makes it easy to see which graph is which. ▶

Label each graph with its a-value.

Stationary points occur where $f'(x) = 0$, i.e., at values of x for which $3x^2 + a = 0$. If $a > 0$, this equation has *no* solutions. The picture agrees: Three of the graphs—those with $a > 0$—have no stationary points. If $a = 0$, the equation $f'(x) = 3x^2 = 0$ has just one solution, at $x = 0$. Similarly, if $a < 0$, then $f'(x) = 3x^2 + a = 0$ has two solutions, at $x = \pm\sqrt{-a/3}$. ▶

Again, the graphs agree.

The simple formula $f''(x) = 6x$ tells all about concavity. Regardless of the value of a, $f''(0) = 0$, $f''(x) < 0$ for $x < 0$, and $f''(x) > 0$ for $x > 0$. Graphically, this means (as the picture shows) that each of the preceding graphs has an inflection point at $x = 0$, where the direction of concavity changes. ◼

Antiderivatives

We've just seen how to find the derivative function f' for certain functions f. Now we'll consider the opposite problem, called **antidifferentiation**:

Given a function f, find another function F such that $F' = f$.

Recall that the function F is called an antiderivative of f.

For polynomials and power functions, finding antiderivatives is not much harder than finding derivatives. (Of course, we need to stay clear about which problem we're tackling at any given time!)

EXAMPLE 5 Find antiderivative functions Q and S for $q(x) = x^2$ and $s(x) = \sqrt{x}$. Are your answers unique?

Solution Since $\left(x^3\right)' = 3x^2$, $Q(x) = x^3/3$ is a natural guess. It's easy to check this guess using the constant rule for derivatives:

$$Q'(x) = \left(\frac{1}{3}x^3\right)' = \frac{1}{3} \cdot 3x^2 = x^2.$$

Finding an antiderivative of s takes a little more thought. The power rule for derivatives suggests that we try a function of the form $S(x) = kx^{3/2}$—but for what constant k? Differentiating gives the answer: For the desired k,

$$S'(x) = k \cdot \frac{3}{2} \cdot x^{1/2} = x^{1/2}.$$

Therefore, $k = 2/3$; the function $S(x) = 2x^{3/2}/3$ works.

These answers aren't (quite) unique. For example, each of the functions

$$\frac{x^3}{3} + 1, \qquad \frac{x^3}{3} - 5, \qquad \text{and} \qquad \frac{x^3}{3} - \pi^2 + e - \ln\left(15 + \sin^2(3)\right)$$

is an antiderivative of q. So is *any* function of the form $x^3/3 + C$, for any constant C. ◼

Antiderivatives Aren't (Quite) Unique

The previous example illustrates an important general principle:

> *If F is an antiderivative for f, then so is F + C, for any constant C.*

Interpret this graphically.

We'll return to this question in the next chapter.

The reason is simple: Any two functions that differ by an additive constant have the same derivative. ◄ The implication runs the other way, too: ◄

> *If $F'(x) = G'(x)$ for all x, then $F(x) = G(x) + C$ for some constant C.*

For the particular function $q(x) = x^2$, these two facts imply just what the preceding example suggests: (i) For any constant C, $Q(x) = x^3/3 + C$ is an antiderivative; and (ii) *every* antiderivative has this form.

Finding Antiderivative Formulas

We've seen how to differentiate any polynomial in x: Use Theorem 1 to differentiate each power x^k; then combine results, using the sum and constant rules for derivatives. ◄

The powers x^k need not be integers!

We can *antidifferentiate* combinations of powers in much the same way. The power rule for derivatives, read "backwards," gets us started.

> **Theorem 4** (Power Rule for Antiderivatives) For any constants C and k ($k \neq -1$), $\dfrac{x^{k+1}}{k+1} + C$ is an antiderivative of x^k.

Proof. With derivative rules, the proof is an easy computation:

$$\frac{d}{dx}\left(\frac{x^{k+1}}{k+1} + C\right) = \frac{(k+1)x^k}{k+1} + 0 = x^k,$$

just as we claimed. ■

The Exception: $k = -1$. Notice that the preceding theorem can't possibly hold for $k = -1$, since division by zero isn't allowed. Nevertheless, the exceptional function $f(x) = 1/x$ *has* a well-behaved antiderivative, though not of the usual power form.

> **Fact** For $x > 0$, the natural logarithm function $\ln(x)$ is an antiderivative of $1/x$.

We'll revisit it soon.

We don't yet have the tools to prove this remarkable fact, ◄ but the following graphs make it plausible.

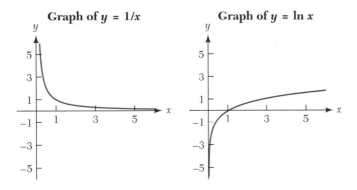

Graph of $y = 1/x$ **Graph of $y = \ln x$**

Antiderivatives of Combinations

The sum and constant rules for derivatives run in reverse: Sums and constant multiples behave just as well for antiderivatives as for derivatives.

EXAMPLE 6   Earlier we differentiated $p(x) = 3x^5 + 7x^4 - x^2/3 + 11$. Now let's antidifferentiate p.

Solution   Again we rend p asunder, antidifferentiate the fragments separately, and reassemble. An antiderivative of $3x^5$ is $x^6/2$; an antiderivative of $7x^4$ is $7x^5/5$; an antiderivative of $x^2/3$ is $x^3/9$; an antiderivative of 11 is $11x$. Thus, for any constant C,

$$\frac{1}{2}x^6 + \frac{7}{5}x^5 - \frac{1}{9}x^3 + 11x + C$$

is an antiderivative of p. The result is easy to check for correctness: Differentiating the preceding expression really does give p. ▶   ■   *See for yourself!*

Easier One Way than the Other.   The concepts of derivative and antiderivative, like those of parent and child, are nothing more than different views of the same functional relationship. Neither concept is any more (or any less) difficult than the other. Moreover, for power functions and polynomials, finding symbolic derivatives and antiderivatives is about equally easy.

It may be surprising, then, that finding *derivative* formulas for elementary functions is, as a rule, much easier than finding *antiderivative* formulas. We'll polish off the derivatives-of-elementary-functions problem quite quickly; indeed, all the ideas are routine enough to fit comfortably into relatively modest computer software. Finding antiderivative formulas, on the other hand, is a greater challenge for both humans and computers. Indeed, most elementary functions have antiderivatives that aren't elementary, i.e., they can't be written as formulas in the usual sense.

Guess and Check.   With the power rule, finding antiderivatives of polynomials and other power functions is easy, and checking answers is easier still. There's a moral:

When finding antiderivatives, guess freely—but check answers carefully.

EXAMPLE 7   Consider the function $f(x) = x^3 - x^2 - 2x$, which follows graphically at left. Four antiderivatives of f, labeled F, G, H, and K, follow at right. Find a formula for each antiderivative.

Graph of $f(x) = x^3 - x^2 - 2x$

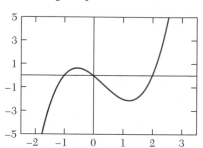

Four antiderivatives of $f(x) = x^3 - x^2 - 2x$

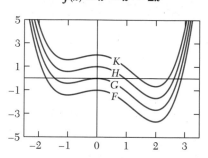

Functions that differ by a constant have the same derivative.

Solution   First notice the resemblance among F, G, H, and K. Graphically, all four graphs are vertical translations of each other; symbolically, all four functions differ from each other by a constant. This is as it should be, since $F' = G' = H' = K' = f$. ◄

Now we find formulas. Theorem 4 (plus a bit of guessing and checking) shows that every antiderivative of f has the form

$$\frac{x^4}{4} - \frac{x^3}{3} - x^2 + C,$$

Check this by differentiation.

for some constant C. ◄   In particular, F, G, H, and K all have this symbolic form; they differ only in the value of C. For example, the graph shows that $F(0) = -1$; thus,

$$-1 = F(0) = C \implies F(x) = \frac{x^4}{4} - \frac{x^3}{3} - x^2 - 1.$$

For similar reasons, G, H, and K correspond to $C = 0$, $C = 1$, and $C = 2$, respectively.   ■

The Power Rule and the Binomial Theorem (Optional)

See any college algebra book for a more complete treatment.

The binomial theorem of elementary algebra ◄ is an explicit formula for an expression of the form $(a+b)^n$, where n is any positive integer. The theorem says, in brief,

$$(a + b)^n = a^n + na^{n-1}b + \frac{n(n-1)}{2}a^{n-2}b^2 + \cdots + nab^{n-1} + b^n.$$

To an algebraist, what the ellipsis ($\ldots$) stands for is the heart of the theorem. For us, only the first few terms matter. Example 8 shows why and how.

EXAMPLE 8   Let $y = x^n$, where n is any positive integer. Show that $dy/dx = nx^{n-1}$.

Solution   By definition,

$$\frac{dy}{dx} = \lim_{h \to 0} \frac{(x + h)^n - x^n}{h}.$$

Applying the binomial theorem in the numerator gives

$$\frac{dy}{dx} = \lim_{h \to 0} \frac{\left(x^n + nx^{n-1}h + \frac{n(n-1)}{2}x^{n-2}h^2 + \cdots\right) - x^n}{h}$$

$$= \lim_{h \to 0} \left(nx^{n-1} + h \cdot \text{stuff}\right)$$

$$= nx^{n-1},$$

as claimed. The key idea is that all terms after the first two involve at least the second power of h and so these terms vanish in the limit—even after we divide by the original denominator. ∎

BASICS

For each function f in Exercises 1–10, find an algebraic expression for f', then plot graphs of the expressions for f and f' on the same axes to check your answer for reasonableness.

1. $f(x) = 3x^2$
2. $f(x) = 7x^{-4}$
3. $f(x) = 4x^5 + 3x^2 - x$
4. $f(x) = 4\sqrt{x} + 1/x^3$
5. $f(x) = 3x^{-1/2} + 4x^{3/2}$
6. $f(x) = -7x^{1/4}$
7. $f(x) = x^3 + \sqrt[3]{x}$
8. $f(x) = 4\sqrt{x^3} - 5x^{-3/2}$
9. $f(x) = 1/\sqrt[3]{x^2} + 5$
10. $f(x) = 4\sqrt[5]{x^3} + 5/\sqrt[3]{x^4}$

11–20. Find the second derivative of each function in Exercises 1–10. Then sketch graphs of the expressions for f and f'' on the same axes to check your answer for reasonableness.

21–30. Find two antiderivatives of each function in Exercises 1–10.

31. When an oil tank is drained for cleaning, there are $V(t) = 100{,}000 - 4000t + 40t^2$ gallons of oil left in the tank t minutes after the drain valve is opened.
    (a) At what average rate does oil drain from the tank during the first 20 minutes?
    (b) At what rate does oil drain out of the tank 20 minutes after the drain valve is opened?
    (c) Explain what $V''(t)$ says about the rate at which oil is draining from the tank.

32. Let $f(x) = e^x$. Use graphs to explain why $f'(x) \neq xe^{x-1}$.

Graph of f

33. Let f be the function shown above, F be an antiderivative of f, and suppose that $F(0) = 10$. (The function f has a local maximum at $x = 1.14$, a point of inflection at $x = 4.67$, and a local minimum at $x = 8.19$.)
    (a) What is the slope of F at $x = 4$? Justify your answer.
    (b) Where is F increasing? Justify your answer.
    (c) Where in the interval $[0, 5]$ does F achieve its largest value? Its smallest value? Justify your answers.

34. Use the limit definition of the derivative function to find f' for each function f below.
    (a) $f(x) = (x + 1)^2$
    (b) $f(x) = 2x^3$
    (c) $f(x) = \sqrt{x + 3}$
    (d) $f(x) = \dfrac{1}{x + 5}$

FURTHER EXERCISES

Find the derivative of each function in Exercises 35–43.

35. $f(x) = (x + 2)^3$
36. $f(x) = 4(x - 1)^5$
37. $f(x) = (x + 4)^{-1}$
38. $f(x) = (x - 2)^{100}$
39. $f(x) = 3(x + 1)^{-2}$
40. $f(x) = \sqrt{x - 1}$
41. $f(x) = \sqrt[3]{x + 7}$
42. $f(x) = 1/\sqrt{x + 4}$
43. $f(x) = 6(x + 4)^{-5/3}$

44–57. Find the second derivative of each of the functions in Exercises 35–43.

58–71. Find an antiderivative of each of the functions in Exercises 35–43.

72. Show that $(1 + x)^r \geq 1 + rx$ when $x \geq 0$ and $r \geq 1$. [**HINT:** Use the racetrack principle.]

73. (a) Over which intervals is $f(x) = (1000 - x)^2 + x^2$ increasing? Decreasing?   [**HINT:** $(1000 - x)^2 = (x - 1000)^2$.]

(b) Use part (a) to decide whether 1000^2 is bigger or smaller than $998^2 + 2^2$. Check your answer using a calculator.

(c) Generalize your result from part (a) to the function $f(x) = (c - x)^n + x^n$, where c is any positive number and n is an *even* positive integer. Use this result to decide whether $10,000^{100}$ is bigger or smaller than $9000^{100} + 1000^{100}$.

74. Let f and F be the functions defined in Exercise 33.

(a) At which values of x, if any, does F have points of inflection? Explain.

(b) Where is F concave down? Justify your answer.

(c) Estimate the value of $F''(5)$.

(d) Explain why F'' achieves its minimum value over the interval $[0, 10]$ at $x = 4.67$.

(e) Where in the interval $[0, 10]$ does F'' achieve its largest value? Justify your answer.

75. Let f and F be the functions defined in Exercise 33, and let L be the line tangent to F at $x = 0$.

(a) Find an equation for L.

(b) Does L lie above or below the curve $y = F(x)$ at $x = 1$? Justify your answer.

76. Let F be the function defined in Exercise 33.

(a) Is it possible that $F(5) = 5$? Explain.

(b) Is it possible that $F(5) = 15$? Explain.

(c) Is $F(10) - F(5) < 0$? Explain.

(d) Is $F(10) - F(7) > 6$? Explain.

77. Let f and F be the functions defined in Exercise 33.

(a) Let g be the line through the points $(2, 7)$ and $(6, 2)$. This line lies below f on the interval $[2, 5]$, so $g(x) < f(x)$ on this interval. Use this fact to compute a lower bound on the value of $F(5) - F(3)$. [**HINT:** Use an antiderivative of g.]

(b) Let h be the line through the points $(2, 8)$ and $(6, 3)$. This line lies above f on the interval $[2, 6]$, so $h(x) > f(x)$ on this interval. Use this fact to compute an upper bound on the value of $F(5) - F(3)$.

78. The function F graphed below is an *antiderivative* of a function f. F has a root at $x = 8.8$ and stationary points at $x = 3.2$, $x = 6$, and $x = 13.3$; F has points of inflection at $x = 4.5$ and $x = 10.5$. [**NOTE:** Although the questions that follow ask about f and f', graphs of f and f' are not shown.]

(a) Where is f positive? Justify your answer.

(b) How many roots does f have in the interval $[0, 15]$? Where are they located?

(c) Rank the five values -2, $f(3)$, $f(6)$, $f(9)$, and $f(14)$ in increasing order.

(d) Where is f increasing? Justify your answer.

(e) Where does f achieve its maximum value in the interval $[0, 15]$? Justify your answer.

(f) Rank the five values 0, $f'(3)$, $f'(6)$, $f'(10)$, and $f'(13)$ in increasing order.

Graph of F

79. Find an equation of the line tangent to the curve $y = x^3 - 6x^2$ at its point of inflection.

80. Suppose that f is a function such that $f(0) = 0$ and f' is the function shown below. Let F be an antiderivative of f. [**NOTE:** The graph is of f', not f or F.]

Graph of f'

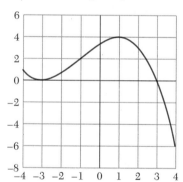

(a) Is F concave up or concave down at $x = -3$? At $x = 0$? At $x = 1$? At $x = 3$? Justify your answers.

(b) Is F increasing, decreasing, or stationary at $x = -3$? At $x = 0$? At $x = 1$? At $x = 3$? Justify your answers.

(c) Where in the interval $[-4, 3]$ does F achieve its minimum value? Justify your answer.

(d) How many stationary points does F have in the interval $[-4, 3]$? Where are they located? Justify your answer.

(e) How many points of inflection does F have in the interval $[-4, 4]$? Where are they located? Justify your answer.

81. (a) How are your answers to Exercise 79 affected if the condition $f(0) = 0$ is changed to $f(0) = 20$?

(b) How are your answers to Exercise 79 affected if the condition $f(0) = 0$ is changed to $f(0) = -15$?

82. Suppose that $y(x)$ is proportional to x^3. Show that $y'(x)$ is proportional to x^2.

83. How many points of inflection does the graph of $y = 2x^6 + 9x^5 + 10x^4 - x + 2$ have? Where are they located?

84. Let $f(x) = 3x^2 - 8x^{-2}$. Evaluate $\lim\limits_{h \to 0} \dfrac{f(2+h) - f(2)}{h}$.

85. Find a symbolic expression for the nth derivative of each of the following functions.
    (a) $f(x) = 1/x$ (b) $f(x) = \sqrt{x}$

86. Let $p(x) = x^n + a_1 x^{n-1} + a_2 x^{n-2} + \cdots + a_{n-1}x + a_n$.
    (a) Evaluate $\dfrac{d^n}{dx^n} p(x)$.
    (b) Show that if k is an integer such that $1 \le k \le n$, $p^{(k)}(0) = k!a_k$.

87. Suppose that g is an nth degree polynomial. Explain why $\dfrac{d^{n+1}}{dx^{n+1}} g(x) = 0$.

88. Let $f(x) = x^n$ where $n \ge 2$ is an integer.
    (a) Show that f is concave up on the interval $(0, \infty)$.
    (b) What can be said about the concavity of f on the interval $(-\infty, 0)$. [**HINT:** Does it matter whether n is even?]

89. The function $f(x) = \dfrac{(x + 0.00001)^8 - x^8}{0.00001}$ is closely approximated by the function $g(x) = ax^b$ for appropriate values of a and b (i.e., for any x, $|f(x) - g(x)|$ is, in some sense, a "small" number). What are these values? Explain why g approximates f so well.

90. The figure below shows the graph of $f(x) = \sqrt{x}$ and its tangent line at $x = a$. Find the value of a.

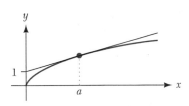

91. Find values of a and b so that the line $2x + 3y = a$ is tangent to the graph of $f(x) = bx^2$ at $x = 3$.

92. Find the value of k for which the graph of $f(x) = 2x^2 + k/x$ has a point of inflection at $x = -1$.

93. Let f be a function defined by $f(x) = Ax^2 + Bx + C$ with the following properties: (i) $f(0) = 2$; (ii) $f'(2) = 10$; (iii) $f''(10) = 4$. Find the value of $A + B + C$.

94. Find the value of b for which the function $y = x^4 + bx^2 + 8x + 1$ has both a horizontal tangent line and a point of inflection at the same value of x.

95. Find c so that the line $y = 4x + 3$ is tangent to the curve $y = x^2 + c$.

96. A fly is crawling from left to right along the edge of a curved barrier on the xy-plane whose shape is the graph

of $y = 7 - x^2$. A spider waits at the point $(4, 0)$. Find the distance between the two creatures when they first see each other. [**HINT:** Draw a picture.]

97. The arithmetic mean of n numbers $0 \le a_1 \le a_2 \le \cdots \le a_n$ is $A_n = (a_1 + a_2 + \cdots + a_n)/n$; their geometric mean is $G_n = (a_1 \cdot a_2 \cdots a_n)^{1/n}$.
    (a) Use algebra to show that $G_2 \le A_2$. [**HINT:** $a_1 \cdot a_2 = \frac{1}{4}\left((a_1 + a_2)^2 - (a_1 - a_2)^2\right)$.]
    (b) Show that $G_2 \le A_2$. [**HINT:** Let $f(a_2) = (a_1 + a_2)/2 - \sqrt{a_1 a_2}$. Use the racetrack principle to show that $f(a_2) \ge 0$.]
    (c) Show that $G_n \le A_n$ when $n \ge 2$. [**HINT:** Show that $f(a_n) = A_n - G_n \ge 0$.]

98. Let m, n be positive integers, and a, b be real numbers.
    (a) Show that $x^n - a^n = (x - a)\left(x^{n-1} + x^{n-2}a + x^{n-3}a^2 + \cdots + xa^{n-2} + a^{n-1}\right)$ by multiplying out the expression on the right.
    (b) Let $f(x) = x^n$. Use part (a) to show that $f'(a) = \lim\limits_{x \to a} \dfrac{x^n - a^n}{x - a} = na^{n-1}$.
    (c) Let $g(x) = x^{-n}$. Use part (b) to show that $g'(a) = -na^{-n-1}$. [**HINT:** $\left(x^{-n} - a^{-n}\right)/(x - a) = (a^n - x^n)/\left(x^n a^n (x - a)\right)$.]
    (d) Let $h(x) = x^{m/n}$. Use part (a) to show that $h'(a) = \dfrac{m}{n} a^{(m-n)/n}$.
    $$\left[\textbf{HINT:} \quad \frac{x^{m/n} - a^{m/n}}{x - a} = \frac{\left(x^{1/n}\right)^m - \left(a^{1/n}\right)^m}{\left(x^{1/n}\right)^n - \left(a^{1/n}\right)^n}.\right]$$
    (e) Let $k(x) = x^{-m/n}$. Show that $k'(a) = -\dfrac{m}{n} a^{-(m+n)/n}$.

99. In the previous exercise you showed that if r is a rational number, then $\dfrac{d}{dx}(x^r) = rx^{r-1}$. In this exercise we extend this result to the case where r is a real number.
    Let r be a real number and p, q be rational numbers such that $0 < p < r < q$.
    (a) Assume that $x > 0$. Explain why
    $$\frac{x^p - 1}{x - 1} < \frac{x^r - 1}{x - 1} < \frac{x^q - 1}{x - 1}.$$
    [**HINT:** Consider the cases $0 < x < 1$ and $x > 1$ separately.]
    (b) Use part (a) to show that $p \le \lim\limits_{x \to 1} \dfrac{x^r - 1}{x - 1} \le q$.
    (c) Explain why part (b) implies that $\lim\limits_{x \to 1} \dfrac{x^r - 1}{x - 1} = r$.
    (d) Suppose that c is an arbitrary real number and $a \ne 0$. Show that
    $$\frac{x^c - a^c}{x - a} = \left(\frac{w^c - 1}{w - 1}\right) a^{c-1} = -w^c \left(\frac{w^{-c} - 1}{w - 1}\right) a^{c-1}.$$
    (e) Use parts (c) and (d) to show that $(x^c)' = cx^{c-1}$ where c is any real number.

3.2 Using Derivative and Antiderivative Formulas

Explicit formulas for derivatives and antiderivatives can be immensely useful when applying calculus to real problems. This section illustrates just a few of many such applications.

Designer Graphs

Using derivative formulas, we can fit graph fragments together to produce given shapes. (Doing so is useful, for instance, in computer-assisted drawing applications.)

EXAMPLE 1   The figure below shows a line segment and a piece of a parabola.

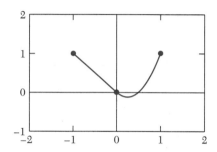

Bits of a line segment and a parabola

Find an equation for the parabola.

Solution   A parabola's equation has the form $p(x) = a + bx + cx^2$. We need to find numerical values for a, b, and c. The picture shows (at least) three features that we'll use to find the needed values:

i. The parabola passes through $(0, 0)$, so $p(0) = 0$. But $p(0) = a + b \cdot 0 + c \cdot 0^2 = a$, so $a = 0$. Thus, $p(x) = bx + cx^2$, with b and c still to be found.

Look closely to be sure you agree!

ii. At $x = 0$ the parabola and the line segment have the *same slope*. ◄   A look at the segment shows that this slope is -1. Therefore, $p'(0) = -1$. We'll use this information to find b. For any x, $p'(x) = b + 2cx$, so $p'(0) = b$. Therefore $b = -1$, and so $p(x) = -x + cx^2$. We still need to find c.

iii. The parabola passes through $(1, 1)$, so $p(1) = 1$. But $p(x) = -x + cx^2$, so $p(1) = -1 + c$; for $p(1) = 1$ to hold, we must have $c = 2$.

Try plotting both $p(x)$ and $y = -x$. Is the picture "right"?

Thus, $p(x) = -x + 2x^2$. ◄   ■

Modeling Motion: Acceleration, Velocity, and Position

A falling object has, at any moment, an acceleration a, a velocity v, and a height h. All three quantities are natural functions of time: If t denotes time (measured in seconds after some reference time $t = 0$) and height is measured in feet above some reference point, then

$$h'(t) = v(t) \qquad \text{and} \qquad v'(t) = a(t).$$

Equivalently, h is an antiderivative of v and v is an antiderivative of a.

According to Newton's laws of motion, an object in free fall, i.e, affected only by gravity, ▶ has a *constant* (vertical) acceleration, about -32 feet per second per second. In symbols: $a(t) = -32$. ▶

Since v is an antiderivative of a,

$$v(t) = -32t + C$$

for some constant C. Which constant? It depends on the setting—because $v(0) = C$, ▶ C represents the **initial velocity** (i.e., the velocity at time 0). Therefore, we write

$$v(t) = -32t + v_0.$$

Antidifferentiating again gives a formula for $h(t)$:

$$h(t) = -16t^2 + v_0 t + C.$$

Again $h(0) = C$, so we write

$$h(t) = -16t^2 + v_0 t + h_0,$$

where h_0 represents **initial height**.

See Section 2.1 for more on free fall.

The acceleration is negative because the object falls.

Plug $t = 0$ into the formula for t.

EXAMPLE 2   A cannonball is shot upward, with initial height 5 feet and initial velocity 288 feet per second. ▶ Assuming free-fall conditions, how high does the cannonball rise? How long does it remain airborne?

This translates to around 200 mph, which is possible for an old-fashioned cannon.

Solution   From the preceding information (with $v_0 = 288$ and $h_0 = 5$), the velocity and height functions have these formulas:

$$v(t) = -32t + 288 \qquad \text{and} \qquad h(t) = -16t^2 + 288t + 5.$$

The ball is highest when $v(t) = 0$; solving this equation gives $t = 9$ seconds. Thus, the cannonball rises to $h(9) = 1301$ feet; ▶ it stays airborne until $h(t) = 0$. Solving *this* equation gives $t \approx 18.02$ seconds. ■

About 1/4 mile.

Maximum–Minimum Problems and the Derivative

Derivative formulas, if they're simple enough, can make locating maxima and minima of f an easy matter. The key idea is that, for a differentiable function f, *local* maxima and minima can occur only at stationary points, i.e., where $f'(x) = 0$. In searching for the largest and smallest values of f on an interval, therefore, the only natural candidates are stationary points and, in some cases, endpoints. To put it more precisely:

Let f be differentiable on an interval I. Maximum and minimum values of f on I occur, if at all, only at stationary points of f or at endpoints (if any) of I.

Although this problem can be solved without calculus, with calculus it's easier.

EXAMPLE 3   Let $f(x) = x^2 - x$ and $I = [0, 2]$. Where on I does f take maximum and minimum values? ▶

Solution   Since $f'(x) = 2x - 1$, there's just one stationary point, at $x = 1/2$. The graph of f is a parabola opening upward, with vertex at $x = 1/2$. Therefore, for x in $[0, 2]$, f takes its minimum value at $x = 1/2$ (the stationary point of f) and its maximum value at $x = 2$ (the right endpoint of I).   ■

Real-World Maxima and Minima.   In real-life problems, the first—and often hardest—step is translation into mathematical language.

EXAMPLE 4   A gardener wants to enclose 1000 square feet of land, using as little fencing as possible. The garden is to be rectangular; one side, unfenced, lies along a river. How much fencing is needed? (**NOTE**: This problem and others like it appear in more detail in Appendix D.)

Solution   Here's the setup:

Let x and y denote the lengths shown. (By assumption, $xy = 1000$, so $y = 1000/x$.) If L denotes the total length of fence, then

$$L = 2x + y = 2x + \frac{1000}{x}.$$

We want to minimize L.

Plot your own.

A plot of L ◄ shows a minimum somewhere near $x = 20$. With the derivative we can locate this stationary point precisely:

$$L'(x) = 2 - \frac{1000}{x^2} = 0 \iff x = 10\sqrt{5} \approx 22.3607.$$

For this value of x, $y = 1000/x = 20\sqrt{5}$, and $L = 40\sqrt{5} \approx 89.44$. In agricultural terms: *The best dimensions for a garden, in feet, are $10\sqrt{5} \times 20\sqrt{5}$. The garden needs $40\sqrt{5}$ feet of fence.* (More realistically, the garden is about $22.5' \times 45'$ and needs about 90 feet of fence.)   ■

BASICS

Where is each of the following functions increasing? Decreasing? Find *all* local maxima and minima. (Express your answers exactly—e.g., as $\sqrt{3}$—rather than in decimal form.)

1. $f(x) = x^3 - 3x^2 + 5$
2. $f(x) = -8x^9 + 9x^8 - 7$
3. $g(x) = 7x^9 - 18x^7 + 63$
4. $h(x) = 5x^6 + 6x^5 - 15x^4$

In Exercises 5–8, find the maximum and the minimum value of the function over the given interval.

5. $f(x) = x^3 - 3x + 5$;   [0, 2]
6. $f(x) = 1 + 2x^3 - x^4$;   [-1, 2]
7. $f(x) = \sqrt{x} - 3/x$;   [0.25, 1]
8. $f(x) = x^3 - x^2$;   [-2, 0]
9. Find the maximum and the minimum value of $f(x) = x^4 - 3x^2 + 6$ over each interval below.
   (a) [-2, 2]   (c) [-1, 2]
   (b) [-4, -2]   (d) [-1, 1]
10. A particle moves on the x-axis in such a way that its position at time $t > 0$ is given by $x(t) = 3t^5 - 25t^3 + 60t$. For which values of t is the particle moving to the left?
11. A particle moves along a straight line so that its velocity is given by $v(t) = t^2$. How far does the particle travel between $t = 1$ and $t = 3$?

FURTHER EXERCISES

12. Find the maximum and minimum values of the derivative of $f(x) = 6x^2 - x^4$ on the interval $[-2, 2]$.
13. For which value(s) of x does the slope of the line tangent to the curve $y = -x^3 + 3x^2 + 1$ take on its largest value? Justify your answer.
14. Suppose that $q(x)$ is a polynomial of degree $n \geq 2$. Explain why q has at most $n - 1$ local extrema.
   [**HINT**: $q'(x)$ is a polynomial of degree $n - 1$.]
15. Let $f(x) = ax^3 + bx^2 + cx + d$.
   (a) Show that f has either no local extrema or a local maximum and a local minimum.
   (b) Show that if they exist, the local extrema of f are located at $x = \dfrac{-b \pm \sqrt{b^2 - 3ac}}{3a}$.
16. Is there a real number x such that $4x^3 - x^4 = 30$? Prove your answer.
17. Let p be the parabola defined by $p(x) = ax^2 + bx + c$ where a, b, and c are constants. Show that the vertex of the parabola is at $x = -b/2a$.
18. (a) Find constants a and b so that the polynomial $p(x) = x^3 + ax + b$ has a local minimum at $(2, -9)$.
   (b) Draw a graph of p.
19. (a) Find constants a, b, and c so that the quadratic polynomial $p(x) = ax^2 + bx + c$ is zero at $x = 1$, decreasing when $x < 2$, and increasing when $x > 2$.
   (b) Draw a graph of p.
20. (a) Find a cubic polynomial $q(x) = ax^3 + bx^2 + cx + d$ that has a local maximum at $(0, 2)$ and a local minimum at $(5, 0)$.
   (b) Draw a graph of q.
21. For which values of the constant k does the graph of $y = x^3 - 3x^2 + k$ have just one x-intercept? Two x-intercepts? Three x-intercepts? Justify your answers.
22. For which values of k, if any, does the function $f(x) = (8x + k)/x^2$ have a local minimum at $x = 4$?
23. The altitude $h(t)$ of a rocket from its moment of launch until its fall to Earth is given by the formula $h(t) =$ $-t^4 + 8t^3 + 375$. (Altitude is measured in feet above sea level; time from launch is measured in seconds.)
   (a) From what altitude is the rocket launched?
   (b) When does the rocket fall into the ocean?
   (c) What is the rocket's maximum altitude?
   (d) What is the rocket's maximum upward velocity?
   (e) What is the rocket's maximum downward velocity?
   (f) What is the rocket's maximum upward acceleration?
   (g) What is the rocket's maximum downward acceleration?
24. An object moves along a straight line in such a way that it is slowing down with an acceleration of -8 meters per second per second. The object has an initial velocity of 14 meters per second. Find the distance the object travels during the first 3 seconds.
25. Suppose that the brakes are applied on a car traveling 50 mph and that this gives the car a constant negative acceleration of 20 ft/sec^2. How long will it take the car to come to a stop? How far will the car travel before stopping?
26. A car traveling at a constant speed of 80 mph along a straight highway fails to stop at a stop sign. Three seconds later, a highway patrol car starts from a point of rest at the stop sign and maintains a constant acceleration of 8 ft/sec^2. How long will it take the patrol car to overtake the speeding automobile? How far from the stop sign will this occur? What is the speed of the patrol car when it overtakes the automobile?
27. Suppose that a rectangle has its base on the x-axis and its two upper corners on the curve $y = 2(1 - x^2)$.
   (a) What is the maximum perimeter of such a rectangle?
   (b) Show that there is no minimum perimeter rectangle.
      [**HINT**: The sides of a rectangle must have nonzero length.]
28. Find the shortest distance from the curve $xy = 4$ to the origin.   [**HINT**: Minimize the *square* of the distance function.]

29. What is the area of the largest rectangle whose base is on the x-axis and whose upper corners are on the curve $y = 12 - x^2$?

30. The cost of fuel for a luxury liner is proportional to the square of the speed v. At a speed of 10 km/hour, fuel costs \$3000/hour. Other fixed costs (e.g., labor) amount to \$12,000/hour. Assuming that the ship is to make a trip of total distance D, find the speed that minimizes the cost of the trip. Does this speed depend on D?

31. An open box is made by cutting squares of side w inches from the four corners of a sheet of cardboard that is 24 inches by 32 inches and then folding up the sides. What should w be to maximize the volume of the box?

32. An open box with a capacity of 36,000 cubic inches is to be twice as long as it is wide. The material for the box costs \$0.10 per square foot. What are the dimensions of the least expensive box? How much does it cost?

33. Coffee mugs can be manufactured for 60¢ each. At a price of \$1 each, 1000 can be sold. For each penny the price is lowered, 50 more mugs can be sold. What price maximizes profit? How many mugs are sold at this price?

3.3 Derivatives of Exponential and Logarithm Functions

Algebraic versus Transcendental Functions

Using the limit definition to find the derivative of a function with an algebraic formula can be messy, but it's usually routine.

EXAMPLE 1 If $f(x) = \sqrt{x + 1}$, find $f'(x)$.

Solution Here's the limit computation:

$$
\begin{aligned}
f'(x) &= \lim_{h \to 0} \frac{f(x + h) - f(x)}{h} \\
&= \lim_{h \to 0} \frac{\sqrt{x + h + 1} - \sqrt{x + 1}}{h} \\
&= \lim_{h \to 0} \frac{\sqrt{x + h + 1} - \sqrt{x + 1}}{h} \cdot \frac{\sqrt{x + h + 1} + \sqrt{x + 1}}{\sqrt{x + h + 1} + \sqrt{x + 1}} \\
&= \lim_{h \to 0} \frac{(x + h + 1) - (x + 1)}{h \cdot (\sqrt{x + h + 1} + \sqrt{x + 1})} \\
&= \lim_{h \to 0} \frac{1}{\sqrt{x + h + 1} + \sqrt{x + 1}} \\
&= \frac{1}{2\sqrt{x + 1}}.
\end{aligned}
$$

We used an algebraic trick in the third line; otherwise the calculation was straight-forward. ◄ ∎

Was this computation really necessary? Could we have used, instead, the (known) derivative of $\sqrt{x}$?

Nonalgebraic functions—those without algebraic formulas—are called **transcendental**. Exponential, logarithm, and trigonometric functions are all transcendental.

Finding derivatives of transcendental functions poses a special challenge. Without algebraic formulas, the limits that define derivatives usually don't succumb to the usual algebraic tricks. We'll need other methods. In another sense, transcendental functions are like all others: Given the graph of any function f, we can graphically estimate the behavior of f'. For each of the following functions, we begin with a graphical approach.

Assuming the Basics. Throughout this section we assume—usually without comment—basic properties of the exponential and logarithm functions. For quick reviews, see Appendices E and F.

Derivatives of Exponential Functions: Graphical Evidence

The Natural Exponential Function

In Example 6, page 112, we conjectured, from graphical evidence, this remarkable property of the natural exponential function: *If $f(x) = e^x$, then $f'(x) = e^x$.*

Does something similar hold for exponential functions with other bases? Here are graphs of several:

Graphs of three exponentials

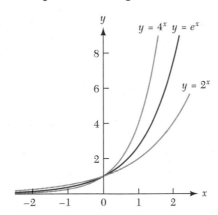

Observe these features.

   Similar Slope Behavior   Slopes of all three graphs behave similarly. As we move from left to right on any graph, slopes start small (but positive) and increase through larger and larger positive values.

   Common Point, Different Slopes   The three graphs cross—but have different slopes—at $x = 0$. The e^x-graph has slope 1 at $x = 0$; hence, at $x = 0$ the graphs of $y = 2^x$ and $y = 4^x$ have slopes respectively less than and greater than 1.

Derivatives at $x = 0$: A Closer Look. All graphs of the form $y = b^x$ pass through the point $(0, 1)$—but with different slopes. Zooming in on this point shows shows how these slopes (i.e., the respective derivatives at $x = 0$) differ from each other:

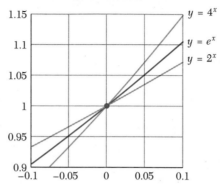

Graphs of three exponentials: a closer look

Notice:

- At this magnification, each exponential curve looks like a straight line—namely, its tangent line at $x = 0$.

Make these estimates!

- Eyeball estimates of the slopes of these lines ◄ say:

$$\left.\frac{d}{dx}2^x\right|_{x=0} \approx 0.7; \qquad \left.\frac{d}{dx}e^x\right|_{x=0} \approx 1; \qquad \left.\frac{d}{dx}4^x\right|_{x=0} \approx 1.4.$$

The middle estimate is crucial. The property of having slope 1 at the origin is sometimes used to define the base e. ◄

There are other ways to define e; we'll see one later.

> **Definition** The number e is the base for which the curve $y = b^x$ passes through $(0, 1)$ with slope 1.

Derivatives of Exponentials: Again, Analytically

With graphical evidence as intuition, we're ready to sharpen our knowledge of derivatives of exponential functions, using the limit definition of derivative. We'll show:

> **Theorem 5** If $f(x) = e^x$, then $f'(x) = e^x$. If $g(x) = b^x$, then $g'(x) = b^x \ln b$.

Before plunging into calculations, let's observe some facts.

- The first formula is a special case of the second. (Set $b = e$.)
- The second formula says that, for any positive base b, $g'(x) = kg(x)$, where $k = \ln b$ is a constant. Thus, every exponential function g has this important property:

  The instantaneous rate of change of g is proportional to the value of g.

  Many real-life varying quantities (populations, undisturbed bank balances, masses of radioactive minerals, comprehensive fees at St. Lena College, etc.) exhibit this property. That's why exponential functions are essential tools in mathematical modeling.
- If $g(x) = 2^x$, then $g'(x) = 2^x \ln 2 \approx 0.693 \cdot 2^x$. If $h(x) = 4^x$, then $h'(x) = 4^x \ln 4 \approx 1.386 \cdot 4^x$. This result agrees with what we saw graphically.

Calculating the Derivative of b^x

Let $g(x) = b^x$. We'll show that $g'(x) = b^x \ln b$.

Step 1: Why $g'(x) = kg(x)$.   The computation starts simply:

$$g'(x) = \lim_{h \to 0} \frac{g(x+h) - g(x)}{h}$$

$$= \lim_{h \to 0} \frac{b^{x+h} - b^x}{h} = \lim_{h \to 0} b^x \cdot \frac{b^h - 1}{h}$$

$$= b^x \cdot \lim_{h \to 0} \frac{b^h - 1}{h} = b^x \cdot k.$$

Two remarks are needed:

- Pulling b^x outside the limit on the last line was legal because the limit depends on h, not on x.
- If the last limit—we called it k—exists, then we've shown what we claimed in Step 1: $(b^x)' = kb^x$.

Step 2: Why k Exists.   Does the limit k exist? What's k's numerical value?
    The key is to interpret the limit k as the slope of the line tangent to the g-graph at $x = 0$. Since $g(x) = b^x$,

$$k = \lim_{h \to 0} \frac{b^h - 1}{h} = \lim_{h \to 0} \frac{b^h - b^0}{h} = \lim_{h \to 0} \frac{g(h) - g(0)}{h}.$$

The last form is familiar by now—it's the *definition* of the slope of the tangent line at $x = 0$. We saw graphically that such tangent lines exist; their slopes depend on b.

Step 3: Why $k = \ln b$.   We'll show in a moment that the numerical value of k (i.e., the slope of the line tangent to $y = b^x$ at $x = 0$) is $\ln b$. First, let's see some numerical evidence for the case $b = 10$.

Our claim, numerically, is that

$$\lim_{h \to 0} \frac{b^h - 1}{h} = \ln 10 \approx 2.3026.$$

A machine did the work. Let's tabulate values of the limit expression for small inputs h: ◀

The slope at $x = 0$ of $y = 10^x$—numerical estimates						
h	-0.01	-0.001	-0.0001	0.0001	0.001	0.01
$(10^h - 1)/h$	2.2763	2.2999	2.3023	2.3029	2.3052	2.3293

The numbers for $b = 10$ look convincing. To prove that $k = \ln b$ for *any* positive base b, we'll reduce the situation to the best understood case, $b = e$. A key

See Appendix E for more details.

relationship holds ◀ between $f(x) = e^x$ and $g(x) = b^x$:

$$g(x) = b^x = e^{x \ln b} = f\big((\ln b) \cdot x\big).$$

In graphical language, g is obtained from f by a *horizontal compression*, with factor

We studied horizontal squashing and stretching in Section 1.2.

$\ln b$. ◀ By assumption, the f-graph has slope 1 at $x = 0$. The g-graph, therefore, has slope $\ln b$ at $x = 0$. We're finished.

Derivatives of Logarithm Functions

To find the derivative of the logarithm function $f(x) = \log_b x$, we'll use what we know about its inverse $f^{-1}(x) = b^x$. Here, in advance, are our conclusions. Notice again the "naturality" of e:

T h e o r e m 6 For any positive base $b \neq 1$,

$$\left(\log_b x \right)' = \frac{1}{x} \cdot \frac{1}{\ln b}.$$

If $b = e$, then

$$\left(\ln x \right)' = \frac{1}{x}.$$

We'll show these facts geometrically; a formal proof would amount only to translating everything into symbolic form.

A Picture "Proof"

As we saw, for example, in Section 1.5.

The key idea is that, for any base $b \neq 1$, the functions b^x and $\log_b x$ are *inverses*. Geometrically, this relationship means ◀ that either graph can be obtained from the other by reflection through the line $y = x$. The following figure, which shows the case $b = e$, illustrates a crucial point: how such a reflection affects tangent lines.

Inverse functions and their derivatives

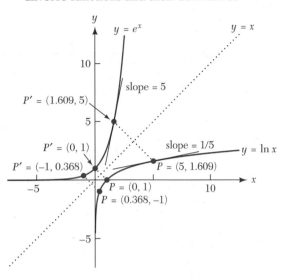

Notice the following features of the graph.

Symmetric Points   A point $P = (x, y)$ lies on one graph if and only if the point $P' = (y, x)$ lies on the other. Several such pairs are shown.

Symmetric Tangent Lines   Consider the lines tangent to the two graphs at corresponding points P and P'. (One such pair of tangent segments is shown.) Like the graphs themselves, these tangent lines are symmetric with respect to the line $y = x$. It follows ▶ that *the slopes of the tangent lines at P and P' are reciprocals.*   *This is the key fact!*

Reciprocal Slopes   The slope of the tangent line at $P' = (\ln a, a)$ is the derivative of the function $y = e^x$ evaluated at $x = \ln a$. We saw earlier that $(e^x)' = e^x$. Therefore, the tangent line at P' has slope $e^{\ln a} = a$. It follows that the tangent line at $P = (a, \ln a)$ has slope $1/a$, just as the theorem predicts.

So far, we've explained that the theorem holds if $b = e$. For *any* base b, $\log_b x = \ln x / \ln b$. Applying the constant rule for derivatives gives the general derivative formula.

Antiderivatives of Exponential and Logarithm Functions

The derivative rules for exponential functions are readily reversed, so the following theorem closely resembles Theorem 5.

> **Theorem 7**   Let C be any constant and b any positive base.   If $f(x) = e^x$, then $F(x) = e^x + C$ is an antiderivative of f. If $g(x) = b^x$, then $G(x) = \dfrac{b^x}{\ln b} + C$ is an antiderivative of g.

Notice:

For exactly the same reason!

Its Own Antiderivative, Naturally The natural exponential function e^x is not only its own derivative but its own antiderivative. ◀

Other Bases The second antiderivative formula is best checked by differentiation. Notice how we use the constant rule:

$$\left(\frac{b^x}{\ln b} + C \right)' = \frac{1}{\ln b} \cdot \left(b^x \right)' = \frac{1}{\ln b} \cdot b^x \ln b = b^x.$$

Antiderivatives of logarithm functions—even the natural logarithm—are a little more complicated. For the moment, we'll just *state* a formula.

> **Fact** Let C be any constant. If $f(x) = \ln x$, then $F(x) = x \ln x - x + C$ is an antiderivative of f.

This claim needs some explanation.

Rabbit from a Hat? The preceding antiderivative formula seems to have been plucked from thin air. Where did it come from? Why is it right? *Is it right?* Tools to be developed soon will let us show—by differentiation— that the formula is correct. ◀ Showing how one might *produce* such a formula will have to wait a little longer.

Also see the exercises at the end of this section.

Graphical Evidence Formal proofs must wait, but graphical evidence for the antiderivative formula is available right now. Here's some:

Antiderivatives of $y = \ln x$

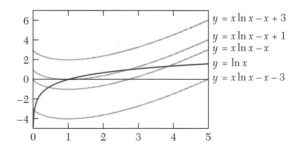

The picture shows graphs of $y = x \ln x - x + C$ for various values of C. ◀ Each such graph looks plausibly like an antiderivative of $y = \ln x$ (bold), as it should.

The antiderivatives differ from each other only by a vertical shift; all have the same derivative.

EXAMPLE 2   Find an antiderivative of $f(x) = 3e^x - 2 \ln x$. How many are there?

We're using the sum and constant rules for derivatives.

Solution   The theorems give antiderivatives of e^x and $\ln x$. Combining them appropriately ◀ tells us that $3e^x - 2 \cdot (x \ln x - x)$ is one suitable antiderivative. So, therefore, is any function of the form

$$F(x) = 3e^x - 2 \cdot (x \ln x - x) + C,$$

where C is any constant. ∎

A Sample Problem

Here's a problem with ingredients from this section.

EXAMPLE 3   Let $f(x) = e^x = \exp(x)$. Find a linear function L and a quadratic function Q that closely approximate f near $x = 0$, as follows:

a. Find constants a and b so that the linear function $L(x) = a + bx$ has the same value and first derivative as f at $x = 0$—i.e., so that $L(0) = f(0)$ and $L'(0) = f'(0)$.

b. Find constants a, b, and c for which the quadratic function $Q(x) = a + bx + cx^2$ has the same value and first *two* derivatives as f at $x = 0$—i.e., so that $Q(0) = f(0)$, $Q'(0) = f'(0)$, and $Q''(0) = f''(0)$.

c. Plot f, L, and Q on the same axes. How closely do L and Q approximate f on the interval $[-0.5, 0.5]$?

Solution   Observe for later reference that $f(x) = f'(x) = f''(x)$; in particular, $f(0) = f'(0) = f''(0) = 1$.

a. Let $L(x) = a + bx$. We need values for a and b. Note first that $L(0) = a$. To have $L(0) = f(0) = 1$, we must have $a = 1$. Similarly, $L'(x) = b$. To have $L'(0) = b = f'(0) = 1$, we must have $b = 1$. Thus, $L(x) = 1 + x$.

b. Let $Q(x) = a + bx + cx^2$. We need values for a, b, and c. As in part (a),

$$Q(0) = f(0) = 1 \quad \Longrightarrow \quad a = 1;$$

$$Q'(0) = f(0) = 1 \quad \Longrightarrow \quad b = 1.$$

Thus, $Q(x) = 1 + x + cx^2$, and we need only find c. Since $Q''(x) = 2c$, ▶   *Check this!*
we have

$$Q''(0) = 2c = f''(0) = 1 \Longrightarrow c = 1/2.$$

Thus, $Q(x) = 1 + x + x^2/2$.

c. Here are graphs of all three functions (the exponential function is bold):

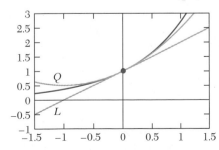

Two approximations to $y = e^x$

Use a calculator to check these numbers.

The picture shows that near $x = 0$, both L and Q approximate f well—Q even better than L. A close look reveals that, for x in the interval $[-0.5, 0.5]$, ◄

$$|f(x) - L(x)| \le f(0.5) - L(0.5) \approx 0.149;$$

$$|f(x) - Q(x)| \le f(0.5) - Q(0.5) \approx 0.024. \qquad \blacksquare$$

BASICS

For each function f in Exercises 1–15, find an algebraic expression for f'. Then plot graphs of the expressions for f and f' on the same axes to check your answer for reasonableness.

1. $f(x) = 2e^x + \pi$
2. $f(x) = e^5 - 4e^x$
3. $f(x) = e^x + x^e + e$
4. $f(x) = 2e^{x+1}$
5. $f(x) = 2^x + x^2 + 2$
6. $f(x) = 3^x$
7. $f(x) = 2 \cdot 3^{x-1}$
8. $f(x) = e^\pi + \pi^x + x^\pi$

9. $f(x) = (\ln 5)\, e^x$
10. $f(x) = (\ln 5)^x$
11. $f(x) = x^{\ln 5}$
12. $f(x) = (\ln 5)\, 5^x$
13. $f(x) = (\ln 5)\, x^5$
14. $f(x) = -2 \ln x$
15. $f(x) = 3 \log_2 x$

16–30. Find the second derivative of each function in Exercises 1–15. Then sketch graphs of the expressions for f and f'' on the same axes to check your answer for reasonableness.

31–45. Find an antiderivative of each function in Exercises 1–15.

46. What is the slope of the curve $y = 3^x$ at $x = 0$?

47. What is the slope of the curve $y = \log_3 x$ at $x = 1$?

48. On January 1, 1950, the population of Boomtown was 50,000. Since then the size of the population has been accurately modeled by the function $P(t) = 50{,}000(0.98)^t$,

where t is the number of years since January 1, 1950.

(a) What was the population of Boomtown on January 1, 1990?

(b) At what rate was the population of Boomtown changing on January 1, 1990? Was it increasing or decreasing then?

49. The position of a particle on the x-axis at time $t > 0$ seconds, is $x(t) = \ln t$ meters.

(a) Find the average velocity of the particle over the interval $1 \le t \le e$.

(b) Find the instantaneous velocity of the particle at time $t = e$.

50. Find an equation of the line tangent to the curve $y = e^x$ at $x = 0$.

51. Let $f(x) = \sqrt{x} - \ln x$.

(a) Show that f achieves its (global) minimum value at $x = 4$.

(b) Show that f has an inflection point at $x = 16$.

52. Let $f(x) = \ln x - e^{x-1}$.

(a) Show that f achieves its (global) maximum value when $x = 1$.   [**HINT:** $e^{x-1} = e^x / e$.]

(b) Show that f is concave down everywhere on its domain.

FURTHER EXERCISES

53. Let f, L, and Q be as in Example 3.

(a) Find the value of the constant d for which the cubic function $C(x) = Q(x) + dx^3$ has the properties $C(0) = f(0)$, $C'(0) = f'(0)$, $C''(0) = f''(0)$, and $C'''(0) = f'''(0)$.

(b) Plot f, L, Q, and C on the same axes over the interval $[-3, 3]$.

(c) Find an interval over which $|f(x) - L(x)| < 0.1$.

(d) Find an interval over which $|f(x) - Q(x)| < 0.1$.

(e) Find an interval over which $|f(x) - C(x)| < 0.1$.

54. Let $f(x) = \ln x$.

(a) Find values of the constants a and b so that the linear function $L(x) = a + b(x - 1)$ has the properties $L(1) = f(1)$ and $L'(1) = f'(1)$.

(b) Find the value of the constant c for which the quadratic function $Q(x) = L(x) + c(x - 1)^2$ has the properties $Q(1) = f(1)$, $Q'(1) = f'(1)$, and $Q''(1) = f''(1)$.

(c) Find the value of the constant d for which the cubic function $C(x) = Q(x) + d(x - 1)^3$ has the properties

$C(1) = f(1)$, $C'(1) = f'(1)$, $C''(1) = f''(1)$, and $C'''(1) = f'''(1)$.

(d) Graph f, L, Q, and C on the same axes over the interval $[0.1, 3]$.

(e) Find an interval over which $|f(x) - L(x)| < 0.1$.

(f) Find an interval over which $|f(x) - Q(x)| < 0.1$.

(g) Find an interval over which $|f(x) - C(x)| < 0.1$.

55. Which of the following functions are solutions of the differential equation $f''(x) = f(x)$? Which are solutions of $f'''(x) = f'(x)$?

(i) $f(x) = \sqrt{3}$

(ii) $f(x) = x + 2$

(iii) $f(x) = e^x$

(iv) $f(x) = 2^x$

(v) $f(x) = \ln x$

56. Let n be a positive integer.

(a) Find an expression for $\dfrac{d^n}{dx^n} e^x$.

(b) Find an expression for $\dfrac{d^n}{dx^n} \ln x$.

57. Suppose that $b > 0$. Find an antiderivative of $\log_b x$. [**HINT:** Use the Fact on p. 204.]

58. (a) Use the Fact on p. 204 to show that $(x \ln x)' = 1 + \ln x$.
    (b) Suppose that $b > 0$. Evaluate $(x \log_b x)'$.

59. Let $f(x) = \ln(kx)$ where k is a positive constant.
    (a) Show that $f'(x) = x^{-1}$. [**HINT:** $\ln(kx) = \ln k + \ln x$.]
    (b) Find an antiderivative of f.

60. Let $f(x) = b^x$.
    (a) Find all values of $b > 0$ such that $f'(0) > 0$.
    (b) Find all values of $b > 0$ such that $f'(1) > f(1)$.

61. Let $g(x) = \log_b x$.
    (a) Find all values of $b > 0$ such that $g'(1) > 1$.
    (b) Find all values of $b > 0$ such that $g'(1) > g(1)$.

62. Let $f(x) = e^{-x} = \exp(-x)$. Show that $f'(x) = -e^{-x}$. [**HINT:** $e^{-x} = (1/e)^x$.]

63. Find the derivative of $f(x) = \ln(1/x)$. [**HINT:** $\ln(1/x) = -\ln x$.]

64. Let $f(x) = 2e^{x+3}$.
    (a) Explain how a graph of f can be obtained from a graph of the natural exponential function using *only* vertical stretching.
    (b) Use part (a) to find a symbolic expression for $f'(x)$.

65. Let $f(x) = 3e^{x+2}$.
    (a) Explain how a graph of f can be obtained from a graph of the natural exponential function using *only* horizontal translation.
    (b) Use part (a) to find a symbolic expression for $f'(x)$.

66. Find the x-intercept of the line tangent to the curve $y = 2^x$ at $x = 0$. [**HINT:** Draw a picture.]

67. Does there exist a real number a such that the line tangent to the curve $y = e^x$ at $x = a$ passes through the origin? If so, find it. If not, explain why there is no number with this property.

68. Does there exist a real number a such that the line tangent to the curve $y = x - \ln x$ at $x = a$ passes through the origin? If so, find it. If not, explain why there is no number with this property.

69. Give an example of a function f with the property that $f'(x) \neq f(x)$ but $f''(x) = f'(x)$.

70. Let $f(x) = Ae^x$ and $g(x) = e^{x+k}$ where A and k are constants.
    (a) Show (by differentiation) that $f'(x) = f(x)$.
    (b) Show (by differentiation) that $g'(x) = g(x)$.
    (c) Find the value of A that makes $f(0) = 3$.
    (d) Find the value of k that makes $g(0) = 3$.
    (e) Show that if $A > 0$, there is a value of k such that $f(x) = g(x)$ for all x.
    (f) Show that if $A < 0$, there is a value of k such that $f(x) = -g(x)$ for all x.
    (g) Find a function h such that $h'(x) = h(x)$ and $h(0) = -2$.

71. Let $f(x) = e^{kx} = \exp(kx)$ and $g(x) = Af(x)$ where A and k are constants.

(a) Explain why $f(x) = b^x$ for some positive number b. What is b?
(b) Use part (a) and Theorem 5 to show that $f'(x) = ke^{kx} = kf(x)$.
(c) Use part (b) to show that $g'(x) = kg(x)$.
(d) Find a function h such that $h'(x) = 3h(x)$ and $h(0) = 2$.
(e) Find a function h such that $h'(x) = -2h(x)$ and $h(0) = 3$.

72. (a) Let A be a constant. Show that $y(x) = 5 + Ae^x$ is a solution of the differential equation $y' = y - 5$.
    (b) Find the solution of the differential equation $y' = y - 5$ with the property $y(0) = 6$.
    (c) Find the solution of the differential equation $y' = y - 5$ that passes through the point $(1, 0)$.

73. Let $f(x) = \ln|x|$, $g(x) = x^{-1}$, and
$$h(x) = \begin{cases} -2 + \ln x & \text{if } x > 0 \\ 3 + \ln(-x) & \text{if } x < 0. \end{cases}$$
    (a) What is the domain of f? Of g?
    (b) Plot graphs of f, g, and h on the same set of axes.
    (c) Explain why f is an antiderivative of g. [**HINT:** f is an even function.]
    (d) Show that h is an antiderivative of g (i.e., show that $h'(x) = g(x)$ for all $x \neq 0$).

74. Let $f(x) = e^{kx}$ and $g(x) = e^{-kx}$; k is a constant.
    (a) Show that $f''(x) = k^2 f(x)$ and $g''(x) = k^2 g(x)$.
    (b) Let $h(x) = Af(x) + Bg(x)$ where A and B are any real numbers. Show that $h''(x) = k^2 h(x)$.
    (c) Let h be the function defined in part (b). Choose values of A and B so that $h''(x) = 4h(x)$, $h(0) = 2$, and $h'(0) = 8$.

75. Using the result of the previous exercise find a function h such that $h''(x) = 4h(x)$ for all x, $h'(0) = 4$, and $h(0) = 4$.

76. Suppose that A is a constant. Verify that $x(t) = 1 + t + Ae^t$ is a solution of the differential equation $x' = x - t$.

77. Suppose that A is a constant. Verify that $y(x) = Ae^x - x - 1$ is a solution of the differential equation $yy' = y^2 + xy$.

78. Suppose that A and B are constants. Verify that $y(x) = A \ln x + B + x$ is a solution of the differential equation $xy'' + y' = 1$.

79. Suppose that A and B are constants. Verify that $y(x) = Ax^{-1} + B + \ln x$ is a solution of the differential equation $x^2 y'' + 2xy' = 1$.

80. Let $f(x) = x - e \ln x$.
    (a) Prove that $f(\pi) > 0$. [**HINT:** Show that f has a global minimum at $x = e$.]
    (b) Use part (a) to show that $e^\pi > \pi^e$.

81. (a) Show that the distance between the point $(0, 0)$ and a point (a, b) on the curve $y = e^{x/2}$ is least when a is the solution of the equation $2a + e^a = 0$. [**HINT:** Minimize the square of the distance.]
    (b) Estimate the distance in part (a).

82. Let $f(x) = 2^x$ and $g(x) = 7x/3 + 1$. What is the slope of the line tangent to the curve $y = f(x)$ at the point in $[0, 3]$ where the vertical distance between f and g is the greatest?

83. Find the area of the smallest rectangle that has one side on the positive x-axis, one side on the negative y-axis, a corner at the origin, and a corner on the curve $y = \ln x$. [**HINT**: Use Exercise 58.]

84. (a) Find the minimum value of $f(x) = e^x - 1 - x$.
    (b) Explain why part (a) shows that $e^x \geq 1 + x$ for all x.
    (c) Use the racetrack principle to show that $e^x \geq 1 + x$ for all x.   [**HINT**: Consider the cases $x > 0$ and $x < 0$ separately.]
    (d) Show that $e^x \geq 1 + x + x^2/2$ for all $x \geq 0$.

85. (a) Show that $(1 + 1/n)^n \leq e$ for any integer $n \geq 1$. [**HINT**: Let $x = 1/n$ and use part (b) of the previous exercise.]
    (b) Show that $e \leq (1 + 1/n)^{n+1}$ for any integer $n \geq 1$. [**HINT**: Proceed as in part (a) but with $x = -1/(n + 1)$.]
    (c) Let $a_n = (1 + 1/n)^n$, $b_n = (1 + 1/n)^{n+1}$. Show that $\lim_{n \to \infty}(b_n - a_n) = 0$ [**NOTE**: It is a fact that $\lim_{n \to \infty}(1 + 1/n)^n = e$, but we don't yet have the tools necessary to prove this fact.]

86. (a) Show that $e^x > 2x$ for all $x \geq 1$.   [**HINT**: $e^x > 2$ for all $x \geq 1$.]
    (b) Use part (a) to show that $e^x > x^2$ for all $x \geq 1$.

87. Show that $1 - 1/x \leq \ln x \leq x - 1$ when $x > 0$. [**HINT**: Consider the cases $x \geq 1$ and $0 < x \leq 1$ separately.]

88. (a) Use the racetrack principle to show that $0 \leq \ln x \leq \sqrt{x} - 1/\sqrt{x}$ if $x \geq 1$.
    (b) Show that $\sqrt{x} - 1/\sqrt{x} \leq \ln x \leq 0$ if $0 < x \leq 1$.
    (c) Use part (b) to show that $\lim_{x \to 0^+} x \ln x = 0$.

89. (a) Prove that $\lim_{x \to \infty} \ln x/x = 0$.   [**HINT**: Use the previous exercise.]
    (b) Prove that $\lim_{x \to \infty} \ln x/\sqrt{x} = 0$.   [**HINT**: $\ln x/\sqrt{x} = 2\ln\left(\sqrt{x}\right)/\sqrt{x}$.]
    (c) Prove that if r is a positive real number, then $\lim_{x \to \infty} \ln x/x^r = 0$.

90. In this exercise we will prove that $\lim_{x \to \infty} x^r/e^x = 0$ for any real number r.
    (a) Prove that $2 \ln x - x < 0$ for all $x > 0$.
    (b) Use part (a) to prove that $\lim_{x \to \infty}(r \ln x - x) = -\infty$ if $r \geq 1$.   [**HINT**: If $r \geq 1$, then $r^2 \geq r$.]
    (c) Use part (b) to prove that $\lim_{x \to \infty} x^r/e^x = 0$ if $r \geq 1$.   [**HINT**: $x^r = e^{r \ln x}$.]
    (d) Use part (c) to prove that $\lim_{x \to \infty} x^r/e^x = 0$ if $r < 1$.

91. (a) Assume that $x \geq 1$. Use the inequality $\ln x \leq x - 1$ to show that $x \ln x - x \leq x^2/2 - x - 1/2$. [**HINT**: Use Exercise 58 and the racetrack principle.]
    (b) Prove that $(x - 1)/2 \leq \ln x$ if $1 \leq x \leq 3$. [**HINT**: Show that $f(x) = \ln x - (x - 1)/2$ is decreasing on the interval $[1, 2]$ and increasing on the interval $[2, 3]$.]

(c) Use the inequality in part (b) to find a quadratic polynomial p with the property that $p(x) \leq x \ln x - x$ if $1 \leq x \leq 3$.

92. (a) Use the fact that $1/x \leq 1$ if $x \geq 1$ to show that $\ln x \leq x - 1$ if $x \geq 1$.
    (b) Use the fact that $1/x^2 \geq 1/x$ if $0 < x \leq 1$ to show that $1 - 1/x \leq \ln x$ if $0 < x \leq 1$.

93. (a) Assume that $x \geq 1$. Explain how the relation $\ln x \leq 2(\sqrt{x} - 1)$ can be derived from the relation $1/x \leq 1/\sqrt{x}$.
    (b) Assume that $0 < x \leq 1$. Explain how the relation $\ln x \geq 2(1 - 1/\sqrt{x})$ can be derived from the relation $1/x \leq 1/x^{3/2}$.

94. The derivative of a logarithmic function $f(x) = \log_b x$ is proportional to $1/x$. In fact, the constant of proportionality is $f'(1)$—the slope of the line tangent to the graph of f at $x = 1$. Justify each step in the following derivation of this result.   [**HINT**: In line (iv), $w = h/x$.]

(i) $\dfrac{d}{dx} \log_b x = \lim\limits_{h \to 0} \dfrac{\log_b(x + h) - \log_b x}{h}$

(ii) $= \lim\limits_{h \to 0} \dfrac{\log_b\left(\dfrac{x + h}{x}\right)}{h}$

(iii) $= \lim\limits_{h \to 0} \dfrac{\log_b\left(1 + \dfrac{h}{x}\right)}{h}$

(iv) $= \lim\limits_{w \to 0} \dfrac{\log_b(1 + w)}{xw}$

(v) $= \dfrac{1}{x}\left(\lim\limits_{w \to 0} \dfrac{\log_b(1 + w)}{w}\right)$

(vi) $= \dfrac{f'(1)}{x}$.

95. The equation $a^x = 1 + x$ has at least one solution in the interval $[0, \infty)$ for every $a > 0$. For which values of $a > 0$ does this equation have a nonzero solution? For these values of a, how many nonzero solutions are there?

96. Find all the solutions of the equation $2^x = 2x$. Prove that you have found all the solutions.   [**HINT**: Use the racetrack principle.]

97. The equation $b^x = bx$ has at least one solution in the interval $[0, \infty)$ for every $b > 0$. For which values of $b > 0$ does this equation have more than one solution?

Evaluate the limits in Exercises 98–101 by recognizing each as the definition of a derivative.

98. $\lim\limits_{t \to 0} \dfrac{e^t - 1}{t}$

99. $\lim\limits_{x \to 0} \dfrac{2^{3+x} - 2^3}{x}$

100. $\lim\limits_{s \to 0} \dfrac{\log_5(25 + s) - 2}{s}$

101. $\lim\limits_{z \to e} \dfrac{\ln z - 1}{z - e}$

102. Evaluate $\lim\limits_{x \to \pi} \dfrac{e^x - e^\pi}{x^e - \pi^e}$.

3.4 Derivatives of Trigonometric Functions

We've already studied the trigonometric functions and some of their derivatives informally, using graphical techniques. In this section we take a more systematic point of view. (To review trigonometric basics, see Appendix G.)

Derivatives of the Sine and Cosine Functions—Graphical Evidence

Graphs suggest almost irresistibly that $(\sin x)' = \cos x$. ▶

We've already succumbed to the temptation, e.g., in Section 2.3.

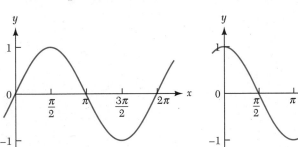

Graph of $f(x) = \sin x$ **Graph of $f'(x) = \cos x$**

It's equally tempting to guess that $(\cos x)' = -\sin x$, as the following graphs show.

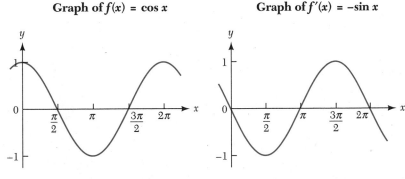

Graph of $f(x) = \cos x$ **Graph of $f'(x) = -\sin x$**

Convincing as they are, graphs aren't proofs. Rigor requires that we deal directly with the limit that defines the derivative.

Derivatives of the Sine and Cosine Functions—Analytic Proofs

Differentiating the Sine Function. To show rigorously that $(\sin x)' = \cos x$, we need—somehow—to calculate the limit

$$\frac{dy}{dx} = \lim_{h \to 0} \frac{\sin(x+h) - \sin(x)}{h}.$$

It appears in Appendix G.

Algebra isn't enough. We need the addition formula for sines: ◄

$$\sin(x + h) = \sin(x)\cos(h) + \cos(x)\sin(h).$$

Check each step.

Now we can do some algebra: ◄

$$\lim_{h \to 0} \frac{\sin(x + h) - \sin(x)}{h} = \lim_{h \to 0} \frac{(\sin(x)\cos(h) + \cos(x)\sin(h)) - \sin(x)}{h}$$

$$= \lim_{h \to 0} \left((\sin x) \cdot \frac{\cos h - 1}{h} + (\cos x) \cdot \frac{\sin h}{h} \right)$$

$$= (\sin x) \cdot \lim_{h \to 0} \frac{\cos h - 1}{h} + (\cos x) \cdot \lim_{h \to 0} \frac{\sin h}{h}.$$

They depend on x, not h; that's why we could pull them outside the limits.

"Nontrivial" is the mathematician's euphemism for "hard."

(As far as the limit is concerned, $\sin x$ and $\cos x$ are constants. ◄   )

To finish the computation we need two nontrivial ◄ facts:

$$\lim_{h \to 0} \frac{\cos h - 1}{h} = 0 \quad \text{and} \quad \lim_{h \to 0} \frac{\sin h}{h} = 1.$$

If we assume these facts for the moment, the conclusion follows easily:

$$\frac{d}{dx}\sin x = (\sin x) \cdot \lim_{h \to 0} \frac{\cos h - 1}{h} + (\cos x) \cdot \lim_{h \to 0} \frac{\sin h}{h}$$

$$= (\sin x) \cdot 0 + (\cos x) \cdot 1$$

$$= \cos x.$$

Two Challenging Limits.   Each of the limits in the preceding calculation can be interpreted as a certain derivative.

- The limit

$$\lim_{h \to 0} \frac{\cos h - 1}{h} = \lim_{h \to 0} \frac{\cos h - \cos 0}{h - 0}$$

defines the derivative of the cosine function at $x = 0$. The cosine graph is horizontal there, so the limit is zero.

- A similar argument applies to

$$\lim_{h \to 0} \frac{\sin h}{h} = \lim_{h \to 0} \frac{\sin h - \sin 0}{h - 0}.$$

The limit *defines* the derivative of the sine function at $x = 0$. In Example 4, page 152, we saw convincing graphical and numerical evidence that this derivative is 1. For a more rigorous symbolic proof, see Appendix H.

Differentiating the Cosine Function.   To show that $(\cos x)' = -\sin x$, we could mimic the argument just given for the sine function. Instead, we'll deduce the derivative from what we know already about the sine function.

As their graphs show, the sine and cosine functions differ by a horizontal translation of $\pi/2$ units. In symbols:

$$\cos x = \sin(x + \pi/2) \quad \text{and} \quad \cos(x + \pi/2) = -\sin x.$$

It follows from these facts ▶   that the cosine graph has the same slope at x as the   *Think about them!*
sine graph has at $x + \pi/2$. Using the preceding identities and the fact—which we
just showed—that $(\sin x)' = \cos x$, we can express all this symbolically:

$$\frac{d}{dx}\cos x = \frac{d}{dx}\sin(x + \pi/2) = \cos(x + \pi/2) = -\sin x.$$

Derivatives of Other Trigonometric Functions

Each of the remaining trigonometric functions (tangent, secant, cotangent and cosecant) is an algebraic combination of the sine and cosine functions. Not surprisingly,
so is each derivative. With ingenious algebra we could calculate, from scratch, the
derivative of each of these functions. It's more efficient, though, to wait ▶   until we   *The wait will be short. In*
develop explicit rules for derivatives of products and quotients. To allay unbearable   *the meantime, see the*
suspense, here's a graphical peek at what's to come: *exercises at the end of this*
section.

Graph of $f(x) = \tan x$ **Graph of $f'(x) = \sec^2 x$**

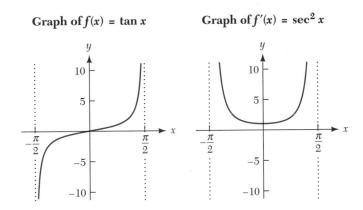

As it should, the second graph represents the slope behavior of the first. It amounts
to circumstantial evidence—but only that—for the fact that $(\tan x)' = \sec^2 x$.

Antiderivatives of Sines and Cosines

With just a little care for signs, ▶   our derivative formulas can be reversed to give   *That's signs, not sines.*
antiderivatives. Here are the details. They're easy to check by differentiation.

> **T h e o r e m   8**   Let C be any constant.   If $f(x) = \sin x$, then
> $F(x) = -\cos x + C$ is an antiderivative of f. If $g(x) = \cos x$, then
> $G(x) = \sin x + C$ is an antiderivative of g.

E X A M P L E   1   Find an antiderivative of $f(x) = 3\sin x - 4\cos x$. How many
antiderivatives are there?

We're using the sum and constant rules.

Solution   Theorem 8 gives antiderivatives of $\sin x$ and $\cos x$. Combining them appropriately ◄   tells us that $-3\cos x - 4\sin x$ is one suitable antiderivative. So, therefore, is *any* function of the form

$$F(x) = -3\cos x - 4\sin x + C,$$

Plot both f and F. Does the picture show what it should?

where C is any constant. ◄   ∎

A Sample Problem

Here's a problem with ingredients developed in this section.

EXAMPLE 2   Here's a graph of $f(x) = x + \cos x$. Two interesting points are bulleted.

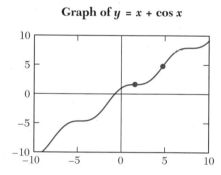

Graph of $y = x + \cos x$

Are there any local maxima or minima? Are there any inflection points?

Would zooming help? If a plotting device is at hand, see for yourself.

Solution   The graph appears to have flat spots, such as at the left bullet, but are they local maxima, local minima, or just flat spots? The picture alone doesn't tell us for sure. ◄

With derivatives we can settle the question easily. Since $f(x) = x + \cos x$, $f'(x) = 1 - \sin x$ and $f''(x) = -\cos x$. Thus,

$$f'(x) = 0 \iff \sin x = 1 \iff x = \frac{\pi}{2} + 2k\pi;$$

$$f''(x) = 0 \iff \cos x = 0 \iff x = \frac{\pi}{2} + k\pi.$$

In particular, the first bulleted point is both a stationary point and an inflection point. In addition, f has stationary points at intervals of 2π and inflection points at intervals of π. (The right bullet is another inflection point.)

Notice, too, that $f'(x) = 1 - \sin x \geq 0$ for all x. Thus, the first derivative never changes sign, and the stationary points are nothing but that—they are neither local maxima nor local minima.   ∎

Further Transcendental Meditations (Optional)

Hyperbolic Functions.   Exponential, logarithmic, and trigonometric functions are not the only useful transcendental functions. Others include the **hyperbolic sine** and **hyperbolic cosine** functions, denoted by **sinh** and **cosh**, respectively, and defined by

$$\sinh(x) = \frac{e^x - e^{-x}}{2} \qquad \text{and} \qquad \cosh(x) = \frac{e^x + e^{-x}}{2}.$$

Hyperbolic functions arise naturally in various engineering applications. Under appropriate conditions, for instance, the cosh function models the shape of a hanging cable. ▶

Plot cosh *on your calculator to see what this means.*

As the definitions show, sinh and cosh are near relatives of the natural exponential function. Less obvious but equally important are links between hyperbolic and trigonometric functions. For example, sinh and cosh—just like sin and cos—can be combined to form such other hyperbolic functions as tanh and sech. More details on hyperbolic functions appear in the exercises at the end of this section.

Recognizing Transcendental Functions.   Proving rigorously that a function is transcendental can be difficult. Consider the sine function, for example. We defined the sine function using the unit circle, not any algebraic formula. Nevertheless, the sine function might conceivably have some algebraic formula.

In fact, the sine function has *no* algebraic formula, and neither does any other trigonometric, logarithm, or exponential function. The following observations don't *prove* that the sine function is transcendental, but they give some idea of what's involved.

- Could $\sin x = p(x)$ for some polynomial $p(x)$? No. A polynomial of degree n has at most n roots; the sine function has infinitely many roots.

- Could the sine function be a rational function? That is, could $\sin x = p(x)/q(x)$ for some polynomials p and q? No again, for the same reason. A rational function has only finitely many roots.

- The sine function is 2π-periodic. Of the raw algebraic functions—powers, roots, and so on—none is periodic. Creating a periodic function from nonperiodic components would be difficult.

BASICS

For each function f in Exercises 1–4, find an algebraic expression for f', then plot graphs of the expressions for f and f' on the same axes to check your answer for reasonableness.

1. $f(x) = 2 \sin x$
2. $f(x) = 3 \cos x$
3. $f(x) = 3 \sin x - 2 \cos x$
4. $f(x) = 2 \cos x - 3 \sin x$

5–8. Find the second derivative of each function in Exercises 1–4. Then sketch graphs of the expressions for f and f'' on the same axes to check your answer for reasonableness.

9–12. Find an antiderivative of each function in Exercises 1–4.

13. Let $f(x) = x - \cos x$.
    (a) Does f have any stationary points in the interval $[0, 2\pi]$? If so, where?
    (b) Where in the interval $[0, 2\pi]$ is f increasing?
    (c) Find the maximum and minimum values of f over the interval $[0, \pi]$.
    (d) Where is f concave down?
    (e) Where is f increasing most rapidly? Evaluate f' there.

14. Repeat the previous exercise with $f(x) = x + \cos x$.

In Exercises 15–20, draw a graph of the function over the interval $[0, 2\pi]$. Label local extrema, global extrema, and points of inflection with both coordinates. (Use exact values rather than decimal approximations for the coordinates.)

15. $f(x) = \dfrac{1}{2}x + \sin x$

16. $g(x) = x + \sin x$

17. $h(x) = \dfrac{3}{2}x + \sin x$

18. $f(x) = \sin x + \cos x$

19. $g(x) = 2\sin x - \sqrt{3}x$

20. $h(x) = \sqrt{3}\sin x + 3\cos x$

FURTHER EXERCISES

21. Let $f(x) = \sin x$.
    (a) Find values of the constants a and b so that the linear function $L(x) = a + bx$ has the properties $L(0) = f(0)$ and $L'(0) = f'(0)$.
    (b) Find the value of the constant c for which the quadratic function $Q(x) = L(x) + cx^2$ has the properties $Q(0) = f(0)$, $Q'(0) = f'(0)$, and $Q''(0) = f''(0)$.
    (c) Find the value of the constant d for which the cubic function $C(x) = Q(x) + dx^3$ has the properties $C(0) = f(0)$, $C'(0) = f'(0)$, $C''(0) = f''(0)$, and $C'''(0) = f'''(0)$.
    (d) Plot f, L, Q, and C on the same axes over the interval $[-4, 4]$.
    (e) Find an interval over which $|f(x) - L(x)| < 0.1$.
    (f) Find an interval over which $|f(x) - Q(x)| < 0.1$.
    (g) Find an interval over which $|f(x) - C(x)| < 0.1$.

22. Repeat the previous problem with $f(x) = \cos x$.

23. The range R of a projectile whose speed is v meters per second and whose angle of elevation is x radians is given by $R = \left(v^2/g\right)\sin(2x)$ where $g \approx 9.8$ meters per second per second is the acceleration due to gravity.
    (a) What angle of elevation gives the projectile maximum range? Justify your answer.
    (b) What is the maximum range of the projectile?

24. Let k be a real number. Find a trigonometric identity for $\cos(x + k)$ by differentiating both sides of the identity $\sin(x + k) = \sin x \cos k + \cos x \sin k$.

25. Let $f(x) = \cos x + \frac{1}{2}\cos(2x)$.
    (a) Determine where f has local maxima and minima over the interval $[0, 5]$.
    (b) Estimate where f has inflection points in the interval $[0, 5]$.

26. Repeat the previous problem using the function $f(x) = \sin x - \frac{1}{2}\cos(2x)$.

27. Find a function that satisfies each set of conditions below.
    (a) $f'(z) = \cos(2z)$, $f(\pi/2) = 1$
    (b) $f'(x) = 2x^3 + \sin(4x)$, $f(0) = 1$
    (c) $g'(w) = \sin(w + 2)$, $g(0) = 3$

28. Let $f(x) = \cos(kx)$ and $g(x) = \sin(kx)$; k is a constant.
    (a) Show that $f''(x) = -k^2 f(x)$ and $g''(x) = -k^2 g(x)$.
    (b) Let $h(x) = Af(x) + Bg(x)$, where A and B are any real numbers. Show that $h''(x) = -k^2 h(x)$.
    (c) Let h be the function defined in part (b). Choose values of A and B so that $h''(x) = -4h(x)$, $h(0) = 2$, and $h'(0) = 8$.

29. Let $f(x) = A\cos x + B\sin x$ where A and B are constants.
    (a) Show that $f''(x) = -f(x)$.
    (b) Show that $f^{(4)}(x) = f(x)$.
    (c) Find a function f such that $f''(x) = -f(x)$, $f'(\pi/3) = -1$, and $f(\pi/3) = \sqrt{3}$.

30. Let $f(x) = \sin(x^2)$. Show that $f'(x) \neq \cos(x^2)$ in a way that your classmates would find convincing. [**HINT:** Compare graphs.]

31. Let $f(x) = 2 - \sin x$ and $g(x) = 1 + \cos x$.
    (a) Find the maximum vertical distance between the curves $y = f(x)$ and $y = g(x)$ over the interval $[0, 2\pi]$.
    (b) Find the slopes of the lines tangent to f and g at the point in the interval $[0, 2\pi]$ where the vertical distance between f and g is greatest.

32. Let $f(x) = \cos x$ and $g(x) = \sqrt{3}\sin x$.
    (a) Find the maximum vertical distance between the curves $y = f(x)$ and $y = g(x)$ over the interval $[0, \pi]$.
    (b) Find the slopes of the lines tangent to f and g at the point in the interval $[0, \pi]$ where the the vertical distance between f and g is greatest.

33. Suppose that two sides of a rectangle are on the coordinate axes and that a corner of the rectangle lies on the curve $y = 2\cos x$, $0 \leq x \leq \pi/2$. Is it possible that the perimeter of the rectangle is $3\pi/2$? If so, find the coordinates of the corners of a rectangle with these properties. If not, find the coordinates of the corners of the rectangle with the greatest possible perimeter.

34. Suppose that one side of a rectangle is on the x-axis and that two corners of the rectangle lie on the curve $y = 2\sqrt{2}\sin x$, $0 \leq x \leq \pi/2$. Among all such rectangles, find the perimeter of the one with maximum perimeter.

35. Suppose that two sides of a rectangle are on the coordinate axes and that a corner of the rectangle lies on the curve $y = k\cos x$, where $\pi/3 \leq k \leq \pi/2$.
    (a) Show that k can be chosen so that the rectangle with maximum perimeter is a square.
    (b) Estimate the value of k in part (a).

36. (a) What is the 50th derivative of $f(x) = \cos x$?
    (b) What is the 51st derivative of $g(x) = \sin x$?
    (c) What is the 52nd derivative of $g(x) = \sin x$?

37. Show that $\dfrac{d^n}{dx^n}\sin x = \sin(x + n\pi/2)$ when n is an integer such that $n \geq 1$. [**HINTS:** $\sin(x + \pi/2) = \cos x$; $\sin(x + \pi) = -\sin x$.]

38. Suppose that the line tangent to the curve $y = \sin x$ at $x = z$ passes through the origin. Show that $\tan z = z$.

39. (a) Use the fact that $(\tan x)' = \sec^2 x$ to show that $x = 0$ is the only solution of the equation $x = \tan x$ in the interval $(-\pi/2, \pi/2)$.
    (b) Suppose that $k \geq 1$ is an integer. Explain why the equation $x = \tan x$ has exactly one solution between $k\pi$ and $(2k+1)\pi/2$.

40. Show that $\cos x + \sin x \leq 1 + x$ for all $x \geq 0$.

41. Show that $2 + \sin x - \cos x \leq e^x$ if $x \geq 0$.
    [**HINT:** Use the previous exercise and the fact that $1 + x \leq e^x$ for all $x \geq 0$.]

42. Let $f(x) = \sin x$.
    (a) Find an equation of the line tangent to f at $x = 0$.
    (b) Find an equation of the line tangent to f at $x = \pi/3$.
    (c) Which of the tangent lines in parts (a) and (b) yields a better approximation of $\sin(\pi/6)$? How could you have used f'' to predict which tangent line would yield the better approximation?

43. Let $f(x) = 2x/\pi$.
    (a) Prove that $f(x) \leq \sin x$ when $0 \leq x \leq \pi/2$.
    (b) Show that the function $g(x) = \sin x - f(x)$ achieves its maximum value over the interval $[0, \pi/2]$ at the point where the slope of the line tangent to the sine function is $2/\pi$.

44. (a) Use the inequality $\cos x \leq 1$ to prove that $\sin x \leq x$ for all $x \geq 0$.
    (b) Use part (a) to show that $1 - x^2/2 \leq \cos x$ for all $x \geq 0$.
    (c) Explain why the inequality in part (b) is true for all x. [**HINT:** $\cos x$ is an even function.]

45. (a) Use the result in part (b) of the previous exercise to prove that $x - x^3/6 \leq \sin x$ if $x \geq 0$.
    (b) Use part (a) to prove that $\cos x \leq 1 - x^2/2 + x^4/24$.
    (c) Prove that $\sin x \leq x - x^3/6 + x^5/120$ if $x \geq 0$.

Evaluate the limits in Exercises 46–49 by recognizing each as the definition of a derivative.

46. $\displaystyle\lim_{x \to \pi/2}\dfrac{\cos x}{x - \pi/2}$

47. $\displaystyle\lim_{x \to \pi}\dfrac{\sin x}{x - \pi}$

48. $\displaystyle\lim_{h \to 0}\dfrac{\sin(3\pi/2 + h) + 1}{h}$

49. $\displaystyle\lim_{h \to 0}\dfrac{2\cos(5\pi/4 + h) + \sqrt{2}}{h}$

50. Evaluate $\displaystyle\lim_{h \to 0}\dfrac{\cos(x - h) - \cos(x + h)}{h}$.

51. Let $f(x) = \sin(x + k)$ where k is a constant.
    (a) Use the limit definition of the derivative to show that $f'(x) = \cos(x + k)$.
    (b) Interpret the result in part (a) graphically. [**HINT:** How is the graph of f related to the graph of the sine function?]

(c) Let $g(x) = \cos x$. Use the result in part (a) to show that $g'(x) = -\sin x$. [**HINT:** $\cos x = \sin(x + \pi/2)$.]

(d) Let $g(x) = \cos(x + k)$. Use the result in part (a) to show that $g'(x) = -\sin(x + k)$. [**HINT:** $\cos(x + k) = \sin(x + (k + \pi/2))$.]

(e) Let $h(x) = 3\sin(x - 2)$. Use the results in parts (a) and (d) to find $h'(x)$, $h''(x)$, and an antiderivative of h.

52. (a) Let $f(x) = \sin(kx)$ where k is a constant. Provide a justification for each step in the following proof that $f'(x) = k\cos(kx)$. [**HINT:** In line (iii), $w = kh$.]

(i) $\displaystyle\frac{d}{dx}\sin(kx) = \lim_{h \to 0}\frac{\sin(k(x + h)) - \sin(kx)}{h}$

(ii) $\displaystyle= \lim_{h \to 0}\frac{\sin(kx + kh) - \sin(kx)}{h}$

(iii) $\displaystyle= \lim_{h \to 0}k \cdot \frac{\sin(kx + w) - \sin(kx)}{w}$

(iv) $\displaystyle= \lim_{w \to 0}k \cdot \frac{\sin(kx + w) - \sin(kx)}{w}$

(v) $\displaystyle= k \cdot \left(\lim_{w \to 0}\frac{\sin(kx + w) - \sin(kx)}{w}\right)$

(vi) $\displaystyle= k\cos(kx).$

(b) Use the result in part (a) to find the derivative of $g(x) = 3\sin(2x)$.

53. Let $f(x) = \cos(kx)$ where k is a constant.
    (a) Use the limit definition of the derivative to prove that $f'(x) = -k\sin(kx)$.
    (b) Use the result in part (a) to find the derivative of $g(x) = 2\cos(3x)$.

54. Let $f(x) = \tan x$. Justify each step in the following proof that $f'(x) = \sec^2 x$.

(i) $\displaystyle\frac{d}{dx}\tan x = \lim_{h \to 0}\frac{\tan(x + h) - \tan x}{h}$

(ii) $\displaystyle= \lim_{h \to 0}\frac{\dfrac{\sin(x + h)}{\cos(x + h)} - \dfrac{\sin x}{\cos x}}{h}$

(iii) $\displaystyle= \lim_{h \to 0}\frac{\cos x \sin(x + h) - \sin x \cos(x + h)}{h\big(\cos x \cos(x + h)\big)}$

(iv) $\displaystyle= \lim_{h \to 0}\frac{\cos(-x)\sin(x + h) + \sin(-x)\cos(x + h)}{h\big(\cos x \cos(x + h)\big)}$

(v) $\displaystyle= \lim_{h \to 0}\frac{\sin h}{h\big(\cos x \cos(x + h)\big)}$

(vi) $\displaystyle= \lim_{h \to 0}\left(\frac{\sin h}{h} \cdot \frac{1}{\cos x \cos(x + h)}\right)$

(vii) $\displaystyle= \frac{1}{\cos^2 x}$

(viii) $\displaystyle= \sec^2 x.$

55. Let $f(x) = \cot x$. Use the limit definition of the derivative to show that $f'(x) = -\csc^2 x$.

56. Let $f(x) = \sec x$. Use the limit definition of the derivative to show that $f'(x) = \sec x \tan x$.

57. Let $f(x) = \csc x$. Use the limit definition of the derivative to show that $f'(x) = -\csc x \cot x$.

58. (a) Show (algebraically) that the hyperbolic cosine (cosh) is an even function.
    (b) Is the hyperbolic sine function (sinh) even, odd, or neither? Justify your answer.

59. (a) Show that $\left(\sinh x\right)' = \cosh x$.
    (b) Show that $\left(\cosh x\right)' = \sinh x$.

60. (a) Plot the hyperbola defined by the equation $x^2 - y^2 = 1$.
    (b) Show that $\cosh^2 x - \sinh^2 x = 1$.

61. Show that $\cosh(x + y) = \sinh x \cosh y + \cosh x \sinh y$.

62. Show that $\sinh(x + y) = \cosh x \cosh y + \sinh x \sinh y$.

63. Show that $(\cosh x + \sinh x)^k = \cosh(kx) + \sinh(kx)$ for any number k.

64. Suppose that $\sinh x = 4/5$. Evaluate $\cosh x$.

65. (a) Show that $y = \sinh x$ satisfies the differential equation $y'' = y$, and the initial conditions $y(0) = 1$ and $y'(0) = 1$.
    (b) Show that $y = \cosh x$ satisfies the differential equation $y'' = y$, and the initial conditions $y(0) = 1$ and $y'(0) = 0$.

66. The hyperbolic tangent function *tanh* is defined by the equation $\tanh x = \sinh x / \cosh x$.
    (a) Plot $\tanh x$.
    (b) Suppose that $\tanh x = 3/4$. Evaluate $\cosh x$ and $\sinh x$.

3.5 New Derivatives from Old: The Product and Quotient Rules

The theme of producing new functions from old runs throughout calculus. In Section 1.6 we discussed various operations—algebraic operations, composition, inversion, and so on—that build new functions from old. In this section and the next, we'll see how *derivatives* of these new functions are related to derivatives of the old functions.

The derivative of a built-up function f reflects (not surprisingly) both the *materials* from which f is built and the *operations* used to combine them. We've already seen this principle in action in simple cases.

EXAMPLE 1　　Let $f(x) = x^2 + 3 \sin x$. What's $f'(x)$? Why?

We discussed them in Section 3.1.

Solution　　According to the sum and constant rules, ◀

$$f'(x) = \left(x^2 + 3 \cdot \sin x\right)' = \left(x^2\right)' + 3 \cdot (\sin x)' = 2x + 3 \cos x.$$

If we let $g(x) = x^2$ and $h(x) = \sin x$, $f(x) = g(x) + 3 \cdot h(x)$, so the same result looks like this:

$$f'(x) = \left(g(x) + 3 \cdot h(x)\right)' = g'(x) + 3 \cdot h'(x).$$

In this (especially simple) case, f' is built from g' and h' *exactly* as f is built from g and h. ∎

Not every situation is so simple. Derivatives of products and quotients of functions need more care than sums and constant multiples. Sadly, the "naïve" rules for products and quotients—$(fg)' = f'g'$ and $(f/g)' = f'/g'$—are false.

EXAMPLE 2 Show by examples that the naïve product and quotient rules are false.

Solution Almost any functions f and g will do. If, say, $f(x) = x$ and $g(x) = 1$, then $f(x)g(x) = x$ and $f(x)/g(x) = x$. But

$$\big(f(x) \cdot g(x)\big)' = \big(x\big)' = 1 \neq f'(x) \cdot g'(x) = 1 \cdot 0 = 0.$$

The "false quotient rule" also produces nonsense:

$$\left(\frac{f(x)}{g(x)}\right)' = (x)' = 1 \neq \frac{f'(x)}{g'(x)} = \frac{1}{0}. \qquad\blacksquare$$

The (True) Product and Quotient Rules

Following, stated as theorems, are the *true* product and quotient rules. Before considering *why* they hold, we'll illustrate how to use them.

In each theorem, u and v are differentiable functions. ▶

For some reason, it's traditional in calculus to use u and v (not f and g) in this context. We won't buck tradition.

Theorem 9 (**The Product Rule**) If $p(x) = u(x) \cdot v(x)$, then

$$p'(x) = \big(u(x) \cdot v(x)\big)' = u'(x) \cdot v(x) + u(x) \cdot v'(x).$$

Equivalently, $(uv)' = u'v + uv'$.

Theorem 10 (**The Quotient Rule**) If $q(x) = u(x)/v(x)$, then

$$q'(x) = \left(\frac{u(x)}{v(x)}\right)' = \frac{v(x)u'(x) - u(x)v'(x)}{v(x)^2}.$$

Equivalently, $(u/v)' = \dfrac{vu' - uv'}{v^2}$.

EXAMPLE 3 (**Antiderivatives of ln x**) In Section 3.3 we claimed, but didn't prove, ▶ that functions of the form $F(x) = x \ln x - x + C$ are antiderivatives of the natural logarithm function. Prove it now.

We gave graphical evidence. See page 204.

Solution With the product rule, nothing could be easier. Watch:

$$(x \ln x - x + C)' = 1 \cdot \ln x + x\,\frac{1}{x} - 1 = \ln x,$$

just as claimed. $\qquad\blacksquare$

EXAMPLE 4 Let $p(x) = x^2 \sin x$ and $q(x) = x^2/\sin x$. Find p' and q'.

Solution In the preceding notation, $u(x) = x^2$ and $v(x) = \sin x$. Thus, $u' = 2x$ and $v' = \cos x$. The rest is "pattern-matching":

$$p'(x) = (uv)' = u'v + uv' = 2x \sin x + x^2 \cos x;$$

$$q'(x) = (u/v)' = \frac{vu' - uv'}{v^2} = \frac{(\sin x)2x - x^2 \cos x}{(\sin x)^2}.$$

(Some algebraic simplification of answers is possible, but since our main point concerns the derivatives, we'll leave these answers as they are.) ∎

EXAMPLE 5   The exponential function is its own derivative. What's the derivative of $r(x) = \exp(-x)$.

Solution   Recall some basic exponential algebra:

$$r(x) = \exp(-x) = e^{-x} = \frac{1}{e^x}.$$

We leave some algebraic details to you.

Now use the quotient rule: ◄

$$r'(x) = \left(\frac{1}{\exp(x)}\right)' = \frac{-e^x}{(e^x)^2} = -\frac{1}{e^x} = -e^{-x} = -\exp(-x).$$

∎

EXAMPLE 6   Let f be a differentiable function; let $g = 1/f$. Find a formula for g' in terms of f and f'.

Solution   We *could* apply the quotient rule to $1/f$, but we'll take another tack. First notice that $1 = g(x)f(x)$ for all sensible inputs x. Differentiating both sides of this equation ◄ gives

Use the product rule on the right.

$$0 = g'f + gf' = g'f + \frac{f'}{f}.$$

Solving for g' gives the answer:

$$\left(\frac{1}{f}\right)' = g' = -\frac{f'}{f^2}.$$

∎

Derivatives of the "Other" Trigonometric Functions

In Section 3.4 we showed carefully that $(\sin x)' = \cos x$ and $(\cos x)' = -\sin x$. The quotient rule makes easy work of finding derivatives of the remaining trigonometric functions. We'll do one such computation, leaving the rest as exercises.

EXAMPLE 7   Find and simplify the derivative of $\tan x$.

Solution   Use the quotient rule:

$$(\tan x)' = \left(\frac{\sin x}{\cos x}\right)' = \frac{\cos^2 x + \sin^2 x}{\cos^2 x} = \frac{1}{\cos^2 x}.$$

To avoid fractions, the answer is usually written as $\sec^2 x$. ∎

The full story of trigonometric derivatives fits snugly in a table:

Derivatives of Trigonometric Functions					
Function	$\sin x$	$\cos x$	$\tan x$	$\cot x$	$\sec x$	$\csc x$
Derivative	$\cos x$	$-\sin x$	$\sec^2 x$	$-\csc^2 x$	$\sec x \tan x$	$-\csc x \cot x$

Evidence for the Product Rule

A formal proof of the product rule ▶ involves difference quotients. Here we'll give two less formal plausibility arguments.

One appears in Appendix H.

Exhibit A: It Works for Linear Functions. Suppose that $f(x) = ax + b$ and $g(x) = cx + d$. We'll differentiate $f \cdot g$ twice—with and without the product rule.

First, notice that

$$f(x) \cdot g(x) = (ax + b) \cdot (cx + d) = acx^2 + (ad + bc)x + bd.$$

The last expression is a polynomial in x. Differentiating it directly gives

$$(fg)' = 2acx + (ad + bc).$$

Happily, the product rule agrees:

$$(fg)' = f'g + fg' = a \cdot (cx + d) + (ax + b) \cdot c = 2acx + ad + bc.$$

These computations show that the product rule gives the right answer for *linear* functions f and g. Most functions, of course, aren't linear. However, every differentiable function is "almost linear" in the sense that its graph looks straight if viewed at sufficiently small scale. ▶

A rigorous proof of the product rule can be constructed from this idea.

Exhibit B: It Works in Practice. Both U.S. population and U.S. per-capita income vary with time; we'll denote them by $p(t)$ and $i(t)$ respectively. The product function $w(t) = p(t)i(t)$ describes the *total annual U.S. income*—another function of t. Let's ask:

How fast was total annual income growing at $t = 1990$?

In 1990, the population of the United States was about 250 million and was increasing at the rate of about 3 million per year. Per-capita income was about $15,000 and was growing at about $1000 per year. In symbols:

$$p(1990) = 250 \text{ million};\qquad p'(1990) = 3 \text{ million};$$
$$i(1990) = 15{,}000;\qquad i'(1990) = 1000.$$

To answer the question, observe first that growth in w comes from two sources: rising population and rising per-capita income. Over a year, rising population produces about 3 million new people, each earning about $15,000, or a total of about

$$p'(1990) \cdot i(1990) = 3 \text{ million} \cdot 15{,}000 = 45 \text{ billion}$$

dollars of new income. Similarly, rising personal income accounts for about

$$p(1990) \cdot i'(1990) = 250 \text{ million} \cdot 1000 = 250 \text{ billion}$$

dollars of new income. In all, therefore, income rises at the rate of

$$p'(1990) \cdot i(1990) + p(1990) \cdot i'(1990) = 45 \text{ billion} + 250 \text{ billion}$$
$$= 295 \text{ billion}$$

dollars per year—just as the product rule predicts.

Evidence for the Quotient Rule

Quotients are similar to products; in the same sense, the quotient rule follows directly from the product rule.

EXAMPLE 8   Deduce the quotient rule from the product rule. (Assume that the derivative of the quotient exists.) ◄

It does exist; showing that it does requires a proof similar to the one for the product rule in Appendix H.

The function q is defined wherever $v(x) \neq 0$.

Use the product rule.

Solution   Let u and v be differentiable functions, and let $q(x) = u(x)/v(x)$ be the quotient function. ◄   We want $q'(x)$.

By definition, $qv = u$. Differentiating both sides ◄ of this equation gives

$$u' = (qv)' = q'v + qv' = q'v + \frac{u}{v}v' = u'.$$

Solving this last equation for q' gives the quotient rule:

$$q' = \frac{vu' - uv'}{v^2}$$ ■

Mixed Examples

Functions built from *several* products and quotients need several applications of the rules.

EXAMPLE 9   Let u, v, and w be differentiable functions. Find a formula for $(uvw)'$.

Solution   Used carefully, the product rule is enough:

$$
\begin{aligned}
(uvw)' &= ((uv) \cdot w)' & \text{[by grouping]} \\
&= (uv)' \cdot w + (uv) \cdot w' & \text{[product rule]} \\
&= (u'v + uv') \cdot w + uvw' & \text{[product rule]} \\
&= u'vw + uv'w + uvw'.
\end{aligned}
$$

Notice the pleasing symmetry of the result: Each term involves one derivative.   ■

EXAMPLE 10   Differentiate $x\,e^x \sin x$ and $x/(e^x \sin x)$.

Solution   By the rule just derived,

$$\left(x\,e^x \sin x \right)' = 1 \cdot e^x \cdot \sin x + x \cdot e^x \cdot \sin x + x \cdot e^x \cdot \cos x.$$

The second expression requires the quotient and product rules:

$$
\begin{aligned}
\left(\frac{x}{e^x \sin x} \right)' &= \frac{(e^x \sin x)(x)' - x(e^x \sin x)'}{(e^x \sin x)^2} \\
&= \frac{(e^x \sin x) - x(e^x \sin x + e^x \cos x)}{(e^x \sin x)^2} & \text{[product rule]} \\
&= \frac{(\sin x)(1 - x) - x \cos x}{e^x \sin^2 x} & \text{[simplified].}
\end{aligned}
$$ ■

Sometimes other ingredients, such as constants, are involved.

EXAMPLE 11   Differentiate $e^{2x}\sin(3x)$.

Solution   In the preceding section we saw how to handle constants:

$$\left(e^{2x}\right)' = 2e^{2x}; \qquad \left(\sin(3x)\right)' = 3\cos(3x).$$

Now, by the product rule,

$$\left(e^{2x}\sin(3x)\right)' = 2e^{2x}\sin(3x) + 3\cos(3x)e^{2x}. \qquad \blacksquare$$

Guessing and Checking Antiderivatives: A Sample Problem

Finding derivatives is relatively easy. With techniques from this section, we can handle a great many functions. Finding *antiderivatives* is often harder. Making a guess—and then checking it—is often an effective technique.

EXAMPLE 12   Find antiderivatives for

$$f(x) = x^2\cos x + 2x\sin x \quad \text{and} \quad g(x) = \frac{x\cos x - \sin x}{x^2}.$$

Solution   The basic forms of f and g suggest the results of applying the product and quotient rules, respectively. With this in mind, it's natural to guess

$$F(x) = x^2\sin x + C \quad \text{and} \quad G(x) = \frac{\sin x}{x} + C$$

as possible antiderivatives. ▶ With the product and quotient rules, it's easy to check our guesses. Just as we hoped,

In each, C stands for any constant.

$$F'(x) = (x^2\sin x)' = x^2\cos x + 2x\sin x;$$

$$G'(x) = \left(\frac{\sin x}{x}\right)' = \frac{x\cos x - \sin x}{x^2}. \qquad \blacksquare$$

BASICS

In Exercises 1–4 find the derivative of the function in two ways: (1) using the product rule, and (2) by multiplying out before differentiating.

1. $h(x) = x^3(x^2 - 4)$
2. $h(x) = (x + 4)^2$

3. $h(x) = e^x \cdot e^{2x}$
4. $h(x) = \tan x \cdot \csc x$

In Exercises 5–8, find the derivative of the function in two ways: (1) using the quotient rule, and (2) algebraically simplifying, then differentiating.

5. $h(x) = \dfrac{x^2}{x^3}$

6. $h(x) = \dfrac{1 - x^2}{1 + x}$

7. $h(x) = \dfrac{1}{e^x}$

8. $h(x) = \dfrac{\cot x}{\csc x}$

9. Let $h(x) = f(x) \cdot g(x)$, and $j(x) = f(x)/g(x)$. Fill in the missing entries in the table below using the information about f and g given in the following table and the definitions of h and j.

x	$f(x)$	$f'(x)$	$g(x)$	$g'(x)$	$h'(x)$	$j'(x)$
-2	1	-1	-3	4		$-1/9$
-1	0	-2	1	1	-2	
0	-1	2	-2	1		
1	2	-2	-1	2	6	
2	3	-1	2	-2		1

In Exercises 10–25, find an algebraic expression for f', then plot graphs of the expressions for f and f' on the same axes to check your answer for reasonableness.

10. $f(x) = x \sin x$

11. $f(x) = x^2 \cos x$

12. $f(x) = \sqrt{x} \ln x$

13. $f(x) = x^3 e^x$

14. $f(x) = \sin x \cos x$

15. $f(x) = (x^2 + 3)e^x$

16. $f(x) = x^{3/2} \sin x$

17. $f(x) = \dfrac{\sin x}{x^2}$

18. $f(x) = \dfrac{\cos x}{x^2 + e^{2x}}$

19. $f(x) = \dfrac{(x + 1)^2}{x^2 + 1}$

20. $f(x) = \dfrac{\sin x}{\cos(2x)}$

21. $f(x) = \cot(3x)$

22. $f(x) = \sec(x + 3)$

23. $f(x) = 3 \csc(2x)$

24. $f(x) = 4 \tan(x + 5)$

25. $f(x) = x^3 e^{2x} \ln x$

In Exercises 26–37, find an antiderivative of the function.

26. $g(x) = \sec^2 x$

27. $g(x) = 3 \sec x \tan x$

28. $g(x) = \sec^2(x + 1)$

29. $g(x) = \csc^2(2x)$

30. $g(x) = 2x \cos x - x^2 \sin x$

31. $g(x) = (x + 1)e^x$

32. $g(x) = e^{2x} \sin x - 2e^{2x} \cos x$

33. $g(x) = \dfrac{xe^x - e^x}{x^2}$

34. $g(x) = 2xe^x + x^2 e^x$

35. $g(x) = \dfrac{\sin x}{x} + \ln x \cos x$

36. $g(x) = e^x \sin x + e^x \cos x$

37. $g(x) = e^x \sin x - e^x \cos x$

38. Suppose that f and g are differentiable functions and that $h(x) = f(x)g(x)$. Find symbolic expressions involving only numerical constants and derivatives of f and g for each of the following.

(a) $h''(x)$

(b) $h'''(x)$

FURTHER EXERCISES

39. Let $f(x) = xe^x$.

(a) Find values of the constants a and b so that the linear function $L(x) = ax + b$ has the properties $L(0) = f(0)$ and $L'(0) = f'(0)$.

(b) Find the value of the constant c for which the quadratic function $Q(x) = L(x) + cx^2$ has the properties $Q(0) = f(0)$, $Q'(0) = f'(0)$, and $Q''(0) = f''(0)$.

(c) Plot f, L, and Q on the same axes over the interval $[-3, 3]$.

(d) Find an interval over which $|f(x) - L(x)| < 0.1$.

(e) Find an interval over which $|f(x) - Q(x)| < 0.1$.

40. Let $f(x) = \dfrac{x}{3 + x^2}$.

(a) Find values of the constants a and b so that the linear function $L(x) = a(x - 1) + b$ has the properties $L(1) = f(1)$ and $L'(1) = f'(1)$.

(b) Find the value of the constant c for which the quadratic function $Q(x) = L(x) + c(x - 1)^2$ has the properties $Q(1) = f(1)$, $Q'(1) = f'(1)$, and $Q''(1) = f''(1)$.

(c) Graph f, L, and Q on the same axes over the interval $[-3, 3]$.

(d) Find an interval over which $|f(x) - L(x)| < 0.1$.

(e) Find an interval over which $|f(x) - Q(x)| < 0.1$.

41. Suppose that f is the function shown below.

Graph of f

(a) Let $h(x) = x^2 f(x)$. Evaluate $h'(2)$ and $h'(4)$.

(b) Let $m(x) = f(x)/(x^2 + 1)$. Evaluate $m'(0)$.

42. Suppose that $f(1) = 2$ and f' is the function shown below. Let $m(x) = x^3 f(x)$.

Graph of f'

(a) Evaluate $m'(1)$.
(b) Show that m is increasing at 2.
(c) Estimate $m''(1)$.

43. Suppose that $f(0) = 3$ and f' is the function shown below. Let $g(x) = e^x f(x)$.

Graph of f'

(a) Evaluate $g'(0)$.
(b) Is g increasing at $x = 1$? At $x = 2$? At $x = -2$? Justify your answers.
(c) Estimate $g''(0)$.
(d) Is g concave up at $x = 1$? At $x = 2$? Justify your answers.

44. Let f be the function described in the previous problem, and let $g(x) = \sin(x) f(x)$.
(a) Evaluate $g'(0)$.
(b) Is g increasing at $x = \pi/2$? At $x = \pi/2$? Justify your answers.
(c) Estimate $g''(0)$.
(d) Is g concave up at $x = \pi/2$? Justify your answer.

45. Let f be the function described in the previous problem, and let $g(x) = f(x) \cdot f(x) = f(x)^2$.
(a) Evaluate $g'(0)$.
(b) Is g increasing at $x = 1$? At $x = -1$? Justify your answers.
(c) Estimate $g''(0)$.
(d) Is g concave down at $x = 1$? Justify your answer.

46. A line is drawn that is tangent to the curve $y = x^2$ at a point in the first quadrant. This tangent line, along with the curve $y = x^2$ and the x-axis form a region whose area is 9. Find the coordinates of the point of tangency.

47. Let $f(x) = xe^x$. Find values of the parameters A and B so that $F(x) = (Ax + B)e^x$ is an antiderivative of f.

48. Let $g(x) = x^2 e^x$. Find values of the parameters A, B, and C so that $G(x) = (Ax^2 + Bx + C)e^x$ is an antiderivative of g.

49. Let $h(x) = e^{2x} \sin x$. Find values of the parameters A and B, so that $H(x) = e^{2x}(A \cos x + B \sin x)$ is an antiderivative of h.

50. Consider the function $f(x) = \dfrac{\ln x}{x}$ on the interval $(1, \infty)$.
(a) Where does f achieve its maximum value? What is this value?
(b) Find all inflection points in the graph of f.

51. Let $f(x) = \dfrac{x + \sin x}{\cos x}$.
(a) Find $f'(x)$.
(b) Find an equation of the line tangent to the graph of f at $x = 0$.

52. Let $f(x) = e^x \cos x$.
(a) At which value of x in the interval $[0, \pi]$ does f achieve its maximum value? Its minimum value?
(b) At which value(s) of x in the interval $[0, \pi]$ does f have an inflection point?

53. Find the maximum and minimum values of $g(x) = e^{-x} \sin x$ over the interval $[0, \infty)$.

54. Let $f(x) = 2x + x \cos x + \sin x$.
(a) Show that f has a local minimum at $x = (2k + 1)\pi$, where $k = 0, 1, 2, \ldots$.
(b) Show that f has a local maximum at $x = -(2k + 1)\pi$, where $k = 0, 1, 2, \ldots$.

55. Suppose that $N(t)$, the number of bacteria in a culture at time t, is proportional to $25 + te^{-t/20}$.
(a) At what time during the interval $0 \le t \le 100$ is the number of bacteria the smallest?
(b) At what time during the interval $0 \le t \le 100$ is the number of bacteria the greatest?
(c) At what time during the interval $0 \le t \le 100$ is the rate of change of the number of bacteria a minimum?

56. What is the area of the largest triangle that can be formed in the first quadrant by the x-axis, the y-axis, and a line tangent to the graph of $y = e^{-x}$?

57. (a) Show that $x + x^{-2} \geq 2$ if $0 < x \leq 1$.
    [**HINT**: Use the racetrack principle.]
    (b) Show that $\sin x + \tan x \geq 2x$ if $0 \leq x < \pi/2$.
    [**HINT**: Use part (a) to show that $\cos x + \sec^2 x \geq 2$ if $0 \leq x < \pi/2$.]

58. Show that $\dfrac{x}{1+x} \leq \ln(1+x) \leq x$ for all real numbers $x \geq 0$. [**HINT**: Use the racetrack principle.]

59. Suppose that f is a differentiable function, $g(x) = \left(f(x)\right)^2$, and $h(x) = \left(f(x)\right)^3$.
    (a) Use the product rule to show that $\left(g(x)\right)' = 2f(x)f'(x)$. [**HINT**: $\left(f(x)\right)^2 = f(x) \cdot f(x)$.]
    (b) Use part (a) and the product rule to find an expression for $\left(h(x)\right)'$.

60. Suppose that f is a differentiable function and $g(x) = \left(f(x)\right)^2$.
    (a) Show that every stationary point of f is a stationary point of g.
    (b) If x_0 is a local maximum point of f, is x_0 also a local maximum point of g? Explain.

61. (a) Show that $F(x) = x(\ln x - 1)$ is an antiderivative of $f(x) = \ln x$.
    (b) Use the result in part (a) to find an antiderivative of $g(x) = \log_b x$.

62. Let $f(x) = \dfrac{\ln x}{k} - \dfrac{kx}{x+1}$ where k is a constant.
    (a) Show that if $0 < k \leq 2$, f has no local extrema.
    (b) Show that f has local extrema if $k > 2$. Determine how many and where they are located.

63. Evaluate the following limits by recognizing each as the definition of a derivative.
    (a) $\displaystyle\lim_{x \to \pi} \dfrac{x \cos x + \pi}{x - \pi}$
    (b) $\displaystyle\lim_{t \to 0} \dfrac{1}{t} \cdot \left(\dfrac{t^2 - 3t + 4}{5t + 2} - 2 \right)$

64. Suppose that f, g, and h are differentiable functions and that $h(x) = f(x)g(x)$. Show that if $h'(1) \neq 0$, then $\left(f(1)\right)^2 + \left(g(1)\right)^2 > 0$.

65. Suppose that the functions f and g have domain $\mathbb{R}$ and the following properties:
    (i) $f(x + y) = f(x)f(y)$
    (ii) $g(x + y) = g(x)g(y)$
    (iii) $f(0) \neq 0$
    (iv) $g(0) \neq 0$
    (v) $f'(0) = 1$
    (vi) $g'(0) = 1$
    (a) Show that $f(0) = 1$. [**HINT**: Let $x = y = 0$ in (i) and use part (a).]
    (b) Show that $f(x) \neq 0$ for all x. [**HINT**: Let $y = -x$ in (i).]
    (c) Show that $f'(x) = f(x)$ for all x. [**HINT**: Use the definition of the derivative.]
    (d) Let $k(x) = f(x)/g(x)$. Show that k is defined for all real inputs and find an expression for $k'(x)$.
    (e) Use part (d) to show that $f(x) = g(x) = e^x$.

66. (An alternative proof of the product rule.) Suppose that f and g are differentiable functions and that $h(x) = f(x)g(x)$.
    (a) Use the definition of the derivative to show that $\left(\left(f(x)\right)^2\right)' = 2f(x)f'(x)$. [**HINT**: Use the algebraic identity $a^2 - b^2 = (a - b)(a + b)$.]
    (b) Show that
    $$h(x) = \frac{1}{2}\left(\left(f(x) + g(x)\right)^2 - \left(f(x)\right)^2 - \left(g(x)\right)^2 \right).$$
    (c) Use parts (a) and (b) to show that $h'(x) = f(x)g'(x) + f'(x)g(x)$.

3.6 New Derivatives from Old: The Chain Rule

In the last section we differentiated new functions constructed by *algebraic* operations on simpler functions. In this section we continue the same theme with respect to the operation of *composition*.

Derivatives of Compositions

With the sum, constant multiple, product, and quotient rules we can differentiate any *algebraic* combination of elementary functions—even monsters such as

$$\frac{x^2 \sin x}{\tan x} - \frac{\sin x \ln x + \cos x}{x^5 + 1} \cdot \left(2^x + \frac{3^x}{\ln x} \right).$$

The Question

All that remains before we can differentiate *all* elementary function is to handle functions such as

$$\sin\left(x^2\right), \quad (\sin x)^2, \quad \text{and} \quad \sin^3\left(x^2\right)$$

constructed by *composing* simpler functions.

The simpler functions involved here are ▶

$$f(u) = \sin u, \qquad g(u) = u^2, \qquad \text{and} \qquad h(u) = u^3.$$

The "neutral" variable name u avoids conflict with x, which was used differently earlier.

Finding f', g', and h' is (by now) no problem; the real question is how to combine everything to find derivatives of composite functions. In simplest form, our question is:

> *How do we combine f, g, f', and g' to find $(f \circ g)'$?*

The Answer

The **chain rule** ▶ is the missing link. In stating it we assume, as usual, that f and g are "nice enough" (i.e., that f, g, f', and g' have large enough domains to permit the desired operations). ▶

The name is apt: Composed functions are "chained" together in series.

More fine print after the theorem.

T h e o r e m 1 1 (**The Chain Rule**)   Let f and g be well-behaved functions. Then

$$(f \circ g)'(a) = f'(g(a)) \cdot g'(a).$$

This important theorem deserves several closer looks:

The Derivative Exists   The theorem guarantees (among other things) that the derivative $(f \circ g)'(a)$ exists, provided that both $f'(g(a))$ and $g'(a)$ exist.

Why a Product?   The theorem says that the derivative of the composition $f \circ g$ is a certain product of the derivatives f' and g'. Why is a *product* appropriate? Interpreting derivatives as rates of change gives some clue. Suppose, say, that $y = f(u)$, $u = g(x)$, $f'(u) = 3$, and $g'(x) = 2$. This means, in rate language, that y changes three times as fast as u, and u changes twice as fast as x. It sounds reasonable, then, that y should change six times as fast as x. (Additional evidence for the chain rule appears later in this section.)

In other words, the naïve chain rule $(f \circ g)' = f'g'$ is false!

Which Product?   The chain rule involves a certain product of derivatives of f and g. But take care: The factors $f'(g(a))$ and $g'(a)$ show that f' and g' are evaluated for *different inputs*. ◄

Other Notations   In the theorem, we used a specific input, $x = a$, to emphasize the point just made. Since the theorem holds for all sensible inputs a, we can also think of it as a relation among functions. From this point of view, we might prefer to state the chain rule as

$$(f \circ g)'(x) = f'(g(x)) \cdot g'(x)$$

or, equivalently, as

$$(f \circ g)' = (f' \circ g) \cdot g'.$$

A fully rigorous proof of the chain rule runs quickly into delicate technical problems. We'll settle for plausible evidence. First, however, let's look at some examples.

EXAMPLE 1   Differentiate $\sin(x^2)$ and $\sin^2 x$. Are the answers the same? Should they be?

Solution   No and no. The trick is to keep clear exactly what's meant, in context, by f, g, f', and g'. If we let $f(u) = \sin u$ and $g(u) = u^2$, then

$$\sin(x^2) = (f \circ g)(x) \qquad \text{and} \qquad \sin^2 x = (g \circ f)(x).$$

From the chain rule and the derivatives of f and g:

$$\left(\sin(x^2)\right)' = (f \circ g)'(x) = f'\left(g(x)\right) \cdot g'(x) = \cos\left(x^2\right) \cdot 2x.$$

Similarly,

$$\left(\sin^2 x\right)' = (g \circ f)'(x) = g'\left(f(x)\right) \cdot f'(x) = 2\sin x \cdot \cos x.$$

The two answers are different—as they should be. After all, composition isn't commutative. ∎

EXAMPLE 2   Differentiate $\sin(3x)$ and $\sin(x + 3)$, using the chain rule.

Solution   Let $f(u) = \sin u$, $g(x) = 3x$, and $h(x) = x + 3$. Then

$$\left(\sin(3x)\right)' = (f \circ g)'(x) = f'\left(g(x)\right) \cdot g'(x) = \cos(3x) \cdot 3.$$

Similarly,

$$\left(\sin(x + 3)\right)' = (f \circ h)'(x) = f'\left(h(x)\right) \cdot h'(x) = \cos(x + 3) \cdot 1. \quad ∎$$

EXAMPLE 3   Differentiate $\sin^3(x^2)$.

Solution   We need to keep our wits about us here. Let $f(u) = \sin u$, $g(u) = u^2$, and $h(u) = u^3$; then

$$(h \circ f \circ g)(x) = \left(\sin(x^2)\right)^3 = \sin^3(x^2).$$

To handle the "three-deep" composition $h \circ f \circ g$, we group it in the form $h \circ (f \circ g)$ and use the chain rule *twice*:

$$(h \circ (f \circ g))' = (h' \circ (f \circ g)) \cdot (f \circ g)' = (h' \circ (f \circ g)) \cdot (f' \circ g \cdot g').$$

For our particular f, g, and h, we get

$$\left((\sin x^2)^3\right)' = 3 \left(\sin x^2\right)^2 \cdot \left(\sin x^2\right)'$$

$$= 3 \left(\sin x^2\right)^2 \cdot \left(\cos x^2 \cdot 2x\right)$$

$$= 6x \left(\sin x^2\right)^2 \cos x^2. \qquad \blacksquare$$

Working from Outside In: How to Use the Chain Rule. Composing two functions can be thought of as *nesting* one within the other. The use of nested parentheses, as in $f(g(x))$, reflects this point of view: Here f is the "outer" function and g the "inner." In this language, the chain rule says, in effect, to *work from outside in*, one level at a time. More precisely:

> To differentiate $f(g(x))$, first differentiate the outer function, leaving the inner function alone. Then multiply by the derivative of the inner function.

This rule works nicely for *any* level of nesting, as the next example illustrates.

EXAMPLE 4 Differentiate $\sin\big(\sin(\sin x)\big)$.

Solution Carefully watch the de-nesting, step by step:

$$\big(\sin(\sin(\sin x))\big)' = \cos\big(\sin(\sin x)\big) \cdot \big(\sin(\sin x)\big)'$$

$$= \cos\big(\sin(\sin x)\big) \cdot \cos(\sin x) \cdot (\sin x)'$$

$$= \cos\big(\sin(\sin x)\big) \cdot \cos(\sin x) \cdot \cos x. \qquad \blacksquare$$

EXAMPLE 5 Let n be any real number. Differentiate $\sin^n x$; then differentiate $f(x)^n$, where f is any differentiable function.

Solution From the chain rule, ▶

$$\left((\sin x)^n\right)' = n (\sin x)^{n-1} \cdot \cos x.$$

Here the "outer" function is u^n.

The same idea works for *any* "inner" function f:

$$\left(f(x)^n\right)' = n\, f(x)^{n-1} \cdot f'(x).$$

(The last formula is sometimes called the **general power rule**.) $\qquad \blacksquare$

Evidence for the Chain Rule

As we did for the other rules, we'll argue informally.

Exhibit A: It Works for Linear Functions. Let $f(x) = ax + b$ and $g(x) = cx + d$. We'll differentiate $f \circ g$ twice—with and without the chain rule. Notice first that

$$(f \circ g)(x) = a(cx + d) + b = acx + ad + b.$$

The last form shows that $f \circ g$ is another linear function; its derivative is the constant ac.

The chain rule gives the same result. Because $f'(x) = a$ and $g'(x) = c$, it's easy to calculate that $f'(g(x)) \cdot g'(x) = ac$, just as claimed.

Exhibit B: Differentiable Functions Are Almost Linear.   Let f and g be any differentiable functions. Because f and g are differentiable, they're almost linear (as we have remarked before and have seen graphically): Near the point of tangency, f and g are well approximated by their tangent lines. The slope of the g-graph at $x = a$ is $g'(a)$, so near $x = a$, g is well approximated by a linear function l with slope $g'(a)$ (i.e., $g(x) \approx l(x) = g'(a)x + k$, for some constant k; the value of k doesn't matter here).

For convenience, let's write $g(a) = b$. For reasons like the earlier ones, f is well approximated near $x = b$ by a linear function m, this time with slope $f'(b)$. In other words, $f(x) \approx m(x) = f'(b)x + K$ for some other constant K.

From all the foregoing, $(f \circ g)(x) \approx (m \circ l)(x)$. Because l and m are linear, Exhibit A applies. It says that

$$(m \circ l)' = f'(b) \cdot g'(a) = f'(g(a)) \cdot g'(a),$$

just as the chain rule predicts.

The Chain Rule in Leibniz Notation

When we don't want to name functions, for example.

The chain rule is easy to remember, and sometimes convenient to state, ◄ in Leibniz form. Here's what it says, informally (we assume that y and u are well-behaved functions):

Let y be a function of u and u a function of x. Then y is a (composite) function of x; the derivative of y with respect to x is

$$\frac{dy}{dx} = \frac{dy}{du} \cdot \frac{du}{dx}.$$

EXAMPLE 6   Differentiate $y = \sin\left(x^2\right)$ yet again.

Solution   We have $y = \sin u$ and $u = x^2$, so

$$\frac{dy}{dx} = \frac{dy}{du} \cdot \frac{du}{dx} = \cos u \cdot 2x = 2x \cdot \cos x^2.$$

(At the last step we replaced u with x^2.)   ■

Derivatives of Inverses and the Chain Rule

The chain rule can be used to relate the (known) derivative of a function to the (unknown) derivative of its *inverse*.

Recall that if f and g are inverse functions, the equation

$$(f \circ g)(x) = f(g(x)) = x$$

must hold for all x in the domain of g. Differentiating both sides—the left side needs the chain rule—gives

$$(f \circ g)'(x) = f'(g(x)) \cdot g'(x) = 1.$$

With a little algebra, our desired formula now follows:

Fact Let f and g be inverse functions. Then
$$g'(x) = \frac{1}{f'(g(x))}$$
for all x for which the right side is defined.

In the next section we'll use this formula to find derivatives of inverse trigonometric functions. For the moment we'll content ourselves with a simpler example.

EXAMPLE 7 Use the fact that e^x and $\ln x$ are inverses and the fact that $(e^x)' = e^x$ to find (again!) the derivative of $\ln x$.

Solution Let $f(x) = e^x$ (so that $f'(x) = e^x$) and $g(x) = \ln x$. Then by the preceding formula,

$$g'(x) = (\ln x)' = \frac{1}{f'(g(x))} = \frac{1}{e^{\ln x}} = \frac{1}{x}.$$
■

Guessing and Checking Antiderivatives: A Sample Problem

With the chain rule in our tool kit, we can now easily differentiate almost any elementary function. Finding *antiderivatives*, however, is another matter. Making a guess and then checking it is often the best strategy.

EXAMPLE 8 Find antiderivatives for

$$f(x) = x\cos(x^2) \qquad \text{and} \qquad g(x) = \frac{\cos x}{\sin x}.$$

Solution The basic form of f suggests the result of applying the chain rule, perhaps to the expression $\sin(x^2)$. A check shows that our guess is *almost* correct:

$$\left(\sin(x^2)\right)' = \cos(x^2) \cdot 2x = 2x\cos(x^2).$$

To dispose of that extra factor of 2, we simply divide it out. It follows ▶ that, for any constant C,

Check for yourself.

$$F(x) = \frac{\sin(x^2)}{2} + C$$

is an antiderivative of f.

It's a little harder to recognize g as a derivative, but if we notice that the numerator is the derivative of the denominator, we may suspect that a logarithm function is involved. So it is—the guess $G(x) = \ln(\sin x) + C$ turns out to be correct:

$$G'(x) = \left(\ln(\sin x) + C\right)' = \frac{\cos x}{\sin x},$$

as desired. ■

BASICS

1. Let $h(x) = f(g(x))$. Use the information about f and g given in the table below to fill in the missing information about h and h'.

x	$f(x)$	$f'(x)$	$g(x)$	$g'(x)$	$h(x)$	$h'(x)$
1	1	2	4	3		
2	2	1	3	4		
3	4	3	1	2		
4	3	4	2	1		

In Exercises 2–11, assume that g is a function such that $g'(x)$ exists for all x. Find the derivative of the function f. (Answers will involve g and g'.)

2. $f(x) = \big(g(x)\big)^n$

3. $f(x) = e^{g(x)}$

4. $f(x) = \ln\big(g(x)\big)$

5. $f(x) = \sin\big(g(x)\big)$

6. $f(x) = \tan\big(g(x)\big)$

7. $f(x) = g(x^n)$

8. $f(x) = g(e^x)$

9. $f(x) = g(\ln x)$

10. $f(x) = g(\sin x)$

11. $f(x) = g(\tan x)$

In Exercises 12–41, find the derivative of the function f.

12. $f(x) = \big(x^2 + 3\big)^{29}$

13. $f(x) = (1 + e^x)^{33}$

14. $f(x) = \sqrt{2 + \sin x}$

15. $f(x) = \sin\big(e^{3x}\big)$

16. $f(x) = \cos\big(x^2\big)$

17. $f(x) = \tan\big(\sqrt[3]{x}\big)$

18. $f(x) = \ln\big(4x^2 + 3\big)$

19. $f(x) = \sqrt{x^2 + e^{-3x}}$

20. $f(x) = \cos\big(\sqrt{x} + e^{-x}\big)$

21. $f(x) = \big(2x^{3/4} + 5x^{-1/6}\big)^{12}$

22. $f(x) = \big(x^{3/2} - 4x^{2/3} + 17\big)^{421}$

23. $f(x) = (x + \ln x)^{14}$

24. $f(x) = \ln\big(x^2 + 3\sin x\big)$

25. $f(x) = \sin\big(e^{2x} + \tan(3x)\big)$

26. $f(x) = \cos^3(x^4)$

27. $f(x) = \ln\big(e^{3x} + \cos^2 x\big)$

28. $f(x) = \tan\big(e^{x^2}\big)$

29. $f(x) = \sin^5\big(3x^4\big)$

30. $f(x) = \sqrt{4 + \cos^2(3x)}$

31. $f(x) = \sqrt[3]{\cos\big(4x^2\big)}$

32. $f(x) = \cos^2\big(e^{3x}\big)$

33. $f(x) = x^2\big(4x^3 + 5\big)^6$

34. $f(x) = x\ln\big(2 + \sin(3x)\big)$

35. $f(x) = e^{x^3}\cos\big(4x^5 + 2\big)$

36. $f(x) = \sin^2(x\cos x)$

37. $f(x) = \sqrt{3 + x\cos x}$

38. $f(x) = e^{x\sin x}$

39. $f(x) = \tan\big(x^2\ln x\big)$

40. $f(x) = \cos\big(xe^{2x}\big)$

41. $f(x) = x^2\sin\big(e^x\ln x\big)$

FURTHER EXERCISES

42. Let $h(x) = f\left(\dfrac{g(x)}{x^2 + 1}\right)$. If $f(1) = 3$, $f(2) = 5$, $f'(1) = 7$, $f'(2) = 11$, $g(1) = 2$, and $g'(1) = 4$, what is $h'(1)$?

43. Let $f(x) = \ln(2 + \cos x)$.
   (a) What is the domain of f?
   (b) At which values of x does f have local maxima? Local minima?
   (c) At which values of x does f have inflection points?

44. Suppose that f is the function shown to the right and that $g(x) = f(x^2)$.
   (a) For which values of x is $g'(x) = 0$?
   (b) Is g increasing or decreasing at -1?
   (c) Is g' positive or negative over the interval $(\sqrt{2}, 5)$?

Graph of f

45. Suppose that $f(1) = 2$, that f' is the function shown below, and that $k(x) = f(x^3)$. Evaluate $k'(-1)$.

Graph of f'

In Exercises 46–48, f and g are the functions defined by the graphs below.

Graph of f

Graph of g

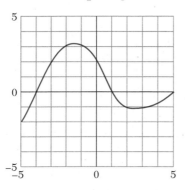

46. Let $h(x) = (f \circ g)(x)$.
    (a) Evaluate $h(-2)$, $h(1)$, and $h(3)$.
    (b) Is $h'(-1)$ positive, negative, or equal to zero? Justify your answer.
    (c) Estimate $h'(-2)$, $h'(1)$, and $h'(3)$.
    (d) What values of x correspond to stationary points of $h(x)$?

47. Let $j(x) = f(x^4)$. Is j increasing or decreasing at $x = -1$? Justify your answer.

48. Let $k(x) = f(e^{-x})$.
    (a) Determine whether k is increasing, decreasing, or stationary at $x = -1$, $x = 0$, and $x = 1$. Justify your answers.
    (b) Show that $x = -\ln 2$ is a stationary point of k and determine whether it corresponds to a local maxima or a local minima. Justify your answer.

49. Let $h(x) = (f \circ g)(x)$ and $j(x) = f(x) \cdot g(x)$. Fill in the missing entries in the table below.

x	$f(x)$	$f'(x)$	$g(x)$	$g'(x)$	$h(x)$	$h'(x)$	$j(x)$	$j'(x)$
-1	3	2	1		0	$-1/2$	3	
0	0	$1/2$	-1	1			0	$-1/2$
1		-5	0			2	0	

In Exercises 50–61, find an antiderivative of the function f.

50. $f(x) = xe^{x^2}$
51. $f(x) = 2x \cos\left(x^2\right)$
52. $f(x) = x^2 \sin\left(2x^3\right)$
53. $f(x) = x\left(x^2 + 3\right)^4$
54. $f(x) = x^3\left(2x^4 + 6\right)^5$
55. $f(x) = \dfrac{3x}{4 + x^2}$

56. $f(x) = \dfrac{x}{\sqrt{1 + x^2}}$
57. $f(x) = x \sin\left(1 - x^2\right)$
58. $f(x) = \dfrac{e^{\sqrt{x}}}{\sqrt{x}}$
59. $f(x) = e^{3x} \sec^2\left(e^{3x}\right)$
60. $f(x) = e^{\cos x} \sin x$

61. $f(x) = e^x \cos\left(e^x\right) \sin\left(e^x\right)$

In Exercises 62–65, find an antiderivative of the function f.
[**HINT:** $\left(\ln(f(x))\right)' = f'(x)/f(x)$.]

62. $f(x) = \dfrac{3x^2 - 5}{x^3 - 5x}$
63. $f(x) = \dfrac{2x + 3}{x^2 + 3x + 5}$

64. $f(x) = \dfrac{e^x}{1 + e^x}$
65. $f(x) = \dfrac{\cos x}{2 + \sin x}$

66. Let f be an even function and g be an odd function.
    (a) Show that f' is an odd function. [**HINT:** $f(-x) = f(-1 \cdot x)$.]
    (b) Show that g' is an even function.
    (c) Show that f'' is an even function. [**HINT:** Use part (b).]
    (d) Is g'' even, odd, or neither? Justify your answer.

67. Suppose that f is a differentiable function. If f is periodic with period P, is f' also periodic? Explain.

68. Some values of the derivative of a function f are shown in the table below:

x	0	1	2	3	4	5	6
$f'(x)$	4	-3	2	0	-1	-6	1

   [**NOTE:** No explicit formula for f is given!]
   (a) Let $g(x) = f(x+3)$. For which values of x can you evaluate $g'(x)$? Evaluate $g'(x)$ for these values of x.
   (b) Let $h(x) = 3f(x)$. For which values of x can you evaluate $h'(x)$? Evaluate $h'(x)$ for these values of x.
   (c) Let $j(x) = f(3x)$. For which values of x can you evaluate $j'(x)$? Evaluate $j'(x)$ for these values of x.
   (d) Let $k(x) = f(x) - 3$. For which values of x can you evaluate $k'(x)$? Evaluate $k'(x)$ for these values of x.

69. Suppose that f is a function such that $f'(x) = x \cos x$.
   (a) Let $g(x) = f(x + \pi)$. Evaluate $g'(0)$.
   (b) Let $h(x) = f(2x)$. Evaluate $h'(\pi/2)$.
   (c) Let $j(x) = 2f(x) + 3$. Find an expression for $j'(x)$.

70. Suppose that F is an antiderivative of a function f.
   (a) Find an antiderivative of $g(x) = f(x) + 3$.
   (b) Find an antiderivative of $h(x) = f(x + 3)$.
   (c) Find an antiderivative of $j(x) = 3f(x)$.
   (d) Find an antiderivative of $k(x) = f(3x)$.

71. Let k be a constant and f a differentiable function. If $g(x) = f(x + k)$, then $g'(x) = f'(x + k)$.
   (a) Justify (with a short sentence) each step in the following proof of this result:

   (i) $g'(x) = \lim\limits_{h \to 0} \dfrac{g(x + h) - g(x)}{h}$

   (ii) $\quad = \lim\limits_{h \to 0} \dfrac{f(x + k + h) - f(x + k)}{h}$

   $\quad = \lim\limits_{h \to 0} \dfrac{f(w + h) - f(w)}{h}$

   (iii) $\quad$ where $w = x + k$

   (iv) $\quad = f'(w)$

   (v) $\quad = f'(x + k)$.

   (b) Explain this result geometrically.

72. Let k be a constant and f a differentiable function. If $g(x) = f(kx)$, then $g'(x) = kf'(kx)$. Justify (with a short sentence) each step in the following proof of this result: [**HINT:** In line (iv), $w = kh$.]

   (i) $g'(x) = \lim\limits_{h \to 0} \dfrac{g(x + h) - g(x)}{h}$

   (ii) $\quad = \lim\limits_{h \to 0} \dfrac{f(kx + kh) - f(kx)}{h}$

   (iii) $\quad = \lim\limits_{w \to 0} k \cdot \dfrac{f(kx + w) - f(kx)}{w}$

   (iv) $\quad = kf'(kx)$

73. Suppose that f is a function such that $f(\pi/2) = 17$ and $f'(x) = 2e^{x \cos x}$.
   (a) Find an equation of the line tangent to the graph of f at $x = \pi/2$.

   (b) Let $L(x)$ be the line tangent to the graph of f at $x = \pi/2$. Is $L(2) < f(2)$? Justify your answer.

74. (a) Suppose that f is a function such that $f'(x) > 0$ for all x. If $g(x) = f(f(x))$, must g be increasing for all x? Explain your answer.
   (b) Suppose that f is a function such that $f'(x) < 0$ for all x. If $g(x) = f(f(x))$, must g be decreasing for all x? Explain your answer.

75. Use the identity $\dfrac{f(x)}{g(x)} = f(x) \cdot \big(g(x)\big)^{-1}$ together with the product and chain rules to derive the quotient rule.

76. Let $h = f \circ g$. Show that $h''(a) = f''\big(g(a)\big)\big(g'(a)\big)^2 + f'\big(g(a)\big)g''(a)$.

77. Let $f(x) = (1 + e^x)^{-1}$.
   (a) Evaluate $\lim\limits_{x \to \infty} f(x)$.
   (b) Evaluate $\lim\limits_{x \to -\infty} f(x)$.
   (c) Explain why f is invertible.
   (d) What are the domain and range of f^{-1}?

78. Let $f(x) = \sin(|x|)$.
   (a) What is the domain of f? Of f'?
   (b) Find an expression for $f'(x)$.

79. Suppose that $f'(x) = g(x)$ and that $h(x) = x^2$. Express $\big(f(h(x))\big)'$ in terms of g and x.

80. Suppose that f and g are functions such that
   (i) $f(0) = 0$
   (ii) $g(0) = 1$
   (iii) $f'(x) = g(x)$ for all x
   (iv) $g'(x) = -f(x)$ for all x
   (a) Show that $h(x) = \big(f(x)\big)^2 + \big(g(x)\big)^2 = 1$ for all x. [**HINT:** Show that $h(0) = 1$ and $h'(x) = 0$ for all x.]
   (b) Let $k(x) = \big(F(x) - f(x)\big)^2 + \big(G(x) - g(x)\big)^2$ where F and G are another pair of functions that satisfy the properties (i)–(iv). Show that $k'(x) = 0$ for all x.
   (c) Use part (b) to show that $F(x) = f(x) = \sin x$ and that $G(x) = g(x) = \cos x$ for all x.

81. Let $f(x) = \sin x + \cos x$ on the interval $[-3\pi/4, \pi/4]$.
   (a) Show that f has an inverse.
   (b) Find an expression for $\big(f^{-1}\big)'(x)$.
   [**HINT:** $(\cos x - \sin x)^2 = 2 - (\sin x + \cos x)^2$.]

82. Let $g(x) = f(x^2)$ where f is a function with the following properties:
   (i) f, f', and f'' have domain $\mathbb{R}$
   (ii) $f(0) = 1$
   (iii) $f'(x) > 0$ for all $x > 0$
   (iv) $f'(x) < 0$ for all $x < 0$
   (v) f is concave down on $(-\infty, 0)$
   (vi) f is concave up on $(0, \infty)$
   (a) Where does g have local minima?
   (b) Where is g concave up?

83. Evaluate $\lim\limits_{x \to \pi} \dfrac{5^{\sin x} - 1}{x - \pi}$ exactly (i.e., as a value of an elementary function).

3.7 Implicit Differentiation

Implicitly vs. Explicitly Defined Functions

The equation $y = -3x + 5$ defines y *explicitly* as a function (we'll call it f) of x; for an input x, the output is given by the explicit rule $f(x) = -3x + 5$.

Consider, by contrast, the equivalent equation $6x + 2y = 10$. Although it doesn't display y explicitly as a function of x, the new equation defines y *implicitly* as a function of x. Solving for y gives the same explicit formula we had before:

$$6x + 2y = 10 \iff 2y = -6x + 10 \iff y = -3x + 5.$$

For the linear equation $6x + 2y = 10$, it was easy to find an explicit formula for y. Things aren't always so simple. More complicated equations are much harder— or even impossible—to solve explicitly for y. Even simple equations, moreover, may not define y uniquely as a function of x, because the graph of an *equation* need not be the graph of a *function*.

EXAMPLE 1   Does the equation $x - y^2 = 0$ define y as a function of x?

Solution   No. Solving algebraically for y shows why:

$$x - y^2 = 0 \implies y^2 = x \implies y = \pm\sqrt{x}.$$

In effect, the equation defines *two* functions of x: $f_1(x) = \sqrt{x}$ and $f_2(x) = -\sqrt{x}$, each defined for nonnegative inputs x. The picture agrees:

Graph of $x - y^2 = 0$

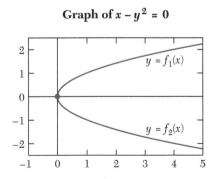

The "sideways" parabola is not the graph of a function, because each value of x corresponds to *two* values of y. ▶   The upper and lower halves of the graph do define functions of x—the same functions (f_1 and f_2) we wrote explicitly before. ■

It fails the vertical line test.

EXAMPLE 2   Does the equation $5x^2 - 6xy + 5y^2 = 16$ implicitly define y as one function of x? As several functions of x?

Solution   Here's the graph, an ellipse. The points $(2, 2)$ and $(1, -1)$ are bulleted.

Graph of $5x^2 - 6xy + 5y^2 = 16$

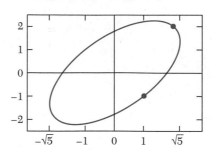

Like the preceding graph, this one defines two different functions y of x, corresponding to the "upper" and "lower" halves of the ellipse. (The upper half, passing through $(2, 2)$, defines a function g_1 for which $g_1(2) = 2$. The lower half defines another function g_2, for which $g_2(1) = -1$. Below we'll find derivatives at these points.) The domain of each function is the interval $[-\sqrt{5}, \sqrt{5}]$. Solving for y is harder this time; we won't bother. ∎

Implicit Functions, Implicit Derivatives

All three of the equations $6x + 2y = 10$, $x = y^2$, and $5x^2 - 6xy + 5y^2 = 16$ implicitly define one or more functions of x. How can we differentiate such functions?

If we can solve *explicitly* for y, there's no problem. For $x = y^2$, for instance,

$$f_1(x) = \sqrt{x} \implies f_1'(x) = \frac{1}{2\sqrt{x}};$$

$$f_2(x) = -\sqrt{x} \implies f_2'(x) = -\frac{1}{2\sqrt{x}}.$$

What if we can't (or don't want to) solve explicitly for y? In that case, we use **implicit differentiation**, a technique for differentiating an *equation* without first solving it for one of the variables. Here's the idea for an equation in x and y:

> **Implicit differentiation.** *Differentiate both sides of the equation, treating y as an (implicitly defined) function of x; use the chain rule. The resulting equation involves x, y, and dy/dx; it can be solved, if necessary, for dy/dx.*

We'll illustrate with examples.

EXAMPLE 3   Let $x = y^2$. Find dy/dx by implicit differentiation.

Solution   We differentiate both sides of the equation, treating y as a function of x:

$$x = y^2 \implies \frac{d}{dx}(x) = \frac{d}{dx}(y^2) \implies 1 = 2y\frac{dy}{dx}.$$

Note the use of the chain rule: Since y is a function of x, $(y^2)' = 2yy'$. Solving for dy/dx is easy:

$$1 = 2y\frac{dy}{dx} \implies \frac{dy}{dx} = \frac{1}{2y}.$$

This agrees—as it should—with *both* of our earlier answers:

$$y = \sqrt{x} \implies y' = \frac{1}{2\sqrt{x}} = \frac{1}{2y};$$

$$y = -\sqrt{x} \implies y' = -\frac{1}{2\sqrt{x}} = \frac{1}{2y}.$$ ■

EXAMPLE 4   As the graph in Example 2 shows, the equation $5x^2 - 6xy + 5y^2 = 16$ defines y implicitly as a function of x near the point $(2, 2)$. Find dy/dx in terms of x and y. What is its value at the point $(2, 2)$? At $(1, -1)$? What do these values mean?

S o l u t i o n   First we differentiate our equation with respect to x: ▶

Don't forget: y is a function of x, so the chain rule applies.

$$5x^2 - 6xy + 5y^2 = 16 \implies 10x - 6y - 6x\frac{dy}{dx} + 10y\frac{dy}{dx} = 0.$$

(The middle term, $-6xy = -6x \cdot y$, is a product, so we used the product rule.)
    Now we solve for dy/dx:

$$10x - 6y - 6x\frac{dy}{dx} + 10y\frac{dy}{dx} = 0 \implies \frac{dy}{dx}(-6x + 10y) = 6y - 10x$$

$$\implies \frac{dy}{dx} = \frac{6y - 10x}{10y - 6x}.$$

At $(2, 2)$, therefore, $dy/dx = -1$; at $(1, -1)$, $dy/dx = 1$. ▶   Geometrically, these results mean that the curve shown has slopes -1 and 1, respectively, at the bulleted points $(2, 2)$ and $(1, -1)$. ▶ ■

Check these simple calculations.

Convince yourself that these results are graphically plausible.

Implicit Differentiation: Pros and Cons.   Implicit differentiation neatly avoids the hard problem of solving complicated equations explicitly for y. As usual, there's a price to pay for such convenience: Derivative formulas obtained via implicit differentiation usually involve both x and y.

A Sample Problem

Here's a problem that applies ideas of this section.

EXAMPLE 5   Use implicit differentiation to find points on the unit circle $x^2 + y^2 = 1$ at which the tangent line has slope 1.

S o l u t i o n   Implicit differentiation with respect to x gives ▶

Check the easy steps.

$$x^2 + y^2 = 1 \implies 2x + 2y\frac{dy}{dx} = 0 \implies \frac{dy}{dx} = \frac{-x}{y}.$$

Thus, $dy/dx = 1$ if and only if $y = -x$, i.e., where the line $y = -x$ intersects the unit circle. From a picture ▶ it's easy to see that these intersections occur at the points $(-\sqrt{2}/2, \sqrt{2}/2)$ and $(\sqrt{2}/2, -\sqrt{2}/2)$. ■

Draw your own!

BASICS

1. This problem refers to the equation $5x^2 - 6xy + 5y^2 = 16$ and its graph—see Example 2 (page 233) and Example 4 (page 235).

   (a) Find the two points (x, y) on the ellipse at which the tangent line is horizontal.

   (b) Find the points (x, y) on the ellipse at which the tangent line is vertical. (At these points, the derivative dy/dx doesn't exist.)

2. This problem refers to the equation $5x^2 - 6xy + 5y^2 = 16$ and its graph—see Example 2 (page 233) and Example 4 (page 235).

   (a) The graph appears to be symmetric about the origin. To see that it really *is* symmetric, show that if (x_0, y_0) is on the graph, then so is the "opposite" point $(-x_0, -y_0)$.

   (b) Let (x_0, y_0) be any point on the graph. The previous part says that the "opposite" point is $(-x_0, -y_0)$. Show that the tangent lines at these two points have the same slope. (This is further evidence of the graph's symmetry.)

   (c) The line $y = -x$ intersects the ellipse at two points. Find these two points; then find the slope at each point.

3. All parts of this exercise are about the equation $y^2 - x^2 = 1$ and its graph, called a **hyperbola**.

   (a) The equation $y^2 - x^2 = 1$ defines two different functions $y = f_1(x)$ and $y = f_2(x)$. Find formulas for these functions.

   (b) Plot the equation $y^2 - x^2 = 1$. [**HINT:** One way is to plot both functions f_1 and f_2.]

   (c) Use implicit differentiation to find dy/dx.

   (d) Find the points on the curve $y^2 - x^2 = 1$ at which the tangent line is horizontal.

   (e) For any constant $c > 0$, the line $x = a$ intersects the hyperbola at two points. (Draw a picture to convince yourself.) Find these two points and the slope of the hyperbola at each.

   (f) What happens in the previous part as $c \to \infty$?

4. Repeat the previous problem, but use the hyperbola $y^2 - 2x^2 = 1$.

5. This problem refers to the curves with equations $x^2 + xy + y^2 = 1$ and $x^2 - xy + y^2 = 1$, shown.

   (a) Find dy/dx for each of the curves.

   (b) Decide which curve is which. [**HINT:** One way is to compare the derivatives at a common point, such as $(0, 1)$.)]

   (c) Draw tangent lines to the curves at the four intersection points. Find the slope of each one.

   (d) Find and label all points at which each curve is horizontal.

   (e) Find and label all points at which each curve is vertical.

   (f) Draw the line $y = x$ into the picture. Use derivatives to show that this line intersects both curves perpendicularly.

Graphs of $x^2 + xy + y^2 = 1$ and $x^2 - xy + y^2 = 1$

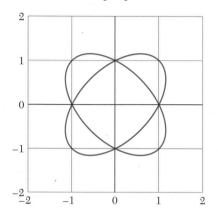

6. The equation $2x + 3y = 6$ implicitly defines y as a function of x.

   (a) By solving for y, find an *explicit* formula for y as a function of x. Use the result to find dy/dx.

   (b) Find dy/dx by implicit differentiation; compare your result with the previous part.

7. The equation $4x^2 y - 3y = x^3$ implicitly defines y as a function of x.

   (a) Use implicit differentiation to find dy/dx.

   (b) Write y as an explicit function of x and compute dy/dx directly.

8. The equation $x^2 + y^2 = 1$ defines *two* different functions of x:

   $$x^2 + y^2 = 1 \implies y = \pm\sqrt{1 - x^2}.$$

   (a) Find dy/dx by implicit differentiation.

   (b) Find dy/dx (directly) for each of the functions $y = \pm\sqrt{1 - x^2}$. Compare your results with those found in part (a).

Each of the equations in Exercises 9–18 defines a curve in the xy plane. Plot the curve, then use implicit differentiation to find an expression for dy/dx.

9. $x^2 + 2y^2 = 4$

10. $4x^2 - 9y^2 = 36$

11. $y^2 - x^2 = x^2 y^2$

12. $y^2(2 - x) = x^3$

13. $x^3 + y^3 = 3xy$

14. $y^4 = y^2 - x^2$

15. $x^{2/3} + y^{2/3} = 1$
16. $\left(x^2 + y^2 - 2y\right)^2 = x^2 + y^2$
17. $\left(x^2 + y^2\right)^2 = x^2 - y^2$
18. $\left(x^2 + y^2\right)^2 = 2xy$

19. Plot the curve defined by the equation $y^2 = x^3$ and find equations of the *two* lines tangent to the curve at $x = 1$.
20. Plot the curve defined by the equation $y^4 = 4y^2 - x^2$ and find equations of the *four* lines tangent to the curve at $x = 1$.

3.8 Inverse Trigonometric Functions and Their Derivatives

The natural logarithm and exponential functions are inverses; each "undoes" the other. In symbols: ▶

$$\ln(e^x) = x \quad \text{and} \quad e^{\ln x} = x.$$

The first equation holds for all x, the second only for $x > 0$.

In this section we find and discuss inverses for several trigonometric functions.

Nomenclature: "Arc" or "Inverse"?

Two common forms are used to denote inverse trigonometric functions:

   $\arcsin x$, $\arccos x$, *and* $\arctan x$ *are synonyms for* $\sin^{-1} x$, $\cos^{-1} x$, *and* $\tan^{-1} x$.

Each form has its advantages. The first makes explicit the connection with arcs on the unit circle; the second reminds us that function inversion is involved. We'll use both. ▶

Don't confuse $\sin^{-1} x$ with $1/\sin x$. They're entirely different.

A Technical Problem and Its Solution: Restricting Domains

To be invertible, a function must be one-to-one; graphically speaking, it must pass the "horizontal line test." Alas, *no* trigonometric function has this property. Far from it—trigonometric functions, being 2π-periodic, repeat themselves on intervals of length 2π, hitting each output value *infinitely often*. The next example illustrates the problem.

EXAMPLE 1   By any sensible definition, ▶ $\arcsin(0.5)$ should be a number whose sine is 0.5. Find such a number. How many are there?

We'll give a formal definition later.

Solution   The following picture shows four inputs x for which $\sin x = 0.5$.

Values of x for which $\sin x = 0.5$

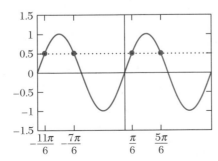

The bulleted points show that

$$0.5 = \sin\left(-\frac{11\pi}{6}\right) = \sin\left(-\frac{7\pi}{6}\right) = \sin\left(\frac{\pi}{6}\right) = \sin\left(\frac{5\pi}{6}\right).$$

Moreover, this list is far from exhaustive; infinitely many more solutions of the equation $\sin x = 0.5$ exist outside our viewing window.

Which of these numbers deserves to be called arcsin(0.5)? None especially, but if *forced* to choose, we might reasonably select $\pi/6$, the solution nearest to zero. Any scientific calculator would agree. According to a TI-82 calculator, for instance,

$$\sin^{-1}(0.5) = 0.5235987756 \approx \frac{\pi}{6}. \qquad \blacksquare$$

Surgery on Domains

The problem just seen—choosing from among reasonable candidates—arises because the trigonometric functions aren't one-to-one. Our solution is crude but effective: Amputate the offending part of each function's domain, and thus *make* the function one-to-one. Graphs will show us where to cut.

Inverting the Sine Function

The sine function is increasing, and therefore one-to-one, on the interval $[-\pi/2, \pi/2]$. (There are, of course, many other intervals on which the sine function is one-to-one. Among all such intervals, this one, centered at 0, seems most natural.) The following picture shows both the ordinary sine curve (drawn dotted) and the special piece (drawn solid) in question here:

Restricting the sine function

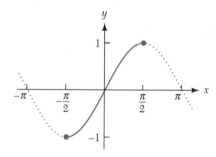

Restricted to this domain, the sine function is invertible. Its inverse is defined formally as follows:

> **Definition** For x in $[-1, 1]$, $y = \sin^{-1} x$ (or $\arcsin x$) is defined by the conditions
>
> $$x = \sin y \qquad \text{and} \qquad -\frac{\pi}{2} \le y \le \frac{\pi}{2}.$$
>
> In words: $\arcsin x$ is the (unique) angle between $-\pi/2$ and $\pi/2$ whose sine is x.

Note the following features.

Graphical Symmetry The graphs of $\sin x$ (restricted!) and $\sin^{-1} x$ show the expected reflective symmetry:

Sine and arcsine

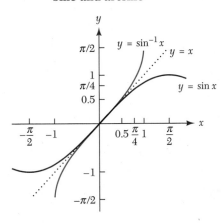

Domain and Range The sine function (as we've restricted it) has domain $[-\pi/2, \pi/2]$ and range $[-1, 1]$. The inverse sine function has it the other way around: Its domain is $[-1, 1]$, and its range is $[-\pi/2, \pi/2]$.

Collapsing Equations By the definition of inverse function, these equations hold:

$$\sin\left(\sin^{-1} x\right) = x \quad \text{if } -1 \le x \le 1;$$

$$\sin^{-1}\left(\sin x\right) = x \quad \text{if } -\pi/2 \le x \le \pi/2$$

Inverting the Cosine Function

The cosine function ▶ becomes one-to-one, and therefore invertible, when restricted to the interval ▶ $[0, \pi]$. This understood, the graphs and formal definition follow naturally.

The situation for $\cos x$ differs only in details from that for $\sin x$.

Restricting $\cos x$ to $[-\pi/2, \pi/2]$ doesn't work. Do you see why not?

Restricting the cosine function

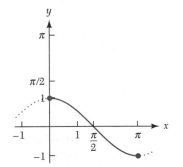

Graphs of cosine and arccosine

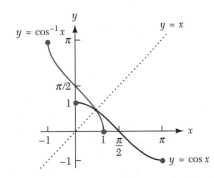

> **Definition** For x in $[-1, 1]$, $y = \cos^{-1} x$ (or $\arccos x$) is defined by the conditions
>
> $$x = \cos y \quad \text{and} \quad 0 \le y \le \pi.$$
>
> In words: $\arccos x$ is the (unique) angle between 0 and π whose cosine is x.

Note the following features.

Graphical Symmetry The graphs of $\cos$ and $\cos^{-1}$ show the usual symmetry for inverse functions—they're symmetric with respect to the line $y = x$.

Domain and Range We restricted the cosine function to have domain $[0, \pi]$ and range $[-1, 1]$. The *inverse* cosine function therefore has domain $[-1, 1]$ and range $[0, \pi]$.

Collapsing Equations As with the arcsine, two equations express the inverse relationship between $\cos$ and $\arccos$:

$$\cos\left(\cos^{-1} x\right) = x \qquad \text{if } -1 \le x \le 1;$$
$$\cos^{-1}\left(\cos x\right) = x \qquad \text{if } 0 \le x \le \pi.$$

Inverting the Tangent Function

The tangent function is one-to-one if restricted to any of its "branches." We'll use the middle branch, through the origin:

The tangent function: one-to-one on the middle branch

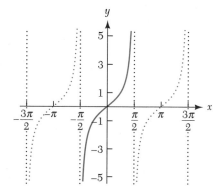

The graph shows this. Do you agree?

Unlike the sine and cosine functions, the tangent function hits *all* real numbers as outputs. ◄ The arctangent function therefore accepts *all* real inputs.

> **Definition** Let x be any real number. Then $y = \tan^{-1} x$ (or $y = \arctan x$) means that
>
> $$x = \tan y \qquad \text{and} \qquad -\pi/2 < y < \pi/2.$$

Note the following features.

Symmetry   The domain of $\arctan x$ (i.e., the range of $\tan x$) is infinite, so any picture of its graph is incomplete. The following figure conveys this idea. Notice the usual symmetry between a function and its inverse.

The tangent function and its inverse

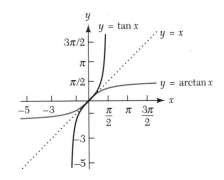

Infinite Domain   The tangent function (as restricted) has domain $(-\pi/2, \pi/2)$ and range $(-\infty, \infty)$. The arctangent function turns things around: It has range $(-\pi/2, \pi/2)$ and domain $(-\infty, \infty)$.

Other Inverse Trigonometric Functions

All six trigonometric functions have inverses (as your calculator knows). As with $\arcsin x$, $\arccos x$, and $\arctan x$, defining $\operatorname{arcsec} x$, $\operatorname{arccsc} x$, and $\operatorname{arccot} x$ requires care with domains and ranges. Among these other functions, the arcsecant is most useful; it appears occasionally in antiderivatives.

For any legal ▶ input t, $\sec t = 1/\cos t$. The simplest definition of $\operatorname{arcsec} x$— based on $\arccos x$—reflects this fact:

There's trouble if $\cos t = 0$.

> **Definition** For any x with $|x| \geq 1$,
>
> $$\operatorname{arcsec}(x) = \sec^{-1}(x) = \cos^{-1}(1/x).$$

In particular,

$$y = \operatorname{arcsec}(x) = \arccos\left(\frac{1}{x}\right) \implies \cos y = \frac{1}{x} \implies \sec y = x,$$

as we'd expect. ▶

Plot your own copy of the arcsecant function; experiment with various domains. See the exercises at the end of this section for more on the arcsecant function.

Inverse Trigonometric Functions and the Unit Circle

We defined the sine, cosine, and tangent functions in terms of the unit circle. Their inverses can be interpreted in the same way. The following examples show how.

EXAMPLE 2   A unit circle follows; its arclengths—starting from the point $(1, 0)$—are in radians. Use the picture to estimate $\arctan 2$ and $\arcsin(0.5)$.

Arctangent, arcsine, and the unit circle

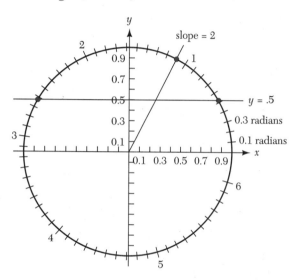

Don't recall? See Appendix G.

Solution   By definition, $\arctan 2$ is an angle (or "arc") whose tangent is 2. Recall ◄   that the tangent of any angle is the *slope* of the line that angle determines. Therefore, we want an angle—the one between $-\pi/2$ and $\pi/2$—that corresponds to a slope of 2. To find this angle, we draw a line from the origin, with slope 2, and observe the radian measure of the angle it determines. The preceding picture shows that the line determines an arc of radian measure about 1.1; this is our estimate for $\arctan 2$. ◄

According to a TI-82 calculator, $\arctan 2 \approx$ 1.107148718.

Reading $\sin^{-1}(0.5)$ from the unit circle is similar. By definition, $\sin t$ is the y-coordinate of the point determined by an arc of length t on the unit circle. Therefore, $\arcsin(0.5)$ is the (one and only) angle between $-\pi/2$ and $\pi/2$ (i.e., to the right of the y-axis) whose corresponding y-coordinate is 0.5. To find this angle, we draw the line $y = 0.5$ (as shown in the preceding picture). It crosses the circle at a bulleted point, the end of an arc with radian measure about 0.52; therefore, $\arcsin(0.5) \approx 0.52$. ◄   ■

This estimate is surprisingly good: The "right" answer is about 0.5235987756.

Combining Ordinary and Inverse Trigonometric Functions

In practice, inverse trigonometric functions are often combined with ordinary trigonometric functions. Diagrams like the following one can help demonstrate how the process works.

Relating inverse trigonometric functions: pictorial aids

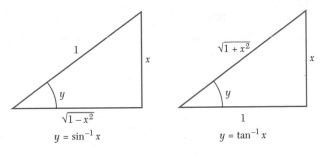

$$y = \sin^{-1} x \qquad\qquad y = \tan^{-1} x$$

The first figure shows (among other things) that

$$\cos(\sin^{-1} x) = \frac{\text{adjacent}}{\text{hypotenuse}} = \sqrt{1 - x^2}.$$

From the second figure, similarly,

$$\sin(\tan^{-1} x) = \frac{\text{opposite}}{\text{hypotenuse}} = \frac{x}{\sqrt{1 + x^2}}.$$

Derivatives of Inverse Trigonometric Functions

We know derivatives of the ordinary trigonometric functions; with the chain rule, we can find derivatives of their inverses. After tabulating the derivatives, we'll derive them.

Derivatives of ordinary trigonometric functions are other trigonometric functions. Surprisingly, inverse trigonometric functions have *algebraic* derivatives:

Derivatives of Inverse Trigonometric Functions				
Function	$\arcsin x$	$\arccos x$	$\arctan x$	$\operatorname{arcsec} x$
Derivative	$\dfrac{1}{\sqrt{1 - x^2}}$	$\dfrac{-1}{\sqrt{1 - x^2}}$	$\dfrac{1}{1 + x^2}$	$\dfrac{1}{|x|\sqrt{x^2 - 1}}$

Observe these features.

Restricted Domains   The functions $\arcsin x$, $\arccos x$, and $\operatorname{arcsec} x$ have restricted domains. So, therefore, must their derivatives. ▶

Not Much Difference   The derivatives of $\arcsin x$ and $\arccos x$ differ only in sign. Therefore, $(\arcsin x + \arccos x)' = 0$ for all legal inputs x. This is no accident, of course. For all x in $[-1, 1]$,

$$\arcsin x + \arccos x = \frac{\pi}{2}.$$

(The graphs of $\arcsin x$ and $\arccos x$ say the same thing.) ▶

For more detail, see the exercises at the end of this section.

Graph both on the same axes to see why.

Graphical Evidence We'll demonstrate by calculation *why* these formulas hold. Graphs show convincingly *that* they hold. Here, for instance, are graphs of the arctangent function and its claimed derivative:

Graph of $f(x) = \arctan x$ and
$f'(x) = 1/(1 + x^2)$

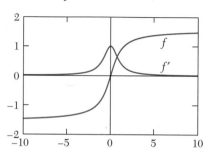

EXAMPLE 3   Explain why $(\arctan x)' = \dfrac{1}{1 + x^2}$.

Solution   If $f(x) = \tan x$ and $g(x) = \arctan x$, then f and g are inverses, and $f'(x) = \sec^2 x$. Unraveling the derivative formula above gives

$$(\arctan x)' = \frac{1}{f'(g(x))} = \frac{1}{\sec^2(\arctan x)}.$$

Despite first appearances, the result is correct. We can simplify the last quantity using the trigonometric identity $\sec^2 t = 1 + \tan^2 t$:

$$\frac{1}{\sec^2(\arctan x)} = \frac{1}{1 + \tan^2(\arctan x)} = \frac{1}{1 + x^2}. \qquad \blacksquare$$

EXAMPLE 4   Explain why $(\arcsin x)' = \dfrac{1}{\sqrt{1 - x^2}}$.

Solution   We play the same game as before, with slightly different twists. If $f(x) = \sin x$ and $g(x) = \arcsin x$, then $f'(x) = \cos x$, so ◄

Watch for the following combination of the cosine and arcsine functions.

$$(\arcsin x)' = \frac{1}{f'(g(x))}$$

$$= \frac{1}{\cos(\arcsin x)}$$

$$= \frac{1}{\sqrt{1 - \sin^2(\arcsin x)}}$$

$$= \frac{1}{\sqrt{1 - x^2}}. \qquad \blacksquare$$

The remaining derivative formulas can be found similarly.

BASICS

Find the numbers in Exercises 1–6 exactly (in radians).

1. $\arcsin(1)$
2. $\arcsin\left(\sqrt{3}/2\right)$
3. $\arccos(-1)$
4. $\arccos\left(-\sqrt{2}/2\right)$
5. $\arctan(1)$
6. $\arctan\left(\sqrt{3}\right)$

Use the unit circle on page 242 to estimate the numbers in Exercises 7–10.

7. $\arcsin(0.8)$
8. $\arccos(0.6)$
9. $\arctan(0.4)$
10. $\arctan(3)$

11. This table gives samples of what one calculator knows (and doesn't know) about inverse trigonometric functions:

Inverse Trigonometric Function Values

x	−2	−1	−0.5	−0.25
arcsin x	−1.57	−0.52	−0.25	0.00
arccos x	3.14	2.09	1.82	1.57
arctan x	−1.11	−0.79	−0.46	−0.24
x	0	0.25	0.5	1
arcsin x	0.25	0.52	1.57	
arccos x	1.31	1.04	0.00	
arctan x	0.00	0.24	0.46	0.79
x	2	5	100	
arcsin x				
arccos x				
arctan x	1.11	1.37	1.56	

(a) Can the gaps in the table above be filled in? Why or why not? Answer using the word "domain."

(b) In each column, the entries for $\arcsin x$ and $\arccos x$ add to 1.57. Surely this can't be coincidence! What's going on?

12. (a) The function $\operatorname{arcsec} x$ accepts as inputs only values of x for which $|x| \geq 1$. Why is this restriction necessary? [**HINT**: What's the domain of $\arccos x$?]

(b) What's the *range* of $\operatorname{arcsec} x$? [**HINT**: What's the range of $\arccos x$?]

Write each of the following as an algebraic expression. For example, $\cos(\arcsin x) = \sqrt{1 - x^2}$.

13. $\sin(\arccos x)$
14. $\tan(\arcsin x)$
15. $\sin(\arctan x)$
16. $\cos(\arctan x)$

In Exercises 17–28, find the derivative of the function f.

17. $f(x) = \arctan(2x)$
18. $f(x) = \arctan\left(x^2\right)$
19. $f(x) = \sqrt{\arcsin x}$
20. $f(x) = \arcsin(\sqrt{x})$
21. $f(x) = \arcsin(e^x)$
22. $f(x) = e^{\arctan x}$
23. $f(x) = \arctan(\ln x)$
24. $f(x) = \arccos(2x + 3)$
25. $f(x) = \operatorname{arcsec}(x/2)$
26. $f(x) = \arcsin x / \arccos x$
27. $f(x) = x^2 \arctan\left(\sqrt{x}\right)$
28. $f(x) = \ln(2 + \arcsin x)$

In Exercises 29–41, find an antiderivative of the function f. (Guess and check.)

29. $f(x) = 1/\left(1 + x^2\right)$
30. $f(x) = 1/\sqrt{1 - x^2}$
31. $f(x) = 2/\left(1 + 4x^2\right)$
32. $f(x) = 3/\left(9 + x^2\right)$
33. $f(x) = 1/\sqrt{9 - x^2}$
34. $f(x) = 1/\sqrt{1 - 4x^2}$
35. $f(x) = e^x/\left(1 + e^{2x}\right)$
36. $f(x) = x/\sqrt{1 - x^4}$
37. $f(x) = (2\arctan x)/(1 + x^2)$
38. $f(x) = (\cos x)/\left(1 + \sin^2 x\right)$
39. $f(x) = e^{\arcsin x}/\sqrt{1 - x^2}$
40. $f(x) = 1/\left(x\left(1 + (\ln x)^2\right)\right)$
41. $f(x) = 1/\left(\left(1 + x^2\right)\arctan x\right)$

FURTHER EXERCISES

42. Find 4 values of x for which $\cos x = 0.5$.

43. Use a calculator to find 4 values of x (2 positive, 2 negative) for which $\tan x = 2$.

44. The graph of the tangent function has *vertical* asymptotes. Where are they? Does the graph of the *arctangent* function have asymptotes? If so, where? Why? [**HINT**: Look carefully at the graphs of the tangent function and its inverse. Explain what you see in terms of reflection across the line $y = x$.]

45. The equation arcsin (sin *x*) = *x* suggests that the graph of *y* = arcsin (sin *x*), over any interval, might be the straight line *y* = *x*. It isn't (see below).

Graphs of *y* = arcsin (sin *x*) and *y* = *x*

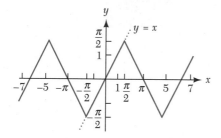

(a) Explain why arcsin(sin 5) ≠ 5.
(b) For which values of *x* *does* the equation arcsin(sin *x*) = *x* hold?
(c) Let $f(x) = \arcsin(\sin x)$. Show that $f'(x) = \dfrac{\cos x}{|\cos x|}$.
(d) How is the result in part (c) related to the graph shown above?

46. (a) What is the domain of the derivative of the arcsine function? Give a geometric explanation for why it is not the same as the domain of the arcsine function.
(b) What is the domain of the derivative of the arctangent function? Is it the same as the domain of the arctangent function?
(c) What is the domain of the derivative of the arcsecant function? Is it the same as the domain of the arcsecant function?

47. Let $f(x) = \arctan x$.
(a) Plot graphs of *f* and *f'* on the same axes.
(b) Evaluate $\lim_{x\to\infty} f(x)$, $\lim_{x\to-\infty} f(x)$, $\lim_{x\to\infty} f'(x)$, and $\lim_{x\to-\infty} f'(x)$. What do these results imply about the graph of the arctangent function?
(c) Is *f* even, odd, or neither?
(d) Show that *f* is increasing on $(-\infty, \infty)$. How does the graph of *f'* reflect this fact?
(e) Where is *f* concave up? Concave down? Does *f* have any inflection points? If so, where?

48. Let $f(x) = \arcsin x$.
(a) Plot graphs of *f* and *f'* on the same axes.
(b) Evaluate $\lim_{x\to 1^-} f(x)$, $\lim_{x\to-1^+} f(x)$, $\lim_{x\to 1^-} f'(x)$, and $\lim_{x\to-1^+} f'(x)$. What do these results imply about the graph of the arcsine function?
(c) Is *f* even, odd, or neither?
(d) Show that *f* is increasing on $(-1, 1)$. How does the graph of *f'* reflect this fact?
(e) Where is *f* concave up? Concave down? Does *f* have any inflection points? If so, where?
(f) How does the graph of *f'* reflect the information found in part (e)?

49. Let $f(x) = \arccos x$.
(a) Plot graphs of *f* and *f'* on the same axes.
(b) Evaluate $\lim_{x\to 1^-} f(x)$, $\lim_{x\to-1^+} f(x)$, $\lim_{x\to 1^-} f'(x)$, and $\lim_{x\to-1^+} f'(x)$. What do these results imply about the arccosine function?
(c) Is *f* even, odd, or neither?
(d) Show that *f* is decreasing on $(-1, 1)$. How does the graph of *f'* reflect this fact?
(e) Where is *f* concave up? Concave down? Does *f* have any inflection points? If so, where are they located.
(f) How does the graph of *f'* reflect the information found in part (e)?

50. Let $f(x) = \operatorname{arcsec} x$.
(a) Plot graphs of *f* and *f'* using the plotting window $[-5, 5] \times [0, 5]$. [**NOTE:** If your calculator or computer doesn't recognize the arcsecant function, trick it by plotting $\arccos(1/x)$ instead.]
(b) Evaluate $\lim_{x\to 1^+} f(x)$, $\lim_{x\to-1^-} f(x)$, $\lim_{x\to 1^+} f'(x)$, and $\lim_{x\to-1^-} f'(x)$. What do these results imply about the graph of the arcsecant function?
(c) Evaluate $\lim_{x\to\infty} f(x)$, $\lim_{x\to-\infty} f(x)$, $\lim_{x\to\infty} f'(x)$, and $\lim_{x\to-\infty} f'(x)$. What do these results imply about the graph of the arcsecant function?
(d) Is *f* even, odd, or neither?
(e) Show that *f* is increasing everywhere in the interior of its domain. How does the graph of *f'* reflect this fact?
(f) Where is *f* concave up? Concave down? Does *f* have any inflection points? If so, where are they located?
(g) How does the graph of *f'* reflect the information found in part (f)?

51. Show that $x/(1 + x^2) \le \arctan x \le x$ for all $x \ge 0$.

52. (a) By drawing an appropriately labeled right triangle, show that $\arctan x + \arctan(1/x) = \pi/2$ if $x > 0$.
(b) Use calculus to show that $\arctan x + \arctan(1/x) = \pi/2$ for all $x > 0$. [**HINT:** Start by differentiating both sides of the equation.]
(c) Find an equation similar to the one in part (a) that is valid when $x < 0$.

Use calculus to prove the identities in Exercises 53–57. [**HINT:** See Exercise 52.]

53. $\arcsin(-x) = -\arcsin(x)$

54. $\arccos(-x) = \pi - \arccos(x)$

55. $\arccos x = \pi/2 - \arcsin x$

56. $\arctan\left(x/\sqrt{1 - x^2}\right) = \arcsin x$

57. $\arcsin\left(\dfrac{x - 1}{x + 1}\right) = 2\arctan(\sqrt{x}) - \pi/2$

58. (a) Show that $2\arcsin x = \arccos(1 - 2x^2)$ if $0 \le x \le 1$.
(b) Find an identity similar to the one in part (a) that is valid if $-1 \le x \le 0$.

59. Let $f(x) = \arctan\left(\dfrac{1+x}{1-x}\right)$.

    (a) Assume that $x < 1$. Show that there is a constant C such that $f(x) = C + \arctan x$.
    (b) Use the result in part (a) to evaluate $\lim_{x \to 1^-} f(x)$.
    (c) Evaluate $\lim_{x \to 1^+} f(x)$.

60. (a) Show that $\tan(\operatorname{arcsec} x) = \sqrt{x^2 - 1}$ if $x \geq 1$.
    (b) Find an algebraic expression for $\tan(\operatorname{arcsec} x)$ if $x \leq -1$.

61. (a) Derive the identity

$$\arctan x + \arctan y = \arctan\left(\frac{x+y}{1-xy}\right)$$

from the identity $\tan(x + y) = \dfrac{\tan x + \tan y}{1 - \tan x \tan y}$.

    (b) What conditions must be satisfied by x and y for the identity derived in part (a) to hold?

62. (a) Use the identity in part (a) of the previous problem to show that $\pi/4 = \arctan(1/2) + \arctan(1/3)$.
    (b) Show that $\pi/4 = 2\arctan(1/4) + \arctan(7/23)$. [**HINT**: Use the identity in part (a) of the previous problem twice.]

3.9 Chapter Summary

This chapter is about derivatives and antiderivatives of elementary functions—how to calculate and combine them, how to use them, and why they matter.

The Simplest Derivatives; First Uses. Section 3.1 introduced and used symbolic derivatives in the simplest possible setting: polynomials and power functions. Section 3.2 illustrated a few of the practical uses of derivative formulas.

Derivatives of Transcendental Functions. Sections 3.3 and 3.4 discussed derivatives of exponential, logarithmic, and trigonometric functions—the transcendental functions of elementary calculus. Finding derivatives of **algebraic functions** is—no surprise here—usually a matter of algebra. For **transcendental functions**, finding derivatives takes a little more ingenuity.

Antiderivatives. Differentiation is a two-way street: If f is the derivative of g, then g is an **antiderivative** of f. Throughout the chapter we kept an eye on the inverse problem of finding antiderivatives for given functions. In general, the problem is harder than the "forward" problem of finding derivatives; in Calculus II we'll return to finding antiderivatives.

New Derivatives from Old. Having found derivatives and antiderivatives of (almost) all the basic elementary functions, we turned in Sections 3.5 and 3.6 to combining those derivatives, in the appropriate ways, to find derivatives (and sometimes antiderivatives) of functions built in various ways from the basic elements. The key combinatorial rules are the **product and quotient rules** (for algebraic combinations) and the **chain rule**, for functions built by composition.

Derivatives of Functions Defined Implicitly. Sometimes functions are given implicitly by equations rather than explicitly by formulas. In such cases, the method of **implicit differentiation** applies; we discussed it in Section 3.7.

Inverse Trigonometric Functions. Inverse trigonometric functions ("arc" functions) are defined as inverses—in the sense that the natural log and exponential functions are inverses—of the trigonometric functions. In Section 3.8 we explored these functions and their derivatives, using tools (such as implicit differentiation) developed in earlier sections.

4

Applications of the Derivative

4.1 Differential Equations and Their Solutions

Such as Pythagoras's formula, $a^2 + b^2 = c^2$.

Algebraic equations ◄ describe relations among varying quantities. *Differential equations* go one step—a giant step—further: They describe relations among changing quantities and the rates at which they change. Only by solving differential equations can many real-life phenomena be usefully described.

> *Nature's voice is mathematics; its language is differential equations.*

The idea of a differential equation first appeared, informally, in Section 2.1. Here and in the next section we take a closer look.

Differential Equations: Basic Ideas

Observe the various derivative notations; all are used in practice.

A **differential equation** (**DE**) is any equation that contains one or more derivatives. Each of the following is a DE: ◄

$$f'(x) = 6x + 5; \qquad y' = y; \qquad \frac{dy}{dx} = 5;$$

$$g'(x) = kg(x); \qquad y''(t) = -32.$$

Or more than one.

(The last DE has **order two**, because it involves a second derivative. The others have **order one**.) To **solve** a differential equation means to find a *function* ◄ that satisfies the DE. Any such function is called a **solution** to the DE. The point bears repeating:

> *A solution to a DE is a function—not a number—that satisfies the DE.*

Simple examples will illustrate the ideas and terminology; watch too for some quirks of notation.

> **EXAMPLE 1** Solve the DE $f'(x) = 6x + 5$. How many solutions are there? How are they related to each other?

248

S o l u t i o n   The DE asks for an antiderivative f of the function $6x + 5$. It's easy to check ▶ that for any constant C, the function $f(x) = 3x^2 + 5x + C$ does the job. Thus, this DE has infinitely many solutions; they differ from each other by *additive* constants. ▶

But do so now!

How do the solutions differ from each other graphically?

EXAMPLE 2   Solve the DE $y' = y$. How many solutions are there? How are they related to each other?

S o l u t i o n   Let's guess a solution. The DE $y' = y$ says that the desired function y is its own derivative. The natural exponential function has this property, so $y = e^t$ is a reasonable guess for a solution.

Once we've made it, our guess is easy to check:

$$y(t) = e^t \implies y'(t) = e^t = y(t).$$

Thus $y' = y$, as we wanted, so $y(t) = e^t$ is indeed a solution of the DE.

Are there other solutions? Given the previous example we might try $y(t) = e^t + C$ for a nonzero constant C. A good try, but no dice:

$$y(t) = e^t + C \implies y'(t) = e^t \neq y(t).$$

Other solutions *do* exist, however. For any real constant C, $y(t) = Ce^t$ works: ▶

$$y(t) = Ce^t \implies y'(t) = Ce^t = y(t).$$

Is $C = 0$ OK? Why or why not?

Again, there are infinitely many solutions, but this time they differ from each other by *multiplicative* constants.

Here are some morals from the examples:

No Arguments Here.   DEs are often stated without arguments. For example, all of the DEs

$$y' = y, \qquad y'(x) = y(x), \qquad \text{and} \qquad y'(t) = y(t)$$

say the same thing: *The desired function y is its own derivative.* The input variable name, or **argument**—either x or t in the preceding examples—can safely be assumed or is understood, so it's often omitted. Once gotten used to, the omission of arguments causes little confusion. Omitting inessential function arguments sometimes unclutters and simplifies a DE's appearance. For example, both

$$\frac{d^2y}{dt^2} + 2t\frac{dy}{dt} + t = y(t) \qquad \text{and} \qquad y'' + 2ty' + t = y$$

say the same thing, but the latter form is more economical and (depending on the method) easier to handle.

Variable Names.   In writing DEs the symbol t, rather than x, often denotes the input variable. This choice is made because the letter t suggests *time*. A solution y (also known as $y(t)$), then, is naturally thought of as a quantity that varies with time.

Instead of x, u, v, or any other letter.

The use of t, ◄ however, is *only* an aid to intuition; mathematically speaking, one variable name is as good as another. DEs, moreover, can model physical situations that have nothing to do with time.

Checking Is Easy, Finding Is Hard. Finding solutions to DEs, starting from scratch, can be very difficult. Many thick books treat the subject. *Checking* whether a given function satisfies a DE, by contrast, is usually much easier.

Initial Value Problems

A first-order DE involves only the first derivative.

A unique solution, in math-talk.

Examples 1 and 2 are typical of first-order DEs. ◄ Such DEs usually have infinite families of solutions, in which the family members differ from each other by additive or multiplicative constants. If we also stipulate that a solution function should have a particular output value for a particular input, then we expect only one solution. ◄ This extra stipulation is called an **initial condition**, and the combination of DE and initial condition is known as an **initial value problem** (**IVP**) .

EXAMPLE 3   Solve the IVP

$$f'(x) = 6x + 5; \qquad f(0) = 42.$$

Solution   As we saw, any function of the form $f(x) = 3x^2 + 5x + C$ is a solution of the DE. With the initial condition we can find C:

$$f(0) = 3 \cdot 0^2 + 5 \cdot 0 + C = 42 \Longrightarrow C = 42.$$

Thus $f(x) = 3x^2 + 5x + 42$ is the unique solution to the IVP.   ∎

EXAMPLE 4   Solve the IVP

$$y' = y; \qquad y(0) = 42.$$

Solution   As we saw already, any function of the form $y(t) = Ce^t$ solves this DE. Again, the initial condition forces C to be 42, so $y(t) = 42e^t$ is the unique solution to the IVP.   ∎

Solving DEs and IVPs: Guessing, Checking, and Parameters

Checking a Guess

Given a DE and a guess at a solution, it's straightforward to check whether the guess works. The next example shows what we mean.

EXAMPLE 5   Consider the DE $y' = k(y - T)$, where k and T are constants. Is $y(t) = T + Ae^{kt}$ a solution? Does the value of the constant A matter?

S o l u t i o n   Deciding whether $y(t) = T + Ae^{kt}$ solves the DE is a straightforward calculation—we'll calculate both the left-hand side (LHS) and the right-hand side (RHS) of the DE, using this y, and compare our results:

$$\text{LHS} = y'(t) = k \cdot Ae^{kt};$$

$$\text{RHS} = k(y - T) = k(T + Ae^{kt} - T) = k \cdot Ae^{kt}.$$

The two sides are equal, so y is a solution.

So far the value of A hasn't mattered—y solves the DE for *any* constant A. Had we specified an initial condition, we would have needed a specific value of A. ▶   *See the next example.*

■

From One Solution, Many: Choosing Values of Parameters

The DE $y' = k(y - T)$ involves two constant parameters, k and T. The general solution, $y(t) = T + Ae^{kt}$, involves still another parameter, A.

Such parameters offer an important "leverage" advantage. Assigning them specific values as needed permits us, in effect, to solve infinitely many differential equations "for the price of one." We illustrate with a practical example.

E X A M P L E 6   Freshly poured coffee in a ceramic cup ▶ has a temperature of 190°F. As it cools, the coffee obeys **Newton's law of cooling**: ▶

In a better-insulated cup, say one made of foam, coffee might cool more slowly. (See the exercises.)

   The rate at which an object cools is proportional to the temperature difference between the object and its environment.

Cooling is actually more complicated than this simple—but still useful—model suggests.

Assume that room temperature is a constant 65°F, and that after 2 minutes, the coffee has cooled to 160°F. Find a formula for the temperature at time t.

S o l u t i o n   Under the verbiage of this example lies the DE of Example 5. If $y(t)$ represents temperature at time t, then Newton's law says that for some constant k, ▶

We'll evaluate k soon.

$$y' = k(y - 65), \qquad y(0) = 190, \qquad \text{and} \qquad y(2) = 160.$$

As we saw in Example 5, a solution of this DE has the general form

$$y(t) = 65 + Ae^{kt}$$

for some constants A and k. All that remains is to give them numerical values.

From the initial condition $y(0) = 190$, it follows that

$$190 = y(0) = 65 + Ae^{k \cdot 0} = 65 + A \implies A = 125.$$

Similarly, $y(2) = 160$ means ▶   that

Check the steps.

$$160 = y(2) = 65 + 125e^{2k} \implies e^{2k} = \frac{95}{125}.$$

Take logs of both sides.

Solving the last equation ◀   gives $k = (\ln 95 - \ln 125)/2 \approx -0.13722$. Putting it all together gives (approximately)

$$y(t) = 65 + 125e^{-0.13722t}.$$

Does the answer make common sense? A graph of y over a 50-minute period suggests it does:

Cooling coffee:
$$y(t) = 65 + 125e^{-0.13722t}$$

As the temperature graph shows, the coffee cools quickly at the beginning, when it's hottest, and then more slowly as it nears room temperature.   ∎

Reading and Writing DEs and IVPs: The Language of Change

And should!

Or family of functions.

If $y = y(x)$ is *any* function of x, then the derivative function y' (that is, dy/dx) represents the instantaneous rate of change of y with respect to x. A differential equation, therefore, can ◀   be understood as a highly compressed statement about the *rate* at which some unknown function "grows." To *solve* a differential equation, therefore, means to find a function ◀   that "grows" in the way the DE stipulates. The DE $y' = y$, for instance, is a succinct way of saying:

> *The quantity y grows at a rate equal to y itself. Thought of as a function, y is its own derivative.*

As remarked above, functions of the form $y = Ce^x$ (but no others!) have this important property.

Translating to and from DE Language

That is, in both directions.

An ability to "translate" back and forth ◀   between the symbolic language of DEs and ordinary English statements about changing quantities and derivatives is important, both for understanding what DEs say and for applying them to solve worthwhile problems. The accompanying table illustrates several such translations. In each entry, y represents an unknown function $y(t)$ of a variable t; ◀ k denotes a constant.

We might have used x instead of t, but t naturally suggests time.

In the ordinary sense of the word "fast."

If the input variable t denotes time, then $y = y(t)$ represents a quantity that varies with time; $y' = y'(t)$ (that is, dy/dt) tells how fast ◀ y grows. (Thinking of the input variable as time is intuitively convenient, but it's not necessary.

Sometimes the input variable denotes distance, position, or some other physical quantity.)

Translating between DE Language and English	
DE Version	**English Version**
$y' = 0$	The growth rate of y is zero (i.e., y remains constant).
$y' = 0.06y$	The growth rate of y is *proportional*, with proportionality constant 0.06, to y itself (i.e., y behaves like a bank account drawing 6% interest).
$y' = kt$	The growth rate of y at time t is *proportional* to t.
$y'' = k$	y has constant *second derivative* (i.e., y has constant "acceleration").
$y' = k(y - 65)$	y varies at a rate proportional to the *difference* between y and 65 .

Interpreting DEs in Real Contexts

Each DE in the table above describes how some (unknown) function y varies with time. Does any familiar, real-life quantity $y(t)$ vary in the manner described by each DE?

A DE chosen purely at random need not "model" any familiar real-world phenomenon. However, many simple DEs—including all those in the table above—do describe important, naturally occurring phenomena. The DE $y' = 0$ describes *any* quantity that remains unchanged over time—for instance the position of an immovable object or the value of a financial "deposit" under one's mattress.

In a similar vein, the DE $y' = 0.06y$ describes the value $y(t)$ at time t of a deposit that bears 6% interest. (We'll discuss financial interest in more detail in the next section.)

The DEs $y' = kt$ and $y'' = k$ are closely related; every solution to the first automatically satisfies the second. ▶   In physical language, both DEs describe quantities $y(t)$ with constant acceleration. For instance, $y(t)$ might represent the varying height of a free-falling object, because earth's gravity exerts a constant downward force, and so causes a constant downward acceleration. (Newton's second law of motion says so.)

Do you see why?

The DE $y' = k(y - 65)$, as we saw in Example 6, can be interpreted as an instance of Newton's law of cooling. If $y(t)$ is the temperature at time t of an object in 65°F room, then the DE says (à la Newton) that the object warms or cools at a rate *proportional to the difference between its own temperature and room temperature.*

Interpreting Initial Conditions

A DE such as $y' = y$ tells how fast a varying quantity $y(t)$ grows, but not how large y is at any specific time. An initial condition, such as $y(0) = 1$, adds a point of reference. It tells, in effect, the value from which y starts.

What initial conditions mean in real-world settings depends on the situation. In the various previously described scenarios an initial condition might specify, at some specific time, the value of a bank deposit, the vertical position of a falling object, or the temperature of a cooling object.

With continuous compounding.

Notice that an "initial" condition need not refer specifically to time $t = 0$. In "DE-speak," *any* condition of the form $y(a) = b$ is called an initial condition. In modeling the growth of a bank account, for instance, the DE $y' = 0.06y$ describes a 6% interest rate. ◄   The initial condition $y(1995.753) = 100$ might describe an initial deposit of $100 on October 1, 1995.

Hard Work, but Worth Doing. Solving DEs can be much more complicated than our simple examples suggest. A gigantic literature exists on the theory and methods of solving many sorts of differential equations. This body of literature attests to the depth and subtlety of the subject—but even more to its practical importance.

BASICS

In Exercises 1–10 decide whether the function is a solution of the differential equation given with it.

1. $y(t) = t^2/2$; $y' = t$

2. $y(t) = t^2/2$; $y' = y$

3. $y(t) = -e^t$; $y' = -y$

4. $y(t) = e^{-t}$; $y' = -y$

5. $y(t) = 42e^{-t}$; $y' = -y$

6. $y(t) = Ce^{kt} + A$; $y' = k(y - A)$

7. $y(t) = e^t + \frac{1}{2}t^2$; $y' = y + t$

8. $y(t) = \frac{1}{2}e^{t^2}$; $y' = ty$

9. $y(t) = e^{t^2/2}$; $y' = ty$

10. $y(t) = \frac{1}{2}t^4 + \frac{3}{2}t^2 + \frac{1}{4}$; $y' - \frac{2y}{t} = t^3$

FURTHER EXERCISES

11. This exercise relates to Example 6, which is about coffee cooling in a 65°F room. We found that for coffee in a ceramic cup, starting at 190°F, the solution function $y(t) = 65 + 125e^{-0.13722t}$ reasonably represents the (Fahrenheit) temperature after t minutes.

    (a) How long does the coffee in a ceramic cup remain above 100°F?

    (b) In a better-insulated cup, say one made of foam, coffee stays hot longer. Starting from 190°F, coffee in a foam cup is found to cool to 180°F in two minutes in a 65°F room. Let $z(t)$ represent the temperature of the coffee in a foam cup. Find a formula for $z(t)$.

    (c) How long does coffee in a foam cup stay above 100°F?

    (d) Plot $y(t)$ and $z(t)$ on the same axes.

12. (a) Check by differentiation that for any constant d,
    $P(t) = \dfrac{1}{1 + de^{-t}}$ is a solution of the DE $P' = P(1 - P)$. (This DE is a simple form of the so-called **logistic DE**; it's used to model certain populations.)

    (b) Solve the IVP $P' = P(1 - P)$; $P(0) = 1/2$.
    [**HINT:** Use the initial condition to choose a value for d.]

    (c) Solve the IVP $P' = P(1 - P)$; $P(0) = 2$.

    (d) Plot the solution functions for the previous two parts on the same axes; use the t-interval $0 \le t \le 5$.

13. Verify by direct calculation that if k, C, and d are con-

stants, then the function

$$P(t) = \frac{C}{1 + d\,e^{-k \cdot C \cdot t}}$$

is a solution of the logistic DE $P' = kP(C - P)$. (We'll see the logistic DE in later applications.)

14. When the valve at the bottom of a cylindrical tank is opened, the depth of liquid in the tank drops at a rate proportional to the square root of the depth of liquid. (This fact, known as **Torricelli's law**, is named after Evangelista Torricelli [1608–1647], a contemporary of [and, briefly, secretary to] Galileo.) If $y(t)$ is the liquid's depth t minutes after the valve is opened, Torricelli's law can be expressed as the DE $y' = -k\sqrt{y}$, where k is some positive constant.

    (a) Does the water level drop faster when a tank is full or when it is half full? Explain your answer.

    (b) Verify by differentiation that, for any constants C and k, the function $y(t) = (C - kt)^2/4$ is a solution of Torricelli's DE $y' = -k\sqrt{y}$. (For physical reasons, it's OK to assume that $C - kt \ge 0$.)

    (c) Suppose that for a certain tank, $y(0) = 9$ feet and $y(20) = 4$ feet. Find an equation for $y(t)$. Show that the tank takes 60 minutes to empty entirely.

    (d) Plot $y(t)$ over the interval $0 \le t \le 60$.

    (e) At what time is the water level dropping at a rate of 0.1 feet per minute?

15. Let $h(t)$ represent the height, in meters above ground level, of an object (a helium balloon, for instance, or a cannonball, an airplane, or a toy rocket) at time t seconds. Let $v(t) = h'(t)$ represent the vertical velocity, in meters per second, at time t seconds. Note that $v'(t) > 0$ means the object is rising; when $v'(t) < 0$, the object falls.

    In this situation, the DE $v' = -9.8$ means the same thing as the English statement

    The object has constant downward acceleration of 9.8 meters per second per second.

    (The negative sign appears in the DE because the acceleration is downward.) Translate each of the following DEs into an English statement about the physical situation. (Specify units carefully.)

    (a) $v'(t) = 1$ (c) $v'(t) = -0.01\big(v(t)\big)^2.$
    (b) $h'(t) = -3$

16. Let the situation and notation be as in the previous exercise.
    (a) Solve the IVP $v'(t) = 1$; $v(0) = 0$. Assuming these conditions, what is the object's upward velocity at $t = 10$ seconds? [**HINT**: If v' is a constant function, then v is a linear function.]
    (b) Solve the IVP $h'(t) = -3$; $h(0) = 100$. Find and interpret $h(10)$. [**HINT**: If h' is a constant function, then h is a linear function.]
    (c) Verify by differentiation that for any constant C, $v(t) = 100/(t + C)$ solves the DE $v' = -0.01v^2$.
    (d) Suppose that $v' = -0.01v^2$ and $v(0) = 5$. What's $v(30)$? What's $v(80)$? Interpret your answers in physical language. [**HINT**: See the previous part.]

17. An object moves along an east–west axis. Distance is measured in meters and time in seconds. Let $p(t)$, $v(t)$, and $a(t)$ denote, respectively, the object's position (in meters east of the starting point), eastward velocity (in meters per second), and eastward acceleration (in meters per second per second) at time t. Then $p' = v$ and $p'' = v' = a$.
    In this situation, the English statement

    The object has constant eastward acceleration 10 meters per second per second.

    says the same thing as the first-order DE $v' = 10$. Translate each of the following English statements into a first-order DE involving either v' or p'.
    (a) The object accelerates westward at a constant 5 meters per second per second.
    (b) The object travels eastward at a constant speed of 15 meters per second.
    (c) Because of friction, the object's eastward acceleration is proportional to the velocity. [**NOTE**: The proportionality constant will be negative; explain why.]

    (d) The object's eastward velocity is proportional to the square root of the object's position.

18. In learning theory, a "performance function" $P(t)$ may be used to measure someone's skill at a task (using units appropriate to the task) as a function of the training time t. The graph of a performance function is called a **learning curve**. For many tasks, performance improves quickly at first, but then tapers off (i.e., the rate of learning decreases) later, as the value of $P(t)$ approaches M, some maximal level of performance.
    (a) Explain how the DE $P' = k(M - P)$, where k is a positive constant, describes this situation. (Notice the similarity to Newton's law of cooling!)
    (b) Verify by differentiation that for any constant C, the function $P(t) = M - Ae^{-kt}$ solves the DE above.
    (c) For a certain learning activity, measurements show that $P(0) = 0.1M$ and $k = 0.05$; t is measured in hours. How long does it take to reach 90% of the maximum level of performance?

19. A flu epidemic spreads through a 3000-student college community at a rate proportional to the product of the number of members already infected and the number of those not yet infected. (This product measures the number of possible infectious contacts.)
    Let $P(t)$ represent the number of students infected after t days. Express the statement above as a differential equation.

20. It's a winter day in Frostbite Falls, Minnesota—10 degrees below zero Celsius. Boris and Natasha stop at a convenience store for hot (initially, 90°C) coffee to warm them on the cold, windy walk home. Two types of cups are available: environmentally destructive foam and politically correct cardboard.
    Boris wants foam. "What do I care about the greenhouse effect?" he asks. "Besides, remember the proportionality constant k in Newton's law of cooling—$y'(t) = k \cdot \big(y(t) - C\big)$, where C is the environmental temperature, $y(t)$ is the Celsius temperature of the coffee at time t, and t is measured in minutes? Well, for foam, $k = -0.05$. For cardboard, k is a pathetic -0.08."
    "Do what you want, Boris," says Natasha. "I'd rather save the world. I'm having cardboard. And make mine decaf."
    (a) How long does Boris's coffee stay above 70 degrees? How about Natasha's? How hot is each cup after 5 minutes?
    (b) Redo part (a), assuming that Boris and Natasha drink their coffee in the overheated (25°C) convenience store.
    (c) What value of k is needed to assure that coffee, starting at 90°C, is still at least 70°C after 5 minutes outdoors. [**HINT**: Use the solution function to set u an appropriate equation; then solve for k.]

4.2 More Differential Equations: Modeling Growth

Exponential Growth

Exponential functions have an important property, best expressed in terms of rates:

> *An exponential function grows at a rate proportional to its size, i.e., "with interest."*

We discussed this property in Section 2.9.

Thanks to this property, ◄ exponential functions model many real phenomena that also grow "with interest."

Let's interpret this property in the language of DEs and IVPs. For any constants A and k, let $y(t) = Ae^{kt}$. Two key properties of y matter here: ◄

Check these claims carefully!

$$y'(t) = Ake^{kt} = ky(t) \qquad \text{and} \qquad f(0) = Ae^0 = A.$$

In other words:

> **Theorem 1** For any constants A and k, the exponential function $y = Ae^{kt}$ solves the initial value problem
> $$y' = ky; \qquad y(0) = A.$$

This theorem offers considerable practical horsepower. Just by choosing k and A judiciously, we can solve a surprising number of useful and important DEs.

Modeling Things That Grow "with Interest"

Many real-world quantities grow or "decay" like exponential functions, at rates proportional to their size. The Consumer Price Index, the value of a financial deposit, college fees, ◄ radioactive decay, biological populations—all can be modeled, more or less accurately, by the IVP of Theorem 1.

Recall Uncle Eric, in Section 1.6.

The Federal Deficit: Keeping Rates Straight

Differential equations relate varying quantities to their derivatives, i.e., their instantaneous rates of change. Practical problems, on the other hand, often mention other sorts of rates; for instance, average rates, interest rates, and percentages. Teasing out the derivatives from all this loose talk may take some care.

Politicians say "hence."

A is for "average."

EXAMPLE 1   At time $t = 0$ years, the federal deficit is \$300 billion and growing fast. "*How* fast?" worries Senator Smith. "What will the deficit be 6 years hence, ◄ when I face reelection?" In Senate hearings, Economist A ◄ predicts:

> *The federal deficit will grow 15% each year.*

I is for "instantaneous."

"Balderdash," says Economist I, ◄

> *The federal deficit will grow at an instantaneous rate of 15% per year.*

Do the two economists disagree? How seriously? Does the difference matter?

Solution   Economists A and I agree on one thing: the deficit grows at a rate proportional to itself. In DE terms, if $y(t)$ represents the deficit in billions, t years out, A and I agree that for some constant k,

$$y'(t) = ky(t).$$

They *disagree*, however, on the value of k. Economist A's statement refers to an *average* rate: Over a 1-year interval, the deficit rises 15%. Economist I's statement means something else: At any *instant*, the deficit's rate of increase, if it remained unchanged for a year, would raise the deficit by 15%.

To understand the subtle difference and see its effects, we'll recast the predictions as IVPs. Combined with the initial value ($300 billion) of the deficit, I's claim comes to this:

$$y'(t) = 0.15 \cdot y(t); \qquad y(0) = 300.$$

A says something else:

$$y'(t) = k \cdot y(t); \qquad y(t+1) = 1.15 \cdot y(t); \qquad y(0) = 300.$$

We'll solve both IVPs and compare the results. I's claim fits perfectly into Theorem 1's template. Since $y'(t) = 0.15 \cdot y(t)$ and $y(0) = 300$, it follows that

$$y(t) = 300e^{0.15t}.$$

This is I's model for the deficit.

Solving A's IVP requires one more step. Since $y'(t) = k \cdot y(t)$ and $y(0) = 300$, it follows (from Theorem 1, again) that $y(t) = 300e^{kt}$.

Now we'll find a value for k. Our new formula $y(t) = 300e^{kt}$ gives $y(t+1) = 300e^{k(t+1)}$. Therefore, ▶

Check the algebra carefully.

$$y(t+1) = 1.15 \cdot y(t) \implies 300e^{k(t+1)} = 1.15 \cdot 300e^{kt}$$

$$\implies e^{k} = 1.15$$

$$\implies k = \ln 1.15 \approx 0.13976.$$

Economist A therefore models the deficit with the function $y(t) = 300e^{0.13976t}$.

What does each scenario predict for the next 6 years? Graphs answer best:

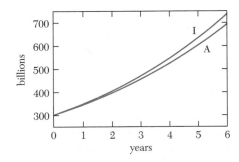

Dueling economists: growth of the deficit

A's and I's predictions diverge as time goes on. At $t = 6$, the difference is

$$300e^{0.15 \times 6} - 300e^{0.13976 \times 6} \approx 43.97 \text{ billion dollars---}$$

real money, even by government standards. ∎

Money in the Bank

Bank interest serves as the most familiar example of growth with interest. Its crucial feature—shared by exponential growth in *all* forms—is that *the rate at which the account's value grows is proportional to the value itself.* The rich don't only get richer—they also get richer *faster.*

Bank interest may even be *too* familiar; some care is needed in translating banking language into mathematical terms. The next example illustrates this.

EXAMPLE 2   A savings account pays a **nominal** annual interest rate of 6%, compounded daily. Model the situation in DE language. How much does the account increase over 1 year?

Solution   If $b(t)$ represents the balance (in dollars) at time t (in years), the preceding terms mean that each day (in a 365-day year), the balance is multiplied by the same small factor:

$$b\left(t + \frac{1}{365}\right) = b(t) \cdot \left(1 + \frac{0.06}{365}\right). \tag{4.1}$$

With initial balance $b(0)$, for instance, here's what happens for the first 2 days:

$$b\left(\frac{1}{365}\right) = b(0) \cdot \left(1 + \frac{0.06}{365}\right); \qquad b\left(\frac{2}{365}\right) = b(0) \cdot \left(1 + \frac{0.06}{365}\right)^2.$$

Over a full year, therefore, the balance increases by the factor

$$\left(1 + \frac{0.06}{365}\right)^{365} \approx 1.0618.$$

The **effective annual interest rate**, therefore, is 6.18%; over a 1-year interval, the balance rises by this percentage.

How do interest rates, nominal or effective, translate into DE language?

Not continuously.

- In reality, the balance $b(t)$ varies discretely, ◄ at regular intervals and in jumps of at least 1 cent. Any DE model, being continuous, must commit *some* error. In practice, such errors are usually small enough to ignore.

- On a time scale measured in years, a single day is effectively an "instant." The *nominal* interest rate can therefore serve as k in the DE of Theorem 1. In other words, the balance $b(t)$ can be taken, with little error, to satisfy the DE

$$b'(t) = 0.06 \cdot b(t).$$

Theorem 1 solves the preceding DE. If the initial balance is b_0, then

$$b'(t) = 0.06 \cdot b(t) \quad \text{and} \quad b(0) = b_0 \Longrightarrow b(t) = b_0 \cdot e^{0.06t}.$$

After 1 year the balance is

$$b(1) = b_0 \cdot e^{0.06} \approx b_0 \cdot 1.0618.$$

Thus, the balance rises by 6.18%, the *effective* interest rate. ■

Interest Rates and Derivatives: Continuous Compounding. The DE $b'(t) = 0.06 \cdot b(t)$ does a good—but not perfect—job of modeling 6% annual interest, compounded daily. The model would do better if interest were compounded hourly or even every minute. The DE represents **continuous compounding**, i.e., the limiting case as the number of compounding periods tends to infinity.

Why does the DE $b'(t) = 0.06 \cdot b(t)$ approximate daily compounding as well as it does? Equation 4.1 and the following computation (for those who wish to follow it) explain. The result shows, too, how the frequency of compounding affects the "goodness" of approximation.

$$b'(t) \approx \frac{b(t + 1/365) - b(t)}{1/365}$$

$$= \left(b(t) \cdot \left(1 + \frac{0.06}{365} \right) - b(t) \right) \cdot 365$$

$$= b(t) \cdot \left(1 + \frac{0.06}{365} - 1 \right) \cdot 365$$

$$= 0.06 \cdot b(t).$$

Radioactive Decay

A sample of carbon-14 (a radioactive isotope of ordinary carbon) decays at time t (in years) at a rate proportional to its weight, with $k = -0.000121$. In DE language,

$$W'(t) = -0.000121 W(t).$$

(The constant k is negative because W shrinks. In banking terms, carbon-14 pays negative interest.)

EXAMPLE 3 If a sample of carbon-14 weighed W_0 pounds in 1991, how much will it weigh in 2991? What did it weigh in 991? How long does it take the sample to lose half its weight? (This period is called the sample's **half-life**.)

That is, t = 0 in 1991.

Solution   If $W(t)$ represents the sample's weight t years after 1991, ◄ then $W(t)$ solves the IVP

$$W'(t) = -0.000121 W(t); \qquad W(0) = W_0.$$

According to Theorem 1, $W(t) = W_0 e^{-0.000121 t}$. Therefore,

$$\text{weight in } 2991 = W(1000) = W_0 e^{-0.121} \approx 0.88603 W_0;$$

$$\text{weight in } 991 = W(-1000) = W_0 e^{0.121} \approx 1.12862 W_0.$$

The sample loses a bit more than 12% of its weight over 1000 years.

The sample's half-life is the value of t for which $W(t) = W_0/2$. To find it, we solve for t:

$$W(t) = \frac{W_0}{2} \iff W_0 e^{-0.000121 t} = 0.5 W_0 \iff -0.000121 t = \ln(0.5)$$

$$\iff t \approx 5728 \text{ years.} \qquad \blacksquare$$

Some Biological Populations Grow Exponentially

We'll soon see why such growth can't last forever.

Certain biological populations grow exponentially—at least for a while. ◄ A fruit fly population in laboratory conditions, for example, might reasonably grow at an instantaneous rate of 5% per day, i.e., at a rate proportional (with $k = 0.05$) to the population itself. In DE language, if the population $P(t)$ starts at P_0, then P satisfies the IVP

$$P'(t) = 0.05 P(t); \qquad P(0) = P_0.$$

Notice the analogy with ordinary interest: Each day's crop of new fruit flies can safely be thought of as "interest" added to the previous day's "balance."

EXAMPLE 4   A population of fruit flies grows at an instantaneous rate of 5% per day. How long does it take for the population to double?

Solution   Given an initial population (i.e., at $t = 0$) of P_0 flies, Theorem 1, on page 256, gives the formula

$$P(t) = P_0 e^{0.05 t}.$$

A little algebra shows that the population doubles in less than 14 days:

$$P(t) = P_0 e^{0.05 t} = 2 P_0 \iff 0.05 t = \ln 2 \iff t \approx 13.8629. \qquad \blacksquare$$

Not All Populations Grow Exponentially: Logistic Growth

A logical, not just biological, impossibility.

Real biological populations can't grow exponentially forever. An initial "deposit" of 100 fruit flies, growing as just described, would number over 8 billion in a year. In a decade, the flies would outnumber Earth's molecules. ◄

Any exponentially-growing population sooner or later exceeds the limits (physical or biological) of its environment. This possibility applied to human population worried the 19th-century clergyman Thomas Malthus, who predicted an early and unpleasant end to the human race. ▶ Whether Malthus was right or wrong, exponentially-growing biological populations do eventually outgrow their environments. How real populations cope with this mathematical certainty (and with other vagaries of real life) varies: Some (e.g., lemmings and Dungeness crabs) have boom–bust cycles; others eat up their environments and die off completely.

Was Malthus right? It's too soon to tell. Some evidence may even suggest that Malthus was too optimistic.

A third (happier) possibility is **logistic growth**. As a population approaches an upper limit C (the environment's **carrying capacity**), growth slows. For our flies, logistic growth might lead to the lower of the following graphs.

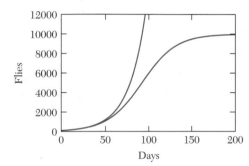

Population growth: exponential vs. logistic

Toward the beginning the two growths are almost identical. This reflects the fact that when populations are small, the effects of crowding are scarcely felt, so logistic growth is essentially exponential.

Like exponential growth, logistic growth is characterized by a rate property, and therefore by a DE (the **logistic DE**):

The population's growth rate is proportional both to the population itself and to the difference between the carrying capacity and the population.

In symbols,

$$P' = kP(C - P),$$

where P represents the population, P' the population's growth rate, C the carrying capacity of the environment, and k a constant of proportionality. ▶

P and P' vary with time; k and C are parameters.

Wise Flies: A Case Study in Logistic Growth

A population of prudent fruit flies, eager to avoid crowding, chooses to reproduce logistically. This is their story.

On a certain day, the fly population is 1000 and growing at the (net) instantaneous rate of 50 flies per day. The environment can support 10,000 flies.

Questions.   Given what we know of growth *rates*, it's natural to wonder about population *amounts*:

- What is the population in 10 days? 100 days? 200 days?
- When is the population 5000? 9000? 9900?
- When does the population grow fastest? What's the population then?

Solving the Logistic DE

Modeling logistic population growth boils down to solving an IVP. If a population has P_0 members at $t = 0$ and grows logistically, then the population function $P(t)$ is a solution of the IVP

$$P' = kP(C - P); \qquad P(0) = P_0.$$

It gets easier with techniques from integral calculus.

Searching systematically for a solution, given our present tools, is a slightly messy process. ◄   Suffice it to say that, for any constants k, C, and d, the function

$$P(t) = \frac{C}{1 + d\, e^{-kCt}}$$

Try checking this; it's messy but not really hard.

solves the logistic DE $P' = kP(C - P)$. ◄   (The constants k and C in the solution are the same as those in the DE; as we'll see, d depends on the initial population.)

What Solutions Mean for Our Flies

To put things in context, let's define:

$$t = \text{time, in days, since the original measurement;}$$

Measuring population in thousands removes pesky zeros from later computations.

$$P(t) = \text{population of flies, in thousands,} \blacktriangleleft \text{ at time } t;$$

$$P'(t) = \text{growth rate, in thousands of flies per day, at time } t.$$

$P'(0) = 0.05$ corresponds to 50 flies/day.

Evaluating the Parameters.   We know three things: $P(0) = 1$, $P'(0) = 0.05$, ◄ and $C = 10$. We'll use what we know to evaluate k and d. Since

$$P'(0) = 0.05 = k \cdot P(0) \cdot (10 - P(0)) = k \cdot 1 \cdot 9,$$

it follows that $k = 0.05/9 \approx 0.00556$.

Now we can evaluate d. By assumption,

$$P(0) = 1 = \frac{10}{1 + d\, e^{-kC0}} = \frac{10}{1 + d},$$

so $d = 9$. Putting everything together gives, at last, an explicit formula for $P(t)$:

$$P(t) = \frac{C}{1 + d\, e^{-kCt}} = \frac{10}{1 + 9\, e^{-0.0556t}}.$$

Here's a graph of P:

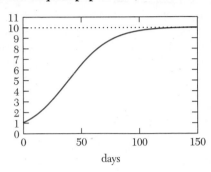

The mosquito population, in thousands

days

Our Questions Answered: What the Graph and Formula Say.   With the formula and the graph we can answer—approximately at least—all the questions posed earlier.

- The graph shows that on day 100 the population was about 9700; room is quickly running out. Day 200 isn't shown, but the graph's asymptotic behavior shows that at $t = 200$ the population should still be just short of 10,000. The formula agrees:

$$P(200) = \frac{10}{1 + 9\,e^{-0.0556\cdot 200}} \approx 9.99867.$$

Thus, rounding to the nearest whole fly (the only reasonable way to round flies) the population is at full capacity.

- Both graph and formula show that the population reaches 5000 (half the upper limit) in about 40 days. Around day 80, the population reaches 9000—90% capacity. Around day 130, 99% capacity is reached. ▶

    All of these numbers can be checked with the formula.

- Intuition says that under logistic conditions a population increases slowly at first (when there are few breeding members), faster as the population rises, and then slowly again it nears the upper limit. The graph agrees. Population grows fastest when the graph points most steeply upward, i.e., around day 40. ▶   At this time the population is about 5000—exactly half the carrying capacity. This isn't a coincidence. Under logistic conditions, it's a disadvantage for a population to be either too large or too small. Thus, intuitively, we'd expect growth conditions to be optimal at the halfway point.

    Take a close look; do you agree?

Afterword: Discrete versus Continuous Growth

All mathematical models more or less simplify the phenomena they describe. The use of DEs and IVPs to model phenomena of growth is no exception to this rule.

One sense in which DEs and IVPs may describe reality imperfectly concerns the difference between discrete and continuous phenomena of change. Human population, for instance, grows discretely, in jumps of at least one unit. Hot coffee in a cool room, by contrast, cools continuously—its temperature takes all intermediate values on the way down.

DEs and IVPs are continuous models of growth. Solutions to DEs and IVPs are continuous functions, so they can't, by definition, describe discrete phenomena exactly. (Neither, for the same reason, can discrete models perfectly describe continuous phenomena.) Such philosophical problems, though undeniable, need not be fatal. In practice, differences between discrete and continuous growth models are often unimportant, especially when the "jumps" are small. It matters little, for instance, whether coffee temperature falls continuously or in tiny increments of 0.0001 degree. Perfection is impossible, but DEs and IVPs can be (and are) used to model growth—even discrete growth—easily, effectively, and accurately.

Discrete Models of Continuous Phenomena. DEs and IVPs are continuous entities that are often used to describe phenomena that vary discretely. Sometimes the relationship goes the other way: Continuous phenomena can be modeled discretely. Movies, digital watches, and electronic computing all depend, in various ways, on discrete approximations of continuous reality.

BASICS

Use Theorem 1, page 256, to solve the IVP's in Exercises 1–4. Check answers by differentiation.

1. $y' = 0.1y$; $y(0) = 100$
2. $y' = -0.0001y$; $y(0) = 1$
3. $y' = (\ln 2) \cdot y$; $y(2) = 4$
4. $y' = ky$; $y(10) = 2y(0)$ [**HINT:** Solve for k using the initial condition.]
5. Suppose that the "wise flies" described in this section had begun by breeding at the net rate of 100 (rather than 50) flies per day. How would things be different? More specifically, answer the following questions, and compare the results with those given on page 263.
   (a) What size will the population be in 10 days? 100 days?
   (b) When is the population 5000 flies? 9000? 9900?
   (c) When does the population grow fastest? What's the population then?

FURTHER EXERCISES

6. First National Bank bank advertises 8% interest, compounded continuously. Second National Bank advertises 10% interest, compounded continuously—but also charges depositors a $100 yearly administrative fee for the privilege of banking there. (For simplicity, assume that the $100 fee is deducted continuously over the full year.) Let $P(t)$ denote the value in dollars of a deposit after t years.
   (a) First National Bank's policy amounts to the DE $P'(t) = 0.08P(t)$ (or just $P' = 0.08P$). Explain why.
   (b) Second National Bank's policy amounts to the DE $P'(t) = 0.1P(t) - 100$ (or just $P' = 0.1P - 100$). Explain why.
   (c) Solve the IVP $P' = 0.08P$, $P(0) = 1000$. If $1000 is deposited now in First National Bank, how much will it be worth in 10 years?
   (d) Solve the IVP $P' = 0.1P - 100$, $P(0) = 1000$. [**HINT:** The DE can be rewritten like that for Newton's law of cooling.]
   (e) If $1000 is deposited now in Second National Bank, how much will it be worth 10 years?

7. A certain menacing biological culture (aka The Blob) grows at a rate proportional to its size. When it arrived unnoticed one Wednesday noon in Chicago's Loop, it weighed just 1 g. By the 4:00 p.m. rush hour it weighed 4 g.

   The Blob has its "eye" on the Sears Tower, a tasty morsel weighing around 3,000,000,000,000 g (i.e., 3×10^{12} g). As soon as it weighs 1000 times as much (i.e., 3×10^{15} g), The Blob intends to *eat* the Sears Tower. By what time must The Blob be stopped? Will Friday's rush-hour commuters be delayed?

8. A mold grows at a rate proportional to the amount present. Initially, its weight is 2 g; after two days, it weighs 5 g. How much does it weigh after eight days?

9. A bacterial culture is placed in a large glass bottle. Suppose that the volume of the culture doubles every hour, and the bottle is full after one day.
   (a) If the culture was placed in the bottle at time $t = 0$ hours, when was the bottle half full?
   (b) Assume that the bottle is "almost empty" when the culture occupies less than 1% of its volume. How long is the bottle "almost empty"?

10. The police guard gave Sara a cold look, but his voice was polite as he directed her to the room she sought. "Don't touch anything, please, Ms. Abrams. The Chief said I had to let you in, but he said to tell you to mind your fingers." "Thank you," Sara replied coolly. "The Chief knows he can trust me." The guard opened his mouth as if to speak, but he merely shook his head and withdrew.

Sara was standing in what appeared to be a combination bedroom and laboratory. A relative had found Dr. Howell's body on the floor of this room that morning. By 9:00 a.m., the coroner had completed his examination; he stated that death was due to a severe blow to the head, and that Dr. Howell had been dead for 36 to 40 hours. It seemed critical to Sara to know exactly when Dr. Howell had died so that she could eliminate certain suspects. But how could she possibly discover exactly when he was killed? Puzzled, she wandered around the small, cluttered room, being careful not to touch anything. The old doctor apparently had been conducting an experiment when he was killed. Sara absentmindedly read from the notebook that was lying open on the bench:

The fungus grows at a rate proportional to its current weight.

"Great," she thought, "I'm here to investigate a murder, and instead I'm getting a biology lesson." At a loss for what else to do, she continued reading.

To exemplify this biological truth, I place the fungus on a scale and record its weight at various times

10 g	5:30 p.m.
12 g	6:15 p.m.
13	

"Hmm," Sara mused, "the poor guy didn't even get to finish the last entry." Sara suddenly frowned in con-

centration. She searched her pockets until she found a pencil stub and an old receipt. When the guard entered the room a few minutes later, Sara had just finished scribbling on the receipt. She smiled as she shoved the receipt and the pencil stub back into her pocket.

"Don't worry," Sara said cheerfully, "I'm leaving. I now know exactly when Dr. Howell was killed." The guard looked sourly at Sara's back as she left the room.

When was Dr. Howell killed? How did Sara know?

11. Oil is pumped continuously from a well at a rate proportional to the amount of oil left in the well. Initially there were 1,000,000 barrels of oil in the well; 6 years later, 500,000 barrels remain.

(a) At what rate was the amount of oil in the well decreasing when there were 600,000 barrels of oil remaining?

(b) It will no longer be profitable to pump oil from the well when there are fewer than 50,000 barrels remaining. When should pumping stop?

12. Human skeletal fragments were brought to a laboratory for carbon dating. Analysis showed that the proportion of ^{14}C to ^{12}C is only 6.25% of the value in living tissue. How long ago did this person die?

13. A tank initially contains 100 gallons of water and 10 pounds of salt, thoroughly mixed. Pure water is added at the rate of 5 gallons per minute, and the mixture is drained off at the same rate. (Assume complete and instantaneous mixing.)

(a) Explain why $S(t)$, the amount of salt in the tank at time t, is the solution of the IVP

$$S' = -\frac{5}{100}S; \quad S(0) = 10.$$

(b) Using Theorem 1, page 256, find a solution of this IVP.

(c) How much salt is left in the tank after 1 hour?

4.3 Linear and Quadratic Approximation; Taylor Polynomials

In this section we consider how to find and use polynomial functions that approximate other (non-polynomial) functions. What does it mean for a function f to approximate another function g? How can we choose "good" polynomial approximations? How are derivatives involved?

We'll start with the simplest cases: linear and quadratic polynomials. Then we'll describe how to find polynomials of *any* degree, called **Taylor polynomials**, that approximate a given function.

> **Why Bother?** Often in mathematics and applied mathematics one wants to approximate one function—say, *f*—with another—say, *g*. Why bother? For several possible reasons: If *f* is complicated, poorly understood, or otherwise inconvenient, it's useful to replace *f* with a simpler, better behaved, more tractable, or better understood *g*.
>   Polynomial functions are simple, convenient, well understood, and easy to use, so it's natural to use them to approximate more complicated functions.

Tangent Lines and Linear Approximation

The line tangent to the graph of a function f at $x = x_0$ can be thought of in several ways. Geometrically, the tangent line is the straight line that best "fits" the graph of f at $x = x_0$. Analytically, the tangent line represents the linear function that best approximates f near $x = x_0$.

EXAMPLE 1   If $f(x) = \sqrt{x}$, then $f(64) = 8$. How closely does the tangent-line function approximate values of $f(x)$ for x near 64?

Verify this.

Solution   Since $f'(x) = 1/(2\sqrt{x})$, $f'(64) = 1/16$. Therefore, the line tangent to the f-graph at $(64, 8)$ has equation ◄

$$\ell(x) = f(64) + f'(64)(x - 64) = 8 + \frac{x - 64}{16} = 4 + \frac{x}{16}.$$

Here are graphs of both f and ℓ:

The tangent approximation at *x* = 64 to √x̄

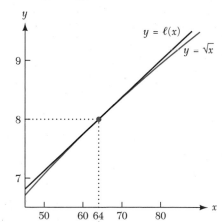

Observe:

- The tangent line hugs the graph of f closely; ℓ should therefore approximate f well for x near $x = 64$—the nearer the better.

But only slightly if x is near 64.

- The graph of f is concave down, so it lies below its tangent line. Therefore, the tangent line function ℓ *overestimates* f ◄   for each $x \neq 64$.   ∎

Constant, Linear, and Quadratic Approximations

How might we approximate *any* function f near *any* convenient "base point" $x = x_0$? Example 1 showed that the tangent line at $x = x_0$ defines a good **linear approximation** to f at x_0. **Constant** and **quadratic approximations** are also possible.

EXAMPLE 2   Which constant function best approximates $f(x) = \sqrt{x}$ near $x = 64$? Which quadratic function?

Solution   Since $f(64) = 8$, the constant function $c(x) = 8$ is the only sensible choice; c has the "right" value at $x = 64$.

How should we choose a good quadratic approximation q? For the linear approximation ℓ we required that

$$\ell(64) = f(64) = 8 \quad \text{and} \quad \ell'(64) = f'(64) = \frac{1}{16}.$$

It's reasonable, then, to ask that a good quadratic approximation q go one step further: ▶

$$q(64) = 8; \qquad q'(64) = \frac{1}{16}; \qquad q''(64) = f''(64) = -\frac{1}{2048}. \quad (4.2)$$

Why is $f''(64) = -1/2048$? Check for yourself.

In other words, our desired quadratic function q should "agree" with f at $x = 64$, both in its value and in its first *two* derivatives.

The easiest way to construct such a q is to write it in the form ▶

$$q(x) = a + b(x - 64) + c(x - 64)^2,$$

Writing q in powers of $(x - 64)$, not x, greatly simplifies some of the algebra.

and then choose the coefficients a, b, and c appropriately. Since we'll soon need them, we first calculate several values and derivatives of q: ▶

$$q(x) = a + b(x - 64) + c(x - 64)^2 \implies q(64) = a;$$
$$q'(x) = b + 2c(x - 64) \implies q'(64) = b;$$
$$q''(x) = 2c \implies q''(64) = 2c.$$

Watch carefully—here's where the form of q pays off.

Choosing a, b, and c is now easy:

Choosing a.   Equation 4.2 requires that $q(64) = 8$. But $q(64) = a$, so $a = 8$.

Choosing b.   Equation 4.2 requires that $q'(64) = 1/16$. But $q'(64) = b$, so $b = 1/16$.

Choosing c.   Equation 4.2 requires that $q''(64) = -1/2048$. But $q''(64) = 2c$, so $c = -1/4096$. ▶

Where did the factor of 2 come from?

Putting everything together gives

$$q(x) = 8 + \frac{x - 64}{16} - \frac{(x - 64)^2}{4096}$$

as the best quadratic approximation to f at $x = 64$.

Graphs of c, ℓ, q, and f show what's happening:

**Constant, linear, and quadratic
approximations to $f(x) = \sqrt{x}$**

Observe:

- All three approximating functions (c, ℓ, and q) have the right value at $x = 64$.
- The graphs of ℓ and q (but not of c!) are tangent to the f-graph (i.e., have the same slope) at $x = 64$.

And much better than c.
- The function q fits f better than ℓ ◄ near $x = 64$. The fact that $q''(64) = f''(64)$ means that q and f have the same concavity at $x = 64$.

How well, numerically, do c, ℓ, and q approximate f? A table gives some idea:

Three Approximations to $f(x) = \sqrt{x}$							
x	50	63	63.9	64	64.1	65	80
$c(x)$	8.00000	8.00000	8.00000	8	8.00000	8.00000	8.00000
$\ell(x)$	7.12500	7.93750	7.99375	8	8.00625	8.06250	9.00000
$q(x)$	7.07715	7.93726	7.99375	8	8.00625	8.06226	8.93750
$f(x)$	7.07107	7.93725	7.99375	8	8.00625	8.06226	8.94427

The numbers agree with the graphs: q approximates f best, especially near $x = 64$.

■

Definitions

Linear and quadratic approximations can be chosen for any well-behaved function f, at any base point x_0. The formal definition summarizes, in general language, what we did above for $f(x) = \sqrt{x}$ and $x_0 = 64$.

Definition Let f be any function for which $f'(x_0)$ and $f''(x_0)$ exist. The linear approximation to f, based at x_0, is the linear function

$$\ell(x) = f(x_0) + f'(x_0)(x - x_0).$$

The quadratic approximation to f, based at x_0, is the quadratic function

$$q(x) = f(x_0) + f'(x_0)(x - x_0) + \frac{f''(x_0)}{2}(x - x_0)^2.$$

EXAMPLE 3 Let $f(x) = 10^x$. Find ℓ and q, the linear and quadratic approximations to f, based at $x = 3$. Use each to estimate $f(3.1)$.

Solution We need two derivatives and a value for f, all at $x = 3$. Calculation gives $f(x) = 10^x$, $f'(x) = 10^x \ln 10$, and $f''(x) = 10^x (\ln 10)^2$, so ▶   *Check our work.*

$$f(3) = 10^3 = 1000; \qquad f'(3) = 1000 \ln 10 \approx 2302.5851;$$

$$f''(3) = 1000(\ln 10)^2 \approx 5301.8981.$$

The definitions say:

$$\ell(x) = f(3) + f'(3) \cdot (x - 3) \approx 1000 + 2302.5851(x - 3);$$

$$q(x) = \ell(x) + \frac{f''(3)}{2} \cdot (x - 3)^2 \approx 1000 + 2302.5851(x - 3) + 2650.9491(x - 3)^2.$$

Numerical calculation shows:

$$\ell(3.1) \approx 1230.2585; \qquad q(3.1) \approx 1256.7680; \qquad f(3.1) \approx 1258.9254.$$

The following graphs, like the preceding numbers, show how much better q does than ℓ in fitting f:

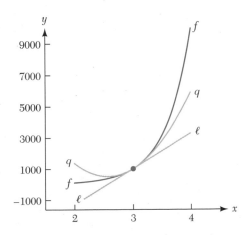

Linear and quadratic approximation to $f(x) = 10^x$

The picture shows why ℓ fits f so poorly: f bends sharply away from ℓ, even quite near $x = 3$. The second derivative, f'', explains why. Any linear function ℓ has zero second derivative. Our target function f, by contrast, has a huge second derivative: $f''(3) \approx 5000$. Because f is so drastically nonlinear, it is badly approximated by any straight line—even the best-fitting one. ∎

Taylor Polynomials

Linear and quadratic approximations ℓ and q are built to match f as well as possible at x_0; more precisely, ℓ and f have the same value and *first* derivative at x_0; q and f have the same value and first *two* derivatives at x_0.

Still better agreement is possible if polynomials of degree 3, 4, and higher are used to approximate f. An important example illustrates this idea.

> **Made to Order.**   The constant, linear, and quadratic approximations are sometimes called **zeroth-order**, **first-order**, and **second-order approximations** to f near x_0. In each case "order" refers to the number of derivatives of f with which a particular approximation agrees at x_0.

EXAMPLE 4   Find a fifth-degree polynomial p_5 that agrees with $f(x) = e^x$ to order 5 at $x_0 = 0$—i.e., such that

$$p_5(0) = f(0); \qquad p_5'(0) = f'(0); \qquad p_5''(0) = f''(0);$$
$$p_5^{(3)}(0) = f^{(3)}(0); \qquad p_5^{(4)}(0) = f^{(4)}(0); \qquad p_5^{(5)}(0) = f^{(5)}(0).$$

Solution   We need values for several derivatives of f, all at $x = 0$. The

That's why we chose this f!

fact that $f(x) = e^x$ makes calculating derivatives very, very easy: ◄

$$f(x) = e^x \implies f(0) = f'(0) = f''(0) = f^{(3)}(0) = f^{(4)}(0) = f^{(5)}(0) = 1.$$

We want a polynomial p_5 with the same derivatives at $x = 0$. To begin, we write $p_5(x)$ in standard form:

$$p_5(x) = a_0 + a_1 x + a_2 x^2 + a_3 x^3 + a_4 x^4 + a_5 x^5,$$

For each i, a_i is the coefficient of the ith power x^i.

Check these easy but important calculations.

The coefficients a_i are still to be found. ◄

Finding a_0 through a_5 is surprisingly easy. The key idea is that the a_i are closely related to derivatives of p_5 at $x = 0$. Specifically: ◄

$$
\begin{aligned}
p_5(x) &= a_0 + a_1 x + a_2 x^2 + a_3 x^3 + a_4 x^4 + a_5 x^5 &&\implies p_5(0) = a_0; \\
p_5'(x) &= a_1 + 2a_2 x + 3a_3 x^2 + 4a_4 x^3 + 5a_5 x^4 &&\implies p_5'(0) = a_1; \\
p_5''(x) &= 2a_2 + 3 \cdot 2a_3 x + 4 \cdot 3a_4 x^2 + 5 \cdot 4a_5 x^3 &&\implies p_5''(0) = 2a_2; \\
p_5'''(x) &= 3 \cdot 2a_3 + 4 \cdot 3 \cdot 2a_4 x + 5 \cdot 4 \cdot 3a_5 x^2 &&\implies p_5'''(0) = 6a_3; \\
p_5^{(4)}(x) &= 4 \cdot 3 \cdot 2a_4 + 5 \cdot 4 \cdot 3 \cdot 2a_5 x &&\implies p_5^{(4)}(0) = 24a_4; \\
p_5^{(5)}(x) &= 5 \cdot 4 \cdot 3 \cdot 2a_5 &&\implies p_5^{(5)}(0) = 120a_5.
\end{aligned}
$$

The pattern of coefficients should now be clear: *For each i, $p_5^{(i)}(0) = i!\,a_i$.* ▶   In other words:

The symbol $i!$ denotes the factorial of i. We will discuss factorials soon.

$$a_0 = p_5(0); \qquad u_1 = p_5'(0); \qquad a_2 = \frac{p_5''(0)}{2!};$$

$$a_3 = \frac{p_5'''(0)}{3!}; \qquad a_4 = \frac{p_5^{(4)}(0)}{4!}; \qquad a_5 = \frac{p_5^{(5)}(0)}{5!}.$$

Now we can find p_5. By design,

$$p_5(0) = f(0) = 1; \qquad p_5'(0) = f'(0) = 1;$$

$$p_5''(0) = f''(0) = 1; \quad \ldots \quad p_5^{(5)}(0) = f^{(5)}(0) = 1.$$

Thus,

$$a_0 = 1; \qquad a_1 = 1; \qquad a_2 = \frac{1}{2!};$$

$$a_3 = \frac{1}{3!}; \qquad a_4 = \frac{1}{4!}; \qquad a_5 = \frac{1}{5!}.$$

Our search is over:

$$p_5(x) = \frac{1}{0!} + \frac{x}{1!} + \frac{x^2}{2!} + \frac{x^3}{3!} + \frac{x^4}{4!} + \frac{x^5}{5!} = 1 + x + \frac{x^2}{2} + \frac{x^3}{6} + \frac{x^4}{24} + \frac{x^5}{120}.$$

How well does p_5 approximate f near $x = 0$? Very well indeed:

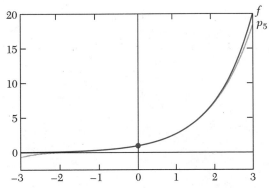

A fifth-order polynomial approximation to $f(x) = e^x$

The graphs of f and p_5 are almost identical near $x = 0$. Compare, for instance, $f(1)$ and $p_5(1)$. Easy numerical calculations show that

$$p_5(1) = 1 + 1 + \frac{1}{2} + \frac{1}{6} + \frac{1}{24} + \frac{1}{120} = \frac{163}{60} \approx 2.717;$$

$$f(1) = e \approx 2.718.$$

Factorials

The repeated product $i \cdot (i-1) \cdot (i-2) \cdots 2 \cdot 1$, denoted by $i!$, is called the **factorial** of i. (By convention, $0! = 1$.) Factorials appear often in mathematics, especially when (as we just saw) an operation is performed repeatedly. Factorials are important enough in practice that many calculators have a factorial key. ◄

Does yours? If so, try it. If not, calculate some factorials from scratch.

The Formal Definition

The polynomial p_5 just constructed is the fifth-order Taylor polynomial for $f(x) = e^x$ based at $x = 0$. Here is the general definition:

> **Definition** (Taylor polynomials) Let f be any function whose first n derivatives exist at $x = x_0$. The **Taylor polynomial** of order n, based at x_0, is defined by
>
> $$p_n(x) = f(x_0) + f'(x_0)(x - x_0) + \frac{f''(x_0)}{2!}(x - x_0)^2$$
> $$+ \frac{f^{(3)}(x_0)}{3!}(x - x_0)^3 + \cdots + \frac{f^{(n)}(x_0)}{n!}(x - x_0)^n.$$

> **Taylor's Idea?**   The idea underlying Taylor polynomials predates Brook Taylor (1685–1731), for whom they are named. According to the author George Simmons, the idea appears in work of John Bernoulli published in 1694. The same versatile Bernoulli is also credited with having discovered—behind the scenes—the famous technical lemma known as l'Hôpital's Rule. (We'll see it later in this book.)

EXAMPLE 5   Let $f(x) = \sqrt{x}$. Find the first three Taylor polynomials p_1, p_2, and p_3, all based at $x_0 = 64$.

Check the last one; we found the others earlier.

Solution   We'll need the value and first three derivatives of f at $x = 64$: ◄

$$f(64) = 8; \qquad f'(64) = \frac{1}{16};$$

$$f''(64) = -\frac{1}{2048}; \qquad f'''(64) = \frac{3}{262144}.$$

The first two are familiar.

Plugging these values into the definition gives our results: ◄

$$p_1(x) = 8 + \frac{x - 64}{16};$$

$$p_2(x) = 8 + \frac{x - 64}{16} - \frac{(x-64)^2}{2! \cdot 2048} = 8 + \frac{x - 64}{16} - \frac{(x-64)^2}{4096};$$

$$p_3(x) = 8 + \frac{x - 64}{16} - \frac{(x-64)^2}{4096} + \frac{3}{3! \cdot 262144}(x - 64)^3$$

$$= 8 + \frac{x - 64}{16} - \frac{(x-64)^2}{4096} + \frac{(x-64)^3}{524288}.$$

More on the Definition

The Taylor polynomial definition deserves a closer look.

Why Powers of $(x - x_0)$? Polynomials usually involve powers of x. ▶ Here, though, powers of $(x - x_0)$ ▶ are more convenient. For instance, the use of powers of $(x - 64)$ in Example 5 simplified our calculations, most of which involved $x = 64$.

x, x^2, x^3, etc.

$(x - x_0)$, $(x - x_0)^2$, $(x - x_0)^3$, etc.

Maclaurin Polynomials Taylor polynomials that happen to be based at $x = 0$ are known as **Maclaurin polynomials**. They look a little simpler than Taylor polynomials: ▶

They're simpler only in appearance. The idea is exactly the same.

$$P_n(x) = f(0) + f'(0)x + \frac{f''(0)}{2}x^2 + \cdots + \frac{f^{(n)}(0)}{n!}x^n.$$

In Example 4 we found the fifth-order Maclaurin polynomial for $f(x) = e^x$.

Order, Degree A Taylor polynomial P_n of order n is chosen so that its value and first n derivatives at $x = x_0$ agree with those of f. The definition shows that P_n also has **degree** n; i.e., it involves powers of $(x - x_0)$ up through the nth power. (If, by chance, $a_n = 0$, then P_n has smaller degree.)

Many Symbols, Just One Variable The definition involves many symbols: a, x, i, x_0, f, and n. While in this thicket, it is helpful to remember that x *is the only variable*. In specific cases, all the other symbols take *numerical* values.

From P_n to P_{n+1} Let f be a function and P_n the Taylor polynomial of order n based at x_0. The next Taylor polynomial, P_{n+1}, has just one more term:

$$P_{n+1}(x) = P_n(x) + \frac{f^{(n+1)}(x_0)}{(n+1)!}(x - x_0)^{n+1}.$$

Approximating Functions with Taylor Polynomials

Taylor polynomials are useful in the approximating of other, more complicated functions. For example, the sine function—which has no algebraic formula—is a good candidate for Taylor approximation.

EXAMPLE 6 Find P_1, P_3, P_5, and P_7, the Taylor (or Maclaurin) polynomials for $f(x) = \sin x$ based at $x = 0$. Plot everything on one set of axes. How closely does each polynomial approximate $\sin 1$?

Solution To find P_7 we need the value and the first through seventh derivatives of $f(x) = \sin x$ at $x = 0$. In order, they are 0, 1, 0, -1, 0, 1, 0, -1. ▶ It follows that

Check this easy calculation.

$$P_7(x) = 0 + 1x + 0x^2 - \frac{1}{3!}x^3 + 0x^4 + \frac{1}{5!}x^5 + 0x^6 - \frac{1}{7!}x^7$$

and, therefore, that

$$P_1(x) = x; \qquad P_3(x) = x - \frac{x^3}{6};$$

$$P_5(x) = x - \frac{x^3}{6} + \frac{x^5}{120}; \qquad P_7(x) = x - \frac{x^3}{6} + \frac{x^5}{120} - \frac{x^7}{5040}.$$

Here are all five graphs:

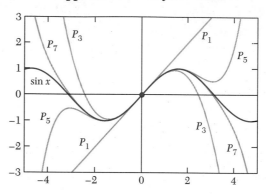

**Several Taylor polynomial
approximations to $f(x) = \sin x$**

The picture shows:

Snug Fits near $x = 0$ All four Taylor polynomials fit the sine curve well near the base point $x = 0$.   Differences appear as one moves away (in either direction) from $x = 0$.

Better and Better   Higher-order Taylor polynomials approximate the sine function better than lower-order ones. P_7, for instance, follows the sine curve much farther than P_1 or P_3.

Standard Shapes   The graphs of P_1 and P_3 have standard shapes for linear and cubic functions. ◄   The wigglier appearance ◄ of P_5 and P_7 reflects their higher degree.

Trace P_3 carefully to see this.

Wiggly is good in sine-curve approximating!

Intuition suggests that P_1, P_3, P_5, and P_7 should give successively closer estimates for $\sin 1 \approx 0.84147$. So they do:

$$P_1(1) = 1; \qquad\qquad P_3(1) = \frac{5}{6} \approx 0.83333;$$

$$P_5(1) = \frac{101}{120} \approx 0.84167; \qquad P_7(1) = \frac{4241}{5040} \approx 0.84147.$$ ■

Bounding the Error of Linear Approximation (Optional)

An Error Bound Formula

How large, numerically, are the errors committed when Taylor polynomials approximate a function f? We'll address the question only for linear approximations. For higher-order approximations the ideas are similar but the calculations are more complicated.

If we compare the results of Example 2 and Example 3, we find that the error of a linear approximations depends on two factors:

1. the distance from x to x_0;

2. the size of the second derivative f'' near x_0; (Large values of $|f''|$ mean that the f-graph bends sharply away from its tangent line; small (i.e., near 0) values of $|f''|$ mean that the f-graph is nearly straight near x_0 and therefore closely approximated by its tangent line.)

The following theorem restates these ideas quantitatively. Throughout, f is a differentiable function, ℓ is the linear approximation based at $x = x_0$, and I is an interval containing x_0.

Theorem 2 (An Error Bound for Linear Approximation) Suppose that the inequality

$$|f''(x)| \leq K$$

holds for all x in I. Then, for all x in I,

$$|f(x) - \ell(x)| \leq \frac{K}{2}(x - x_0)^2.$$

Observe:

The Error Function The function $f(x) - \ell(x)$ on the left is called an **error function**; we'll call it $e(x)$. For a given x, $e(x)$ is the **error** committed by ℓ in approximating f. Ideally, $e(x) \approx 0$.

An Error Bound Formula The inequality

$$|f(x) - \ell(x)| \leq \frac{K}{2}(x - x_0)^2$$

is (appropriately) called an **error bound formula**; it guarantees that the error function is no larger than the *computable* quantity on the right.

It's OK to Overestimate K Any upper bound for $|f''(x)|$ on I can serve as K in the inequality. Choosing K as small as possible improves the error bound, but the inequality holds even for larger values of K. ▶

It's safe to overestimate K but not to underestimate.

Worst-Case Scenario The function $f(x) = K(x-x_0)^2/2$ represents the worst case that the hypothesis allows. For this f,

$$f(x_0) = 0, \qquad f'(x_0) = 0; \qquad \text{and} \qquad f''(x) = K.$$

In this case, the linear approximation is simply $\ell(x) = 0$, and the error ℓ commits is as large as the theorem permits:

$$|f(x) - \ell(x)| = f(x) = \frac{K}{2}(x - x_0)^2.$$

Before proving the error bound theorem, we'll show it in action.

Satisfaction Guaranteed

The practical value of an error bound formula is its guarantee of the accuracy of an approximation.

EXAMPLE 7   We showed that $\ell(x) = 4 + x/16$ is the linear approximation to $f(x) = \sqrt{x}$ at $x = 64$. What accuracy does the error bound formula guarantee if ℓ approximates f on the interval $[50, 80]$? What does the formula guarantee for the interval $[63, 65]$?

Check ours.

Solution   First we need a suitable K. Calculation ◄ shows that

$$f''(x) = \frac{d^2}{dx^2}\left(\sqrt{x}\right) = -\frac{1}{4x^{3/2}} \implies |f''(x)| = \frac{1}{4x^{3/2}}.$$

A moment's thought—or a look at the graph of $|f''|$—shows that $|f''(x)|$ decreases as x increases. For the interval $[50, 80]$, therefore, we may use $K = |f''(50)| \le 0.00071$. ◄   On $[63, 65]$, $K = |f''(63)| \le 0.0005$ works.

$|f''(x)|$ is largest at the left endpoint.

Watch all inequalities carefully.

All ingredients are ready. The theorem says: ◄

- For x in $[50, 80]$, the largest possible error is no more than

$$|\sqrt{x} - \ell(x)| \le \frac{0.00071}{2}(x - 64)^2 \le \frac{0.00071}{2}(80 - 64)^2 \approx 0.09.$$

(We used $x = 80$, not $x = 50$, to make $(x - 64)^2$ as large as possible.) In other words, on $[50, 80]$, ℓ approximates f to about *one-decimal-place accuracy*.

- For x in $[63, 65]$, the largest possible error is even less:

$$|\sqrt{x} - \ell(x)| \le \frac{0.0005}{2}(x - 64)^2 \le \frac{0.0005}{2}(65 - 64)^2 \approx 0.00025.$$

In other words, on $[63, 65]$, ℓ approximates f to about *three-decimal-place accuracy*.

The table on page 268 shows that these error estimates are conservative: ℓ and q behave a little better than the theorem predicts.   ∎

A Proof of the Error Bound Formula

We use f, ℓ, and K as in the theorem. From these data we'll construct two new functions—an error function e and a **bounding function** b, defined as follows:

$$e(x) = f(x) - \ell(x); \qquad b(x) = K\frac{(x - x_0)^2}{2}.$$

The same proof, with minor variations, shows that the inequality holds for $x < x_0$ and in absolute value.

The theorem claims—and we need to show—that for x in I, $|e(x)| \le b(x)$. To simplify things slightly, we'll show only that $e(x) \le b(x)$ for $x \ge x_0$. ◄

To our aid come three key properties of e and b:

$$e(x_0) = 0 = b(x_0); \qquad e'(x_0) = 0 = b'(x_0); \qquad |e''(x)| \le K = b''(x). \quad (4.3)$$

Make them!

We met it first in Section 2.1.

(All three properties of e follow from the construction of ℓ: By design, $\ell(x_0) = f(x_0)$, $\ell'(x_0) = f'(x_0)$, and $e''(x) = f''(x) - l''(x) = f''(x)$. All three properties of b come from direct calculations.) ◄

To finish the proof, we use the racetrack principle. ◄   Recall what it says about *any* functions g and h:

If $g(x_0) = h(x_0)$ and $g'(x) \le h'(x)$ for $x \ge x_0$, then $g(x) \le h(x)$ for $x \ge x_0$.

Let's apply the racetrack principle to e and b. As we saw earlier, $e(x_0) = 0 = b(x_0)$. We'll be finished, says the racetrack principle, if we can somehow show that $e'(x) \le b'(x)$ for $x \ge x_0$. To do so, we apply the racetrack principle *again*, but this time to e' and b' and their derivatives e'' and b''. Equations 4.3 contain everything we need: $e'(0) = b'(0)$ and $e''(x) \le b''(x)$. Thus, the racetrack principle says that $e'(x) \le b'(x)$ for $x \ge x_0$—exactly what we need to finish the proof. ■

BASICS

For each function in Exercises 1–5, first find the Taylor polynomial P_n of order n for the function f with base point x_0. Then plot both f and P_n on the same axes. Choose your plotting window to show clearly the relationship between f and P_n.

1. $f(x) = \dfrac{1}{1-x}$, $n = 3$, $x_0 = 0$
2. $f(x) = \sin x + \cos x$, $n = 4$, $x_0 = 0$
3. $f(x) = \ln x$, $n = 3$, $x_0 = 1$
4. $f(x) = \tan x$, $n = 2$, $x_0 = 0$
5. $f(x) = \sqrt{x}$, $n = 3$, $x_0 = 4$

In Exercises 6–11, find linear and quadratic approximations of each function at $x_0 = 0$. Then, find an interval over which each approximating polynomial makes an error no greater than 0.01.

6. $f(x) = \sin x$
7. $f(x) = \cos x$
8. $f(x) = \tan x$

9. $f(x) = e^x$
10. $f(x) = \arctan^{-1} x$
11. $f(x) = \arcsin^{-1} x$

12. The line tangent to the curve $y = g(x)$ at the point $(3, 5)$ intersects the y-axis at the point $(0, 10)$.
    (a) What is $g'(3)$?
    (b) Estimate $g(2.95)$ using a linear approximation.

FURTHER EXERCISES

13. In Example 2 we worked with the quadratic function $q(x) = a + b(x - 64) + c(x - 64)^2$. Here are more details:
    (a) We claimed that $q'(x) = b + 2c(x - 64)$ and that $q''(x) = 2c$. Verify these claims by differentiation.
    (b) Rewrite q in powers of x, i.e., in the form $q(x) = A + Bx + Cx^2$.
    (c) Rewrite $q'(x) = b + 2c(x - 64)$ in powers of x. Does your answer agree, as it should, with the previous part?

14. This exercise refers to Example 6 (p. 273). Throughout this problem, $f(x) = \sin x$, and $P_n(x)$ is the Taylor (Maclaurin) polynomial of order n based at $x = 0$.
    (a) Every function mentioned in Example 6—f, P_1, P_3, P_5, and P_7—is odd. How is this fact "reflected" in their graphs?
    (b) Find the *even-order* Maclaurin polynomials P_2, P_4, P_6, and P_8 for $f(x) = \sin x$. How are they related to P_1, P_3, P_5, and P_7?

15. Throughout this exercise, $g(x) = \cos x$ and $P_n(x)$ is the Taylor (Maclaurin) polynomial of order n based at $x = 0$. For use below, here are graphs of g, P_0, P_2, P_4, and P_6.

Several Taylor polynomial approximations to $f(x) = \cos x$

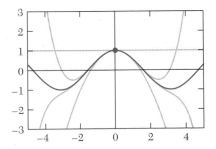

(a) Find formulas for P_0 through P_6, the Maclaurin polynomials through order 6 for g based at $x = 0$. Use your results to label the graphs shown above.
(b) Is g odd, even, or neither? What about the Maclaurin polynomials for g found in part (a)? How do the graphs "reflect" the situation?
(c) In Example 6 we found the Maclaurin polynomials P_1, P_3, P_5, and P_7 for $f(x) = \sin x$. Find the derivatives $P_1'(x)$, $P_3'(x)$, $P_5'(x)$, $P_7'(x)$, and $f'(x)$ of these functions. How are the results related to the rest of this problem?

16. Suppose that f is a function such that $f'(x) = \sin\left(x^2\right)$ and $f(1) = 0$.   [**NOTE:** No explicit formula for f is given.]
   (a) Estimate the value of $f(0.5)$ using a linear approximation.
   (b) Is the estimate you computed in part (a) greater than or less than the exact value of $f(0.5)$? Explain.
   (c) Estimate the value of $f(0.5)$ using a quadratic approximation.

17. Let g be a well-behaved function defined on the interval $[0, 10]$ such that the graph of g passes through the point $(5, 2)$ and the derivative of g is the function sketched below.

Graph of g'

   (a) Estimate $g(0)$ using a linear approximation.
   (b) Estimate $g(8)$ using a quadratic approximation.

18. Suppose that h is a well-behaved function such that $h(2) = 3$, $h'(2) = -2$, and $-2 \le h''(x) \le 1$ for all x in $(0, 4)$.
   (a) Find a linear approximation to h.
   (b) Show that $0 \le h(3) \le 2$.

19. At time $t = 0$ seconds a car passed a reference point, heading eastward at 25 meters per second, with eastward acceleration 2 meters per second per second. Let $p(t)$ be the car's eastward position at time t seconds, in meters east of the reference point. (In particular, $p(0) = 0$.)
   (a) Find the linear approximation function ℓ_p to p based at $t_0 = 0$. Use it to predict where the car will be at time $t = 1$, and where the car was at time $t = -1$.
   (b) Find the quadratic approximation function q_p to p based at $t_0 = 0$. Use it to predict where the car will be at time $t = 1$, and where the car was at time $t = -1$.
   (c) Let $v(t)$ be the car's eastward velocity, in meters per second, at time t. Find the linear approximation ℓ_v to v based at $t_0 = 0$. What does it predict at time $t = 1$ second?
   (d) (Requires the error-bound theorem.) The car's motor and brakes are powerful enough to produce acceleration anywhere from -3 to 3 meters per second per second. Use this fact and Theorem 2, page 275,

to decide how much much error the linear approximation ℓ_p can commit, at worst, over the interval $0 \le t \le 1$.

20. Repeat the previous exercise, but assume that at $t = 0$ the car was heading eastward at 25 meters per second, and had eastward acceleration -2 meters per second per second.

21. At time $t = 0$ seconds an object was dropped from a 100-meter cliff. Its downward acceleration, due to gravity, was 9.8 meters per second per second. Let $h(t)$ be the height in meters of the object above ground level at time t seconds.
   (a) Find the linear approximation ℓ to h, based at $t_0 = 0$. What does it predict for $t = 1$?
   (b) Find the quadratic approximation q to h, based at $t_0 = 0$. What does it predict for $t = 1$?
   (c) Plot the functions ℓ and q on the same axes; use $0 \le t \le 6$. Which model—ℓ or q—seems more realistic? Does either model describe the real situation perfectly? Why or why not?

22. Theorem 2, page 275, gives a bound on the error committed by a linear approximation. Here, without proof, is a similar result for quadratic approximation:

   An error bound for quadratic approximation. *Suppose that the inequality $|f'''(x)| \le K$ holds for all x in an interval I. Then for all x in I,*

   $$|f(x) - q(x)| \le \frac{K}{6}(x - x_0)^3.$$

   Let $f(x) = \cos(x)$.
   (a) Find the quadratic approximation to f based at $x_0 = 0$.
   (b) Use the result above to bound the error committed by q in approximating f on the interval $[-1, 1]$.
   (c) Plot both f and q on the same axes; use $-1 \le x \le q$. How is the picture consistent with what you found in the previous part?

23. Let $f(x) = \ln(x)$.
   (a) Find the quadratic approximation q to f, based at $x_0 = 1$.
   (b) Use the error-bound result from the previous problem to bound the error committed by q in approximating f on the interval $[1/2, 3/2]$.
   (c) Plot both f and q on the same axes; use $1/2 \le x \le 3/2$. How are the graphs consistent with what you found in the previous part?

24. Estimate the value of each of the following expressions using a linear approximation. Then compute the difference between your estimate and the value given by a scientific calculator.
   (a) $\sqrt{103}$ (c) $\tan 31°$
   (b) $\sqrt[3]{29}$ (d) 0.8^{10}

25. For each part of the previous exercise, compare the "actual" approximation error with the theoretical error bound.

26. Let f be a function, and let ℓ and q be the linear and quadratic approximations to f based at x_0. This exercise is about the two functions E_1 and E_2 defined by

    $$E_1(x) = f(x) - \ell(x) \quad \text{and} \quad E_2(x) = f(x) - q(x).$$

    Observe that E_1 and E_2 measure the *difference* between $f(x)$ and either $\ell(x)$ or $q(x)$—i.e., E_1 and E_2 measure the *errors* ℓ and q commit in approximating f. (Ideally, $E_1(x) \approx 0$ and $E_2(x) \approx 0$ when $x \approx 0$.)

    (a) Let $f(x) = e^x$. Find formulas for the functions ℓ, q, E_1, and E_2. (Use the base point $x_0 = 0$.)

    (b) Plot f, ℓ, and q on one set of axes; use $-1 \le x \le 1$.

    How does the picture show that q is a better approximation to f than ℓ?

    (c) Plot E_1 and E_2 on one set of axes; use $-1 \le x \le 1$. How does *this* picture show that q is a better approximation to f than ℓ?

    (d) The two graphs in the previous part should resemble a quadratic and a cubic curve, respectively. Do they? Which resembles which?

    (e) Are the curves in part (d) *really* quadratic and cubic, or do they just look that way?

27. Repeat the previous exercise, but use $f(x) = \ln(x)$, based at $x = 1$.

28. Repeat Exercise 26, but use $f(x) = x^5$, based at $x = 1$.

4.4 Newton's Method: Finding Roots

Consider these problems:

- Let $f(x) = 2x^3 - x - 2$. Find a number r for which $f(r) = 0$.
- Estimate $\sqrt{2}$ with high numerical accuracy.
- Find a number r such that

$$150r = (1 + r)^{31} - 1.$$

Each of these problems involves finding a **root** of some function f, i.e., a number r such that $f(r) = 0$. None can be solved easily or directly by algebraic methods. Newton's method offers a simple and efficient alternative. We'll use it to solve all three problems.

Roots, Graphically

A number r is a root of a function f if $f(r) = 0$. Graphically speaking, a root of f is an x-intercept of the f-graph. Given a graph of f, it's easy to find roots—approximately—just by looking. Following, for instance, are two views of the graph of $f(x) = 2x^3 - x - 2$.

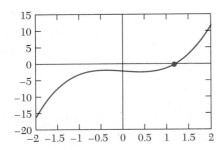

Graph of $f(x) = 2x^3 - x - 2$

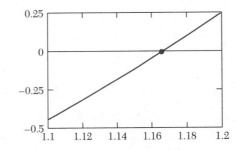

A closer view: $f(x) = 2x^3 - x - 2$

The first graph shows a root somewhere near $x = 1$; the second locates it more precisely, near $x = 1.17$. Further zooming might reveal additional digits. Newton's method offers a more efficient strategy, so we won't bother.

EXAMPLE 1   Graphically estimate a root of $f(r) = (1 + r)^{31} - 1 - 150r$. (Such a root solves the equation $150r = (1 + r)^{31} - 1$, already mentioned.)

Solution   The graph

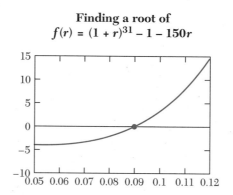

Finding a root of
$$f(r) = (1 + r)^{31} - 1 - 150r$$

shows a root somewhere near $r = 0.09$. A numerical check shows that $f(0.09) \approx -0.0382$, so our guess isn't exact. We'll refine it with Newton's method.   ■

Roots, Algebraically

Roots can sometimes be found algebraically, by solving equations. Unfortunately, such methods often fail; unless a function has an algebraic formula—a simple one—solving algebraically for roots may be hard or impossible. For linear functions, roots are easy to find. ◄

Don't skip the next example; we'll refer to it later.

EXAMPLE 2   A line passes through the point (x_0, y_0) slope m. Find its x-intercept.

Solution   The line has point–slope equation $y - y_0 = m(x - x_0)$. To find the x-intercept, we set $y = 0$ and solve for x:

$$- y_0 = m(x - x_0) \iff x = x_0 - \frac{y_0}{m}. \tag{4.4}$$

(This formula fails, of course, if $m = 0$. The line is horizontal and so has no x-intercept.) ◄   ■

Unless $y_0 = 0$, in which case the line is the x-axis itself.

Newton's Idea: Ride the Tangent Line

Most functions aren't linear and so don't lend themselves easily to algebraic root finding. When algebra fails, Newton's method often succeeds.

Newton's method starts with one approximation to a root and produces another, better approximation. The following picture illustrates one **Newton step**.

Zeroing in on a root: one Newton step

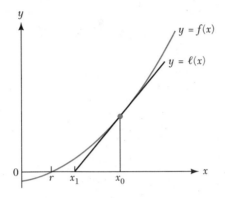

Observe:

- In the picture, r is the sought-after root, x_0 is an **initial guess** for r, and x_1 is an *improved* approximation to r—the result of one Newton step.

- If f happens to be linear, then for any initial guess x_0, the next estimate, x_1, is an *exact* root of f. In other words, for linear functions Newton's method succeeds exactly in only one step.

Repeating the preceding process produces a sequence $x_0, x_1, x_2, x_3, \ldots$ of successive approximations to r. As the picture suggests, successive approximations converge quickly toward the root, r.

Finding a root: several Newton steps

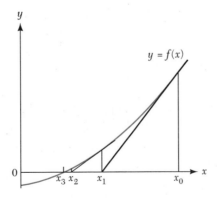

Each approximation is obtained from its predecessor in the same way:

   x_{n+1} *is the x-intercept of the tangent line at* $x = x_n$.

Equation 4.4, with $m = f'(x_0)$, gives an **iteration rule**, i.e., an explicit recipe for moving from one estimate to the next.

Fact (Newton's Iteration Formula) Let f be a function and x_0 an initial guess at a root. For $n \geq 0$,

$$x_{n+1} = x_n - \frac{f(x_n)}{f'(x_n)}. \tag{4.5}$$

Notice:

Iteration   The Fact defines $x_1, x_2, x_2, x_3, \ldots$ **iteratively**; that is, each term is defined by its predecessor. Each successive use of the formula is called an **iteration**.

Found a Root? Then Stop   If x_n happens to *be* a root of f, then $x_{n+1} = x_n$. Here's why:

$$x_{n+1} = x_n - \frac{f(x_n)}{f'(x_n)} = x_n - \frac{0}{f'(x_n)} = x_n.$$

N is for Newton.

The Iteration Function   Consider the new function (we'll call it the **iteration function**) N ◄   defined by $N(x) = x - f(x)/f'(x)$. Then

$$x_1 = N(x_0), \qquad x_2 = N(x_1), \qquad x_3 = N(x_2), \qquad \ldots, \qquad x_{n+1} = N(x_n).$$

Newton's method, then, amounts to applying N *repeatedly* to x_0; r is a root of f if and only if $N(r) = r$. Such an r is called a **fixed point** of the function N.   (The modern theory of dynamical systems studies how inputs behave under repeated applications of a function.)

Aren't Graphs Enough?   Isaac Newton didn't have computer graphics. We do, so why should we bother with Newton's method? Why not just zoom in, as often as necessary, to find roots to any desired accuracy?

One reason is that the process can be tedious, especially if many decimal places of accuracy are required. Moreover, some graphics hardware and software display no more than 8 or 10 digits.

A more important reason concerns efficiency of effort. Whether performed by human or machine, plotting any function often requires dozens, if not hundreds, of function evaluations—an "expensive" proposition. Newton's method is much more efficient. For most functions, successive Newton estimates converge to a root with remarkable speed: Each estimate has about twice as many correct digits as its predecessor.

EXAMPLE 3   We estimated graphically that $f(x) = 2x^3 - x - 2$ has a root near $x = 1.17$. Use Newton's method to do better.

Solution   Since $f(x) = 2x^3 - x - 2$, $f'(x) = 6x^2 - 1$. By the preceding Fact,

$$x_{n+1} = x_n - \frac{f(x_n)}{f'(x_n)} = x_n - \frac{2x_n^3 - x_n - 2}{6x_n^2 - 1}.$$

Check some of these numbers.

If, say, $x_0 = 1$, then ◄

$$x_1 = 1.2, \qquad x_2 \approx 1.16649, \qquad x_3 \approx 1.16537, \quad \text{and} \quad x_4 \approx 1.16537.$$

Our guess, then, is $r \approx 1.16537$.   It checks out numerically (graphically, too): $f(1.16537) \approx -0.00002$.   ∎

Newton's Method in Action

Successful applications of calculus ideas to real problems often depend, sooner or later, on finding roots. Newton's method can simplify the process.

EXAMPLE 4   Which point P on the parabola $y = x^2$ is closest to $(2, 1)$?

Solution   Here's the picture:

On $y = x^2$, P is closest to $(2, 1)$

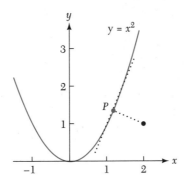

From the picture alone we can only estimate P. However, the picture does suggest the crucial idea: ▶

We'll return to this idea more formally in Section 4.6.

> *The line segment from $(2, 1)$ to P is perpendicular to the tangent line at P.*

We'll translate this condition into an equation, then solve it to find P.

Since P lies on the parabola, its coordinates have the form (r, r^2). Thus, the line tangent to $y = x^2$ at P has slope $2r$, ▶ and the line segment from $(2, 1)$ to P has slope $(r^2 - 1)/(r - 2)$. The perpendicularity condition, therefore, reads symbolically as follows : ▶

*Because
$y = x^2 \implies dy/dx = 2x$.*

Perpendicular lines have negative reciprocal slopes.

$$\frac{r^2 - 1}{r - 2} = -\frac{1}{2r}.$$

Equivalently, $2r^3 - r - 2 = 0$. ▶

This last equation should look familiar—we solved it in Example 3, with Newton's method, and got $r \approx 1.16537$. (The result looks graphically plausible, too.) ■

Convince yourself of the equivalence.

EXAMPLE 5   Calculators "know" that $\sqrt{2} \approx 1.4142136$. Use Newton's method to derive such an estimate.

Solution   Our target, $\sqrt{2}$, is a root of $f(x) = x^2 - 2$. Applied to f, Newton's formula is especially simple: ▶

Check the last algebraic step.

$$x_{n+1} = x_n - \frac{f(x_n)}{f'(x_n)} = x_n - \frac{x_n^2 - 2}{2x_n} = \frac{x_n}{2} + \frac{1}{x_n}.$$

Check some with a calculator.

Starting with $x_0 = 1$, successive estimates ◀ quickly zero in on $\sqrt{2}$:

$$x_1 = 1.5, \quad x_2 \approx 1.4166667, \quad x_3 \approx 1.4142157, \quad x_4 \approx 1.4142136. \qquad ■$$

> **Other Root-Finding Methods.** Newton's method is only one of several common root-finding techniques. The **bisection, or interval-halving, method** (discussed further in Section 4.10) works by isolating a root in smaller and smaller intervals. The **secant method** resembles Newton's tangent method, except that secant lines replace the tangent lines of Newton's method.

Root-Grubbing for Money: IRAs and Newton's Method

U.S. tax law allows some taxpayers to deposit $2000 per year in tax-sheltered individual retirement accounts (IRAs). Doing so raises interesting questions: How much will such an account be worth after 30 years of faithful contributions? What interest rate will produce $300,000 after 30 years?

Expressed as a decimal number, not a percent.

The **compound amount formula** helps answer such questions. If p dollars are deposited yearly at effective annual interest rate r, ◀ then the total value $V(n)$ after n years is given by

$$V(n) = \frac{p}{r}\left((1 + r)^{n+1} - 1\right).$$

How nice? It depends, for example, on inflation and tax rates 30 years hence.

For an IRA paying 6% over 30 years, we'd use $p = 2000$, $r = 0.06$, and $n = 30$. The result is a nice ◀ little nest egg:

$$V(30) = \frac{2000}{0.06}\left((1.06)^{31} - 1\right) \approx \$169{,}603.35.$$

EXAMPLE 6   Riskier investments might yield higher interest rates for one's IRA. What interest rate would ensure a value of $300,000 after 30 years of $2000 contributions?

Solution   We need a value of r for which

$$V(30) = \frac{2000}{r}\left((1 + r)^{31} - 1\right) = 300{,}000.$$

Solving this equation algebraically for r is hard or impossible, so we resort to Newton's method. A little preliminary algebra simplifies the work:

$$\frac{2000}{r}\left((1 + r)^{31} - 1\right) = 300{,}000 \iff \left((1 + r)^{31} - 1\right) - 150r = 0.$$

In Example 1, page 280, we estimated $r \approx 0.09$ as a root of the last equation. With

$$f(r) = \left((1 + r)^{31} - 1\right) - 150r, \qquad f'(r) = 31(1 + r)^{30} - 150,$$

By hand the computations are messy. With a calculator they're easy.

and initial guess $x_0 = 0.09$, Newton's method gives ◀

$$x_1 \approx 0.09014631, \qquad x_2 \approx 0.09014585, \qquad x_3 \approx 0.09014585.$$

After just three Newton steps, the answer is clear: An annual interest rate of 9.015% will suffice. ■

When Things Go Wrong

Newton's method, though impressive when it works, sometimes fails. Several problems can arise.

Horizontal Tangent Lines. If $f'(x_n) = 0$, then the recursion formula

$$x_{n+1} = x_n - \frac{f(x_n)}{f'(x_n)}$$

makes no sense, and the process fails. In graphical terms, the problem is that the tangent line at x_n is horizontal and so never meets the x-axis.

Near-Horizontal Tangent Lines. The preceding problem occurs only rarely in practice; hitting a point of horizontal tangency *exactly* is unlikely. Trouble may still occur, however, if x_n lands too near a root of f'. The next example shows how this can happen.

EXAMPLE 7 Let $f(x) = \sin x$ and $x_0 = 1.3$. Among the roots of f, $r = 0$ lies closest to x_0. What does Newton's method say?

Solution Applied to f, with initial guess $x_0 = 1.3$, Newton's method gives

$$x_1 \approx -2.3021; \qquad x_2 \approx -3.4166; \qquad x_3 \approx -3.1344; \qquad x_4 \approx -3.1416.$$

We found a root but it's the wrong one. The picture shows what happened:

A shallow tangent: finding the wrong root

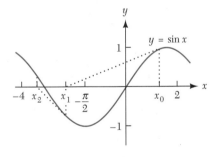

At x_0 the tangent line was too nearly horizontal; riding it to the x-axis took us far away from the expected root. ∎

Cycling. In rare (but interesting) circumstances, successive Newton's approximations cycle, repeating endlessly without converging to a root.

EXAMPLE 8 It's easy to check (by factoring the formula) that the cubic polynomial $f(x) = x^3 - 6x^2 + 7x + 2$ has three roots: $x = 2$, $x = 2 - \sqrt{5} \approx -0.24$, and $x = 2 + \sqrt{5} \approx 4.24$. But what happens if Newton's method is applied to f, with initial guess $x_0 = 1$? *Why* does it happen?

S o l u t i o n   Successive estimates x_0, x_1, x_2, x_3, ... follow the repeating pattern 1, 3, 1, 3, ... The following picture shows why:

Cycling around with Newton's method

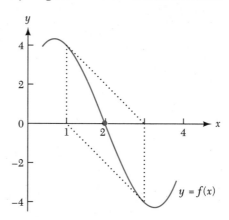

One that won't, however, is $x_0 = 3$. Do you see why not?

By careful planning, we chose the worst possible initial guess x_0. Almost any other initial estimate x_0 would lead successfully to a root. ◄   If $x_0 = 0.999$, for example, Newton's method finds the negative root $2 - \sqrt{5}$. If $x_0 = 1.001$, Newton's method finds the middle root, at $x = 2$. ∎

> **The Butterfly Effect.**   Tiny changes in the initial guess x_0 for Newton's method may cause dramatically different results. In Example 8, for instance, the initial inputs $x_0 = 1.000$, $x_0 = 0.999$, and $x_0 = 1.001$, although almost identical, produced completely different outputs.
>   This phenomenon—small causes producing large effects—occurs often in mathematics and in the real world. It's sometimes called the **butterfly effect**: A butterfly's wing flap in Kansas may eventually trigger a typhoon in Taiwan. Mathematicians use a more prosaic name: **sensitive dependence on initial conditions**.

BASICS

1. By hand, find the first three Newton estimates for $\sqrt{5}$, starting with $x_0 = 2$. Write answers as fractions.
   [**HINT**: First use the Fact on page 282 to write a simple formula for x_{n+1} in terms of x_n.]
2. Use Newton's method with the value of x_0 indicated to compute approximate values of each of the following expressions. (Iterate until successive approximations agree to three decimal places.)
   (a) $\sqrt[3]{100}$; $x_0 = 4.5$ (b) $(13)^{-2}$; $x_0 = 0.01$

3. (a) Use Newton's method to find a solution of the equation $x^5 + 4x - 3 = 0$ in the interval $[0, 1]$ to three-decimal-place accuracy.
   (b) Does the equation in part (a) have any other solutions? Justify your answer.
4. Find a solution of the equation $x \tan x = 1$ in the interval $[0, \pi/2]$ to three-decimal accuracy.

FURTHER EXERCISES

5. For the sample problems on the first page of this section, try to find roots by solving algebraically. (The point is to see that this can't always be done.)

6. Showing your work, derive Equation 4.4, page 280. Use the result to find the x-intercepts of each of the following lines:
   (a) Through $(1, 1)$, with slope 1.
   (b) Through $(1, 2)$, with slope 3.
   (c) The line with equation $y = mx + b$.
   (d) The line tangent to the graph $y = x^2$ at $x = 1$.
   (e) The line tangent to the graph $y = x^2 - 3$ at $x = 1$.

7. Use Equation 4.4 to derive Equation 4.5. Show all work.

8. Here's another way to find the point P on the parabola $y = x^2$ that's closest to $(2, 1)$:
   (a) Let $d(r)$ be the distance from $(2, 1)$ to $P = (r, r^2)$. Write a formula for $d(r)$.
   (b) At the desired value of r, d has a local minimum, so $d'(r) = 0$. Write out and simplify the equation $d'(r) = 0$. Does the result look familiar?

9. Let $f(x) = x^2 - a$.
   (a) Explain why the two numbers x and a/x bracket $\sqrt{a}$.
   (b) Show that Newton's method applied to f gives the iteration function $N(x) = (x + a/x)/2$—i.e., the average of x and a/x.
   (c) What happens if $x_n = \sqrt{a}$?

10. Let $f(x) = x^3 - 2x^2 - 5x + 7 = (x-3)(x-1)(x+2)+1$.
   (a) Explain why it is natural to guess that f has roots near $x = 3$, $x = 1$, and $x = -2$.

   (b) Use Newton's method to estimate the values of the roots of f to four decimal places.

11. Let $f(x) = x^3 - 10x^2 + 22x + 6$.
   (a) Find each of the three roots of f to three decimal places.
   (b) What happens when Newton's method is used with $x_0 = 2$? Explain why this happens.
   (c) Newton's method behaves oddly if $x_0 = 1.39$. How? Why?

12. Let $h(x) = 179.5x^6 - 436.8x^4 + 257.3x^2 - 1$.
   (a) $h(x)$ has three positive roots. Use Newton's method to find each one to three decimal places.
   (b) $h(x)$ has three negative roots. How are they related to the three positive roots you found in part (a)?

13. Find (approximately) the minimum value of $g(x) = 10x^{-2} + 3x^2 + x - 4$ when $1 \le x \le 10$.

14. Let $f(x) = x^2 \sin x^2$ on the interval $[0, 3]$.
   (a) Find (approximately) each of the following numbers to two decimal places using Newton's method:
       (i) The minimum value of f.
       (ii) The maximum value of f.
       (iii) The inflection points of f.
   (b) Sketch a graph of f and label both coordinates of each of the points you found above.

15. Apply Newton's method to $f(x) = x^{1/3}$, starting from $x_0 = 1$. Explain what happens.

16. Apply Newton's method to $f(x) = e^{-x}$, starting from $x_0 = 0$. Explain what happens.

4.5 Splines: Connecting the Dots

How does one draw a curve through several—perhaps many—points in a plane? (**NOTE:** In this section, "curve" means any unbroken line, with or without corners.) The answer depends partly on who does the drawing. For a human artist, it's easy to draw freehand curves that pass through, say, five given points. Indeed, there are many ways to do so. Here are four possibilities:

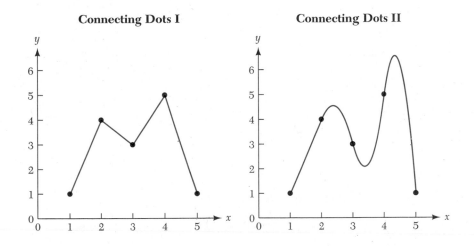

Connecting Dots I   **Connecting Dots II**

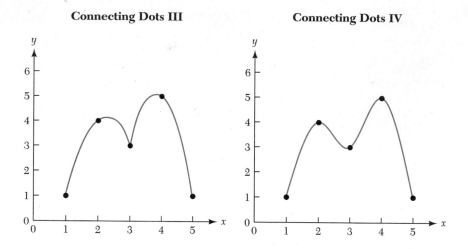

Connecting Dots III   **Connecting Dots IV**

For humans, such ambiguity is no problem. We can easily choose (on aesthetic, physical, or other grounds) one curve through the given points that meets our purposes. Computers, on the other hand, need precise instructions in their own language—mathematics. Calculus provides the tools both to *find* curves that fit data points and to *describe* those curves in terms computers can understand.

Splines: Basic Ideas and Key Words

The idea behind splines is to "tie" small curve segments, called **elements**, together to form a single larger curve, called a **spline**. The points at which elements are tied together (the dots in the preceding pictures) are called **knots**.

Our connect-the-dots pictures illustrate these ideas. Picture I, formed from line segments, shows a **linear spline**. Picture II shows a **quadratic spline**, formed by joining pieces of four parabolas. Picture III shows another (perhaps less pleasing) possible way of joining parabolic arcs. Picture IV shows a **cubic spline**, formed by joining pieces of four cubic arcs. (A cubic arc is defined by a cubic polynomial, $a + bx + cx^2 + dx^3$.)

Linear, quadratic, and cubic splines are formed by joining pieces of lines, parabolas, or cubic curves end to end. Individual lines, parabolas, and cubic arcs are smooth and unbroken, so any new curve built from such pieces is also smooth and unbroken, except perhaps at the knots. With help from derivatives, we'll fit such elements together properly, even at the knots.

Linear Splines: Continuous but Crooked

The simplest splines are formed from straight lines.

EXAMPLE 1   Write a formula for the linear spline in Picture I.

Solution   The curve in question is **piecewise-linear**, i.e., built from several straight-line pieces. The formula for a piecewise-linear function has several parts, one for each piece. The first piece in Picture I, for instance, is part of the line joining the points $(1, 1)$ and $(2, 4)$. This line has slope 3, so its equation in point–slope form ▶ is $y - 1 = 3(x - 1)$, or, equivalently, $y = 3x - 2$. This equation describes only the first piece, where $1 \leq x \leq 2$. Similar reasoning applied to the other spline elements gives the following multiline formula for the spline function S:

Convince yourself of this.

$$S(x) = \begin{cases} 3x - 2 & \text{if } 1 \leq x \leq 2; \\ -x + 6 & \text{if } 2 \leq x \leq 3; \\ 2x - 3 & \text{if } 3 \leq x \leq 4; \\ -4x + 21 & \text{if } 4 \leq x \leq 5. \end{cases}$$

The formula, like the graph, is piecewise-linear: On each piece of its domain the function S is defined by a linear polynomial. ∎

A linear spline, by definition, is a piecewise-linear function that's also continuous. For piecewise-linear functions, continuity holds as long as the linear pieces join properly at their ends.

Quadratic Splines: Smoother

Linear splines, although simple, have an ugly kink at each knot (unless, by pure chance, successive slopes match up). One way to avoid such kinks is to use a quadratic spline, formed by linking parabolic pieces. ▶

Each piece is defined by a quadratic equation.

When Is a Parabola Degenerate?   A quadratic polynomial q has the form $q(x) = a + bx + cx^2$, where a, b, and c are constants. The graph of a quadratic polynomial is almost always a parabola. There's one exception: If $c = 0$, then $q(x) = a + bx + 0x^2 = a + bx$ and q is actually a *linear* polynomial; its graph is a straight line.

To allow for this possibility, it's often best to interpret the word "quadratic" as meaning "having powers no greater than two." Similarly, cubic polynomials involve powers no greater than three. A polynomial that happens to have less than its "advertised" power is sometimes called **degenerate**. (Mathematicians attach no stigma to degeneracy.) From this point of view, a line can be thought of as a degenerate parabola.

Linear splines are crooked because only one possible line segment can join one knot to the next. Quadratic segments are more flexible—literally! Given any two successive knots, we can find a parabola that joins them, with any desired slope at the left knot. Here, for instance, are several parabolas joining $(0, 0)$, and $(1, 1)$,

*See the exercises at the end
of this section for more on
these parabolas.*

each labeled with its slope at $(0, 0)$: ◄

Parabolas joining $(0, 0)$ and $(1, 1)$:
m = slope at $(0, 0)$

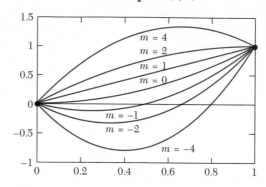

(One "parabola" is actually a line—which one?) By choosing slopes judiciously at each knot, we can iron out the kinks. The next example shows exactly how.

EXAMPLE 2   Use quadratic pieces to form a smooth spline through the same knots as in Example 1:

x	1	2	3	4	5
y	1	4	3	5	1

*From among the infinite
possibilities.*

*Smoothness is an issue only
where two curves join.*

*Remember: A straight line is
a degenerate quadratic.*

*We saw earlier that
$S_1'(2) = 3$.*

Solution   To join each knot to the next, we choose ◄ a parabolic curve. We produce, in other words, a **piecewise-quadratic function**. To avoid kinks, we make sure that the slopes of "incoming" and "outgoing" elements match at each knot. The first spline element, S_1 (from $x = 1$ to $x = 2$), is simplest. Smoothness is no problem at the first knot, ◄ so any quadratic curve joining $(1, 1)$ to $(2, 4)$ will do. The simplest choice, perhaps, is a straight line; ◄ we start with that. We saw in Example 1 that the linear function $S_1(x) = 3x - 2$ works. In particular, $S_1'(2) = 3$.

The second spline element (from $x = 2$ to $x = 3$) is more interesting. We need a quadratic polynomial $S_2(x)$ that joins the second and third knots and also agrees in slope with S_1 at $x = 2$. ◄ In other words, our quadratic polynomial S_2 should satisfy three conditions:

$$S_2(2) = 4; \qquad S_2'(2) = 3; \qquad S_2(3) = 3.$$

How can we find such a polynomial? The following strategy simplifies the algebra considerably. *Write S_2 in powers of $(x - 2)$*, like this:

$$S_2(x) = a + b(x - 2) + c(x - 2)^2.$$

Using powers of $(x - 2)$, rather than powers of x or powers of $(x - 3)$, conveniently "fixes" attention at the current starting point, and makes it easy to choose the constants a, b, and c properly.

Setting $x = 2$ in the preceding expression gives $S_2(2) = a$. Thus, to arrange that $S_2(2) = 4$, we need $a = 4$. Choosing b is similar but starts with a derivative:

$$S_2(x) = a + b(x - 2) + c(x - 2)^2 \implies S_2'(x) = b + 2c(x - 2)$$

$$\implies S_2'(2) = b \implies b = 3.$$

Now we know that $S_2(x) = 4 + 3(x - 2) + c(x - 2)^2$; only c is still to be found. To find it, we use the remaining condition, ▶ $S_2(3) = 3$, as follows:

We've already used the other two conditions, $S_2(2) = 4$ and $S_2'(2) = 3$.

$$3 = S_2(3) = 4 + 3(3 - 2) + c(3 - 2)^2 = 7 + c \implies c = -4.$$

Everything is now in place: ▶

Check for yourself that S_2 does indeed satisfy all three of the preceding requirements.

$$S_2(x) = a + b(x - 2) + c(x - 2)^2$$

$$= 4 + 3(x - 2) - 4(x - 2)^2$$

$$= -18 + 19x - 4x^2.$$

(The last expression above can be found by multiplying out the second-last expression. The last version, although shorter typographically, is not necessarily simpler— the preceding version shows better how the coefficients were found.)

We find the remaining spline elements in just the same way. The next one, S_3, must pass through the third and fourth knots and, to ensure smoothness, must agree in slope with S_2 at $x = 3$. The preceding formula shows that $S_2'(x) = 19 - 8x$, so $S_2'(3) = -5$. Thus, S_3 must satisfy these three conditions:

$$S_3(3) = 3; \qquad S_3'(3) = -5; \qquad S_3(4) = 5.$$

It follows, as before, ▶ that

Convince yourself; only the last coefficient takes much effort.

$$S_3(x) = a + b(x - 3) + c(x - 3)^2 = 3 - 5(x - 3) + 7(x - 3)^2 = 81 - 47x + 7x^2.$$

Similarly, S_4 is chosen to satisfy ▶

Verify that $S_3'(4) = 3$.

$$S_4(4) = 5; \qquad S_4'(4) = S_3'(4) = 9; \qquad S_4(5) = 1.$$

The result ▶ is

The details are left to you; see the exercises at the end of this section.

$$S_4(x) = a + b(x - 4) + c(x - 4)^2$$

$$= 5 + 9(x - 4) - 13(x - 4)^2$$

$$= -239 + 113x - 13x^2.$$

Joining the elements produces this quadratic spline:

$$S(x) = \begin{cases} S_1(x) = 1 + 3(x - 1) & \text{if } 1 \le x \le 2; \\ S_2(x) = 4 + 3(x - 2) - 4(x - 2)^2 & \text{if } 2 \le x \le 3; \\ S_3(x) = 3 - 5(x - 3) + 7(x - 3)^2 & \text{if } 3 \le x \le 4; \\ S_4(x) = 5 + 9(x - 4) - 13(x - 4)^2 & \text{if } 4 \le x \le 5. \end{cases}$$

The picture is worth a careful look.

Following is the graphical result. ◄   The spline S is the bold curve; the quadratic elements S_1 through S_4 are extended beyond the knots to show how they fit together *at* the knots.

**Smoothing the knots:
quadratic spline elements**

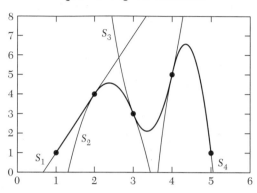

Example 2 illustrates the formal definition.

> **Definition** A quadratic spline is a continuous, piecewise-quadratic function S whose first derivative, S', is also continuous.

This definition may seem forbiddingly technical. Here it means simply that (as we arranged) incoming and outgoing quadratic elements agree at each knot, both in their values and in their first derivatives.

Quadratic Splining: The Process in General

The technique just illustrated works for any collection (x_0, y_0), (x_1, y_1), ..., (x_n, y_n) of knots, ordered from left to right. Here's the drill:

Start   Choose any quadratic curve S_1 joining (x_0, y_0) to (x_1, y_1). (A line segment is one possibility; other choices are possible.) For later use, calculate $m_1 = S_1'(x_1)$.

Fit the Next Parabola   Find a quadratic curve S_2 joining (x_1, y_1) to (x_2, y_2), such that $S_2'(x_1) = m_1$. ◄   To do so, find coefficients a, b, and c so that $S_2(x) = a + b(x - x_1) + c(x - x_1)^2$. Then, for later use, calculate $m_2 = S_2'(x_2)$.

m_1 is the "incoming slope" at (x_1, y_1).

Continue   Proceed as before. For each successive pair of knots (x_i, y_i) and (x_{i+1}, y_{i+1}), choose a quadratic curve S_{i+1} joining the knots, with $S_{i+1}'(x_{i-1}) = m_i$. ◄

m_i is the "outgoing slope" at (x_i, y_i).

How Smooth are Quadratic Splines?

A quadratic spline S has a continuous first derivative. The graph of S therefore has no kinks or corners. The second derivative S'', on the other hand, is *discontinuous* at each knot. ▶ This means, graphically, that the concavity of the S-graph can change abruptly at each knot. A close look at the preceding graph shows this happening. ▶ With cubic splines, we'll smooth things even further.

Unless, by sheer coincidence, successive elements have equal second derivatives at a knot.

Imagine riding a roller coaster along a quadratic spline. What would you feel at each knot?

Counting the Ways: Degrees of Freedom. There is just one line through two given points, or through one point with given slope. Either way, a straight line is completely determined by two conditions.

The reason, algebraically, is that the generic linear equation $y = a + bx$ involves *two* parameters, a and b; different choices of a and b give different lines. In mathematical language, we have **two degrees of freedom** in choosing a line; specifying two points (or one point and a slope) "uses up" both degrees of freedom.

By contrast, a quadratic polynomial $y = a + bx + cx^2$ involves *three* parameters. Thus, three degrees of freedom are available when we choose quadratics. The practical result is that a quadratic spline element—a parabola—can be chosen to satisfy three conditions. Two degrees of freedom are used to hit the left and right knots, but the third degree of freedom remains available. Our method "spends" the extra degree of freedom in specifying a derivative at the left knot, thus matching incoming and outgoing slopes.

Cubic spline elements, of the form $y = a + bx + cx^2 + dx^3$, offer four degrees of freedom. As with linear and cubic elements, hitting the knots costs two degrees of freedom. We'll spend the remaining two on matching two derivatives of incoming and outgoing elements at the left knot.

Cubic Splines: Smoother Still

A **cubic spline**, by definition, is a piecewise-cubic function S for which the first two derivatives, S' and S'', are continuous. Like linear and quadratic splines, cubic splines are constructed by piecing elements together, with each element spanning two knots. With cubic equations ▶ we can ensure that, at each knot, incoming and outgoing elements agree in their values and in their first two derivatives. ▶ By avoiding abrupt changes in concavity, cubic elements produce smoother, more pleasing splines than quadratic elements can achieve.

Cubic equations offer four degrees of freedom, one for each coefficient.

With quadratic splines we could manage just one derivative.

EXAMPLE 3   Construct a cubic spline through the same knots:

x	1	2	3	4	5
y	1	4	3	5	1

Solution   Smoothness is no problem at the first knot. In theory, therefore, we can start by choosing *any* cubic curve S_1 that joins $(1, 1)$ to $(2, 4)$. Infinitely many possibilities exist—any values for $S_1'(1)$ and $S_1''(1)$ are possible. In practice, some choices give better results than others. Moreover, the resulting spline depends strongly on how the first element is chosen. Here, for instance, are two different cubic splines through the same knots:

Two cubic splines through the same knots

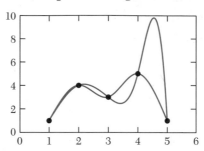

The two splines differ only in how the first element is chosen. As the picture shows, a tiny difference in the first element produces huge differences later on.

Faced with such exquisite sensitivity to initial choices, mathematicians have found systems for choosing the first cubic spline element. ◄ The **natural spline method**, for instance, arranges things so that $S''(x_0) = 0 = S''(x_n)$, i.e., so that S has zero concavity at the first and last knots. (The "tamer" curve in the preceding picture is a natural spline.) Full details on how natural splining works are beyond the scope of this discussion. ◄ We'll simply remark that the unlikely-looking cubic

Or for choosing all the elements simultaneously.

Numerical analysis courses discuss natural and other splining methods.

$$S_1(x) = 1 + \frac{123}{28}(x-1) - \frac{39}{28}(x-1)^3$$

Check that $S_1(1) = 1$ and $S_1(2) = 4$.

can be used to start things off. ◄ For impending use, we notice that $S_1'(2) = 3/14$ and $S_1''(2) = -117/14$.

For the next spline element, from $x = 2$ to $x = 3$, we need a cubic polynomial $S_2(x)$ that satisfies four conditions: ◄

The arithmetic will soon get messy. Bear with us.

$$S_2(2) = 4; \qquad\qquad S_2'(2) = \frac{3}{14} = S_1'(2);$$

$$S_2''(2) = -\frac{117}{14} = S_1''(2); \qquad S_2(3) = 3.$$

As before, writing S_2 in powers of $(x-2)$, as in

$$S_2(x) = a + b(x-2) + c(x-2)^2 + d(x-2)^3,$$

simplifies our search.

We choose the coefficients a and b exactly as in Example 2. The preceding expression shows ◄ that $S_2(2) = a$, so $a = 4$. Similarly,

Just plug in $x = 2$.

$$S_2'(x) = b + 2c(x-2) + 3d(x-2)^2 \implies S_2'(2) = b \implies b = \frac{3}{14}.$$

To find c, we differentiate again:

$$S_2''(x) = 2c + 6d(x-2) \implies S_2''(2) = 2c \implies c = -\frac{117}{28}.$$

Now we know that $S_2(x) = 4 + \frac{3}{14}(x-2) - \frac{117}{28}(x-2)^2 + d(x-2)^3$; we need only find d. To do so, we use the last remaining condition, $S_2(3) = 3$: ▶

Check the arithmetic if you like, but be sure to check the logic.

$$3 = S_2(3) = 4 + \frac{3}{14}(3-2) - \frac{117}{28}(3-2)^2 + d(3-2)^3 \implies d = \frac{83}{28}.$$

We're finished at last:

$$S_2(x) = 4 + \frac{3}{14}(x-2) - \frac{117}{28}(x-2)^2 + \frac{83}{28}(x-2)^3.$$

For later reference, ▶ straightforward calculation shows that $S_2'(3) = 3/4$ and $S_2''(3) = 66/7$.

We'll need these numbers to fit the next element.

We find the remaining spline elements in the same way. The next one, S_3, must satisfy four conditions:

$$S_3(3) = 3; \qquad\qquad S_3'(3) = \frac{3}{4} = S_2'(3);$$

$$S_3''(3) = \frac{66}{7} = S_2''(3); \qquad S_3(4) = 5.$$

When all the dust settles, calculations like these reveal that our desired spline has the following piecewise-cubic form: ▶

It isn't necessary to check all calculations; the idea's the thing.

$$S(x) = \begin{cases} S_1(x) = 1 + \frac{123}{28}(x-1) - \frac{39}{28}(x-1)^3 & \text{if } 1 \le x \le 2; \\[2mm] S_2(x) = 4 + \frac{3}{11}(x-2) - \frac{117}{28}(x-2)^2 + \frac{83}{28}(x-2)^3 & \text{if } 2 \le x \le 3; \\[2mm] S_3(x) = 3 + \frac{3}{4}(x-3) + \frac{33}{7}(x-3)^2 - \frac{97}{28}(x-3)^3 & \text{if } 3 \le x \le 4; \\[2mm] S_4(x) = 5 - \frac{3}{14}(x-4) - \frac{159}{28}(x-4)^2 + \frac{53}{28}(x-4)^3 & \text{if } 4 \le x \le 5. \end{cases}$$

■

Postscript and Confession

The general idea of linear, quadratic, and cubic splines—our main interest in this section—is simple. Elements of the desired type are linked end to end; as many derivatives as possible are made to agree at knots.

The necessary calculations, although routine, can be tedious and repetitive, especially when dozens or hundreds of knots are involved. In real-life practice, such calculations are left to machines. Specialized methods that extend the ideas mentioned here are used to speed up and simplify the machines' work.

BASICS

1. What does it mean for a function to be "piecewise-linear?" Explain briefly in words.

2. Describe (i.e., give a formula for) a piecewise-linear function whose graph passes through the points $(0,0)$, $(1,1)$, and $(2,4)$.

3. A linear spline is given by the multiline formula—one element is missing:

$$S(x) = \begin{cases} -2x + 7, & \text{if } 1 \le x \le 2 \\ \vdots & \vdots \\ -3x + 13, & \text{if } 3 \le x \le 4 \\ x - 3, & \text{if } 4 \le x \le 5 \end{cases}$$

Draw the spline; then fill in the missing line of the formula.

FURTHER EXERCISES

4. (a) Which upper-case letters of the alphabet could be graphs of piecewise-linear functions?

   (b) For two of the letters you chose in the previous part, describe (i.e., give a formula for) a piecewise-linear function whose graph has the right shape. (There are several ways to do this.)

5. Verify the last three lines of the multiline formula for the linear spline S in Example 1.   [**HINT:** For each piece, start with the point–slope form of the line.]

6. In Example 2 we claimed that

   $$S_4(x) = 5 + 9(x - 4) - 13(x - 4)^2 = -239 + 113x - 13x^2$$

   satisfies the three conditions (i) $S_4(4) = 5$; (ii) $S_4'(4) = 9$; (iii) $S_4(5) = 1$. Verify that all three conditions really hold.

In Exercises 7–10, find a quadratic polynomial that satisfies the three given conditions. Then plot the polynomial on an appropriate interval to *show* that it has the desired properties.

7. $Q(0) = 0$; $Q'(0) = 3$; $Q(1) = 5$.

8. $Q(0) = 0$; $Q'(0) = 3$; $Q(10) = 5$.

9. $Q(1) = 0$; $Q'(1) = 3$; $Q(3) = 0$.

10. $Q(-17.3) = 11.1$; $Q'(-17.3) = -1.7$; $Q(-17.1) = 11.5$.

11. The picture on page 290 shows several quadratic curves, each joining $(0, 0)$ and $(1, 1)$, each labeled with its slope at $x = 0$. Find a quadratic equation for each curve.

12. Let (x_0, y_0) and (x_1, y_1) be any two points with $x_0 \neq x_1$, and let m_0 be any real number. Consider the quadratic function q defined by

   $$q(x) = y_0 + m_0(x - x_0) + \frac{y_1 - y_0 - m_0(x_1 - x_0)}{(x_1 - x_0)^2}(x - x_0)^2.$$

   (a) Find $q(x)$ in the case $(x_0, y_0) = (2, 4)$, $(x_1, y_1) = (3, 3)$, $m_0 = 3$. Then check that $q(2) = 4$, $q(3) = 3$, and $q'(2) = 3$.

   (b) Plot your function q from the previous part. How does the graph show that $q(2) = 4$, $q(3) = 3$, and $q'(2) = 3$?

   (c) Show (work with the general formula above) that $q(x_0) = y_0$, $q(x_1) = y_1$, and $q'(x_0) = m_0$. What do these properties mean about the graph of q?

   (d) Give an example in which $q(x)$ turns out to be a linear function. Interpret your example in graphical terms.

13. In each part below, find a quadratic spline S through the three knots

x	0	1	2
y	1	-2	3

   (a) Choose the first element by assuming that $S'(0) = -3$.

   (b) Choose the first element by assuming that $S'(0) = 0$.

   (c) The graphs of the two splines described in (a) and (b) follow. Which is which? Why?

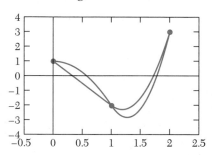

Two quadratic splines through the same knots

14. (For algebra buffs.) Let (x_0, y_0) and (x_1, y_1) be any two points with $x_0 \neq x_1$, and let d_1 and d_2 be any real numbers. Consider the cubic function c defined by

   $$c(x) = y_0 + d_1(x - x_0) + \frac{1}{2}d_2(x - x_0)^2 + \frac{y_1 - y_0 - d_1(x_1 - x_0) - \frac{1}{2}d_2(x_1 - x_0)^2}{(x_1 - x_0)^3(x - x_0)^3}$$

   Show by (careful!) calculation that c satisfies the four conditions (i) $c(x_0) = y_0$; (ii) $c'(x_0) = d_1$; (iii) $c''(x_0) = d_2$; (iv) $c(x_1) = y_1$. (These four conditions mean that c is the "general" spline element joining (x_0, y_0) to (x_1, y_1) with first two derivatives d_1 and d_2 at (x_0, y_0)).

In Exercises 15–18, use the formula in the previous exercise (even if you didn't do the previous exercise) to find a cubic function with the given properties. Plot each result.

15. $c(0) = 0$; $c'(0) = 0$; $c''(0) = 0$; $c(1) = 0$.

16. $c(0) = 0$; $c'(0) = 0$; $c''(0) = 0$; $c(1) = 1$.

17. $c(0) = 1$; $c'(0) = 0$; $c''(0) = -1$; $c(1) = 1/2$.

18. $c(0) = 1$; $c'(0) = 1$; $c''(0) = 1$; $c(1) = 8/3$.

19. In each part below, find a cubic spline S through the three knots

x	0	1	2
y	1	-2	3

   (a) Choose the first element by assuming that $S'(0) = -3$, $S''(0) = 0$.

   (b) Choose the first element by assuming that $S'(0) = 0$, $S''(0) = 0$.

(c) Below are graphs of the two splines you just found. Which is which? Why?

Two cubic splines through the same knots

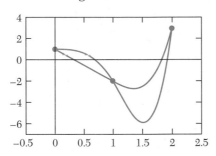

20. By definition, a quadratic spline is a piecewise-linear function whose derivative is also continuous. In fact, the derivative of a quadratic spline is a linear spline. Verify this in the case of Example 2 by writing a (multiline) formula for S'. Then plot both S' and S on the same axes. [**HINT:** To find S', differentiate S line by line.]

4.6 Optimization

Optimization and the Derivative

Optimization is a two-dollar word for the general problem of finding maximum or minimum values of functions. By now it should be clear that the derivative is a useful optimization tool: Anywhere f' is zero is a natural candidate for a maximum or minimum of f. Several subtleties can arise, however. A root of f' might, for instance, correspond to a local maximum, a local minimum, or neither. A maximum, in turn, could be global or merely local.

To help sort things out, we'll collect and summarize the language, the techniques, and a few of the subtleties involved in optimization problems.

Stationary Points and Critical Points. Let f be a function and f' its derivative. As we've seen, f increases where $f' > 0$ and decreases where $f' < 0$. Therefore, any point x_0 at which f' changes sign is naturally of interest.

Barring bizarre circumstances, ▸ f' can change sign at x_0 in only two ways:

We do bar them.

Either $f'(x_0) = 0$ or $f'(x_0)$ doesn't exist.

If the first condition holds, x_0 is a **stationary point** of f. If *either* condition holds, then x_0 is a **critical point** of f.

EXAMPLE 1 The function $f(x) = |x|$ ▸ has no stationary points, because $f'(x)$ is never zero. The critical point $x_0 = 0$ (where the derivative fails to exist) is a local minimum point of f. ∎

Draw (or imagine) the graph of f.

Local vs. Global Extreme Values. Local maximum and minimum values of a function may or may not be global. ▸ The function $f(x) = x/2 + \sin(x)$, for example, has many local maxima and minima on $(-\infty, \infty)$, but none is global—this

Formal definitions of these words appear in Section 2.3.

f is unbounded above and below on $(-\infty, \infty)$. The graph agrees:

Local but not global maxima and minima:
$$f(x) = x/2 + \sin x$$

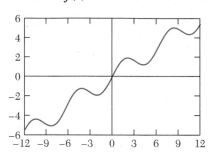

Which Functions Have Global Maxima and Minima? Continuous Functions on Closed Intervals.   As we just saw for $f(x) = x/2 + \sin(x)$ on $(-\infty, \infty)$, a function may have neither a maximum nor a minimum on a given interval. All is not lost. The **extreme value theorem** ◄   guarantees that a continuous function f on a closed interval $[a, b]$ must assume both a global maximum and a global minimum.

We'll study it carefully in Section 4.10.

Evidently a continuous function.

Another look at the preceding graph shows that this property holds for $f(x) = x/2 + \sin(x)$ ◄   on the closed interval $[-12, 12]$. In this case, f takes its largest and smallest values (approximately ± 5.5) at the interval's endpoints, $x = \pm 12$.

Finding Maxima and Minima of f on $[a, b]$.   If f is continuous on a *closed* interval $[a, b]$, then global maxima and minima of f on $[a, b]$ ◄   must exist. The following Fact is a key to finding them.

The extreme value theorem guarantees that they exist.

> **Fact** A continuous function f on a closed interval $[a, b]$ can assume its maximum and minimum values only at critical points in (a, b) or at endpoints of $[a, b]$.

(We will state and prove an equivalent fact in the first lemma on page 335.) The Fact says that the only possible candidates for maximum and minimum points of f are (1) the endpoints a and b, (2) roots of f', and (3) points at which f' doesn't exist. In practice, this slate of candidates is usually short. Choosing among them is usually easy, with or without computer graphics.

EXAMPLE 2   Find the maximum and minimum values of $f(x) = x/2 + \sin(x)$ on $[0, 3]$.

Solution   First we find critical points. Since f' exists everywhere, it's enough to find roots of f'. An easy computation does the trick:

$$f'(x) = \frac{1}{2} + \cos x = 0 \iff \cos x = -\frac{1}{2}.$$

Right? See for yourself.   The only such x in $[0, 3]$ is $x = 2\pi/3 \approx 2.094$. ◄

According to the preceding Fact, the maximum and minimum values of f on $[0, 3]$ can occur only at 0, 3, or $2\pi/3$. All that remains is to check f at these values:

$$f(0) = 0;$$

$$f\left(\frac{2\pi}{3}\right) = \frac{\pi}{3} + \frac{\sqrt{3}}{2} \approx 1.91;$$

$$f(3) = \frac{3}{2} + \sin 3 \approx 1.64.$$

The conclusion for f on $[0, 3]$ is now clear: f assumes its minimum value, 0, at $x = 0$ and its maximum value, 1.91, at $x = 2\pi/3$. ■

Finding Maxima and Minima on Any Interval. The preceding Fact applies only to *closed* intervals. On an arbitrary interval, f may or may not assume global maximum and minimum values. In practice, however, it's usually clear either from context or from graphics whether extreme values exist and how to find them. Critical points are still key.

EXAMPLE 3   Which point on the line $y = -2x + 2$ is closest to the origin?

Solution   Theory temporarily aside, it's intuitively clear that *some* point on the line minimizes the distance in question. To find it, let's define the function

$$f(x) = \text{distance from } (0, 0) \text{ to } (x, -2x + 2)$$

as follows; $f(x)$ is the dotted distance.

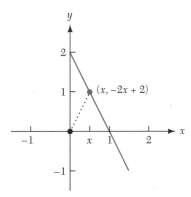

Which point on $y = -2x + 2$ is closest to $(0, 0)$?

An explicit formula for f is easy ▶ to find:

Use the distance formula in the plane.

$$f(x) = \sqrt{x^2 + (-2x + 2)^2} = \sqrt{5x^2 - 8x + 4}.$$

Intuition (or a graph) says that f has just one local (and therefore global) minimum; theory tells us to look for a stationary point. Finding one involves symbolic calculus ▶ and algebra:

Use the chain rule.

$$f'(x) = \frac{10x - 8}{2\sqrt{5x^2 - 8x + 4}} = 0 \iff 10x - 8 = 0 \iff x = \frac{4}{5}.$$

Convince yourself that the answer is reasonable.

Thus, the minimum distance occurs at the point $(4/5, 2/5)$; ◄ the distance itself is $\sqrt{(4/5)^2 + (2/5)^2} \approx 0.8944$. (A slightly different approach is to minimize the *square* of the distance; this eliminates square roots and simplifies the algebra.)   ■

Objective Functions, Constraint Equations, and Implicit Derivatives

Many optimization problems have the same general form. Some convenient jargon will help us describe it.

Objective Function: Describes the quantity to be maximized or minimized. In Example 3, for instance, we minimized the distance D from the origin to a point (x, y) on a certain line. This distance—the objective function—has the form $D(x, y) = \sqrt{x^2 + y^2}$. (As written here, the objective function depends on *two* variables, x and y. In Example 3 we rewrote the objective function using just one variable, x.)

Constraint Equation: Describes a condition that must be satisfied by the variables in an optimization problem. In Example 3, we required (i.e., "constrained") (x, y) to lie on the line $y = -2x + 2$; this is our constraint equation, or simply the **constraint**.

Constrained Optimization Problem: Any optimization problem that (like Example 3) involves maximizing or minimizing some objective function subject to one or more constraint equations.

To solve Example 3, we used the constraint equation to rewrite the objective function to depend on only one variable: $f(x) = \sqrt{5x^2 - 8x + 4}$. Then we used ordinary differentiation to minimize f. Implicit differentiation offers a slightly different (and sometimes simpler) approach to optimizing an objective function with a constraint. Two easy examples will illustrate. ◄

Watch the use of technical language, too.

> **EXAMPLE 4**   Two nonnegative numbers sum to 10. How large can their *product* be?

> **Solution**   Let x and y be the two numbers. Then our **objective function** is $P = xy$. Not just any x and y will do; the **constraint equation** says that $x + y = 10$. Legal values of x are in the interval $[0, 10]$. Note that if either $x = 0$ or $x = 10$, then $P = 0$. Thus, P must be largest at some critical value of x, where $dP/dx = 0$. We find it by differentiation.

We *could* use the constraint $x + y = 10$ to solve for y and then rewrite P as a function of x alone. Instead, we differentiate both the objective function and the constraint equation

$$P = xy; \qquad x + y = 10$$

Check the result for yourself.

implicitly with respect to x (i.e., treating y as an implicit function of x). Here are the results: ◄

$$\frac{dP}{dx} = x\frac{dy}{dx} + y; \qquad 1 + \frac{dy}{dx} = 0.$$

At a critical point x, $dP/dx = 0$. This makes, in all, three equations (including the constraint equation) that must hold at a critical point:

$$0 = x\frac{dy}{dx} + y; \qquad 1 + \frac{dy}{dx} = 0; \qquad x + y = 10.$$

Substituting the second equation into the first gives

$$\frac{dy}{dx} = -1 \implies 0 = -x + y \implies x = y.$$

Thus, the objective function is greatest when the two factors x and y are equal. Since $x + y = 10$, $x = y = 5$ and $P = 25$. ∎

The answer isn't too surprising. Neither is the answer to the following (closely related) problem.

EXAMPLE 5   Two nonnegative numbers multiply to 25. How small can their *sum* be? ▶

First make a guess. (The preceding example should help.)

Solution   The new objective function and constraint equation are, respectively,

$$S = x + y \qquad \text{and} \qquad xy = 25.$$

Differentiating both of these equations implicitly with respect to x and setting $dS/dx = 0$ gives a familiar result:

$$\frac{dS}{dx} = 1 + \frac{dy}{dx} = 0; \qquad y + x\frac{dy}{dx} = 0.$$

It follows—by the same calculation as in the preceding example—that at a critical point, $x = y$. Thus (just as before), $x = y = 5$ at the critical point, and so $S = 10$ is minimum. ∎

The Best Pop Cans: An Extended Example

Real-world calculus optimization problems usually involve translating English prose into—and out of—mathematical language. Once a problem is translated into mathematical terms, more than one solution avenue may be open. Mathematical solutions, once found, require interpretation in the original context.

The next example, chosen to illustrate the preceding points, has to do with the design of ordinary 12-ounce (355-ml) aluminum soft-drink containers. Do such cans have the right dimensions? What does "right" mean? Why do other products come packed in cans of other shapes?

EXAMPLE 6   An ordinary pop can has a volume of 355 cm³. Find the dimensions (radius r and height h) that minimize the surface area of such a can.

Solution   We'll offer two solutions, the first using ordinary derivatives and the second using *implicit* derivatives.

Solution 1:   The can's volume V and surface area A are given by

$$V = 355 = \pi r^2 h; \qquad A = 2\pi rh + 2\pi r^2.$$

From the first equation, $h = 355/(\pi r^2)$, so

$$A = 2\pi rh + 2\pi r^2 = 2\pi r \cdot \frac{355}{\pi r^2} + 2\pi r^2 = \frac{710}{r} + 2\pi r^2.$$

Therefore,

$$\frac{dA}{dr} = -\frac{710}{r^2} + 4\pi r = 0 \implies 710 = 4\pi r^3$$

$$\implies r = \left(\frac{710}{4\pi}\right)^{1/3} \approx 3.837.$$

Not a local maximum or just a flat spot.

It's easy to check (either graphically or by taking another derivative) that A has a local minimum ◄ here. Thus, the can's optimal dimensions are $r \approx 3.84$ cm and $h = 355/\left(\pi r^2\right) \approx 7.67$ cm $= 2r$. In particular, the optimal can has a *square* profile!

Solution 2:   We start with the same equations: $A = 2\pi rh + 2\pi r^2$ and $V = \pi r^2 h = 355$. Differentiating both equations implicitly with respect to r gives

$$\frac{dA}{dr} = 2\pi h + 2\pi r \frac{dh}{dr} + 4\pi r;$$

$$\frac{dV}{dr} = 2\pi rh + \pi r^2 \frac{dh}{dr} = 0 \implies \frac{dh}{dr} = -\frac{2h}{r}.$$

Now we substitute the last result into the expression for dA/dr and simplify:

$$\frac{dA}{dr} = 2\pi h + 2\pi r \left(-\frac{2h}{r}\right) + 4\pi r = 0 \implies h = 2r.$$

Thus, the optimal can has a square profile.   ■

EXAMPLE 7   Example 6 showed that, to minimize surface area, a pop can should have a square profile. Why don't real pop cans have this shape?

Solution   Minimizing surface area does not necessarily minimize the *cost* of a can. Surface area ignores, for instance, the extra cost of the can's top, which is thicker than the sides. ◄   ■

Look carefully at an aluminum pop can. Is the top obviously thicker?

EXAMPLE 8   It's reasonable to estimate that the top of a can is three times as thick as the sides and bottoms. Choose h and r to minimize the amount of aluminum in the can.

Solution   For simplicity, let's pretend that the top is made of three layers of aluminum and still minimize total area A. The argument runs as before, except that now

$$A = \text{sides} + \text{bottom} + \text{top} = 2\pi rh + \pi r^2 + 3\pi r^2 \implies A = 710/r + 4\pi r^2.$$

Now

$$\frac{dA}{dr} = \frac{-710}{r^2} + 8\pi r = 0 \implies r = \left(\frac{710}{8\pi}\right)^{1/3} \approx 3.046.$$

Therefore, the optimal dimensions are now $r \approx 3.046$ cm and $h = 12.182$ cm.   ■

EXAMPLE 9   What are the dimensions of a real pop can? Are they "right"?

Solution   Measuring a real can gives $h \approx 12.5$ cm and $r \approx 3.1$ cm. These values correspond well to the numbers computed before—they're a bit greater, since pop cans have some room to spare. Moreover, the tops of pop cans do seem to be about three times as thick as the sides. ∎

On Beyond Soda Pop

Derivatives are good for more than designing aluminum cans for carbonated beverages. They work with noncarbonated beverages, too.

EXAMPLE 10   Fruit juice is often sold in 6-ounce ($\approx$ 168-cm^3) *steel* cans. Assuming that the top of such a can is no thicker than the sides, find the dimensions that minimize surface area. Are the results realistic?

Solution   Mimicking the foregoing work shows, as in Example 6, that $h = 2r$ for the optimal can. From this it follows that $r \approx 2.99$ cm and $h \approx 5.98$ cm.

Actual juice cans have a more elongated profile than these numbers suggest, perhaps in order to fit the hand. ∎

BASICS

1. Which point on the line $y = -2x + 2$ is closest to $(0, 1)$? [**HINT**: See Example 3 (page 299).]

2. A triangle has legs on the positive x- and y-axes, and its hypotenuse passes through the point $(2, 1)$. Which such triangle has the smallest area?

3. This exercise offers another approach to Example 3 (page 299). This time, let $g(x) =$ square of the distance from $(0, 0)$ to $(x, -2x + 2) = x^2 + (2 - 2x)^2$. (Instead of minimizing the distance from the origin to the line, we'll minimize the square of the distance.)
   (a) Find the minimum value of $g(x)$ and the value of x for which it occurs.
   (b) Is the answer to the previous part consistent with what was found in Example 3 (page 299)? Why or why not?

4. This exercise has to do with finding a point P on the right half of the parabola $y = 1 - x^2$ that's closest to the origin.
   (a) Plot the parabola carefully in the rectangle $[0, 1] \times [0, 1]$. (Use equal size units on both axes.) Then use your plot to estimate the coordinates of P.
   (b) Use calculus to find the coordinates of P. [**HINT**: Mimic either the previous exercise or Example 3 (page 299).]
   (c) Show that the line from the origin to P is perpendicular to the line that's tangent to the parabola at P. How does this property appear on your graph?

5. Redo the previous exercise, but use the parabola $y = 1/4 - x^2$. What accounts for the difference between this exercise and the previous one?

FURTHER EXERCISES

6. Do Example 4 (page 300) without using implicit differentiation. Instead, use the equations $P = xy$ and $x + y = 10$ to rewrite P as a function of only one variable, x. Then maximize P over an appropriate interval.

7. Do Example 5 (page 301) without using implicit differentiation. Instead, use the equations $S = x + y$ and $xy = 25$ to rewrite S as a function of only one variable, x. Then minimize S.

8. Suppose that a projectile (e.g., a thrown baseball or a kicked football) leaves the origin with initial velocity v_0, is aimed along the line $y = mx$, and is influenced only by gravity. Then the projectile's path, or "trajectory,"

satisfies the equation

$$y = mx - \frac{g}{2v_0^2}(1 + m^2)x^2,$$

where g is the constant (downward) acceleration due to gravity. (Numerically, $g \approx 9.8$ meters/second$^2 \approx 32$ feet/second^2.)
   (a) A football is kicked with initial velocity $v_0 = 30$ meters per second and initial slope $m = 1$. (Use $g = 9.8$ meters/second^2.) How high does the football rise? Where does it land?
   (b) Plot the football's trajectory, labeling important points found in the previous part.

9. Use the trajectory formula of Exercise 8 as needed.
   (a) A baseball is thrown with initial velocity $v_0 = 128$ feet per second, with initial slope $m = 1/2$. (Use $g = 32$ feet per second^2.) How high does the baseball rise? Where does it land?
   (b) Another baseball is thrown with initial velocity $v_0 = 128$ feet per second, but with initial slope $m = 2$. How high does this baseball rise? Where does it land?
   (c) Plot the trajectories of both baseballs, labeling the important points found in the previous parts.

10. In the situation of Exercise 8, the projectile has range $R = \dfrac{2v_0^2}{g} \dfrac{m}{1 + m^2}$. (The range is the horizontal distance covered before landing.)
    (a) Use the formula of Exercise 8 to show why the range formula above holds.
    (b) Suppose that v_0 is fixed, but m is variable. What value of m maximizes the range?

11. To find the point on the line $x + y = 1$ that's closest to the origin, we can minimize the objective function $D = x^2 + y^2$ (the square of the distance from (x, y) to $(0, 0)$) subject to the constraint $x + y = 1$. Differentiating gives two equations that must hold at a critical point:

    $$\frac{dD}{dx} = 2x + 2y\frac{dy}{dx} = 0 \quad \text{and} \quad 1 + \frac{dy}{dx} = 0.$$

    Plugging the second equation into the first gives $x = y$. Thus the minimum distance occurs where both $x = y$ and $x + y = 1$, i.e., at the point $(1/2, 1/2)$.

    In each part that follows, find the point on the given line that's closest to the origin. In all parts, a, b, and c are nonzero constants.
    (a) $x + y = c$ (c) $ax + by = c$
    (b) $ax + y = c$

12. In each part, a farmer wants to fence a rectangular area as inexpensively as possible. Assume that fencing materials cost \$1 per foot.
    (a) Suppose that \$40 is available for the project. How much area can be enclosed? [**HINT:** Let x and y be the two dimensions of the fence, so that the area A is xy and the perimeter P is $2x + 2y$. Imitate the method of Example 4 (page 300) to maximize the objective $A = xy$ subject to the constraint $2x + 2y = 40$.]
    (b) Suppose that 100 square feet must be enclosed. What is the least possible cost? [**HINT:** Minimize $P = 2x + 2y$ subject to the constraint $xy = 100$.]
    (c) Discuss the relation between the two parts above. Did the two approaches give the same result?

13. In each part that follows, a farmer wants to fence a rectangular area as inexpensively as possible. Fences must run east–west and north–south. Assume that fencing ma-

terials cost \$h per foot in the east–west direction and \$v per foot in the north-south direction.
    (a) Suppose that a total budget \$b is available for the project. How much area can be enclosed? What are the best dimensions for the fence? [**HINT:** Let x and y be the east–west and north–south dimensions of the fence, respectively. Then the area is $A = xy$ and the total cost is $C = 2hx + 2vy$. Maximize $A = xy$ subject to $C = 2hx + 2vy = b$.]
    (b) Suppose that a total area of a square feet must be enclosed. What is the least possible cost? What are the best dimensions for the fence? [**HINT:** Minimize $C = 2hx + 2vy$ subject to $A = xy = a$.]
    (c) Are the two parts above consistent with each other? In each case, how much is spent for east–west fence? How much is spent for north–south fence?

14. A rectangle has its base on the x-axis, a vertex on the y-axis, and a vertex on the curve $y = e^{-x^2}$.
    (a) What choice of vertices gives the largest area?
    (b) Show that one of the vertices found in part (a) is at an inflection of the curve.

15. Show that the optimal fruit juice can has the dimensions claimed in Example 10.

16. Soup cans are to be manufactured to contain a given volume V. No waste is involved in cutting the material for the vertical sides, but the top and bottom (circles of radius r) are cut from squares that measure $2r$ units on a side. Thus the area A of material consumed for a can of height h is $A = 2\pi rh + 8r^2$. Find the ratio of height to diameter for the most economical can (i.e., the can requiring the least material). [**HINT:** Use the fact that $V = \pi r^2 h$ to express A as a function of r.]

17. A piece of wire 100 cm long is going to be cut into several pieces and used to construct the skeleton of a rectangular box with a square base.
    (a) What are the dimensions of the box with the largest volume?
    (b) What are the dimensions of the box with the largest surface area?

18. Federalist Express limits the size of a mailable parcel: The longest side plus the girth (the perimeter of a cross-section perpendicular to the longest side) may not exceed 108 inches.
    (a) Which mailable cube has largest volume?
    (b) Find the dimensions of the mailable rectangular parcel with a square base that has the largest volume. [**HINT:** Consider two cases, depending on which side is longest.]
    (c) Among mailable right circular cylinders, which has the largest volume? [**HINT:** Again, consider two cases.]
    (d) Give an example of a mailable parcel with larger volume than a nonmailable parcel.

19. It's 10:30 p.m. Your snowmobile is out of gas and you are 3 miles due south of a major highway. The nearest service station is on the highway 6 miles east; it closes at midnight. You can walk 4 miles per hour on roads, but only 3 miles per hour through snowy fields. Can you make it to the service station? What route is best?

20. The water depth in Moon River x miles downstream from Hard Rock is $D(x) = 20x + 10$ feet; the width of the river is $W(x) = 10(x^2 - 8x + 22)$ feet. To create Moon Lake, a dam is to be built downstream from Hard Rock. For engineering reasons, the dam cannot be more than 130 feet high.

    (a) For which values of x is $0 \le D(x) \le 130$?

    (b) How far downstream can the dam be built? If the dam were constructed at this point, how wide would it be? How high?

(c) What are the dimensions of the widest dam that could be constructed?

(d) What are the dimensions of the narrowest dam that could be constructed?

(e) If the cost of building the dam is proportional to the product of the length and the depth of the dam, where should the cheapest dam be located?

21. A truck driving over a flat interstate at a constant speed of 50 mph gets 4 miles to the gallon. Fuel costs $1.19 per gallon. The truck loses a tenth of a mile per gallon in fuel efficiency for each mile per hour increase in speed. Drivers are paid $27.50 per hour in wages and benefits. Fixed costs for running the truck are $11.33 per hour. A trip of 300 miles is planned. What speed minimizes operating expenses?

4.7 Calculus for Money: Derivatives in Economics

Mathematics has been applied to finance for a very long time. A 3000-year-old Babylonian problem (cited in *A History of Mathematics*, by Carl Boyer) goes as follows.

> *Ten brothers received 1;40 minas of silver, and brother over brother received a constant difference. If the eighth brother received 6 shekels, find out how much each earned.*

(A mina is 60 shekels. The Babylonians used a base-60 number system, so the notation 1;40 means 1 mina plus 40 shekels, or 100 shekels.)

Much has changed since Old Babylon. Calculus and economics were invented, and base 60 disappeared from common use. ▶ Financial applications of mathematics, however, remain very much alive, and calculus—especially the derivative—is among modern economists' first tools of choice. In this section we hint briefly at one application of calculus to economics.

Ten-brother families are rare nowadays, too.

Derivatives in Economics

Economics is full of *rates*—rates of interest, tax, inflation, unemployment, return on investment, and many more. Most rates can somehow be thought of as derivatives. Naturally, therefore, many economic calculations involve calculus with derivatives.

Economic Functions: Functions of What?

To apply calculus, we need functions with which to work. Economics offers plenty. Here are just a few examples, informally described: ▶

$HW(t)$ = average U.S. hourly manufacturing wage at time t;

$U(t)$ = number of unemployed U.S. workers at time t;

$ND(t)$ = cumulative U.S. national debt at time t;

$C(q)$ = seller's total cost of q bushels of soybeans;

$p(q)$ = price per bushel at which q bushels of soybeans can be sold.

See the exercises at the end of this section for more details.

The first three of these functions vary with *time*, so their derivatives measure rates with respect to time. To say that $U'(1993.5) = -50,000$, for instance, could mean that halfway through 1993 the unemployment population was dropping at an instantaneous rate of 50,000 people per year. ◄

Notice the units of U′: people/year.

The functions C and p are a little different: For them the independent variable represents *quantity* (here, the quantity of soybeans), not time. Thus, their derivatives measure rates of change with respect to quantity, not time. To say, for instance, that $C'(2000) = 5$ might mean that at 2000 bushels the seller's total cost is increasing at the rate of $5 per bushel. ◄   To put it another way, the 2001st bushel costs the seller $5. ◄

Notice the units of C′: dollars/bushel.

Read further for more on comparing these two interpretations of the derivative.

Discrete versus Continuous Phenomena

Technically speaking, the cost and price functions C and p accept only *whole-number* inputs q: Grain traders sell only whole bushels. A variable that takes only isolated values (e.g., integers) is called a **discrete variable**; a **continuous variable** can take *any* real value. Although many financial functions involve discrete variables, it's common to approximate them—as we will later in this section—with functions of continuous variables. Doing so makes the powerful tools of calculus available.

> **Modeling Reality Mathematically.**   To model any phenomenon (e.g., soybean sales on the Chicago Board, snowshoe hare and fox populations in the Arctic, heating requirements in Minnesota, demand for hard Montana wheat, or the spread of AIDS) mathematically means to *describe* or *represent* the phenomenon quantitatively, in mathematical language.
>
> Every mathematical model simplifies, to some extent, the phenomenon it describes. Economic models, for example, often involve "pretending" that a discrete variable is continuous. A useful model is simple enough to permit calculations but detailed enough to capture essential qualities of the phenomenon studied.
>
> Mathematical modeling involves three steps: (i) *describing* the phenomenon in mathematical terms, using functions, variables, equations, and other mathematical ingredients; (ii) *deducing* mathematical consequences of the model; and (iii) *interpreting* mathematical results in terms of the phenomenon studied.

Mathematical economists use all sorts of functions. For simplicity, in this section we treat only functions that (like C and p) vary with quantity. Our first example uses several such functions in context.

EXAMPLE 1   Lotta Mays, an experienced grain trader, wonders how many bushels of soybeans to buy—and then sell—on the Chicago Board of Trade. Lotta knows that the per-bushel price p (in dollars) she can get for her beans depends on the quantity q (in bushels) she wants to sell: Too many beans on the market drives prices down. Here's her best guess for a **price function**:

$$p(q) = 8 - \frac{q}{1000}.$$

If Lotta sells q bushels at $p(q)$ dollars each, she'll raise a total of $q \cdot p(q)$ dollars. Thus, her **total revenue function** has the formula ◄

The variable q is in bushels, and R is in dollars.

$$R(q) = q \cdot p(q) = q\left(8 - \frac{q}{1000}\right) = 8q - \frac{q^2}{1000}.$$

Before selling any beans, Lotta must buy some. As a large wholesale customer, she can buy any amount of beans for $5 a bushel—$q$ bushels cost $5q$ dollars. In addition, Lotta has fixed expenses of $1000 for buying and selling her beans (handling fees, taxes, commissions, etc.). Lotta's **total cost function** therefore has the formula

$$C(q) = 5q + 1000.$$

How many bushels should Lotta buy and sell? At what price will she sell them? How much will she make?

Solution   Lotta wants to maximize her **profit**—the difference between revenue and cost—on the transaction. Like R and C, the total profit P depends on q:

$$P(q) = R(q) - C(q) = 8q - \frac{q^2}{1000} - (5q + 1000) = 3q - \frac{q^2}{1000} - 1000.$$

Lotta first plots P, R, and C together:

Cost, revenue, and profit functions

The bulleted point on the P-graph is clearly of interest: It suggests that the maximum possible profit—a little more than $1000—occurs for $q = 1500$. The other two bulleted points are also important. They mark the spot at which the R-graph is *farthest above* the C-graph, i.e., where revenue most exceeds cost.

Symbolic calculations agree with the graphs. Differentiating our formulas for C, R, and P gives

$$C'(q) = 5; \qquad R'(q) = 8 - \frac{2q}{1000}; \qquad P'(q) = 3 - \frac{2q}{1000}.$$

Setting $P'(q) = 0$ ▶   gives $q = 1500$—as the graph shows. At this optimal quantity,   *Check our work.*

$$C(1500) = 8500; \qquad R(1500) = 9750;$$

$$P(1500) = 1250; \qquad p(1500) = 6.5.$$

All is now revealed: Lotta earns the maximum profit, $1250, by selling 1500 bushels at $6.50 per bushel.   ■

On the Margin: Derivatives of Revenue, Cost, and Profit

The term "marginal" will be explained soon.

The derivatives R', C', and P' from Example 1, though simple, deserve a look: ◄

Marginal cost, marginal revenue, and marginal profit functions

Again, $q = 1500$ is the quantity of interest. As the bullets show,

$$C'(1500) = R'(1500) = 5; \qquad P'(1500) = R'(1500) - C'(1500) = 0.$$

In other words, at the critical value $q = 1500$, C and R increase at the same rate, and P is stationary.

Mathematicians usually interpret derivatives as rates of change. The statement $P'(1000) = 1$ means, in rate language, that after selling 1000 bushels, the trader is profiting at the rate of \$1 per bushel.

Economists often interpret derivatives such as R', C', and P' as **marginal** quantities. In this language, the statement $P'(1000) = 2$ means that, after selling 1000 bushels, the trader makes \$2 profit on the *next bushel*. (The 1001st bushel is "on the margin.") For any q, **marginal profit at** q is defined as the additional profit to be made on the *next* (i.e., the $(q + 1)$st item. The fact that (as we found earlier) $P'(1500) = 0$ means, in the marginal sense, that zero profit will be made

This would be a good time to quit the business.

on the 1501st item. ◄

Marginal cost and **marginal revenue** have similar meanings. In Example 1 we saw that $C'(1500) = 5 = R'(1500)$. In marginal language, this means that the 1501st bushel is a wash: It costs Lotta \$5 and sells for the same amount.

Here's the general marginal idea for *any* function f that depends on quantity q:

Marginal f is the increase in f due to the $(q + 1)$st item.

Marginals and Instantaneous Rates: Economists versus Mathematicians

How different are the marginal and rate interpretations of the derivative? Are mathematicians and economists really at odds?

Not seriously. In practice, rates and marginals seldom differ by much. A little symbolic algebra shows why. Suppose that $P(q)$ describes the profit to be made on q items. Then the marginal profit function $MP(q)$ is defined by

$$MP(q) = \text{profit on } (q + 1)\text{st item} = P(q + 1) - P(q).$$

The derivative function $P'(q)$, by contrast, is defined by the limit

$$P'(q) = \lim_{h \to 0} \frac{P(q + h) - P(q)}{h}.$$

Setting $h = 1$ in the last expression gives

$$P'(q) \approx \frac{P(q+1) - P(q)}{1} = MP(q).$$

Thus, $P'(q)$ and $MP(q)$ are at least approximately equal; mathematicians and economists are on the same wavelength.

In practice, the difference between MP and P' is seldom worth fretting over. One reason is that P, P', and MP are all mere models of reality and so are inherently inexact (although *not* inherently useless). Another reason is that values of q are usually large, so setting $h = 1$ in the difference quotient gives a good approximation. If, say, P is the profit function in the example, and $q = 2000$, then (as easy calculations show)

$$P'(2000) = -1; \qquad MP(2000) = P(2001) - P(2000) = -1.001.$$

The tiny difference isn't worth a fight.

Varying Parameters

Economic strategies often depend sensitively on various parameters—the wholesale price of soybeans, the prime interest rate, and the rate of inflation—that change unpredictably from day to day. As parameters change, so do the results of economic calculations. The *form* of the calculations, however, may remain essentially the same.

EXAMPLE 2   Lotta Mays (see Example 1) finds a new wholesale supplier of soybeans at $4.50 a bushel. What should she do?

Solution   Lotta's price and revenue functions p and R are unchanged, but her cost and profit functions C and P are new. Here are full details:

$$p(q) = 8 - \frac{q}{1000}; \qquad R(q) = q \cdot p(q) = 8q - \frac{q^2}{1000};$$
$$C(q) = 4.5q + 1000; \qquad P(q) = R(q) - C(q) = 3.5q - \frac{q^2}{1000} - 1000.$$

We maximize P just as before:

$$P'(q) = 3.5 - \frac{2q}{1000} = 0 \Longrightarrow q = 1750.$$

Since $P(1750) = 2062.5$ and $p(1750) = 6.25$, we have this result: Lotta should now sell 1750 bushels at $6.25 a bushel; she'll make $2062.50. (Notice the bottom line: Lotta passed *half* her 50-cent-per-bushel savings on to her customers.) ∎

Treating the wholesale price as a parameter—without a specific value—can solve a whole lotta problems at once.

EXAMPLE 3   The wholesale price of beans seems to change every day; Lotta's sick of working the same problem again and again. How many beans should she sell if the price to her is b dollars a bushel? At what price? How much will she make or lose?

> **Solution**   For any value of b, the functions look like this:
>
> $$p(q) = 8 - \frac{q}{1000}; \qquad R(q) = q \cdot p(q) = 8q - \frac{q^2}{1000};$$
>
> $$C(q) = b \cdot q + 1000; \qquad P(q) = R(q) - C(q) = (8 - b)q - \frac{q^2}{1000} - 1000.$$
>
> We'll maximize P in the usual way:
>
> $$P'(q) = 8 - b - \frac{2q}{1000} = 0 \implies q = 4000 - 500b.$$
>
> For this quantity the price formula gives
>
> $$p(4000 - 500b) = 8 - \frac{4000 - 500b}{1000} = 4 + \frac{b}{2}.$$
>
> A little algebra shows, too, that
>
> $$P(4000 - 500b) = 15{,}000 - 4000b + 250b^2.$$
>
> Lotta has her general result: If beans wholesale for b dollars a bushel, she should sell at $4 + b/2$ dollars per bushel. Doing so, she'll sell $4000 - 500b$ bushels, for a total profit of $15{,}000 - 4000b + 250b^2$ dollars.   ∎

BASICS

1. This exercise is about several functions described informally in this section. In each part, first specify reasonable units for the input and output variables. Then give a plausible domain and range for the function, and draw a graph that could represent the function in question. (No special knowledge of economics is required—if in doubt, guess!)

   (a) $HW(t)$ = average U.S. hourly manufacturing wage at time t

   (b) $U(t)$ = number of unemployed U.S. workers at time t

   (c) $ND(t)$ = cumulative U.S. national debt at time t

2. What does "marginal tax rate" mean? Is this use of "marginal" consistent with those in marginal cost, marginal revenue, and marginal profit?

FURTHER EXERCISES

3. In Example 1 Lotta Mays guessed that the formula

   $$p(q) = 8 - \frac{q}{1000}$$

   might reasonably describe the selling price per bushel of q bushels of soybeans.

   (a) Plot p; choose *xrange* and *yrange* to make sense in context. What are the units on each axis?

   (b) State a sensible domain for the function p.

   (c) Using your answer from the previous part, state a plausible range for the function p.

   (d) The function p above is decreasing. Why would Lotta choose a decreasing function? Explain briefly in words.

   (e) The function p above is linear. Lotta chose a linear function purely for simplicity, but is that a good choice? Would you expect a real price function to be linear? If so, why? If not, sketch a more convincing price function and explain why it is a more realistic choice.

4. This exercise refers to Example 2.

   (a) On one set of axes, plot the functions C, R, and P in Example 2. (A similar picture appears in Example 1.) What do the results show graphically about Lotta's maximum possible profit?

   (b) Calculate $C'(q)$, $R'(q)$, and $P'(q)$. Use the results to find the marginal cost, marginal revenue, and marginal profit at $q = 1500$. What does it mean, in context, that the marginal profit at $q = 1500$ is positive?

   (c) Find a value of q for which the marginal profit at q is negative. What are the marginal revenue and marginal cost at this q?

5. Consider again the situation in Example 1; use the functions C, p, R, and P given there.

   (a) Suppose that Lotta wants only to break even. What's the smallest quantity of soybeans she can sell? What's the greatest quantity? Explain your answers. [**HINT**: Look first at the P-graph on page 307.]

   (b) Suppose that Lotta wants to maximize her total revenue—regardless of profit. How many bushels should she sell? At what price? What is her bottom line?

6. Redo the previous exercise in the situation of Example 2.

7. In Example 3 we concluded: If beans sell wholesale for b dollars per bushel, Lotta should sell $4000 - 500b$ bushels, at $4 + b/2$ dollars per bushel, for an optimal profit of $15,000 - 4000b + 250b^2$ dollars.

   (a) What does our conclusion mean if $b = 5$? If $b = 4.5$? Do the results agree with earlier work?

   (b) What's Lotta's best strategy if $b = 7$?

   (c) Not all values for b make realistic sense. In this context, the sensible values for b lie between $b = 0$ and $b = 8$. (Prices can't be negative, and our price function says that nobody buys above $8.) Among these possible values of b, what's the largest one for which Lotta can at least break even?

8. A lightning rod manufacturer has monthly overhead costs of $6000. Materials cost $1.00 per unit. If no more than 4500 lightning rods are made in a month, labor cost is $0.40 per unit; for each unit over 4500, the manufacturer must pay time-and-a-half for labor. The manufacturer can sell 4000 units per month at $7.00 per unit. Monthly sales are expected to rise by 100 for each $0.10 reduction in price. How many lightning rods should be manufactured each month to maximize profit?

9. A successful retail store must control its inventory. Too much inventory results in excessive interest costs, extra warehouse fees, and the danger of obsolescence. Too small an inventory requires more paperwork for reordering, extra delivery charges, and increases the likelihood of shortages.

   If microwave ovens are ordered in batches of size n, on average $n/2$ ovens will be in stock at any time. An appliance store estimates that it costs $25 to hold an oven in stock for a year.

   The cost of ordering n ovens is $250 plus $2.50 for each oven.

   Assume that the store sells 1000 microwave ovens each year and that it always orders the ovens in "lots" of size n. What lot size will result in the smallest total yearly cost? Is your answer reasonable? Why or why not?

4.8 Related Rates

Related Amounts, Related Rates. Calculus was invented to describe and predict phenomena of change: planetary motion, objects in free fall, varying populations, etc. In many practical applications, several related quantities vary *together*; naturally, the rates at which they vary are also related to each other. With calculus we can describe and calculate such **related rates.**

EXAMPLE 1 A spacecraft travels directly toward Mars. Describe several amounts that vary with time. Interpret their derivatives as rates of change.

Solution Many quantities vary with time, including the spacecraft's distance from Earth, the spacecraft's distance from Mars, the distance and direction from Earth to Mars, the amount of fuel consumed, and so on. These quantities are related to each other in various ways; at any time t, for instance, the fourth quantity depends on the first two.

Each such quantity is a function of time; its derivative, therefore, is the rate of change of the given quantity with respect to time. For the preceding quantities, the derivatives represent, respectively, the rocket's velocity away from Earth, its velocity toward Mars, Earth's velocity relative to Mars, and the rate of fuel consumption. ▶ These rates, like their corresponding amounts, are related to each other. ■

Each rate's units depend on corresponding units of amount.

The following (more mundane) example illustrates the technique (as well as the idea) of related rates.

EXAMPLE 2   A point P moves from left to right along the curve $y = x^2$, at a constant *horizontal* speed of 3 units/second. ◄   How fast does the y-coordinate of P increase at the moment when P passes through $(1, 1)$? When P passes $(2, 4)$?

The shadow cast by P on the x-axis moves to the right at 3 units per second. Can you picture the situation?

Solution   As P moves, many different quantities vary with time. The two of interest here are x and y, the coordinates of P. Written as functions, they are

$$x(t) = x\text{-coordinate of } P \text{ at time } t;$$

$$y(t) = y\text{-coordinate of } P \text{ at time } t.$$

Because P moves along the parabola $y = x^2$, the functions x and y satisfy the equation

$$y(t) = \big(x(t)\big)^2. \tag{4.6}$$

Differentiating both sides of Equation 4.6 with respect to t gives

$$\frac{dy}{dt} = 2x(t)\frac{dx}{dt}, \tag{4.7}$$

or, more compactly, $y' = 2xx'$. With Equation 4.7 we can answer our questions. By assumption, $dx/dt = 3$, for all t. At the point $(1, 1)$, $x = 1$, so

$$\frac{dy}{dt} = 2x(t)\frac{dx}{dt} = 2 \cdot 1 \cdot 3 = 6 \text{ units per second.}$$

When $(x, y) = (2, 4)$,

$$\frac{dy}{dt} = 2x(t)\frac{dx}{dt} = 2 \cdot 2 \cdot 3 = 12 \text{ units per second.}$$

Think again about the situation. Is it obvious that y should grow faster and faster?

Notice, finally, that although x increases at the constant rate of 3 units per second, y increases faster and faster. ◄   ■

What We Just Did: Related Rates in a Nutshell.   The generic related rates problem asks for one rate of change, given information about one or more others. *Solving* related rates problems can be hard, but the underlying idea is simple:

> *Start with an equation relating two or more quantities that vary with time. Differentiate both sides with respect to time, using the chain rule. The resulting equation relates the rates at which the quantities vary. Use it to solve for the desired rate.*

As usual with story problems, the first step—identifying the quantities in question and setting up the original equation—is often the hardest.

Textbook authors love related-rates problems; the supply is almost inexhaustible. Many old favorites involve right triangles and the Pythagorean rule (sometimes thinly disguised).

EXAMPLE 3   (**Pythagoras, Newton, and the State Patrol**)   It's spring break at Huxley College; the annual migration to Fort Lauderdale has begun.

Hovering 1000 feet directly above the southbound interstate is the helicopter state patrol, with Officer Ingkvist on the radar gun.

A car with Huxley decals passes underneath and continues south. Jason and Medea, fresh from their calculus midterm, aren't watching the speedometer. Officer Ingkvist fires. In moments, the results are in: At the moment of firing the car was 2000 feet from the helicopter and moving away at 85 feet per second. "Got 'em!" mutters Ingkvist under her breath.

What happens next?

Solution   In traffic court, Jason and Medea deny all wrongdoing. "The speed limit is 65 mph," they tell the judge. "We were clocked at 85 feet per second. But a simple calculation shows that 65 mph is 95 feet per second. Hence, clearly, ▶ we couldn't have been speeding." "Watch your language," replies the judge, "but what you say makes sense. Explain yourself, Ingkvist. Use that blackboard. Be clear. And cite your sources."

Under duress, even students say "hence" and "clearly."

Officer Ingkvist rises calmly; she's been through this before. Do judges *never* learn? She draws this diagram:

Officer Ingkvist's diagram

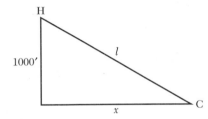

"In the diagram, your honor, H is the helicopter, C is the car, l is the straight-line distance from H to C, x is the horizontal distance from H to C, and 1000 feet is the constant height of the helicopter. The amounts x and l vary with time, of course. Pythagoras says, however, that at *any* time t, x and l satisfy the equation $l^2 = x^2 + 1000^2$. Newton says that differentiating this relation between l and x with respect to t relates the rates of change of l and x, as follows:

$$l^2 = x^2 + 1000^2 \implies 2l \cdot \frac{dl}{dt} = 2x \cdot \frac{dx}{dt} \implies \frac{dx}{dt} = \frac{l}{x} \cdot \frac{dl}{dt}.$$

At the time in question, dl/dt was 85 and l was 2000, so ▶ x was $\sqrt{2000^2 - 1000^2} = 1000\sqrt{3}$. At that moment, therefore, Jason and Medea's southward speed was

Here officer Ingkvist used the Pythagorean equation.

$$\frac{dx}{dt} = \frac{l}{x} \cdot \frac{dl}{dt} = \frac{2000}{1000\sqrt{3}} \cdot 85 \approx 98.1495457622 \text{ feet per second.}$$

Any questions?"

"Just one," says the judge. "Who are Pythagoras and Newton, and why won't they testify in person?"

"Never mind," say Jason and Medea. "We'll pay the ticket."

BASICS

1. The width of a rectangle is increasing at a rate of 2 cm/sec, and its length is increasing at a rate of 3 cm/sec. At what rate is the area of the rectangle increasing when its width is 4 cm and its length is 5 cm?

2. Digging in your backyard, you accidentally break a town water line. Water, bubbling up at a rate of 1 in^3/sec, is forming a circular pond of depth $\frac{1}{2}$ inch in your backyard. How quickly is this pond covering your lawn? (That is, how quickly is the surface area of this pond increasing?)

3. Two cars leave an intersection simultaneously. One travels north at 30 mph, the other travels east at 40 mph. How fast is the distance between them changing after 5 minutes?

FURTHER EXERCISES

4. Boyle's Law states that if the temperature of a gas remains constant, then the pressure P and the volume V of the gas satisfy the equation $PV = c$, where c is a constant. If the volume is decreasing at the rate of 10 in^3 per second, how fast is the pressure increasing when the pressure is 100 lbs/in^2 and the volume is 20 in^3?

5. Two aircraft are in the vicinity of the Erehwon air traffic control center in Neverland. Pachyderm flight 1003 from Celesteville to Paris is approaching from the south, and Peterpan flight 366 from London to Wendytown is approaching from the east. Both are at an altitude of 33,000 feet on paths that will take them directly over the air traffic control center. The Pachyderm flight is 36 nautical miles from the center and approaching it at a rate of 410 knots. The Peterpan flight is 41 nautical miles from the center and approaching it at a rate of 455 knots. [NOTE: One **nautical mile** is equal to 1852 meters. One **knot** is one nautical mile per hour.]
   (a) How close will the planes come to each other? Will they violate the FAA's minimum separation requirement of 5 nautical miles?
   (b) How many minutes before the time of closest approach?

6. Sand is being dumped on a pile in such a way that it always forms a cone whose radius equals its height. If the sand is being dumped at a rate of 10 cubic feet per minute, at what rate is the height of the pile increasing when there is 1000 cubic feet of sand in the pile?

7. A tanker accident has spilled oil in Pristine Bay. Oil-eating bacteria are gobbling 5 ft^3/hour. The oil slick has the form of a circular cylinder. When the radius of the cylinder is 500 ft, the thickness of the slick is 0.01 ft and decreasing at a rate of 0.001 ft/hr. At what rate is the area of the slick changing at this time? Is the area of the slick increasing or decreasing?

8. Oil is leaking from an ocean tanker at the rate of 5000 liters per minute. The leakage results in a circular oil slick. The depth of the oil varies linearly from a maximum of 5 cm at the point of leakage to a minimum of 0.5 cm at the outside edge of the slick. How fast is the radius of the slick increasing 4 hours after the tanker started leaking? [NOTE: 1 liter = 1000 cm^3]

9. A spherical iron ball 8 in. in diameter is coated with a layer of ice of uniform thickness.
   (a) If the ice melts at the rate of 10 in^3/min, how fast is the thickness of the ice decreasing when it is 2 in thick?
   (b) How fast is the outer surface area of the ice decreasing at this time?

10. Bazooka Joe is blowing a spherical bubble gum bubble. Let V be the volume of gum in the bubble, R the inside radius of the bubble, and T the thickness of the bubble. V, T, and R are functions of time t.
   (a) Write a formula for V in terms of T and R. [HINT: Draw a picture.]
   (b) Assume that the amount of bubble gum in the bubble is not changing. What is $V'(t)$?
   (c) After 5 minutes of blowing, the bubble is 3 feet in diameter and 0.01 feet thick. If the inside radius of the bubble is expanding at a rate of 0.5 feet per minute, how fast is the thickness changing? [HINT: Remember that the volume of gum in the bubble does not change over time.]

11. At a certain moment, one bicyclist is 4 miles east of an intersection, traveling toward the intersection at the rate of 9 miles/hour. At the same time a second bicyclist is 3 miles south of the intersection, traveling away from the intersection at the rate of 10 miles/hour. Is the distance between the bicyclists increasing or decreasing at that moment? At what rate?

12. A radio transmitter is located 3 miles from a straight section of interstate highway. A truck is traveling away from the transmitter along the highway at a speed of 65 miles per hour. How fast is the distance between the truck and the transmitter increasing when they are 5 miles apart?

13. A baseball diamond is a square 90 ft on a side. A runner travels from home plate to first base at 20 ft/sec. How fast is the runner's distance from second base changing when the runner is halfway to first base?

14. A baseball diamond is a square 90 ft on a side. A batter hits a ball along the third base line at 100 ft/sec and runs to first base at 25 ft/sec.
    (a) At what rate is the distance between the ball and first base changing when the ball is halfway to third base?
    (b) At what rate is the distance between the ball and the player changing when the ball is halfway to third base?

15. A highway patrol plane flies one mile above a straight section of rural interstate highway at a steady ground speed of 120 miles per hour. The pilot sees an oncoming car and, with radar, determines that the line-of-sight distance from the plane to the car is 1.5 miles and that this distance is decreasing at a rate of 136 miles per hour. Should the driver of the car be given a ticket for speeding? Tell it to the judge. (Assume the judge knows calculus.)

16. A coffee filter has the shape of an inverted cone. Water drains out of the filter at a rate of 10 cm³/min. When the depth of water in the cone is 8 cm, the depth is decreasing at 2 cm/min. What is the ratio of the height of the cone to its radius?

17. An inverted cone has height 10 cm and radius 2 cm. It is partially filled with liquid, which is oozing through the sides at a rate proportional to the area of the cone in contact with the liquid. Liquid is also being poured in the top of the cone at a rate of 1 cm³/min. When the depth of the liquid in the cone is 4 cm, the depth is decreasing at a rate of 0.1 cm/min. At what rate must liquid be poured into the top of the cone to keep the liquid at a constant depth of 4 cm?

18. The minute and hour hands on the face of a town clock are 7 feet and 5 feet long, respectively. How fast is the distance between the tips of the hands increasing when the clock reads 9:00?

19. A swimming pool is 15 feet wide, 40 feet long, 3 feet deep at one end, and 10 feet deep at the other end (see the sketch). Water is being added to it at the rate of 25 cubic feet per minute.
    (a) How fast is the water level rising when there is 2000 cubic feet of water in the pool?

    (b) How fast is the water level rising when there is 3000 cubic feet of water in the pool?

20. A rocket rises vertically from a point on the ground that is 100 m from an observer at ground level. The observer notes that the angle of elevation is increasing at a rate of 12° per second when the angle of elevation is 60°. Find the speed of the rocket at that instant.

21. A Ferris wheel, 50 feet in diameter, revolves at the rate of 10 radians per minute. How fast is a passenger rising when the passenger is 15 feet higher than the center of the Ferris wheel and is rising?

22. The ends of a horizontal trough 10 ft long are isosceles trapezoids with lower base 3 ft, upper base 5 ft, and altitude 2 ft. If the water level is rising at a rate of 0.25 inches per min. when the depth is 1 ft, how fast is water entering the trough?

23. A railroad bridge is 20 ft above, and at right angles to, a river. A person in a train traveling at 60 mph passes over the center of the bridge at the same instant that a person in a motor boat traveling 20 mph passes under the center of the bridge. How fast are the two people separating 10 seconds later?

24. A ladder, 12 feet long, is leaning against a wall. If the foot of the ladder slides away from the wall along level ground, what is the rate of change of the top of the ladder with respect to the distance of the foot of the ladder from the wall when the foot is 6 ft from the wall?

25. An airplane flying east at 400 mph goes over a certain town at 11:30 a.m., and a second plane, flying northeast at 500 mph goes over the same town at noon. How fast are they separating at 1:00 p.m.?

4.9 Parametric Equations, Parametric Curves

A point P wanders about the xy-plane, tracing its path as it goes. As P travels, its coordinates $x = f(t)$ and $y = g(t)$ are functions of time. In this situation t is called a **parameter**, ▶ f and g are **coordinate functions**, and the figure traced out by P is a **parametric curve**. ▶
    Our first example illustrates the idea and some specialized vocabulary.

Here the parameter is a variable. In other situations, parameters are often constants.

Most plotting programs and graphing calculators handle parametric curves. Try yours.

EXAMPLE 1   At any time t with $0 \leq t \leq 10$, the coordinates of P are given by the **parametric equations**

$$x = t - 2\sin t; \qquad y = 2 - 2\cos t.$$

What curve does P trace out? Where is P at $t = 1$? In which direction is P moving?

Solution   The simplest (but most tedious) way to draw a curve is to calculate many points (x, y), plot each one, and "connect the dots." The first step works as usual: For many inputs t, calculate the corresponding x and y. Here's a sampler of results, ◄   rounded to two decimals:

Check several entries.

Parametric Plot Points for $x = t - 2\sin t$, $y = 2 - 2\cos t$											
t	0	0.1	0.2	0.3	0.7	0.8	0.9	1.0	...	9.8	9.9	10
x	0	−0.10	−0.20	−0.29	−0.59	−0.63	−0.67	−0.68	...	10.53	10.82	11.09
y	0	0.01	0.04	0.09	0.47	0.61	0.76	0.92	...	3.86	3.78	3.68

Here's the result, a curve C:

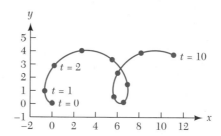

The parametric curve
$x = t - 2\sin t, y = 2 - 2\cos t, 0 \leq t \leq 10$

Notice:

- Points corresponding to *integer* values of t are bulleted. At $t = 1$, P has coordinates $(-0.68, 0.92)$; at this instant, P is heading almost due north.

- The full curve C is not the graph of a function $y = f(x)$; some x-values correspond to more than one y-value. ◄   Certain *pieces* of C, however, do define such functions. (The part of C from $t = 2$ to $t = 5$ does so.)

- The picture shows the x- and y-axes *but no t-axis*: t-values are indicated only by the bulleted points. In most parametric curves t-values don't appear at all. ◄

- The bullets on the graph appear at equal *time* ◄   intervals but not at equal *distances* from each other, because P speeds up and slows down as it moves. ◄ We'll soon see how to calculate the speed of a parametric curve at a point.

- If t measures time, we can visualize C dynamically, as a curve traced by a moving point. Curves defined by ordinary equations in x and y are *static* objects, lying passively on the page. ■

$x = 6$, for instance.

Most graphing calculators don't show t-values graphically, but some calculate t-values numerically.

In this example, t represents time.

When is P moving fastest? Slowest?

A Sampler of Parametric Curves

Parametric curves come in mind-boggling variety. Any choice of two equations $x = f(t)$ and $y = g(t)$ and a t-interval produces a parametric curve. Surprisingly often, the result is beautiful, useful, interesting, or all three. The next several examples hint at some of the possibilities and at connections between parametric curves and ordinary function graphs.

EXAMPLE 2   (**Every Ordinary Function Graph Can Be Written in Parametric Form**)   To produce a sine curve, for example, we can use

$$x = t; \qquad y = \sin t; \qquad -2\pi \le t \le 2\pi.$$

Here's the graph. Integer multiples of $\pi/2$ are bulleted.

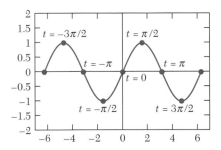

The parametric curve
$x = t,\ y = \sin t,\ -2\pi \le t \le 2\pi$

EXAMPLE 3   (**Parametric Curves May Have Loops, Cusps, Vertical Tangents, and Other Peculiar Features**)   Consider the curve

$$x = 2\cos(t) + 2\cos(4t); \qquad y = \sin t + \sin(4t); \qquad 0 \le t \le 5.$$

Here it is. Bullets show the starting and ending points.

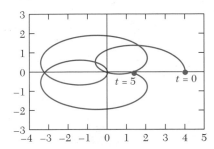

The parametric curve $x = 2\cos(t) + 2\cos(4t)$,
$y = \sin(t) + \sin(4t),\ 0 \le t \le 5$

The curve in Example 3 has this property.

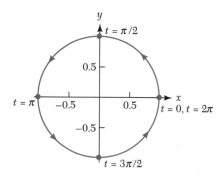

EXAMPLE 4   **(The Unit Circle)**   If a parametric curve's coordinate functions are periodic (i.e., repeat themselves), then the curve's shape reflects this fact. ◄ The simplest and most important such curve is the **unit circle**, often written in xy form as $x^2 + y^2 = 1$. The simplest *parametric* description of the unit circle is as follows:

$$x = \cos t; \qquad y = \sin t; \qquad 0 \le t \le 2\pi.$$

Here's the picture. Integer multiples of $\pi/2$ are bulleted.

The parametric curve
$$x = \cos t, y = \sin t, 0 \le t \le 2\pi$$

Notice the following facts:

- As usual, the parameter t *can* be thought of as time. Alternatively, t can be thought of as the radian measure of the angle determined by the x-axis and the line from the origin to P. Then, for any *angle t*, $(x, y) = (\cos t, \sin t)$ is the point on the unit circle lying t radians counterclockwise from $(1, 0)$.

- Because $x = \cos t$ and $y = \sin t$,

$$x^2 + y^2 = (\cos t)^2 + (\sin t)^2 = 1.$$

This shows that every point on the parametric curve satisfies the Cartesian equation $x^2 + y^2 = 1$ and so—as we intended—lies on the unit circle.

"Reducing" two parametric equations to xy-form, as we just did, is called **eliminating the parameter**. (Doing so is sometimes possible, sometimes not.)

- The circle is traced *once* as t runs from $t = 0$ to $t = 2\pi$. With a larger t-interval, the same circle would be traced repeatedly; with a smaller t-interval, the circle would be traced only partially. ∎

EXAMPLE 5   **(Other Circles)**   The idea in Example 4 extends to *all* circles. If (a, b) is any point in the plane and $r > 0$ is any radius, then the parametric equations

$$x = a + r \cos t; \qquad y = b + r \sin t; \qquad \text{and} \qquad 0 \le t \le 2\pi$$

produce the circle with center (a, b) and radius r. (See the exercises at the end of this section for more details.) ∎

EXAMPLE 6 **(Other Curves with Periodic Coordinate Functions)**
Many curves defined by periodic (i.e., repeating) coordinate functions have striking, beautiful shapes. A typical **Lissajou curve**, for instance, is defined by

$$x = \sin(5t); \qquad y = \sin(6t); \qquad 0 \le t \le 2\pi.$$

(Replacing 5 and 6 with other integers would produce other Lissajou figures.) Our curve looks like this:

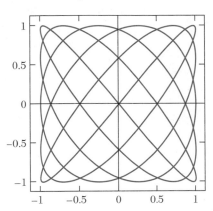

The Lissajou curve
$x = \cos(5t), y = \sin(6t), 0 \le t \le 2\pi$

EXAMPLE 7 **(Same Curve, Different Parametric Equations)** Different pairs of parametric equations may produce exactly the same curve in the xy-plane. In such cases, labeling t-values can make differences appear. For example, setting

$$x = t; \qquad y = t^2; \qquad \text{and} \qquad -1 \le t \le 1$$

produces a parabolic arc. So does

$$x = t^3; \qquad y = t^6; \qquad -1 \le t \le 1.$$

Following are both arcs. The bullets mark 0.25-second time intervals.

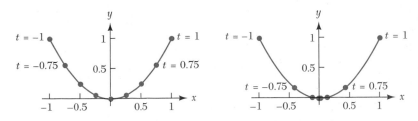

The curve $x = t, y = t^2, -1 \le t \le 1$ **The curve $x = t^3, y = t^6, -1 \le t \le 1$**

The curves are geometrically identical; they differ only in the ways in which they are traced out.

Calculus with Parametric Curves: Speed and Slope

Suppose that a parametric curve C has differentiable coordinate functions $x = f(t)$ and $y = g(t)$. What do the derivatives f' and g' say about the situation?

Speed

As the bullets on some of the foregoing graphs show, speed varies with time.

If t tells time, then $\big(f(t), g(t)\big)$ gives P's *position* in the xy-plane at time t, and it makes sense to consider the speed of P at time t. ◄ Since the derivatives f' and g' tell, respectively, how fast x and y increase, their appearance is no surprise:

> **Definition** Suppose that the position of a point P at time t, $a \le t \le b$, is given by differentiable coordinate functions $x = f(t)$ and $y = g(t)$. Then the speed at time t is given by
>
> $$\text{speed of } P \text{ at time } t = \sqrt{f'(t)^2 + g'(t)^2}.$$

Making Sense. Is the definition "right"? Definitions aren't subject to proof, but they *should* appeal to common sense. Does this one?

Speed, properly defined, should tell how fast the distance traveled by P increases with respect to time. Showing carefully that the preceding definition actually does so involves a rather subtle definition of **arclength**, i.e., distance measured along a curve. We won't define arclength here, but the next example shows that the definition makes good sense for linear coordinate functions.

EXAMPLE 8 At time t seconds, a point P has coordinates $x = at + b$ and $y = ct + d$; a, b, c, and d are constants. How fast is P moving at time t_0?

Solution The definition says that the speed of P at any time t is constant:

$$\sqrt{f'(t)^2 + g'(t)^2} = \sqrt{a^2 + c^2}.$$

Let's see why this result makes sense.

See the exercises at the end of this section for more on this point.

Check our algebra.

Notice first that because P has *linear* coordinate functions, P *moves along a straight line.* ◄ When $t = t_0$, P has coordinates $(x_0, y_0) = (at_0 + b, ct_0 + d)$; by time t_1, P has moved to $(at_1 + b, ct_1 + d)$—a distance of ◄

$$\sqrt{(at_1 + b - at_0 - b)^2 + (ct_1 + d - ct_0 - d)^2} = (t_1 - t_0)\sqrt{a^2 + c^2}$$

units. Therefore, from $t = t_0$ to $t = t_1$, P has

$$\text{average speed} = \frac{\text{distance}}{\text{time}} = \frac{(t_1 - t_0)\sqrt{a^2 + c^2}}{t_1 - t_0} = \sqrt{a^2 + c^2}.$$

This shows that P has the same average speed—$\sqrt{a^2 + c^2}$—over any time interval. Thus, the instantaneous speed of P at any time t_0 is also $\sqrt{a^2 + c^2}$, as the definition says. ∎

Look first at the picture. Try to guess reasonable answers.

EXAMPLE 9 Consider again the parametric curve

$$x = f(t) = t - 2\sin t; \qquad y = g(t) = 2 - 2\cos t$$

of Example 1 on page 316. Find the speed at $t = 3$. When does P move fastest? ◄

S o l u t i o n   The definition says that at any time t,

$$\text{speed} = \sqrt{(f'(t))^2 + (g'(t))^2} = \sqrt{(1 - 2\cos t)^2 + (2\sin t)^2} = \sqrt{5 - 4\cos t}.$$

(The last step uses a trigonometric identity.) ▶   *Check for yourself.*

The rest is easy. At $t = 3$ the speed is $\sqrt{5 - 4\cos 3} \approx 2.99$. The speed formula also shows—and the picture agrees—that the speed $\sqrt{5 - 4\cos t}$ is greatest when $\cos t = -1$, e.g., when $t = \pi$. ■

Slopes on Parametric Curves

On an ordinary, $y = f(x)$-style curve, $f'(x)$ gives the slope at (x, y). Not surprisingly, slopes on parametric curves also involve derivatives.

Not all functions have derivatives; not all parametric curves t have slopes. To avoid needless complications, we'll work only with smooth parametric curves, defined as follows:

> **Definition** The parametric curve C defined by
>
> $$x = f(t); \qquad y = g(t); \qquad a \le t \le b$$
>
> is **smooth** if f' and g' are continuous functions of t, and f' and g' are not simultaneously zero.

(The second requirement says that the moving point P *never stops.*) ▶   For smooth   *See the exercises at the end*
curves, the next theorem tells how to find the slope at a point.   *of this section for more on*
                                                                                    this idea.

> **Theorem 3** Let the smooth parametric curve C be defined as before. If $f'(t) \ne 0$, then the slope dy/dx at the point $(x, y) = (f(t), g(t))$ is given by
>
> $$\frac{dy}{dx} = \frac{g'(t)}{f'(t)} = \frac{dy/dt}{dx/dt}.$$

Before proving the theorem, let's use it.

E X A M P L E   10   Consider the parametric curve

$$x = f(t) = t - 2\sin t; \qquad y = g(t) = 2 - 2\cos t$$

of Example 1, on page 316. Find the slope at $t = 1$. Where is the curve horizontal? Where is it vertical? ▶   *Guess first, by looking at the*
                                                                                    picture.

S o l u t i o n   The theorem says that

$$\text{slope at } t = \frac{g'(t)}{f'(t)} = \frac{2\sin t}{1 - 2\cos t},$$

unless $f'(t) = 0$ (where the denominator is zero). Setting $t = 1$ gives

$$\text{slope at time } 1 = \frac{g'(1)}{f'(1)} = \frac{2\sin 1}{1 - 2\cos 1} \approx -20.88.$$

The picture agrees: At $\big(f(1), g(1)\big) \approx (-0.68, 0.92)$, the curve has large negative slope.

The general slope formula

$$\frac{dy}{dx} = \frac{2\sin t}{1 - 2\cos t}$$

shows that the curve is *horizontal* only when the numerator is zero, i.e., when t is an integer multiple of π. Again the picture agrees. At $t = \pi$, for instance, $(x, y) = (\pi, 4)$. At this point the curve appears to be horizontal.

Otherwise, the slope of C is finite.

The curve C can be *vertical* only where the preceding denominator is zero, ◄ i.e., when $\cos t = 1/2$. One such value is $t = \pi/3 \approx 1.05$; again the picture agrees.
∎

Proof of the Theorem

Remember—we're assuming that f is smooth.

Suppose that $f'(t_0) \neq 0$; then either $f'(t_0) > 0$ or $f'(t_0) < 0$. If $f'(t_0) > 0$, then continuity of f' ◄ implies that $f'(t) > 0$ for t near t_0 and therefore that $x = f(t)$ is *increasing* near $t = t_0$. If $f'(t) < 0$, then (by the same reasoning) $x = f(t)$ is *decreasing* near $t = t_0$. In either case, it follows that if t is near t_0 but $t \neq t_0$, then

Denominators don't vanish unexpectedly.

$f(t) \neq f(t_0)$. This ensures, in turn, that all of the following limit computations make sense. ◄ Notice first that

$$\text{slope} = \left.\frac{dy}{dx}\right|_{t=t_0} = \lim_{t \to t_0} \frac{y - y_0}{x - x_0} = \lim_{t \to t_0} \frac{g(t) - g(t_0)}{f(t) - f(t_0)}.$$

A common sort of trick in proofs of calculus theorems.

Now for a little trick. ◄ We'll divide the preceding numerator and denominator by $(t - t_0)$, take limits of everything in sight, and see what happens. Here goes:

$$\left.\frac{dy}{dx}\right|_{t=t_0} = \lim_{t \to t_0} \frac{\frac{g(t) - g(t_0)}{t - t_0}}{\frac{f(t) - f(t_0)}{t - t_0}} = \frac{\lim_{t \to t_0} \frac{g(t) - g(t_0)}{t - t_0}}{\lim_{t \to t_0} \frac{f(t) - f(t_0)}{t - t_0}} = \frac{g'(t_0)}{f'(t_0)},$$

which is what we wanted to show. (The second-last step is OK because, by assumption, the limit in the denominator is not zero.)
∎

Modeling with Parametric Equations

Parametric equations often help model physical processes that involve quantities that vary in both time and space. Descriptions of **trajectories**—paths taken by projectiles—are one such setting. ◄

Projectiles could be baseballs, BBs, ballistic missiles, etc.

EXAMPLE 11   A major-league fastball leaves the pitcher's hand traveling horizontally, with initial speed 150 ft/sec and initial height 7 feet. Ignoring wind resistance, ◄ describe the ball's trajectory. When, and at what speed, does the ball cross the plate, 60.5 feet away? Is the pitch a strike? If the batter, catcher, and umpire all miss the ball, where does it land?

Major-league batters don't ignore wind resistance. See the exercises at the end of this section for a more realistic example.

Solution   We use an xy-coordinate system with its origin at the pitcher's feet and with home plate at $(60.5, 0)$; the ball starts at $(0, 7)$.

Since we're ignoring wind resistance, the ball, once released, is influenced only by gravity. ▶   Thus, the ball's horizontal acceleration ▶   is zero; its vertical acceleration due to gravity is -32 ft/sec/sec. ▶

If $x = f(t)$, $y = g(t)$, and t measures time in seconds since the ball's release, our information boils down to this:

$$f''(t) = 0; \qquad f'(0) = 150; \qquad f(0) = 0;$$
$$g''(t) = -32; \qquad g'(0) = 0; \qquad g(0) = 7.$$

For more on falling objects, with and without air resistance, see pp. 102–105.

Recall that acceleration is the second derivative of position.

These data lead to simple formulas for f and g. Here is the argument for f:

$$f''(t) = 0 \implies f'(t) = a \qquad \text{and} \qquad f(t) = at + b,$$

Why is the vertical acceleration negative?

where a and b are constants, as yet unknown. Finding them is easy—a combination of the facts that $f(t) = at + b$, $f(0) = 0$, and $f'(0) = 150$ gives $a = 150$ and $b = 0$, so $f(t) = 150t$. A similar argument applies to g, with the result that

$$x = f(t) = 150t; \qquad y = g(t) = 7 - 16t^2.$$

These equations make sense, of course, only while the ball remains airborne, i.e., until $y = 7 - 16t^2 = 0$, or $t = \sqrt{7}/4 \approx 0.661$ seconds. Here's the parametric curve for $0 \le t \le 0.661$:

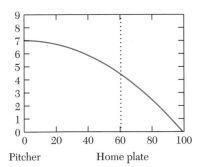

A baseball's trajectory:
$x = 150t, y = 7 - 16t^2, 0 \le t \le 0.661$

Notice the parabolic shape. In fact, every free-falling object either moves vertically or follows a parabolic trajectory. ▶

Using the picture and the formulas, we can answer our original questions. The ball crosses the plate when $x = f(t) = 150t = 60.5$, i.e., when $t = 121/300 \approx 0.4033$ seconds. At this time, ▶

$$\text{speed} = \sqrt{f'(0.4033)^2 + g'(0.4033)^2} \approx 150.55$$

See the exercises for more on parabolic trajectories.

Check the arithmetic.

feet per second. At the same time, the ball's height is $y = g(0.4033) \approx 4.4$ feet—high but in the strike zone for an average batter. Left untouched, the ball hits the ground at $t = 0.661$; at this time, $x = f(t) = 150 \cdot 0.661 \approx 99.2$ feet—almost 40 feet behind home plate.   ■

BASICS

Plot each parametric curve below. (Using a machine is fine, but then draw or copy your own curve on paper.) In each case, mark the direction of travel and label the points corresponding to $t = -1$, $t = 0$, and $t = 1$.

1. $x = t$, $y = \sqrt{1 - t^2}$, $-1 \le t \le 1$
2. $x = t$, $y = -\sqrt{1 - t^2}$, $-1 \le t \le 1$
3. $x = \sqrt{1 - t^2}$, $y = t$, $-1 \le t \le 1$
4. $x = -\sqrt{1 - t^2}$, $y = t$, $-1 \le t \le 1$
5. $x = \sin(\pi t)$, $y = \cos(\pi t)$, $-1 \le t \le 1$

FURTHER EXERCISES

6. Each parametric "curve" below is actually a line segment. In each part, state the beginning point ($t = 0$) and ending point ($t = 1$) of the segment. Then state an equation in x and y for the line each segment determines.
   (a) $x = 2 + 3t$, $y = 1 + 2t$, $0 \le t \le 1$
   (b) $x = 2 + 3(1 - t)$, $y = 1 + 2(1 - t)$, $0 \le t \le 1$
   (c) $x = t$, $y = mt + b$, $0 \le t \le 1$
   (d) $x = a + bt$, $y = c + dt$, $0 \le t \le 1$
   (e) $x = x_0 + (x_1 - x_0)t$, $y = y_0 + (y_1 - y_0)t$, $0 \le t \le 1$
7. A parametric curve $x = f(t)$, $y = g(t)$, $a \le t \le b$ has **constant speed** if—what else?—its speed is constant in t. Which of the curves in Exercises 1–5 have constant speed?
8. Show that if a curve C has linear coordinate functions $x = at + b$ and $y = ct + d$, then C has constant speed. (See the previous exercise.)
9. Consider the curve C shown in Example 1; suppose that t tells time in seconds.
   (a) At which bulleted points would you expect P to be moving quickly? Slowly? Why?
   (b) Use the curve to estimate the speed of P at $t = 3$. (One approach: Estimate how far P travels—i.e., the length of the curve—over the one-second interval from $t = 2.5$ to $t = 3.5$. If d is this distance, then d distance units per second is a reasonable speed estimate at $t = 3$.)
   (c) Use the curve to estimate the speed of P at $t = 6$.
10. Plot the parametric curve
$$x = t^3; \qquad y = \sin t^3; \qquad -2 \le t \le 2.$$
What familiar curve is produced? Why does the result happen?
11. Let (a, b) be any point in the plane, and $r > 0$ any positive number. Consider the parametric equations
$$x = a + r \cos t; \qquad y = b + r \sin t; \qquad 0 \le t \le 2\pi.$$
   (a) Plot the parametric curve defined above for $(a, b) = (2, 1)$ and $r = 2$. Describe your result in words.
   (b) Show by calculation that if x and y are as above, then $(x - a)^2 + (y - b)^2 = r^2$. Conclude that the curve defined above is the circle with center (a, b) and radius r.
   (c) Write parametric equations for the circle of radius $\sqrt{13}$, centered at $(2, 3)$.

(d) What "curve" results from the equations above if $r = 0$?
12. Let a and b be any positive numbers, and let a parametric curve C be defined by
$$x = a \cos t; \qquad y = b \sin t; \qquad 0 \le t \le 2\pi.$$
The resulting curve is called an **ellipse**.
   (a) Plot the curve defined above for $a = 2$ and $b = 1$. Describe C in words. Where is the "center" of C? Why do you think the quantities $2a$ and $2b$ are called the **major and minor axes** of C.
   (b) What curve results if $0 \le t \le 4\pi$? Why?
   (c) Write parametric equations for an ellipse with major axis 10 and minor axis 6.
   (d) Write parametric equations for another ellipse with major axis 10 and minor axis 6.
   (e) Show that for all t, $\dfrac{x^2}{a^2} + \dfrac{y^2}{b^2} = 1$.
   (f) How does the "ellipse" look if $a = b$? How does its xy-equation look?
   (g) How does an ellipse look if $a = 1000$ and $b = 1$?
   (h) How does an ellipse look if $a = 1$ and $b = 1000$?
13. Consider the Lissajou curve in Example 6.
   (a) Label the points corresponding to $t = 0$, $t = 0.1$, and $t = \pi/2$. Add some arrows to the curve to indicate direction.
   (b) How often in the interval $0 \le t \le 2\pi$ does the tracing point P return to $(0, 0)$?
   (c) How would the picture be different if the t-interval $0 \le t \le 4\pi$ were used?
14. We stated in this section that if a parametric curve C has linear coefficient functions $x = f(t) = at + b$ and $y = g(t) = ct + b$, then C is a straight line (or part of a line). This exercise explores that fact.
   (a) Plot the parametric curve $x = 2t$, $y = 3t + 4$, $0 \le t \le 1$. Where does the curve start? Where does it end? What is its shape?
   (b) Eliminate the variable t in the two equations above to find a single equation in x and y for the line of the previous part.
   (c) Find the slope at $t = t_0$ of the parametric curve $x = 2t$, $y = 3t + 4$, $0 \le t \le 1$. Why doesn't the answer depend on t?

15. This exercise pursues the idea that if a parametric curve C has linear coefficient functions $x = f(t) = at + b$ and $y = g(t) = ct + d$, then C is a straight line (or part of a line).

    (a) Consider the parametric curve $x = at + b$, $y = ct + d$, $t_0 \leq t \leq t_1$. Where does the curve start? Where does it end?

    (b) Assume that $a \neq 0$. (Don't assume that $c \neq 0$.) Eliminate the variable t in the two equations above to find one equation in x and y. What line does it describe?

    (c) Assume that $c \neq 0$. (Don't assume that $a \neq 0$.) Eliminate the variable t in the two equations above to find one equation in x and y. What line does it describe?

    (d) What happens if a and c are both zero?

16. This exercise explores the technical assumptions in the definition of smooth curves.

    (a) Consider the "curve" defined for all t by $x = 0$, $y = 0$. Plot C. Does C "deserve" to have a slope at $t = 0$? If so, what slope? If not, why not? Is C smooth in the sense of the definition?

    (b) Consider the curve defined for $-1 \leq t \leq 1$ by $x = t^3$ and $y = t^3$. Plot C. Does C "deserve" to have a slope at $t = 0$? What does Theorem 3 say in this case?

    (c) Consider the curve defined for $-1 \leq t \leq 1$ by $x = t$ and $y = t$. Plot C. Does this C "deserve" to have a slope at $t = 0$? What does Theorem 3 say about slope this time?

17. This exercise concerns Example 11 (p. 322).

    (a) How could the model be made more realistic? What additional information would be needed?

    (b) Use the conditions $g''(t) = -32$, $g'(0) = 0$, and $g(0) = 7$ to show that $g(t) = 7 - 16t^2$, as claimed in the Example.

    (c) Find a formula for $s(t)$, the ball's speed at time t. Plot $s(t)$ over an appropriate interval. How does the graph's shape reflect the physical situation?

18. Use parametric equations to describe the trajectory of a baseball thrown exactly as in Example 11 (p. 322), except that this time the initial velocity is 100 feet per second. Plot the result over an appropriate interval. Is the pitch a strike? At what speed does it cross the plate? (The strike zone is roughly from 1.5 to 4.5 feet above the ground at home plate.)

19. Consider a baseball thrown horizontally, starting from height 7 feet (as in Example 11), but with initial speed s_0.

    (a) Explain briefly why the parametric equations

    $$x = s_0 t, \quad y = 7 - 16t^2$$

    describe the ball's trajectory (ignoring wind resistance).

    (b) When does the ball reach home plate?

    (c) Is the trajectory parabolic? Why?

20. This exercise is again about the situation in Example 11, but takes wind resistance into account. In practice, wind resistance *does* affect a baseball's trajectory. One possible model of wind resistance (we omit the details) leads to the parametric equations

    $$x = \frac{\ln(150 \cdot k \cdot t + 1)}{k}; \quad y = 7 - 16t^2,$$

    where the constant k can be thought of as the ball's "drag coefficient"—the smoother the ball, the lower the value of k. For a typical baseball, $k = 0.003$ is reasonable. With this value of k we get

    $$x = \frac{1000 \ln(0.45t + 1)}{3}; \quad y = 7 - 16t^2.$$

    The resulting trajectory is plotted below; the curve from Example 11 is also shown:

    Baseball trajectories with and without air drag

    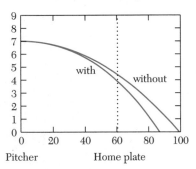

    In each part below, use the graphs to give approximate answers; then use appropriate formulas to improve your results.

    (a) At what time does the air-dragged ball cross the plate? (**NOTE:** The graphs don't show time.) How much longer does it take to reach the plate than the drag-less ball?

    (b) At what height does the air-dragged ball cross the plate?

    (c) At what speed does the air-dragged ball cross the plate?

    (d) Where does the air-dragged ball land?

21. The situation is as in the previous exercise, except that this time the ball is scuffed, so that its "drag coefficient" is 0.005, not 0.003. Plot the new trajectory. Then answer the same questions as in the previous exercise.

4.10 Why Continuity Matters

In Section 2.6 we defined continuity in terms of limits: f is **continuous** at $x = a$ if

$$f(x) \to f(a) \quad \text{as} \quad x \to a.$$

The standard elementary functions—algebraic, exponential, logarithm, and trigonometric functions—are continuous wherever they're defined. Combinations—sums, products, quotients, compositions, etc.—of elementary functions are also continuous wherever they're defined, thanks to the analogous properties of limits. (See, for example, Theorem 2, page 169.)

Indeed, continuity is so much the rule in calculus that its importance may be hard to appreciate. This section aims to remedy that oversight.

The Importance of Being Continuous: The Intermediate Value Theorem and the Extreme Value Theorem

Continuity is clearly a desirable property of a function. The **intermediate value theorem (IVT)** and the **extreme value theorem (EVT)** give two good reasons why. We won't prove either, but we'll indicate why the property each guarantees is practically useful.

Each theorem pertains to a continuous function f on a *closed and bounded* interval. ◄

See the exercises at the end of this section for more on why $[a, b]$ must be closed.

> **Theorem 4** (Intermediate Value Theorem) Let f be continuous on the closed and bounded interval $[a, b]$, and let y be any number between $f(a)$ and $f(b)$. Then, for some input c between a and b, $f(c) = y$.

> **Theorem 5** (Extreme Value Theorem) Let f be continuous on the closed, bounded interval $[a, b]$. Then f assumes both a maximum value and a minimum value somewhere on $[a, b]$.

Existential Philosophy

The EVT and the IVT are **existence theorems**. Each guarantees the existence of (at least) one point in the domain with some desirable property. Neither theorem says *where* in $[a, b]$ these points may fall or *how many* there may be. Such high-toned philosophy may suggest that these theorems, whatever their theoretical importance, have little practical use. That's wrong, as we'll see.

The IVT, Pictorially

The IVT says that if f is continuous on $[a, b]$, then the range of f contains not just $f(a)$ and $f(b)$ but *everything in between*. This means that the graph of a continuous function f is unbroken in the following sense.

Enroute from $(a, f(a))$ to $(b, f(b))$, the graph of f crosses every horizontal line between $y = f(a)$ and $y = f(b)$.

Here's the picture:

The IVT:
f "hits" every intermediate value

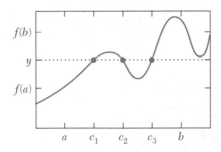

The EVT, Pictorially

The EVT guarantees that if f is continuous on the closed interval $[a, b]$, then f attains both a maximum and a minimum somewhere therein. In the case pictured as follows, these extreme values are attained at $x = x_{max}$ and $x = x_{min}$, respectively.

The EVT:
On $[a, b]$, f assumes extreme values

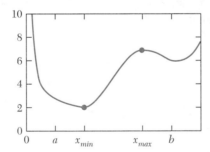

Both hypotheses of the EVT—that f be *continuous* and that the interval be closed—are necessary. If either fails, f need not assume a maximum or a minimum. In the case just pictured, for instance, f does not achieve a maximum on the *open* interval $(0, b)$ even though f is continuous there.

From Abstract Theory, a Concrete Payoff

The IVT is a theoretical result, but it has useful, concrete consequences. One of them, the bisection method, or interval-halving method, is a simple but effective technique for solving equations. First, let's take care of some preliminaries.

A number r is a root (or zero) of a function f if $f(r) = 0$.

EXAMPLE 1 The function $f(x) = x^2 - 1 = (x - 1)(x + 1)$ has *two* roots, at $x = \pm 1$. The function $g(x) = x^2$ has *one* root, at $x = 0$. Since $h(x) = x^2 + 1 \geq 1$ for all x, h has *no* roots. How many roots can any quadratic polynomial have?

Solution   The **quadratic formula**

$$r = \frac{-b \pm \sqrt{b^2 - 4ac}}{2a}$$

finds all the roots—there may be 0, 1, or 2—of any quadratic polynomial $p(x) = ax^2 + bx + c$. ∎

Digging Systematically for Roots

Finding roots by algebraic solution techniques is often difficult or impossible. Even when algebraic methods (such as the quadratic formula) do succeed, finding decimal equivalents may remain a problem. It's clear, for example, that $x = \sqrt{2}$ is a root of $x^2 - 2$, but how can we find an accurate decimal value?

The bisection method will help. It's based on this key fact:

Fact If f is continuous on $[a, b]$, and $f(a)$ and $f(b)$ have opposite signs, then $[a, b]$ contains a root of f. Hence, the midpoint $(a + b)/2$ of $[a, b]$ differs from a root of f by no more than $(b - a)/2$—the "radius" of $[a, b]$.

This fact follows directly from the IVT. Because f is negative at one end of $[a, b]$ and positive at the other, the IVT guarantees that $f(r) = 0$ somewhere in between—i.e., that a root r lies somewhere in $[a, b]$. Lacking further information, the midpoint of $[a, b]$ is as good an estimate of r as any. By applying this fact repeatedly, we can approximate a root as closely as we wish.

EXAMPLE 2   We use the bisection algorithm to estimate $\sqrt{2}$ accurately. Since $\sqrt{2}$ is a root of $f(x) = x^2 - 2$ and certainly lies somewhere in $[1, 2]$, we have all we need to get started.

Solution   We'll converge on an answer, step by step:

Step 1:   Since $f(1) = -1$ and $f(2) = 2$, our Fact says that $m_1 = 1.5$—the midpoint of $[1, 2]$—differs from a root of f by less than 0.5.

Step 2:   Since $f(1.5) = 0.25 > 0$, the Fact applies to the new—and shorter—interval $[1, 1.5]$. Conclusion: $m_2 = 1.25$, the new midpoint, differs from a root by less than 0.25.

Step 3:   Since $f(1.25) = -0.4375 < 0$, our *latest* interval is $[1.25, 1.50]$. Conclusion: $m_3 = 1.375$ differs from a root by less than 0.125.

We can continue these steps as long as we like; in practice, we would continue until we were sure of some predetermined accuracy. (The next few estimates are 1.4375, 1.40625, 1.421875, 1.4140625, 1.41796875, and 1.416015625, not far from $\sqrt{2} \approx 1.41421356237$.) ∎

Satisfaction Guaranteed—Up to a Point

Each step *doubles* the guaranteed precision of our estimate. In other words, the nth midpoint m_n can't possibly miss a root c by more than the radius of the nth interval. More precisely,

$$|m_n - c| < \frac{b - a}{2^n}.$$

On the other hand, there's no guarantee of *ever* finding a root exactly. ▶   In real life, excellent approximations are almost always good enough.

If a and b are rational numbers, then so is each midpoint m_n. Thus, if a root r is irrational, the method cannot ever locate r exactly.

A Bisection Algorithm

The bisection method is an **algorithm**—a finite sequence of explicit instructions which, starting from appropriate inputs, reliably produces a desired output. ("Explicit" means, roughly, "executable by a machine." Computer programs are prime examples of algorithms.)

 To emphasize the algorithmic nature of the bisection method, we'll state it explicitly in **pseudocode**—a sort of generic computer language. (Computerese is printed in the "typewriter" typeface.) Notice that the error tolerance is stated up front, as an input to the program. Once a root is found to within that tolerance, the procedure stops. (Without a stopping rule, a program might run forever.)

> **Input:**   A continuous function f, an interval $[a, b]$, and a tolerance $t > 0$. (For simplicity, we'll assume that $f(a) < 0$ and $f(b) > 0$.)
>
> **Output:**   An approximate root of f, i.e., a number m that differs from a root by less than the tolerance t.
>
> **Step 1:**   (Find midpoint) `m:=(b+a)/2`
>
> **Step 2:**   (Done?) `if b-a<2t then output m; QUIT`
>
> **Step 3:**   (New interval) `if f(m)<0 then a:=m else b:=m`
>
> **Step 4:**   (Start again) `return to step 1`

A Sample Problem

Here's a problem that exploits ideas discussed in this section.

EXAMPLE 3   Let $f(x) = \sqrt{x}$.

a. What does the EVT say about f on the interval $[0, 1]$?

b. Although f is continuous on $[0, \infty)$, it has no maximum value on this interval. Doesn't this contradict the EVT?

c. Although f is continuous on $(0, 1]$, it assumes no minimum value on this interval. Doesn't *this* contradict the EVT?

d. Show that the IVT holds for f on $[0, 4]$ and $y = 1$.

Solution

a. The EVT says that f must achieve a maximum and a minimum somewhere on $[0, 1]$. It does: $f(0) = 0$ is the minimum value and $f(1) = 1$ is the maximum value.

b, c. There's no contradiction in either part. The EVT applies only for closed and bounded intervals. Neither $[1, \infty)$ nor $(0, 1]$ is closed and bounded.

d. The IVT says that f must assume all values between $f(0) = 0$ and $f(4) = 2$. In particular, $y = 1$ is one such value. Sure enough, $f(1) = 1$, as the IVT requires.

■

BASICS

1. Let f and g be the functions shown graphically below. Use the graphs to answer the following questions; no formulas are given (or needed). Notice that f is continuous except at $x = -1$, $x = 1$, and $x = 2$.

Graph of f

Graph of g

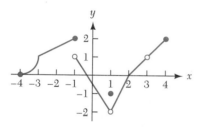

(a) Over what intervals is g continuous?

(b) Which of the following functions are continuous at $x = 2$? (Justify your answers.)

  (i) $f(x) + g(x)$ (iii) $\dfrac{f(x)}{g(x)}$

  (ii) $f(x) \cdot g(x)$ (iv) $\dfrac{g(x)}{f(x)}$

2. Let $f(x) = 1/x$.

(a) What does the EVT say about f on the interval $[0.1, 1]$?

(b) Although f is continuous on $[1, \infty)$, it has no minimum value on this interval. Why doesn't this contradict the EVT?

(c) Although f is continuous on $(0, 1]$, it assumes no maximum value on this interval. Why doesn't this contradict the EVT?

(d) What does the EVT say about f on the interval $[0, 1]$?

(e) Although $f(-1) = -1$ and $f(1) = 1$, there is no value of c for which $f(c) = 0$. Why doesn't this contradict the IVT?

(f) Show that the IVT holds for f on $[0.1, 1]$ and $y = 5$.

3. Use the IVT to show that $f(x) = x^5 + x^3 - 5x + 2$ has a root in $[0, 1]$.

4. Use paper and pencil alone to find the first four midpoint approximations to $\sqrt{3}$, starting with the interval $[1, 2]$. Express your answers as fractions.

5. Use a calculator to find the first six midpoint approximations to a root of $\cos x - x$, starting on $[0, 1]$. Then use graphics or any other method to find the root to within 0.001.

FURTHER EXERCISES

6. Let f and g be the functions defined in Exercise 1. Neither f nor g is continuous on the entire interval $[-4, 4]$, so neither satisfies the hypothesis of the EVT or the IVT. Still, it's possible that either theorem's conclusion might hold.

(a) Does the conclusion of the IVT hold for f on $[-4, 4]$? Explain.

(b) Does the conclusion of the IVT hold for g on $[-4, 4]$? Explain.

(c) Does the conclusion of the EVT hold for f on $[-4, 4]$? Explain.

(d) Does the conclusion of the EVT hold for g on $[-4, 4]$? Explain.

(e) The function g is continuous on $[-4, -1]$, so the EVT holds. What are the maximum and minimum values of g on $[-4, -1]$? Where are they attained?

7. Suppose that f is a continuous function defined for all real numbers and that f achieves a maximum value of 5 and a minimum value of -7. Determine which of the following statements *must* be true, which *might* be true, and which *cannot* be true.

(a) The maximum value of $f(|x|)$ is 7.

(b) The minimum value of $f(|x|)$ is 0.

(c) The maximum value of $|f(x)|$ is 7.

(d) The minimum value of $|f(x)|$ is 5.

8. Suppose that f is a continuous function and $f(3) = -1$. Evaluate $\lim_{x \to 3} f(x)$.

9. Determine which of the following statements *must* be true, which *might* be true, and which *cannot* be true.
   (a) If $f(1) < 0$ and $f(5) > 0$, then there must be a number c in $(1, 5)$ such that $f(c) = 0$.
   (b) If f is continuous on $[1, 5]$, $f(1) > 0$, and $f(5) < 0$, then there must be a number c in $(1, 5)$ such that $f(c) = 0$.
   (c) If f is continuous on $[1, 5]$ and there is a number c in $[1, 5]$ such that $f(c) = 0$, then $f(1) \cdot f(5) \leq 0$.
   (d) If f is continuous and has no roots in $[1, 5]$, then $f(1) \cdot f(5) > 0$.

10. Use the IVT to find an interval in which $f(x) = x^5 + x^3 + 5x + 2$ has a root.

11. Let
$$f(x) = \begin{cases} ax^2 + 3, & \text{if } x < 2 \\ 3x - 5, & \text{if } x \geq 2 \end{cases}.$$
Find the value of a that makes f continuous at $x = 2$.

12. Suppose that f is defined on $[-5, 5]$, that $f(-5) = 2$ and $f(5) = -2$. If f assumes every value in $[-2, 2]$, must f be continuous? Explain. [**HINT:** The answer is *no*. Draw a graphical example that illustrates this.]

13. Use the IVT to prove this fixed point theorem: If f is continuous on $[0, 1]$ and has range contained in $[0, 1]$, then f has a fixed point—i.e., an input x for which $f(x) = x$. [**HINT:** Consider $g(x) = f(x) - x$.]

14. Let
$$f(x) = \begin{cases} x, & \text{if } x \text{ is rational} \\ 0, & \text{otherwise}. \end{cases}$$
Show that f is continuous at 0.

4.11 Why Differentiability Matters; The Mean Value Theorem

Recall the limit definition of derivative, from Section 2.5:

> **Definition** Let f be a function defined near and at $x = a$. Then
> $$f'(a) = \lim_{h \to 0} \frac{f(a + h) - f(a)}{h}.$$

For a given function f and $x = a$, the limit in the definition may or may not exist. If it does exist, we say that f is **differentiable** at $x = a$.

The standard elementary functions—algebraic, exponential, logarithm, and trigonometric functions—are differentiable wherever they're defined. Combinations—sums, products, quotients, compositions, and so on—of elementary functions are also differentiable where they're defined, thanks to the constant, sum, product, quotient, and chain rules.

Like continuity, differentiability is the rule in calculus—so much so, perhaps, that its importance might be overlooked. This section aims to avoid that pitfall.

Differentiable Functions Are Continuous

The absolute value function $g(x) = |x|$ is *continuous* at $x = 0$, but not *differentiable* at $x = 0$. On the other hand, a differentiable function must be continuous.

> **Theorem 6** If f is differentiable at $x = a$, then f is also continuous at $x = a$.

Proof. The theorem is both graphically plausible—it's hard to imagine a discontinuous graph having a tangent line—and easy to prove. To prove the theorem, we need only show that

$$f(x) \to f(a) \quad \text{as} \quad x \to a.$$

(This is the formal definition of continuity of f at $x = a$.) The definition of differentiability at $x = a$ means that, as $x \to a$,

$$\frac{f(x) - f(a)}{x - a} \to f'(a).$$

From this fact it follows easily that as $x \to a$,

$$f(x) - f(a) = \frac{f(x) - f(a)}{x - a} \cdot (x - a) \to f'(a) \cdot 0 = 0,$$

as desired.   ■

Theory and Practice: A Case Study

In later mathematics courses, such as elementary real analysis, the theory of calculus is the main subject.

Calculus is more than a set of powerful methods for solving important problems. It is also a mathematical *theory*—a coherent, rigorously developed body of definitions and theorems. We don't aim here (or elsewhere) to present the theory of calculus in full detail or with full rigor. ◄ This section is, instead, a sort of case study—a close look at one important theorem in its mathematical context.

> **The Guts (and the Beauty) of the Machine.** Using calculus, like driving a car, can be done effectively without perfect or constant attention to the machine's inner workings. Why bother with theory?
>
> One good reason is that ignorance of underlying principles leaves one vulnerable in unexpected situations—no owner's manual can cover everything. A better reason concerns the difference between driver's education and calculus. Driving is a useful skill, but mathematics is a science.
>
> Another good reason is that the theory of calculus is a human creation of unparalleled power and beauty, putting even the internal combustion engine to shame.

The Mean Value Theorem in Context

We encountered them in the preceding section.

The intermediate value theorem and the extreme value theorem ◄ guarantee that a continuous function assumes certain desirable values on a closed interval. The **mean value theorem**, similar in spirit, guarantees that the derivative of a well-behaved function assumes a certain desirable mean value.

> **T h e o r e m 7** (**Mean Value Theorem**) Suppose that f is continuous on the closed interval $[a, b]$ and differentiable on the open interval (a, b). Then, for some c between a and b,
>
> $$f'(c) = \frac{f(b) - f(a)}{b - a}.$$

Taken out of context, the mean value theorem (MVT) is difficult to understand, ▶ let alone appreciate for what it is: one of the two or three "big" theorems of calculus. Most of what follows aims to put the theorem in context. What do its words mean? Why is the name appropriate? What does it say? Are the hypotheses really necessary? What questions does the theorem answer? Why do the answers matter? What makes the MVT a big theorem?

Few statements of any kind make much sense out of context. Ask a politician.

What the MVT Says

The MVT says that, under appropriate hypotheses on f, the derivative f' must assume a certain mean value. Interpreting f' either as an instantaneous rate or as a slope makes everything plausible.

The MVT—A Motorist's View

If a car averages 60 mph over an interval of time, then, at some instant, the speedometer must have read exactly 60. The MVT says the same thing, a little more generally. If $f(t)$ represents a car's position in miles at time t in hours, then (as we saw in Section 2.1) $f'(t)$ represents instantaneous velocity at time t, and $(f(b) - f(a))/(b - a)$ represents average velocity over the interval $a \le t \le b$. In this language, the MVT equation

$$f'(c) = \frac{f(b) - f(a)}{b - a}$$

reads as follows:

> At some time c between a and b, the instantaneous velocity is equal to the average (i.e., mean) velocity over the entire interval.

The MVT—A Graphical View

On the graph of f, each side of the MVT equation represents a *slope*—of the tangent line at $x = c$ and of the secant line from a to b, respectively. The theorem asserts that, for at least one value of c between a and b, these two slopes are equal. In the case shown graphically as follows, *two* such c's exist.

The mean value theorem

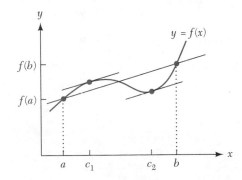

The Hypothesis of Differentiability; Why It Matters

Mathematicians like to state theorems as generally as possible.

The MVT requires that $f'(x)$ should exist for every x in (a, b). (Most often, $f'(x)$ exists at the endpoints, too, but that isn't strictly necessary.) ◄

This differentiability hypothesis can't be diluted. If f isn't differentiable at even *one* point in (a, b), the MVT can fail. The function $f(x) = |x|$ on the interval $[-1, 1]$ illustrates this problem.

Two Natural Questions, Their Answers, and How the MVT Helps

The MVT has important theoretical uses; many amount to putting intuitively reasonable results on a sound logical footing. Here we illustrate two such uses; we'll see others later.

Question 1: When Is a Function Constant?

It's clear either from the definition of derivative or from graphical arguments that the derivative of a constant function is the zero function. In other words,

> If $f(x) = k$ for all inputs x, then $f'(x) = 0$.

Is the converse true? In other words, does our supposed theorem really hold?

> **Theorem 8** If $f'(x) = 0$ for all x in an interval I, then f is constant on I.

Geometric intuition supports the theorem; if the graph of f has a horizontal tangent line everywhere, then it's natural to expect the graph itself to be horizontal. Not surprisingly, the theorem is true. What *is* surprising is that proving this "obvious" fact requires a gun of the MVT's caliber.

We don't care which constant.

Proof.   To show that f is constant, ◄   it's enough to show that for any two inputs $x = a$ and $x = b$, $f(a) = f(b)$. According to the MVT, there's some input c between a and b for which

$$f'(c) = \frac{f(b) - f(a)}{b - a}.$$

We don't know which c works, but here it doesn't matter. By hypothesis, $f'(c) = 0$ for *any* input c. Thus,

$$0 = f'(c) = \frac{f(b) - f(a)}{b - a} \implies f(b) - f(a) = 0 \implies f(b) = f(a).$$

A simple but important consequence follows:

> *Any two antiderivatives of the same function differ by a constant.*

More succinctly stated:

> **Theorem 9** If $F'(x) = G'(x)$ for all x in an interval I, then $F(x) = G(x) + C$ for some constant C.

One would prove this theorem applying Theorem 8 to the difference function $F(x) - G(x)$.

Question 2: When Is a Function Increasing?

Up to now we've often stressed—but never proved—the general principle that the sign of f' determines whether f increases or decreases over an interval. The MVT offers a simple but rigorous proof. First let's make a concise statement:

Theorem 10 Suppose that $f'(x) > 0$ for all x in an interval I. Then f is increasing on I.

Proof.   It's enough to show that, for any a and b in I with $a < b$, $f(a) < f(b)$. By the MVT, there's a c in (a, b) ▶   for which

As before, the location of c doesn't matter.

$$f(b) - f(a) = (b - a) \cdot f'(c).$$

Since both factors on the right must be positive, ▶   so must $f(b) - f(a)$, so $f(b) > f(a)$.

Why?

The MVT and the Racetrack Principle

One version of the racetrack principle ▶   says, in functional language:

We introduced it in Section 2.1, using canine language.

> If $f(a) = g(a)$ and $f'(x) \le g'(x)$ for $x \ge a$, then $f(x) \le g(x)$ for $x \ge a$.

The racetrack principle (like so much else) can be proved by applying the MVT to the "right" function—in this case, to the function $h(x) = g(x) - f(x)$. Let's do so.

The hypotheses say that $h(a) = 0$ and $h'(x) \ge 0$ if $x \ge a$. The MVT assures that for any $x \ge a$, there's a number $c \ge a$ such that

$$h'(c) = \frac{h(x) - h(a)}{x - a} = \frac{h(x)}{x - a}.$$

By hypothesis, moreover, $h'(c) \ge 0$. The rest is easy. ▶   If $x \ge a$, then

But check each step.

$$g(x) - f(x) = h(x) = h(x) - h(a) = h'(c) \cdot (x - a) \ge 0 \cdot (x - a) = 0. \quad \blacksquare$$

Proving the MVT, Step by Step (Optional)

Understood graphically, the MVT may seem so natural as to require no proof at all. ▶   That impression, although understandable, is wrong. A rigorous proof of the MVT involves several steps and appeals to several other results.

For excellent reasons, the same applies to other powerful, important theorems of mathematics.

Two Lemmas

We'll sketch the outline of a proof, based on two preliminary lemmas. ▶

*The term **lemma** denotes an auxiliary result, used to help prove another.*

Lemma 1 Suppose that f assumes either its maximum or its minimum on $[a, b]$ at a point c between a and b. If $f'(c)$ exists, then $f'(c) = 0$.

Proof. We'll show what happens if $f(c)$ is a maximum (the minimum case is similar). By the definition of derivative,

$$f'(c) = \lim_{h \to 0} \frac{f(c+h) - f(c)}{h} = \lim_{h \to 0^-} \frac{f(c+h) - f(c)}{h} = \lim_{h \to 0^+} \frac{f(c+h) - f(c)}{h}.$$

(Because the ordinary limit exists, so must the left-hand and right-hand limits, and all three are equal.) Notice first that since f assumes a maximum at c, the numerator $f(c+h) - f(c)$ in all three limit expressions must be negative. The denominator h, on the other hand, is negative if $h \to 0^-$ and positive if $h \to 0^+$. Therefore,

$$f'(c) = \lim_{h \to 0^-} \frac{f(c+h) - f(c)}{h} = \lim_{h \to 0^-} [\text{positive stuff}] \geq 0.$$

Similarly,

$$f'(c) = \lim_{h \to 0^+} \frac{f(c+h) - f(c)}{h} = \lim_{h \to 0^+} [\text{negative stuff}] \leq 0.$$

The only possible conclusion is that $f'(c) = 0$, as desired.   ■

Lemma 2 (Rolle's Theorem) Suppose that f is continuous on $[a, b]$ and differentiable on (a, b), and that $f(a) = f(b)$. Then, for some c between a and b, $f'(c) = 0$.

Remark. This lemma, known as **Rolle's theorem**, is nothing more (or less) than the MVT in the special case that $f(a) = f(b)$. The picture, too, is similar:

Rolle's theorem

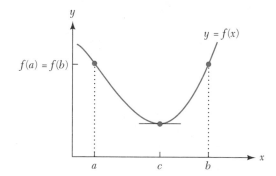

Proof of Rolle's Theorem. Because f is continuous on $[a, b]$, the extreme value theorem guarantees that f assumes both a maximum and a minimum somewhere in $[a, b]$. If both maximum and minimum occur at the endpoints a and b, then, since $f(a) = f(b)$, f must be a constant function. In this case, there's nothing more to do—any c works.

The alternative is that at least one of the maximum and the minimum lies at some c *between* a and b. But, by Lemma 1, $f'(c) = 0$. ■

All Together Now: Proving the MVT

Using Rolle's theorem to prove the MVT requires a technical step. Using f, a, and b, we create a new function to which Rolle's theorem applies.

Proof of the Mean Value Theorem. Given f as in the mean value theorem, consider the linear function l whose graph is the secant line to f from a to b, as shown:

Proving the mean value theorem: f and l

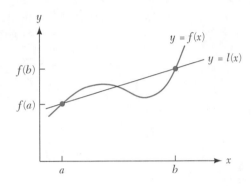

The precise formula for l doesn't matter. What does matter is that

$$l(a) = f(a); \qquad l(b) = f(b); \qquad \text{and} \qquad l'(x) = \frac{f(b) - f(a)}{b - a}.$$

Finally, we consider the difference function $g(x) = f(x) - l(x)$. By design, $g(a) = 0 = g(b)$, so Rolle's theorem applies to g. The rest is bookkeeping: For some c in (a, b),

$$0 = g'(c) = f'(c) - l'(c) = f'(c) - \frac{f(b) - f(a)}{b - a}.$$

We're finished. ■

Sample Problems

Here are two sample problems that show the mean value theorem in action.

EXAMPLE 1 Show, using derivatives, that if $x \geq 0$, then $\sin x \leq x$.

Solution One approach uses the racetrack principle. If $f(x) = \sin x$ and $g(x) = x$, then $f'(x) = \cos x$ and $g'(x) = 1$, so $f'(x) \leq g'(x)$ for all x. By the racetrack principle, $f(x) \leq g(x)$, as desired. ■

EXAMPLE 2 Suppose that the function f is differentiable for all x and that f has n roots. Show that f has *at least* $n - 1$ stationary points.

Solution   We'll show that between any two roots, there's at least one stationary point. Suppose, then, that x_0 and x_1 are roots, so $f(x_0) = 0 = f(x_1)$. By the mean value theorem, there's a c between x_0 and x_1 such that

$$\frac{f(x_1) - f(x_0)}{x_1 - x_0} = f'(c).$$

But since x_0 and x_1 are roots, the left side is zero, so c is a stationary point.   ■

BASICS

1. Throughout this exercise, $f(x) = |x|$.
   (a) Does f satisfy all hypotheses of the MVT on $[a, b] = [-1, 1]$? Explain.
   (b) Does f satisfy the conclusion of the MVT on $[a, b] = [-1, 1]$? Explain.
   (c) Does f satisfy all hypotheses of the MVT on $[a, b] = [1, 2]$? Explain.
   (d) Does f satisfy the conclusion of the MVT on $[a, b] = [1, 2]$? If so, find at least one suitable value of c. How many such c's exist?

2. Let $f(x) = x^7 - x^5 - x^4 + 2x + 1$. Explain why there must exist a point c in the interval $(-1, 1)$ such that the line tangent to the graph of f at the point $(c, f(c))$ has slope 2.

FURTHER EXERCISES

3. Throughout this exercise, $f(x) = x^2$. Because f is differentiable everywhere, it satisfies all hypotheses of the MVT on any interval $[a, b]$.
   (a) What does the MVT say about f on $[a, b] = [-1, 1]$? Find all suitable values of c. How many are there?
   (b) What does the MVT say about f on $[a, b] = [1, 2]$? Find all suitable values of c. How many are there?
   (c) (A little harder) Show that for $f(x) = x^2$ and any interval $[a, b]$, the MVT's number c is the midpoint of $[a, b]$.

4. Let $g(x) = x^{2/3} - 1$. Notice that although $g(-1) = g(1) = 0$ there is no point c in the interval $(-1, 1)$ where $g'(c) = 0$. Why doesn't this contradict Rolle's Theorem?

5. The sketch below shows portions of the graph of a continuous function f.

Graph of f

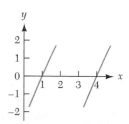

(a) Explain why f must have a root in the interval $(1, 4)$.

(b) Assume that f' is a continuous function. Explain why f' must have a root in the interval $(1, 4)$.

(c) Assume that f' is a continuous function. Explain why the graph of f must have a horizontal tangent line somewhere between $x = 1$ and $x = 4$.

(d) Why is it necessary to assume that f' is continuous in parts (b) and (c)? Sketch a function f that is continuous on the interval $(1, 4)$ for which $f'(x) \neq 0$ anywhere in this interval.

6. Let f be a continuous function that is positive on the interval $[0, 4]$ and differentiable on $(0, 4)$. Answer "yes" or "no" to each of the following questions about f. If you answer "yes" explain how you know that the statement is true. (Your answer should include references to appropriate theorems.) If you answer "no" give an example of a function that shows that the statement is not always true.

(a) Is it always possible to find a c in $[0, 4]$ such that $f(x) \leq f(c)$ for all x in $[0, 4]$?

(b) Is it always possible to find a c in $[0, 4]$ such that the instantaneous rate of change of f at c equals the average rate of change of f over the interval $[0, 4]$?

(c) Is it always possible to find a c in $[0, 4]$ such that the line tangent to the graph of f at $(c, f(c))$ is horizontal?

7. Let f be a function such that $f(0) = 0$ and $\frac{1}{2} \le f'(x) \le 1$ for all values of x.
   (a) Use the Mean Value Theorem to explain why $f(2)$ cannot be equal to 3.
   (b) Find upper and lower bounds for the value of $f(3)$.
   (c) Find upper and lower bounds for the value of $f(-1)$.

8. Use the Mean Value Theorem to show that $|\sin b - \sin a| \le |b - a|$. [**HINT:** $|\cos x| \le 1$ for all x.]

9. Let f be a function that is differentiable everywhere. Furthermore, assume that the line $y = 3x - 1$ is tangent to f at $x = 1$ and that the line tangent to f at $x = 3$ passes through the points $(0, 5)$ and $(4, 1)$.
   (a) Evaluate $f'(1)$ and $f'(3)$.
   (b) Evaluate $f(1)$ and $f(3)$.
   (c) Explain why there must be a number c in the interval $[1, 3]$ such that $f'(c) = 0$.

(d) Suppose that it is known that $f''(x) < 0$ for all x in $[1, 3]$. Explain why this implies that f has a unique maximum value over the interval $[1, 3]$.

10. Suppose that f is differentiable for all values of x, $f(-3) = -3$, $f(3) = 3$, and $|f'(x)| \le 1$ for all x in $[-3, 3]$. Show that $f(0) = 0$.

11. At 1:00 p.m. a truck driver picked up a fare card at the entrance of a tollway. At 2:15 p.m., the trucker pulled up to a toll booth 100 miles down the road. After computing the trucker's fare, the toll booth operator summoned a highway patrol officer who issued a speeding ticket to the trucker. (The speed limit on the tollway is 65 mph.)
   (a) The trucker claimed that he hadn't been speeding. Is this possible? Explain.
   (b) The fine for speeding is \$35 plus \$2 for each mph by which the speed limit is exceeded. What is the trucker's minimum fine?

4.12 Chapter Summary

This chapter offered a potpourri of applications of the derivative andantiderivative. (The sampling is a skimpy one, unfortunately—these ideas have far more uses than any textbook could possibly cover).

Differential Equations. A **differential equation** (**DE**) is any equation that involves derivatives. To **solve** a differential equation means to find a function ▶ that satisfies the DE.

Yes, a function; not a number!

Typical differential equations have many solutions. Every function of the form

$$y = Ce^x,$$

for example, solves the differential equation ▶

$$y' = y.$$

The world's most important DE.

Adding an **initial condition** to a DE creates an **initial value problem**, or **IVP**. An IVP typically has only one solution. The only solution of the IVP ▶

$$y' = y; \qquad y(0) = 1,$$

for instance, is the function

$$y = e^x.$$

The world's most important IVP, perhaps.

Sections 4.1 and 4.2 treated DEs, first briefly in theory and then in more detail, in applications related to population growth.

Linear and Quadratic Approximation. The graph of any well-behaved function is approximated—more or less closely, and near the point of tangency—by its tangent line.This means, in functional language, that f is approximated well by the linear function, called the **linear approximation** to f at x_0, whose graph is the tangent line. Section 4.3 explored the meaning and ramifications of this fact and took matters one step further, to quadratic approximation. In graphical terms, the **quadratic approximation** to f at x_0 is the function whose graph is the parabola that best fits f

at x_0. Going further still, Section 4.3 introduced **Taylor polynomials**, polynomials of any degree that approximate nonpolynomial functions.

Newton's Method. Finding roots of any but the simplest functions—e.g., linear functions—is difficult or impossible algebraically. **Newton's method** (like the bisection method of Section 4.10) offers a numerical alternative. To estimate a root of f near an initial guess x_0, Newton's method "rides the tangent line" from the f-graph at x_0 to the x-axis. The result is seldom *exactly* a root, but repeating Newton's method usually produces—remarkably quickly—a highly accurate approximation to a root. Section 4.4 treated the idea and typical applications of Newton's method—and suggested some ideas for "breaking" it.

Splines. Section 4.5 is about **splining**—fitting together pieces of simple curves (linear, quadratic, or cubic) to formlarger curves of desired shapes. Derivatives are the basic tool: Bymatching one or more derivatives of "incoming" and "outgoing" functions at each knot, curves of any desired smoothness can be constructed.

Optimization. Among the derivative's most important and useful applications is in finding **extreme values**—i.e., maximum and minimum values—of functions. As Section 4.6 illustrated, solving optimization problems usually means identifying an appropriatefunction and then using the derivative to look for and classify its **critical points**. We stressed problems posed in words. Translation of English descriptions into mathematically useful forms is important wherever mathematics is applied.

Economic Applications. Calculus ideas apply to many areas, including economic measurement and prediction. In Section 4.7 we applied derivative ideas (using the **marginal** language favored by economists) to maximize a grain trader's profit on various transactions.

Related Rates. In practical situations, several quantities vary together in time, and the rates at which these quantities vary are themselves related. The method of **related rates** is a technique for discovering such relationships.

A typical problem begins with an equation relating several quantities, all varying with time. Differentiating everything in sight—usually with respect to time— produces a new equation, one that relates the rates at which the quantities vary.

Parametric Equations and Parametric Curves. Many interesting plane curves—e.g., circles—are not graphs of functions. Parametrically defined curves offer much more flexibility. Defining x and y separately as **coordinate functions** of another variable—say t—lets us draw all sorts of interesting,beautiful, and complicated curves. If the coordinate functions are differentiable in t, then calculus ideas apply; we use them to define and calculate slopes and speeds.

Continuity and Differentiability. Sections 4.10 and 4.11 are brief excursions into the theory of calculus. They explored, respectively, the mathematical significance of continuity and differentiability. The most important results are three "value theorems": the **intermediate value theorem**, the **extreme value theorem**, and, most important of all, the **mean value theorem**. All three results are implicit in much of the machinery of calculus.

5

The Integral

5.1 Areas and Integrals

The **tangent-line problem** and the **area problem** are the two main geometric problems of calculus. ▶ As we've seen, the idea of derivative, together with the rules for computing derivatives, solves the tangent-line problem. For an impressive variety of functions, even complicated ones, it's now a routine matter to describe the tangent line at any point on the graph.

The Area Problem and the Integral

The general area problem asks how to measure the area of a plane region. For special regions—rectangles, triangles, squares, circles, trapezoids, and on on—well-known ▶ formulas do the job.

   Area problems are more challenging, and far more interesting, for regions bounded by less familiar curves, such as graphs of functions. The most important area problem of this type is to measure an area like the shaded one in the following picture.

These problems have been "big" for many centuries: The ancient Greeks worked hard and ingeniously at both. Had Greek mathematicians had our algebraic advantages, they would have gone farther than they did.

And long known—the Greeks knew them.

What's the shaded area?

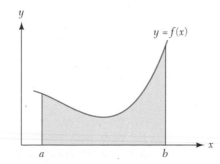

Stated carefully, the problem reads as follows.

Find the area of the region bounded above by the graph of f, below by the x-axis, on the left by the vertical line x = a, and on the right by the vertical line x = b.

The shorthand phrase is less tedious but also less precise. If in doubt, use the full form.

That's a mouthful; we usually say just "the area under the graph of f from a to b." ◄

Signed Area

In this chapter it is often convenient to consider **signed area** as opposed to area in the everyday sense. The adjective "signed" means that *all area below the x-axis counts as negative.* (In the preceding picture, all the shaded area is above the x-axis, so the question of sign doesn't arise.) We'll give more details and examples later.

A compact notation for signed area is even more convenient:

> **Definition** (**The Integral as Signed Area**) Let f be a function defined for $a \le x \le b$. Either of the equivalent expressions
> $$\int_a^b f \quad \text{or} \quad \int_a^b f(x)\, dx$$
> denotes the signed area bounded by $x = a$, $x = b$, $y = f(x)$, and the x-axis.

Observe:

In Words   For either $\int_a^b f$ or $\int_a^b f(x)\, dx$, read "the **integral** of f from a to b." The function f is the **integrand**.

Which Is Better?   Both $\int_a^b f$ and $\int_a^b f(x)\, dx$ denote the same area. The first form looks simpler; for now, we'll usually choose it. The other (dx) form has advantages that will sometimes matter later on. ◄

Differences between the two forms are akin to those between the dy/dx and f′ notations for derivative.

How to Find It?   So far, all we've done is define the symbolic expression $\int_a^b f$ to *stand for* area. The problem of *calculating* areas remains.

EXAMPLE 1   Several areas follow, labeled as integrals. Use familiar area formulas to evaluate each integral.

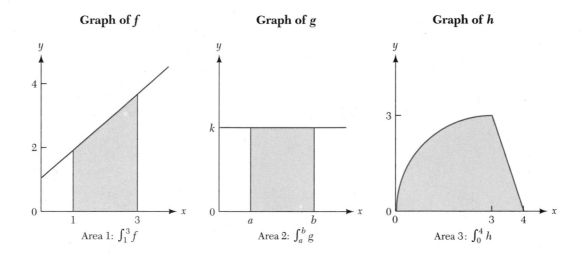

Graph of f   **Graph of g**   **Graph of h**

Area 1: $\int_1^3 f$ Area 2: $\int_a^b g$ Area 3: $\int_0^4 h$

Solution   All three areas are easy to find. Area 1 is a trapezoid, with base 2 and vertical sides 2 and 4; hence, the area ▶ is 6. Area 2 is a rectangle; by the usual formula (area = base × height), its area is $(b-a) \cdot k$. Area 3 is a quarter circle plus a triangle; a close look at these figures reveals a total area of $9\pi/4 + 3/2$, or about 8.569. In integral notation:

A trapezoid with base b and heights h_1 and h_2 has area $b(h_1 + h_2)/2$.

$$\int_1^3 f = 6; \qquad \int_a^b g = k(b-a); \qquad \int_0^4 h = \frac{9\pi}{4} + \frac{3}{2}. \qquad ■$$

Signed Area: Positive and Negative Contributions

Integrals measure *signed area*. In calculating integrals, therefore, area above the graph of f and beneath the x-axis counts as negative. Thus if an integrand f takes negative values in an interval $[a, b]$, then so may the integral $\int_a^b f$.

EXAMPLE 2   Let $f(x) = 1 - x^2$. Find (or estimate) values for the integrals $I_1 = \int_0^2 f$ and $I_2 = \int_{-2}^2 f$.

Solution   The areas in question look like this:

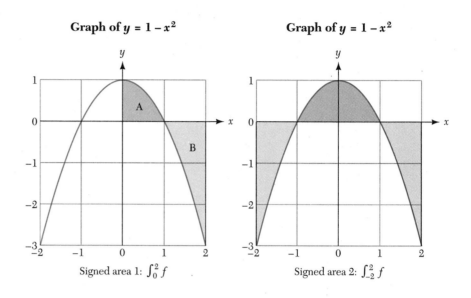

Graph of $y = 1 - x^2$

Signed area 1: $\int_0^2 f$

Graph of $y = 1 - x^2$

Signed area 2: $\int_{-2}^2 f$

Soon it will be easy.

Note that each square has area 1.

For now, computing I_1 and I_2 *exactly* is difficult. ◄ *Estimating* the integrals, on the other hand, is easy. Eyeballing general sizes ◄ suggests that A has area around 2/3; similarly, B has area around 3/2. From these guesses, we estimate (counting B as negative!) that

$$I_1 = \int_0^2 f \approx \frac{2}{3} - \frac{3}{2} = -\frac{5}{6}.$$

It is in the ballpark; the exact answer turns out to be −2/3.

Convince yourself; how is evenness of f involved?

How close is this estimate? We can't yet tell for sure, but it seems to be in the ballpark. ◄ In any event, symmetry shows ◄ that

$$I_2 = 2 \cdot I_1 \approx -\frac{5}{3}. \qquad \blacksquare$$

EXAMPLE 3   Let $g(x) = x^3$. Find or estimate $\int_{-1}^1 g(x)\,dx$ and $\int_0^1 g(x)\,dx$.

Solution   Here's the graph:

Graph of $y = x^3$

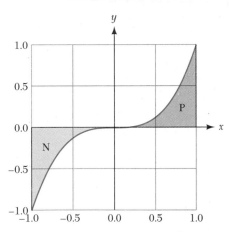

Because of the graph's symmetry ▶   the positive and negative areas P and N cancel.   *Or, equivalently, because g is odd.*
Hence

$$\int_{-1}^{1} g(x)\,dx = 0.$$

(*Any* odd integrand, over *any* interval that's symmetric about $x = 0$, behaves the same way.)

   Symmetry is no help with $\int_0^1 g$. A close look suggests that P's area is about that of one small square, or 0.25. In other words,

$$\int_0^1 g(x)\,dx \approx 0.25.$$

(We'll see soon that—and why—this estimate is actually exact.) ■

Properties of the Integral

Viewing the integral as area makes many of the integral's important properties simple and natural. For example, this picture

Adding areas: $\int_a^b f = \int_a^c f + \int_c^b f$

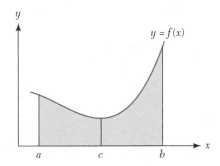

is a convincing illustration of the following simple, but often useful, property of the integral:

If $a < c < b$, then $\int_a^b f = \int_a^c f + \int_c^b f$.

Requiring f and g to be continuous ensures that all the integrals in question exist.

The next theorem collects several similar properties of the integral. ◄

Theorem 1 (New Integrals From Old) Let f and g be continuous functions on $[a, b]$; let k denote a real constant. Then

1. $\int_a^b \big(f(x) \pm g(x)\big)\, dx = \int_a^b f(x)\, dx \pm \int_a^b g(x)\, dx$.

2. $\int_a^b k f(x)\, dx = k \int_a^b f(x)\, dx$.

3. If $f(x) \le g(x)$ for all x in $[a, b]$, then
$\int_a^b f(x)\, dx \le \int_a^b g(x)\, dx$.

4. If $a < c < b$, then $\int_a^b f(x)\, dx = \int_a^c f(x)\, dx + \int_c^b f(x)\, dx$.

Here are some remarks on the theorem:

Like the Derivative The first two of the four properties are akin to properties of the derivative: Like the derivative, the integral behaves well with respect to sums and constant multiples. The following picture illustrates the **constant multiple rule** (the second identity in the theorem) for integrals.

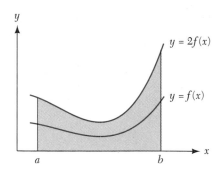

Constant multiple rule for integrals:
why $\int_a^b 2f = 2\int_a^b f$

A Useful Case The third property is often used in the case where either f or g is a constant. Here's one useful version:

Fact (Bounding an Integral) Suppose that, for some numbers m and M, the inequality $m \le f(x) \le M$ holds for all x in $[a, b]$. Then

$$m \cdot (b - a) = \int_a^b m\, dx \le \int_a^b f(x)\, dx \le \int_a^b M\, dx = M \cdot (b - a).$$

A picture makes the point best. ▶

The numbers m and M are lower and upper bounds for f on [a, b].

Bounding areas: why $m(b - a) \leq \int_a^b f \leq M(b - a)$

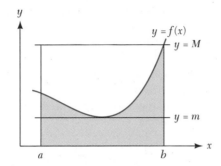

As the picture illustrates, the value of the integral (i.e., the area of the shaded region) lies between the areas of the smaller and the larger rectangles.

The next three examples illustrate uses of the theorem.

EXAMPLE 4   The integral $I = \int_1^3 f(x)\, dx$, where $f(x) = 1/x$, measures the shaded area in the following picture. Use the linear functions g_1, g_2, and g_3 to estimate I.

Three ways to estimate $\int_1^3 f(x)\, dx$

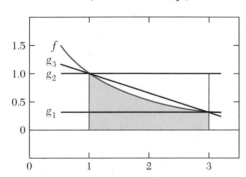

Solution   The shaded area isn't a rectangle or other simple shape, so we can't find I exactly from simple area formulas. We can, however, easily estimate I, using the third part of Theorem 1. As the picture shows, $f(x) \geq g_1(x) = 1/3$ for all x in $[1, 3]$. Therefore, by Theorem 1, ▶

The right-hand integral is easy—it corresponds to a rectangle.

$$\int_1^3 f(x)\, dx \geq \int_1^3 g_1(x)\, dx = \frac{2}{3}.$$

Similarly, $f(x) \le g_2(x) = 1$, for all x in $[1, 3]$, so

$$\int_1^3 f(x)\,dx \le \int_1^3 g_2(x)\,dx = 2.$$

This is an upper bound for I—but not a very good one, as the picture shows. Using g_3 instead of g_2 gives a better upper bound:

$$\int_1^3 f(x)\,dx \le \int_1^3 g_3(x)\,dx = \frac{4}{3}.$$

(The right-hand integral represents a trapezoidal area, so an elementary area formula applies.) Combining these results gives both lower and upper bounds for I: $2/3 \le I \le 4/3$. (By choosing the upper and lower bounding functions more cleverly, we could improve these estimates.) ∎

It is a fact—a remarkable one.

EXAMPLE 5   Using the fact ◄ that $\int_0^\pi \sin x\,dx = 2$, find $\int_0^{\pi/2} \sin x\,dx$ and $\int_0^\pi (3\sin x + 2\cos x)\,dx$.

Draw them to convince yourself!

Solution   Symmetry of the sine and cosine graphs ◄ shows that

$$\int_0^{\pi/2} \sin x\,dx = \frac{1}{2}\int_0^\pi \sin x\,dx = 1; \qquad \int_0^\pi \cos x\,dx = 0.$$

Therefore, by the sum and constant multiple rules,

$$\int_0^\pi (3\sin x + 2\cos x)\,dx = 3\int_0^\pi \sin x\,dx + 2\int_0^\pi \cos x\,dx = 6 + 0 = 6.$$

Does it?

A look at the graph should make the general size of the answer plausible. ◄

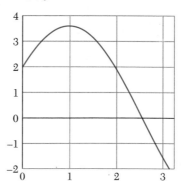

Why $\int_0^\pi (3\sin x + 2\cos x)\,dx = 6$

The "net area" does seem to be about 6. ∎

EXAMPLE 6   For any positive constant b,

$$\int_0^b 1\,dx = b \qquad \text{and} \qquad \int_0^b x\,dx = \frac{b^2}{2}.$$

Explain why. Then use the theorem to find $\int_0^b (Cx + D)\, dx$ for any constants C and D.

Solution   Graphs ▶ of the functions $f(x) = 1$ and $g(x) = x$ over the   *Draw them!* interval $[0, b]$ show that the first two integrals have the claimed values. To finish, we use the sum and constant multiple rules for integrals:

$$\int_0^b (Cx + D)\, dx = \int_0^b (Cx)\, dx + \int_0^b D\, dx$$

$$= C \int_0^b x\, dx + \int_0^b D\, dx$$

$$= C\frac{b^2}{2} + Db. \qquad ■$$

Average Value and the Integral

In the following picture, the rectangle is chosen to have the same area as the shaded region:

Equal areas: average value of a function

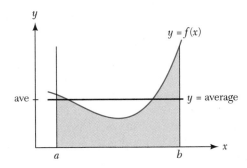

The height of that rectangle is, in a natural way, the **average value** of the function f over the interval $[a, b]$. The following definition states in analytic language what the picture shows geometrically.

> **Definition** Let f be defined on an interval $[a, b]$. The quantity
>
> $$\frac{\int_a^b f(x)\, dx}{b - a}$$
>
> is the average value of f over $[a, b]$.

The ideas of integral and average value are close cousins; understanding either helps with the other. The preceding picture summarizes this two-way relationship. The average value is the *height* of a certain rectangle, the integral is the *area* of the same rectangle.

EXAMPLE 7   The following graphs show speed functions s_A and s_B for two cars A and B.

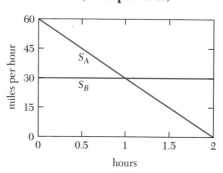

Speeds of cars A and B
(miles per hour)

What's the average value of each function over the interval $[0, 2]$? What do the results mean in "car talk?"

Convince yourself.

Solution   A look at areas shows readily ◀   that

$$\int_0^2 s_A(t)\, dt = 60 = \int_0^2 s_B(t)\, dt.$$

Thus, using the preceding definition, *both* speed functions have average value 30 over the interval $[0, 2]$.

We'll explain exactly why this is true later in this chapter.

The answers make excellent automotive sense. The area under each graph represents the total distance—60 miles—that each car covers in the 2-hour interval. ◀ Dividing the distance by the time elapsed gives an average speed of 30 miles per hour (mph) for each car.   ■

Interpreting the Integral

So far, we've interpreted the integral $\int_a^b f$ mainly in geometric terms. The integral has many other natural and important interpretations, several of which we'll see in this book. Physical motion is one important example.

Speed and Distance

If a function $f(t)$ represents the speed of a moving object at time t, then $\int_a^b f(t)\, dt$ represents the distance traveled by the object over the interval $a \le t \le b$. Why is this true?

As it is for car B in Example 7.

If the speed function $f(t)$ happens to be constant, ◀   say, k—then

$$\int_a^b f(t)\, dt = k \cdot (b - a) = \text{speed} \times \text{time} = \text{distance},$$

as claimed. Remarkably, integrating the speed function gives distance traveled even if $f(t)$ *isn't* constant. Example 7 should make this result believable; in the following sections we'll see exactly why it's true.

Velocity and Net Distance

Speed, by definition, is always positive. Velocity can be either positive or negative, depending on the direction of motion. As a result, integration of a velocity function $v(t)$ gives slightly different information from integration of a speed function. If $v(t)$ tells the velocity of an object moving along a line at time t, then $\int_a^b f(t)\,dt$ represents the *net* distance traveled over the interval $a \le t \le b$. In particular, net distance, like velocity, can be either positive or negative.

Integrating from Right to Left: A Technicality

The English description of $\int_a^b f(x)\,dx$—"the integral of f from $x = a$ to $x = b$"— implies a *direction*: x starts at a and ends at b. ▶ It's usually natural to think of x as moving from left to right. Up to now, we've done just that: In every integral $\int_a^b f$ we've treated so far, we've assumed or stated that $a \le b$. For example, we've discussed

The variable name x isn't sacred; t is another popular choice.

$$\int_0^\pi \sin x\,dx, \quad \text{but not} \quad \int_\pi^0 \sin x\,dx \quad \text{or} \quad \int_\pi^0 (3\sin x + 2\cos x)\,dx.$$

Sometimes the latter sort of "right-to-left" integrals *do* arise. ▶ The following convention, based on signs, handles all the possibilities with a minimum of fuss:

We'll see some in the next section.

$$\int_a^b f(x)\,dx = -\int_b^a f(x)\,dx. \tag{5.1}$$

The convention says (for instance) that, since $\int_0^1 x^3\,dx = 0.25$, ▶

See Example 3.

$$\int_1^0 x^3\,dx = -\int_0^1 x^3\,dx = -0.25.$$

Why does Equation 5.1 make sense? The speed–distance context offers one answer:

If $f(t)$ represents speed at time t, and $a < b$, then $\int_a^b f(t)\,dt$ gives the (positive) distance traveled from time a to time b.

It's physically reasonable, therefore, to consider as negative the distance covered over the reverse interval. ▶

Even if Equation 5.1 seems arbitrary now, it will prove useful in later work.

Integrals and Areas: The Story So Far and a Look Ahead

In this section we defined the integral $\int_a^b f$ as a certain area. Then we observed some properties and uses of the integral, arguing mainly from geometric intuition. What's missing so far is any really practical or effective method of *measuring* areas. In the simplest cases elementary area formulas (for squares, rectangles, and so on) are some help. For all *but* the simplest functions, however, such formulas are useless. In the next three sections we will offer two different practical methods of calculating integrals. In Sections 5.2 and 5.3, we'll relate the new idea of the integral to the older (and better understood) idea of the derivative. This key connection between derivative and integral, known as the **fundamental theorem of calculus**, is among the most important results of mathematics. In Section 5.4 we'll give another definition of the integral, this time as a limit of approximating sums. Logically, the situation resembles that for the derivative, which we first defined informally as the slope of a line and later, more formally, as a limit of differences.

Must Area Exist?   There seems to be little question that areas under simple graphs— lines, parabolas, sinusoids, and so on—exist. The situation is less clear for "not-nice" (e.g., discontinuous) functions, whose graphs may be extremely ragged. Consider, for instance, this graph:

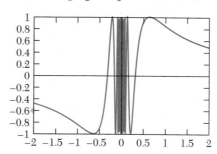

The graph of $y = \sin(1/x)$

Can such a nasty curve sensibly be said to bound area?

The answer turns out (for this curve, anyway) to be yes. Defining area (let alone computing it) in such unpromising circumstances, however, requires special care. In Section 5.4 we'll redefine the integral, using limits, partly in order to handle cases like this one. In practice, there's no need to worry unduly; the vast majority of standard calculus functions behave nicely enough to accord well with our everyday intuition about area.

BASICS

1. Let g be the function shown graphically below. When asked to estimate $\int_1^2 g(x)\,dx$, a group of calculus students submitted the following answers: -4, 4, 45, and 450. Only one of these responses is reasonable; the others are "obviously" incorrect. Which is the reasonable one? Why?

Graph of g

2. Let f be the function shown graphically below.

Graph of f

(a) Which of the following is the best estimate of $\int_0^{40} f(x)\,dx$: $-65, -30, 0, 35, 60$? Justify your answer.
(b) Show that the average value of f over the interval $[10, 30]$ is between -2 and -1.5.

3. The graph of a function f is shown below.

Graph of f

(a) Which of the following is the best estimate of $\int_1^6 f(x)\,dx$: $-24, 9, 20, 38$? Justify your answer.
(b) Find positive integers A and B such that $A \leq \int_3^7 f(x)\,dx \leq B$. Explain how you know that the values of A and B you chose have the desired properties.
(c) $\int_6^8 f(x)\,dx \approx 4$. Does this approximation overestimate or underestimate the exact value of the integral? Justify your answer.
(d) Estimate the average value of f over the interval $[0, 2]$.
(e) Is your estimate in part (d) too large or too small? Justify your answer.

4. For each function below, use a graph of f to evaluate
$$\int_0^2 f(x)\,dx, \quad \int_1^4 f(x)\,dx, \quad \int_{-5}^{-1} f(x)\,dx, \text{ and } \int_{-2}^{3} f(x)\,dx.$$

(a) $f(x) = 3x$ (c) $f(x) = 5 - 2x$
(b) $f(x) = 2x + 5$

5. The graph of a function f (shown below) consists of two straight lines and two one-quarter circles.

Graph of f

Evaluate each of the following integrals.

(b) $\int_0^2 f(x)\,dx$ (f) $\int_0^{15} f(x)\,dx$

(b) $\int_2^5 f(x)\,dx$ (g) $\int_0^{15} |f(x)|\,dx$

(c) $\int_0^5 f(x)\,dx$ (h) $\int_{15}^9 f(x)\,dx$

(d) $\int_5^9 f(x)\,dx$ (i) $\int_{12}^{15} f(x)\,dx$

(e) $\int_4^4 f(x)\,dx$ (j) $\int_9^{12} f(x)\,dx$

6. Suppose that $\int_{-2}^5 f(x)\,dx = 18$, $\int_{-2}^5 g(x)\,dx = 5$, and $\int_{-2}^5 h(x)\,dx = -11$. Evaluate each of the following integrals.

(a) $\int_{-2}^5 \big(f(x) + g(x)\big)\,dx$

(b) $\int_{-2}^5 \big(f(x) + h(x)\big)\,dx$

(c) $\int_{-2}^5 \big(f(x) + g(x) - h(x)\big)\,dx$

(d) $\int_{-2}^5 3f(x)\,dx$

(e) $\int_{-2}^5 -4g(x)\,dx$

(f) $\int_5^{-2} f(x)\,dx$

(g) $\int_{-2}^5 \big(h(x) + 1\big)\,dx$

(h) $\int_0^7 g(x-2)\,dx$

FURTHER EXERCISES

7. Four students disagree on the value of the integral $\int_0^{\pi/2} \cos^8 x\, dx$. Jack argues for $\pi \approx 3.14$, Joan for $35\pi/256 \approx 0.43$, Ed for $2\pi/9 - 1 \approx -0.30$, and Lesley for $\pi/4 \approx 0.79$. Use the graph below to determine who is right. (One *is* right!) Explain how you know that the other values are incorrect.   [**HINT:** $\left(\cos(1/2)\right)^8 < 0.4.$]

Graph of $y = \cos^8 x$

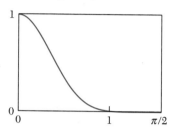

8. The graph of a function h is shown below.

Graph of h

List, from smallest to largest: (i) $h'(5)$; (ii) the average value of h over the interval $[0, 10]$; (iii) the average rate of change of h over the interval $[0, 10]$; (iv) $\int_0^{10} h(x)\, dx$; (v) $\int_0^5 h(x)\, dx$; (vi) $\int_6^{10} h(x)\, dx$; (vii) $\int_5^5 h(x)\, dx$.

9. Estimate each of the quantities (i)–(vi) listed in Exercise 8.

10. Let h be the function shown graphically in Exercise 8 and suppose that $h(t)$ represents the *eastward* velocity (in meters/second) of an object moving along an east-west axis at time t seconds. Estimate each of the following quantities (use appropriate units!) and interpret in the language of velocity, distance, etc.

   (a) $\int_0^{10} h(t)\, dt$

   (b) $\dfrac{1}{10}\int_0^{10} h(t)\, dt$

   (c) $\dfrac{1}{10}\left(h(10) - h(0)\right)$

   (d) $|h(8)|$

   (e) $\int_0^{10} |h(t)|\, dt$

   (f) $\dfrac{1}{10}\int_0^{10} |h(t)|\, dt$

11. Suppose that a car travels on an east-west road with eastward velocity $v(t) = 60 - 20t$ mph at time t hours.

   (a) Evaluate $\int_0^4 v(t)\, dt$. Interpret the answer in car talk.

   (b) Find the average value of $v(t)$ on the interval $[0, 4]$. Show your result graphically.

   (c) Let $s(t)$ be the car's speed at time t. Evaluate $\int_0^4 s(t)\, dt$ and interpret the answer in car talk.

   (d) What is the car's average speed between $t = 0$ and $t = 4$?

12. Sketch the graph of a continuous function f such that $\int_{-2}^3 f(t)\, dt = -10$ and

   (a) $\int_{-2}^3 |f(t)|\, dt = 10.$ (b) $\int_{-2}^3 |f(t)|\, dt \neq 10.$

13. For each function below, use a graph of f to evaluate $\int_0^2 f(x)\, dx$, $\int_1^4 f(x)\, dx$, $\int_{-5}^{-1} f(x)\, dx$, and $\int_{-2}^3 f(x)\, dx$. (a is a positive constant.)

   (a) $f(x) = 1$ (c) $f(x) = -a$

   (b) $f(x) = a$

14. Let f be the function shown graphically in Exercise 5.

   (a) Show that $\int_2^4 f(x)\, dx > 5$.   [**HINT:** Draw the trapezoid with vertices at $(2, 0)$, $\left(2, f(2)\right)$, $(4, 0)$, and $\left(4, f(4)\right)$.]

   (b) Show that $\int_6^9 f(x)\, dx < -9$.

   (c) Is $\int_0^6 f(x)\, dx < 11$?

   (d) Is $\int_2^7 f(x)\, dx$ positive or negative?

15. (a) Evaluate $\int_{-3}^3 (x + 2)\, dx$.

   (b) Evaluate $\int_{-3}^3 |x + 2|\, dx$.

   (c) Evaluate $\int_{-3}^3 \left(|x| + 2\right) dx$.

   (d) Evaluate $\int_{-3}^3 \left(2 - |x|\right) dx$.

16. Evaluate $\int_0^2 f(x)\, dx$, where
$$f(x) = \begin{cases} 1 + x, & \text{if } 0 \leq x \leq 1 \\ 2 - x, & \text{if } 1 < x \leq 2. \end{cases}$$
[**HINT:** Sketch a graph of the integrand.]

17. Evaluate $\int_0^1 \sqrt{1 - (x - 1)^2}\, dx$.   [**HINT:** Sketch a graph of the integrand.]

18. Evaluate $\int_1^3 \left(6 - \sqrt{4 - (x - 3)^2}\, dx\right) dx$.

19. Evaluate $\int_0^3 f(x)\,dx$, where

$$f(x) = \begin{cases} 1 - \sqrt{1 - x^2}, & \text{if } 0 \le x \le 1 \\ 1 + \sqrt{4 - (x-3)^2}, & \text{if } 1 < x \le 3. \end{cases}$$

20. Explain why

$$\int_1^2 x^3\,dx = \int_3^4 (x-2)^3\,dx = \int_{-3}^{-2} (x+4)^3\,dx.$$

21. Suppose that (i) $\int_0^2 f(x)\,dx = 2$; (ii) $\int_1^2 f(x)\,dx = -1$; (iii) $\int_2^4 f(x)\,dx = 7$.

    (a) Evaluate $\int_1^4 f(x)\,dx$.

    (b) Evaluate $\int_0^4 3f(x)\,dx$.

    (c) Evaluate $\int_0^1 f(x)\,dx$.

    (d) Evaluate $\int_0^1 f(x+1)\,dx$.

    (e) Evaluate $\int_0^1 \big(f(x)+1\big)\,dx$.

    (f) Evaluate $\int_2^4 f(x-2)\,dx$.

    (g) Evaluate $\int_2^4 \big(f(x)-2\big)\,dx$.

    (h) Explain why f must be negative somewhere in the interval $[1, 2]$.

    (i) Explain why $f(x) \ge 3$ somewhere in the interval $[2, 4]$.

    (j) Draw the graph of a function f with properties (i)–(iii).

22. Suppose that f is a function such that $\int_0^3 f(x)\,dx = -1$.

    (a) Suppose that f is an *even* function. Explain why $\int_{-3}^3 f(x)\,dx = -2$.

    (b) If f is an *odd* function, what is the value of $\int_{-3}^3 f(x)\,dx$? Explain.

23. Evaluate each of the following integrals using graphical arguments. [**HINT:** See the previous exercise.]

    (a) $\int_0^\pi \cos x\,dx$

    (b) $\int_{\pi/2}^{3\pi/2} \sin x\,dx$

    (c) $\int_{-2}^2 (7x^5 + 3)\,dx$

    (d) $\int_{-1}^1 (4x^3 - 2x)\,dx$

24. Evaluate each of the following integrals using graphical arguments and the fact that $\int_0^\pi \sin x\,dx = 2$.

    (a) $\int_0^{2\pi} \sin x\,dx$

    (b) $\int_0^{2\pi} |\sin x|\,dx$

    (c) $\int_0^\pi (1 + \sin x)\,dx$

    (d) $\int_0^{\pi/2} \sin x\,dx$

    (e) $\int_{-\pi/2}^{\pi/2} \cos x\,dx$

    (f) $\int_0^{\pi/2} (x + \cos x)\,dx$

    (g) $\int_0^{100\pi} \sin x\,dx$

    (h) $\int_0^{100\pi} |\sin x|\,dx$

    (i) $\int_0^{100\pi} \cos x\,dx$

25. Explain why $\displaystyle\int_1^3 \frac{1-x}{x^2}\,dx < \int_1^2 \frac{1-x}{x^2}\,dx$.

26. Is $\int_0^2 (2x^3 - x^2)\,dx$ greater than $\int_0^2 (3x^2 + x)\,dx$? Justify your answer.

27. (a) Explain why $\displaystyle\int_0^{2\pi} (1 + \cos x)\,dx \ge 0$.

    (b) Show that $\displaystyle\int_0^{2\pi} (1 + \cos x)\,dx > 0$. [**HINT:** Consider the intervals $[0, \pi]$ and $[\pi, 2\pi]$ separately.]

28. Let $f(x) = \frac{1}{2}x + \cos x$. Show that $1.3 \le \int_0^3 f(x)\,dx \le 3.5$. [**HINT:** Find the maximum and minimum values of f over the interval $[0, 3]$.]

29. Show that $4.5 \le \displaystyle\int_1^3 e^x \sin x\,dx \le 15$.

30. Show that $0 \le \displaystyle\int_0^\pi x \sin x\,dx \le \pi^2/2$. [**HINT:** $x \sin x \le x$ when $0 \le x \le \pi$.]

31. Show that $\pi/6 \le \displaystyle\int_{\pi/6}^{\pi/2} \sin x\,dx \le \pi/3$.

32. Show that $-\pi/3 \le \displaystyle\int_{2\pi/3}^{\pi} \cos x\,dx \le -\pi/6$.

33. (a) Show that $\pi(1 + 1/e) \le \displaystyle\int_0^{2\pi} e^{\sin x}\,dx \le \pi(1 + e)$. [**HINT:** When $0 \le x \le \pi$, $0 \le \sin x \le 1$. When $\pi \le x \le 2\pi$, $-1 \le \sin x \le 0$.]

    (b) Use the result in part (a) to find upper and lower bounds for $\int_0^{50\pi} e^{\sin x}\,dx$.

34. (a) Show that $\pi/6 \le \displaystyle\int_0^{\pi/2} \cos x\,dx \le 5\pi/12$. [**HINT:** $1/2 \le \cos x \le 1$ when $0 \le x \le \pi/3$.]

    (b) Using ideas similar to those in part (a), find upper and lower bounds on the value of $\int_0^{\sqrt{\pi/2}} \cos(x^2)\,dx$.

35. Show that $0.4 \le \displaystyle\int_0^1 \sin(e^x)\,dx \le 1$.

36. Sketch the graph of a function f with the property that

$$\left| \int_1^5 f(x)\,dx \right| < \int_1^5 |f(x)|\,dx.$$

Explain why the function f you sketched has this property.

37. Show that $\dfrac{\pi}{2} < \displaystyle\int_0^\pi \cos(\sin x)\,dx \le \pi$. [**HINT:** $\cos 1 > \cos(\pi/3) = 1/2$.]

38. (a) Let $y = f(x)$ be the line through the points $(0, \arctan 0)$ and $(1, \arctan 1)$. Explain why $f(x) \leq \arctan x$ when $0 \leq x \leq 1$. [**HINT:** Consider the concavity of the arctangent function.]

   (b) Find an equation of the line tangent to the curve $y = \arctan x$ at $x = 0$.

   (c) Show that $\dfrac{\pi}{8} \leq \displaystyle\int_0^1 \arctan x \, dx \leq \dfrac{1}{2}$.

39. (a) Prove that $2x/\pi \leq \sin x \leq x$ when $0 \leq x \leq 1$.

   (b) Use part (a) to show that $\pi/16 \leq \displaystyle\int_0^{\pi/4} \sin x \, dx \leq \pi^2/32$.

40. (a) Show that $1 + x \leq e^x \leq 1 + 3x$ when $0 \leq x \leq 1$.

   (b) Use part (a) to find upper and lower bounds on $\displaystyle\int_0^1 e^x \, dx$.

41. (a) Suppose that $f(x) \geq 0$ when $0 \leq x \leq 1$. Show that $\displaystyle\int_0^1 x^k f(x) \, dx \leq \int_0^1 f(x) \, dx$ for all integers $k \geq 0$.

   (b) Does the result in part (a) remain valid if the assumption "$f(x) \geq 0$ when $0 \leq x \leq 1$" is not made? Explain.

42. Show that $\displaystyle\int_0^{\pi/4} \cos(2x) \, dx \neq \int_0^{\pi/2} \cos x \, dx$. Which integral is larger? Justify your answer.

43. (a) Show that $\displaystyle\int_0^{\pi/2} \sin^2 x \, dx = \int_0^{\pi/2} \cos^2 x \, dx$.

   [**HINT:** $\cos x = -\sin(x - \pi/2) = \sin(\pi/2 - x)$.]

   (b) Show that $\displaystyle\int_0^{\pi/2} \sin^2 x \, dx + \int_0^{\pi/2} \cos^2 x \, dx = \pi/2$.

   (c) Show that $\displaystyle\int_0^{\pi/2} \sin^2 x \, dx = \pi/4$.

44. (a) Show that $\displaystyle\int_0^{\pi/2} \sqrt{1 + \cos(2x)} \, dx = \sqrt{2} \int_0^{\pi/2} \cos x \, dx$.

   [**HINT:** $2\cos^2 x = 1 + \cos(2x)$ for all x.]

   (b) Does $\displaystyle\int_{\pi/2}^{\pi} \sqrt{1 + \cos(2x)} \, dx = \sqrt{2} \int_{\pi/2}^{\pi} \cos x \, dx$? Justify your answer.

45. Suppose that f is an odd function and continuous everywhere. Explain why the average value of f over any interval $[-a, a]$ is zero.

46. Suppose that f is an even function and continuous everywhere. Explain why the average value of f over the interval $[-a, a]$ is equal to the average value of f over the interval $[0, a]$.

47. Suppose that f is a continuous function and that $-2 \leq f(x) \leq 5$ if $1 \leq x \leq 3$. Show that the average value of f over $[1, 3]$ is between -2 and 5.

48. Suppose that f is a continuous function. If the average value of f over the interval $[0, 1]$ is 2 and the average value of f over the interval $[1, 3]$ is 4, what is the average value of f over the interval $[0, 3]$?

49. Suppose that f is a continuous function. If the average value of f over the interval $[-3, 1]$ is 2 and the average value of f over the interval $[-3, 7]$ is 5, what is the average value of f over the interval $[1, 7]$?

50. Show that the expression $\displaystyle\int_a^b \big(f(x) - c\big)^2 \, dx$ assumes its minimum value when c is the average value of f over the interval $[a, b]$.

51. Suppose that f is continuous on $[0, 1]$. Explain why $\displaystyle\int_0^1 f(x) \, dx = \int_0^1 f(1 - x) \, dx$. [**HINT:** How is the graph of $g(x) = f(1 - x)$ related to the graph of f?]

52. Suppose that f is continuous on $[a, b]$. Show that $\left| \displaystyle\int_a^b f(x) \, dx \right| \leq \int_a^b |f(x)| \, dx$. [**HINT:** $-|f(x)| \leq f(x) \leq |f(x)|$.]

53. Suppose that f and g are continuous on $[a, b]$ and $\displaystyle\int_a^b f(x) \, dx \leq \int_a^b g(x) \, dx$.

   (a) Must $f(x) \leq g(x)$ for every x in $[a, b]$? If so, explain why. If not, give a counterexample.

   (b) Must there be a number c such that $a \leq c \leq b$ and $f(c) \leq g(c)$? If so, explain why. If not, give a counterexample.

54. Archimedes (ca. 250 B.C.E.) proved that the area under a parabolic arch is $2bh/3$, where b is the width of the base of the arch and h is the height. [**NOTE:** Archimedes actually proved a more general theorem about the area of any region cut off from a parabola by a line.]

   (a) Use Archimedes's result to show that $\displaystyle\int_{-a}^{a} x^2 \, dx = 2a^3/3$. [**HINT:** Draw the curve $y = x^2$ and the line $y = a^2$.]

   (b) Use part (a) and the fact that x^2 is an even function to show that $\displaystyle\int_0^a x^2 \, dx = a^3/3$. [**HINT:** $\displaystyle\int_{-a}^{a} x^2 \, dx = \int_{-a}^{0} x^2 \, dx + \int_0^a x^2 \, dx$.]

   (c) Use part (b) to show that $\displaystyle\int_a^b x^2 \, dx = (b^3 - a^3)/3$. [**HINT:** $\displaystyle\int_a^b x^2 \, dx = \int_0^b x^2 \, dx - \int_0^a x^2 \, dx$.]

55. Let $f(x) = 2x + 3$.

   (a) Sketch a graph of f in the viewing window $[1, 2] \times [5, 9]$.

   (b) Evaluate $\displaystyle\int_1^2 f(x) \, dx$. Shade the region of the part (a) graph represented by this integral.

   (c) Show that $f^{-1}(x) = (x - 3)/2$.

   (d) Evaluate $\displaystyle\int_5^9 f^{-1}(x) \, dx$ and shade the region of the part (a) graph represented by this integral.

   (e) Verify that

$$\int_1^2 f(x) \, dx = 2f(2) - f(1) - \int_{f(1)}^{f(2)} f^{-1}(x) \, dx.$$

56. Suppose that f is a continuous function and invertible on $[a, b]$. Then it can be shown that

$$\int_a^b f(x) \, dx = bf(b) - af(a) - \int_{f(a)}^{f(b)} f^{-1}(x) \, dx.$$

(a) Give a graphical proof of this result when $0 < a < b$, and $0 < f(a) < f(b)$. [**HINT:** Draw a sketch of f in the viewing window $[0, b] \times [0, f(b)]$.]

(b) Show that $\displaystyle\int_1^e \ln x \, dx = e - \int_0^1 e^x \, dx.$

(c) Show that $\displaystyle\int_a^b \sqrt{x} \, dx = b^{3/2} - a^{3/2} - \int_{\sqrt{a}}^{\sqrt{b}} x^2 \, dx.$

57. Use the results stated in Exercises 54 and 56 to show that $\int_0^a \sqrt{x} \, dx = 2a^{3/2}/3$.

5.2 The Area Function

In Section 5.1 we defined the integral $\int_a^b f$ to be the signed area of the region from $x = a$ to $x = b$, bounded above or below by the graph of f. For given f, a, and b, this area is a certain fixed *number* (positive, negative, or zero!). In this section we define and study a *function* based on the signed area bounded by the graph of f. This area function, which we'll denote by A_f, is "built" from f, using the integral.

We've played the new-function-from-old game before. In the most important case, we built the derivative function f' from an "original" function f, and then studied what each tells about the other. Here we repeat the process, this time with f and A_f. The relationship between f and A_f will prove remarkably similar to that between f' and f.

Defining the Area Function

First comes the formal definition:

> **Definition** Let f be a function and a any point of its domain. For any input x, the **area function** A_f is defined by the rule
>
> $$A_f(x) = \int_a^x f(t) \, dt = \text{signed area defined by } f, \text{ from } a \text{ to } x.$$

A generic picture illustrates the idea:

The area function: $A_f(x)$ = the shaded area

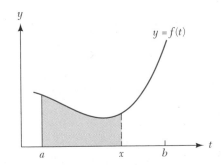

The idea is best understood through pictorial examples. First, three brief remarks.

Variable Names: Why So Many?   Why did we use another variable, t, in the defining expression $A_f(x) = \int_a^x f(t)\,dt$? We did it because the letter x was already taken; we used it to denote the right-hand boundary of the defining region. We introduced t to avoid using x in two different ways.

The Domain of A_f   For which inputs x does the rule $A_f(x) = \int_a^x f$ make sense? Barring bad behavior (e.g., discontinuities) of f, A_f has the same domain as f itself. In particular, $A_f(x)$ makes sense even if x is to the left of a; in this case we use the sign convention ◄ $\int_a^x f = -\int_x^a f$.

We discussed this convention in the preceding section.

The Choice of a   The role of a is to fix one edge (the left edge, in the preceding picture) of the region whose area gives $A_f(x)$. The other edge varies freely. As the definition says, any a in the domain of f is a legal choice. We'll soon see that different choices of a give different (but only slightly different) functions A_f.

EXAMPLE 1   Let $f(x) = 3$ and $a = 0$. Describe the area function $A_f(x) = \int_0^x f(t)\,dt$ for positive inputs x. Does A_f have a simple formula?

A rectangle with height 3 and base x has area 3x.

For now we ignore negative inputs x; we'll return to them in the next example.

Solution   Because $f(x) = 3$, $A_f(x) = 3x$. The first of the following pictures shows why; ◄ in the second, graphs of f and A_f appear together. ◄

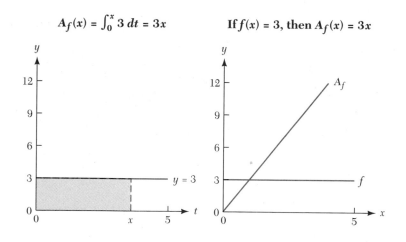

Notice too that $A_f(x) = 3x$ is an *antiderivative* of $f(x) = 3$. This result is no accident, as we'll see. ◄   ∎

We'll return to this important fact in the next example.

EXAMPLE 2   As before, let $f(x) = 3$ and $a = 0$. Discuss the area function $A_f(x) = \int_0^x f(t)\,dt$ for *negative* values of x. Does the formula $A_f(x) = 3x$ still hold?

Solution   Yes—$A_f(x) = 3x$ for *all* values of x. The key step is a careful look at signs. Here are the pictures:

$A_f(x)$ for negative x

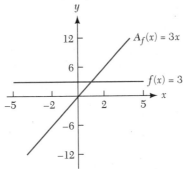

If $f(x) = 3$, then $A_f(x) = 3x$

To make sense of them, recall from the preceding section how we handle right-to-left integrals. Since $x < 0$,

$$\int_0^x f = -\int_x^0 f = -\,(\text{shaded area})\,.$$

Since $x < 0$ and the rectangle has height 3, $3x$ is, just as claimed, *minus* the shaded area. Combining this example with the preceding one leads to this conclusion:

If $f(x) = 3$ and $a = 0$, then $A_f(x) = \int_0^x f = 3x$ for all x. ■

A Happy Moral.   What $\int_a^x f$ means geometrically is clear if $x \geq a$. ▶ If $x < a$, the situation looks trickier; the preceding example was all about this possibility. Working Example 2 took some care with plus and minus signs. When the dust settled, though, the happiest possible outcome was revealed:

The same formula for $A_f(x)$ works for all x, regardless of whether $x < a$ or $x > a$.

This is evidence that our signed area conventions are not just convenient, but "right."

The same principle holds for any well behaved function f. Consequently, we'll seldom worry about (or even mention) the issue of whether $x < a$ or $x \geq a$.

See the first picture in this section if you've forgotten.

EXAMPLE 3   Consider the linear function $f(x) = x$ and the area function $A_f(x) = \int_0^x t\,dt$. What's A_f now?

Solution   As the left-hand graph that follows shows, the region defining $A_f(x)$ for a positive input x is a triangle, with base x and height x. ▶ The formula for A_f is therefore

$$A_f(x) = \frac{\text{base} \times \text{height}}{2} = \frac{x^2}{2}.$$

Graphs of $f(x)$ and $A_f(x)$ appear together at the right.

We skip the case of negative x, for reasons outlined earlier.

$$A_f(x) = \int_0^x t \, dt = x^2/2 \qquad \textbf{If } f(x) = x, \textbf{ then } A_f(x) = x^2/2$$

This is still no accident. Stay tuned.

Again, A_f turns out to be an antiderivative of f. ◄   ■

EXAMPLE 4   **(Another f, same a)**   Consider the new linear function $f(x) = 2 - x$. Find the new area function $A_f(x) = \int_0^x f$. Is A_f again an antiderivative of f?

Notice the different shadings above and below the t-axis.

Solution   The usual area picture applies; area below the x-axis counts as negative. ◄

$$A_f(x) = \int_0^x (2 - t) \, dt \qquad \begin{array}{c} \textbf{If } f(x) = 2 - x, \\ \textbf{then } A_f(x) = 2x - x^2/2 \end{array}$$

With a little care, one can use triangles (or trapezoids, if $x < 2$) to find the net shaded area. On the left-hand of the two preceding graphs, the upper triangle has area 2 and the lower triangle has area $(x - 2)^2/2$. Thus, the net area for given x

Check the algebra.

is ◄

$$A_f(x) = 2 - \frac{(x - 2)^2}{2} = 2x - \frac{x^2}{2}.$$

The exercises at the end of this section pursue this point further.

[**NOTE:** The picture shows a value of x greater than 2. If $0 < x < 2$, the picture is different but the resulting formula is the same.] ◄   The graphs of both f and A_f appear at the right.

Another, perhaps easier, approach to the problem is to use additive properties of the integral ▶ and ideas from earlier examples:

As outlined in Section 5.1.

$$A_f(x) = \int_0^x (2 - t)\, dt = \int_0^x 2\, dt - \int_0^x t\, dt = 2x - \frac{x^2}{2}.$$

Either way, the point is the same:

In all examples so far, the area function A_f is an antiderivative of f. ■

E X A M P L E 5   (**Same f, another a**)   Consider the same linear function, $f(x) = 2 - x$, this time setting $a = 1$ so that $A_f(x) = \int_1^x f$. What's the area function now?

S o l u t i o n   The only difference from the previous example concerns the lower endpoint of integration. Thus, the new function A_f differs from the old one only by the *area of the trapezoidal region* from $x = 0$ to $x = 1$. ▶ In symbols,

This area is 3/2—convince yourself.

$$A_f(x) = \int_1^x f(t)\, dt = \int_0^x f(t)\, dt - \int_0^1 f(t)\, dt$$

$$= 2x - \frac{x^2}{2} - \frac{3}{2}.$$

The graphs support this result:

$$A_f(x) = \int_1^x (2 - t)\, dt$$

If $f(x) = 2 - x$ and $a = 1$, then $A_f(x) = 2x - x^2/2 - 3/2$

Yet again, A_f is an antiderivative of f; ▶ this time, it's the one for which $A_f(1) = 0$. ■

Two antiderivatives of the same f can differ only by a constant, as they do here.

In each of the preceding examples we used an elementary area formula to find an explicit algebraic expression for the area function. In the next example, no such simple formula is available.

E X A M P L E 6   Let $f(x) = 1/x$ and $A_f(x) = \int_1^x f(t)\, dt$. Discuss the area function A_f. Does it look familiar?

S o l u t i o n   The graph that follows on the left shows $f(x)$; the shaded area represents $A_f(3)$.

Graph of $f(x) = 1/x$

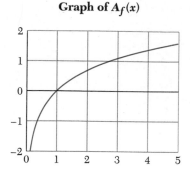

Graph of $A_f(x)$

Values of A_f can be estimated by use of the grid. The shaded area, for example, seems to be a little more than one unit, so $A_f(3)$ is a little more than 1. By plotting such estimates we can produce a rough graph of A_f, such as the one at upper right. ◄

The graph of A_f should look familiar; in fact, it shows the natural logarithm function. We'll explain this striking result in the next section. ■

Note that $A_f(x) < 0$ for $x < 1$. Do you see why?

Properties of A_f : What the Examples Show

In each of the preceding examples, direct calculation showed the area function A_f to be an antiderivative of f. This was no accident:

> **T h e o r e m 2** (**The Fundamental Theorem of Calculus, Informal Version**)   For any well-behaved function f and any base point a, A_f is an antiderivative of f.

In the next section we'll discuss and prove this theorem, which is one of the most important in mathematics. In the rest of this section, we continue to amass evidence for it.

To do so, we list various properties of A_f, each following directly from the definition. Added to the earlier concrete examples, these properties build a strong case for the fundamental theorem of calculus.

Properties of A_f

Let f be a continuous function, a a point of its domain, and $A_f(x) = \int_a^x f(t)\,dt$. Then:

- $A_f(a) = 0$.
- Where f is positive, A_f is increasing.
- Where f is negative, A_f is decreasing.
- Where f is zero, A_f has a stationary point.
- Where f is increasing, A_f is concave up.
- Where f is decreasing, A_f is concave down.

Use, for example, the graphs in Example 5 to convince yourself.

Examples—especially in graphical form—are the best evidence for each claim. ◄
Each claim restates or sharpens the general principle of the fundamental theorem.

BASICS

1. Let $f(x) = x$ and $A_f(x) = \int_0^x f(t)\,dt$. It was shown in the text that $A_f(x) = x^2/2$ if $x > 0$. Show that this formula is correct even if $x < 0$.

2. Let $F(x) = \int_0^x f(t)\,dt$, $G(x) = \int_1^x f(t)\,dt$, and $H(x) = \int_{-2}^x f(t)\,dt$.

   (a) Suppose that $f(x) = 1$. Find symbolic expressions for the functions F, G, and H. Are F, G, and H antiderivatives of f?

   (b) Suppose that $f(x) = 3x$. Find symbolic expressions for the functions F, G, and H. Are F, G, and H antiderivatives of f?

   (c) Suppose that $f(x) = 3x + 1$. Find symbolic expressions for the functions F, G, and H. Are F, G, and H antiderivatives of f?   [**HINT**: Use your answers to parts (a) and (b).]

3. Let $F(x) = \int_0^x f(t)\,dt$, $G(x) = \int_2^x f(t)\,dt$, and $H(x) = \int_{-1}^x f(t)\,dt$.

   (a) Suppose that $f(x) = a$, a positive constant. Find symbolic expressions for the functions F, G, and H. Are F, G, and H antiderivatives of f?

   (b) Suppose that $f(x) = bx$, where b is a positive constant. Find symbolic expressions for the functions F, G, and H. Are F, G, and H antiderivatives of f?

   (c) Suppose that $f(x) = a + bx$, where a and b are positive constants. Find symbolic expressions for the functions F, G, and H. Are F, G, and H antiderivatives of f?

4. Let a be a constant, $f(x) = 2 - x$, and $A_f(x) = \int_a^x f(t)\,dt$.

   (a) Evaluate $A_f(a)$.

   (b) Find a symbolic expression for $A_f(x)$.
   [**HINT**: $\int_a^x (2 - t)\,dt = \int_0^x (2 - t)\,dt - \int_0^a (2 - t)\,dt$.]

   (c) Show that A_f is an antiderivative of f.

   (d) On the same axes, plot f and A_f for $a = 0$, $a = 2$, and $a = -1$.

5. Let $F(x) = \int_a^x f(t)\,dt$ and $G(x) = \int_b^x f(t)\,dt$, where a and b are constants, and f is a continuous function. Show that $G(x) = F(x) + C$, where C is a constant.

6. Suppose that F is an antiderivative of a differentiable function f.

   (a) If F is increasing on $[a, b]$, what is true about f?

   (b) If f is negative on $[a, b]$, what is true about F?

   (c) If f' is positive on $[a, b]$, what is true about F?

   (d) If F is concave down on $[a, b]$, what is true about f'?

   (e) Suppose that G is another antiderivative of f. What relationship exists between F and G?

FURTHER EXERCISES

7. Let $f(x) = \sin x$ and $A_f(x) = \int_0^x f(t)\,dt$.

   (a) It can be shown that $A_f(\pi/2) = 1$. Use this fact (and symmetry) to evaluate $A_f(0)$, $A_f(\pi)$, $A_f(3\pi/2)$, $A_f(2\pi)$, $A_f(-\pi/2)$, $A_f(-\pi)$, $A_f(-3\pi/2)$, and $A_f(-2\pi)$.

   (b) Explain why A_f is a periodic function. [**HINT**: Use the fact that f is 2π periodic.]

   (c) Show that $0 \le A_f(x) \le 2$ for all x.

   (d) Sketch graphs of f and $A_f(x)$ on the same axes over the interval $[-2\pi, 2\pi]$.

8. Use the fact that $\int_0^{\pi/2} \cos x\,dx = 1$ to show that $-1 \le \int_0^x \cos t\,dt \le 1$.

9. Suppose that f is a continuous function and that $\int_0^x f(t)\,dt = \sin(x^2)$.

   (a) Show that $\int_{\sqrt{\pi/2}}^x f(t)\,dt = \sin(x^2) - 1$.

   (b) Find an expression for $\int_{-\sqrt{3\pi/2}}^x f(t)\,dt$.

10. Let $f(x) = 2 - |x|$ and $A_f(x) = \int_0^x f(t)\,dt$.

    (a) Find a symbolic expression for A_f. [**HINT**: Write A_f as a piecewise-defined function.]

    (b) Plot f and A_f on the same axes.

    (c) Where is A_f increasing? What is true about f on these intervals?

    (d) Where is A_f decreasing? What is true about f on these intervals?

    (e) Where does A_f have local extrema? What is true about f at these points?

    (f) Where is A_f concave up? What is true about f on these intervals?

    (g) Where is A_f concave down? What is true about f on these intervals?

    (h) Does A_f have any inflection points? If so, where are they located? What is true about f at these points?

    (i) How would your answers to parts (c)–(h) change if the definition of A_f were changed to $A_f(x) = \int_3^x f(t)\,dt$?

11. Let $A_f(x) = \int_0^x f(t)\,dt$ where f is the function graphed below.

Graph of f

(a) Which is larger: $A_f(1)$ or $A_f(5)$? Justify your answer.

(b) Which is larger: $A_f(7)$ or $A_f(10)$? Justify your answer.

(c) Which is larger: $A_f(-2)$ or $A_f(-1)$? Justify your answer.

(d) Where is A_f increasing?

(e) Explain why A_f has a stationary point at $x = 6$. Is this a local maximum or a local minimum?

(f) Let $F(x) = \int_{-2}^x f(t)\,dt$. Explain why $A_f(x) = F(x) + C$ where C is a negative constant.

(g) Rank the five numbers 0, $A_f(-1) - A_f(-2)$, $A_f(0) - A_f(-1)$, $A_f(1) - A_f(0)$, and $A_f(2) - A_f(1)$ in increasing order.

(h) The five numbers $A_f(4) - A_f(3)$, $A_f(5) - A_f(4)$, $A_f(6) - A_f(5)$, $A_f(7) - A_f(6)$, and $A_f(8) - A_f(7)$ can each be interpreted as the slope of a secant line of A_f. What do these values suggest about the concavity of A_f on the interval $[3, 8]$?

12. Let $A_f(x) = \int_c^x f(t)\,dt$, where c is a constant and f is a continuous function.

(a) Prove that if f is positive on $[a, b]$, then A_f is increasing on $[a, b]$.   [**HINT**: Let y and z be numbers such that $a \le y < z \le b$. Compare $A_f(y)$ and $A_f(z)$.]

(b) Use part (a) to show that if f is negative on $[a, b]$, then A_f is decreasing on $[a, b]$.   [**HINT**: The function g defined by $g(x) = -f(x)$ is positive wherever f is negative.]

(c) Use parts (a) and (b) to show that A_f has a local maximum or a local minium wherever f changes sign.

13. Suppose that $G(x) = \int_1^x g(t)\,dt$. Evaluate each of the following expressions in terms of G.

(a) $\displaystyle\int_1^3 g(u)\,du$ (c) $\displaystyle\int_{-2}^2 g(t)\,dt$

(b) $\displaystyle\int_0^1 g(x)\,dx$

14. Suppose that $F(t) = \int_{-1}^t \sqrt{1 + x^4}\,dx$. Evaluate each of the following in terms of F.

(a) $\displaystyle\int_{-1}^4 \sqrt{1 + u^4}\,du$ (c) $\displaystyle\int_{-2}^3 \sqrt{1 + t^4}\,dt$

(b) $\displaystyle\int_0^{-1} \sqrt{1 + z^4}\,dz$

15. Let $g(x) = \int_1^x f(t)\,dt$ and $h(x) = \int_3^x f(t)\,dt$ where f is a function with the following properties:

(i) $f'(x) \ge 0$ when $0 \le x \le 3$ and $6 \le x \le 7$; $f'(x) \le 0$ when $3 \le x \le 6$.

(ii) $f''(x) \le 0$ when $0 \le x \le 5$; $f''(x) \ge 0$ when $5 \le x \le 7$.

(iii) f has the values shown below

x	0	1	2	3	4	5	6	7
$f(x)$	0	3	5	6	4	0	-2	0

(a) Sketch a graph of f.

(b) Evaluate $g(1)$.

(c) Explain why $h(x) < g(x)$ when $0 \le x \le 7$.

(d) Rank the four values $h(0)$, $g(3)$, $h(4)$, and $g(7)$ in increasing order.

(e) How many roots does g have in the interval $[0, 7]$? Justify your answer.

(f) How many roots does h have in the interval $[0, 7]$? Justify your answer.

16. Let $A_f(x) = \int_c^x f(t)\,dt$, where c is a constant and f is a continuous function.

(a) Prove that if f is increasing on $[a, b]$, then A_f is concave up on $[a, b]$.   [**HINT**: A function g is concave up on an interval if and only if $\big(g(x) - g(a)\big)/(x - a) < \big(g(b) - g(a)\big)/(b - a)$ for every a, b, and x in the interval such that $a < x < b$.]

(b) Use part (a) to show that if f is decreasing on $[a, b]$, then A_f is concave down on $[a, b]$.

(c) Suppose that f has a local maximum at a. Explain why A_f has an inflection point at a.

17. Let $G(x) = \int_{-3}^x f(t)\,dt$ and $H(x) = \int_2^x f(t)\,dt$ where f is the function graphed on the next page.   [**NOTE**: The graph of f is made up of straight lines and a semicircle.]

(a) How are the values of $G(x)$ and $H(x)$ related? Give a geometric explanation of this relationship.

(b) On which subintervals of $[-5, 5]$, if any, is H increasing?

(c) Explain why G has a local minimum at $x = 1$.

(d) Where in the interval $[-5, -2]$ does G achieve its minimum value? Its maximum value? What are these values?

(e) Where in the interval $[-5, 5]$ does H achieve its minimum value? Its maximum value? What are these values?

(f) On which subintervals of $[-5, 5]$, if any, is G concave down?

(g) Where does H have points of inflection?

Graph of f

(c) Let $f(x) = \sqrt{1 - x^2}$ and let $A_f(x) = \int_0^x f(t)\,dt$. Show that A_f is an antiderivative of f.

Graph of $\sqrt{1 - t^2}$

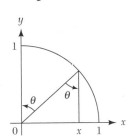

18. Let $F(x) = \int_a^x f(t)\,dt$ and $G(x) = \int_a^x g(t)\,dt$, where f and g are continuous functions. Suppose that $F(a) = G(a)$ and $f(x) \le g(x)$ when $x \ge a$. Show that $F(x) \le G(x)$ when $x \ge a$.

19. Let $F(x) = \int_a^x f(t)\,dt$ and $G(x) = \int_b^x g(t)\,dt$, where f and g are continuous functions. Suppose that $F(c) = G(c)$ and $f(x) \le g(x)$ when $x \ge c$. Show that $F(x) \le G(x)$ when $x \ge c$.

20. (a) Use the figure to show that
$$\int_0^x \sqrt{1 - t^2}\,dt = \frac{1}{2}x\sqrt{1 - x^2} + \frac{1}{2}\arcsin x \text{ if } 0 \le x \le 1.$$
[**HINT:** The area of a circular sector of radius r and angle θ is $r^2\theta/2$.]

(b) Show that $\int_0^x \sqrt{1 - t^2}\,dt = \frac{1}{2}x\sqrt{1 - x^2} + \frac{1}{2}\arcsin x$ if $-1 \le x \le 0$.

21. Let $f(x) = \sqrt{1 - x^2}$ and let $A_f(x) = \int_{-1/2}^x f(t)\,dt$.

(a) Use the result stated in part (a) of the previous exercise to find a symbolic expression for A_f.

(b) Is A_f an antiderivative of f? Justify your answer.

22. Let $a > 0$ be a constant. Use a sketch similar to that given in Exercise 20 to find an expression for $\int_0^x \sqrt{a^2 - t^2}\,dt$ when $-a \le x \le a$. [**HINT:** Check that your answer produces the correct value when $x = a$.]

23. Let $f(x) = x^2$ and $A_f(x) = \int_0^x f(t)\,dt$.

(a) Use the result of Exercise 54 in Section 5.1 to find a symbolic expression for $A_f(x)$.

(b) Is A_f an antiderivative of f? Justify your answer.

24. Let $f(x) = x^2$ and $A_f(x) = \int_{-3}^x f(t)\,dt$.

(a) Use the result of Exercise 54 in Section 5.1 to find a symbolic expression for $A_f(x)$.

(b) Is A_f an antiderivative of f? Justify your answer.

5.3 The Fundamental Theorem of Calculus

What the Theorem Says

In the last section we defined and studied the area function A_f, built from an original function f and a base point a. All the evidence so far points to this conclusion:

A_f *is an antiderivative of* f.

The formal statement of this fact is the most important theorem of our subject.

Here it is, complete with mathematical fine print:

Theorem 3 (The Fundamental Theorem of Calculus, Formal Version)   Let f be a continuous function, defined on an open interval I containing a. The function A_f with rule

$$A_f(x) = \int_a^x f(t)\,dt$$

is defined for every x in I, and

$$\frac{d}{dx}\left(A_f(x)\right) = f(x).$$

We'll say much more about the FTC (our shorthand) and provide a proof at the end of this section. First, some remarks.

Why Continuous? Requiring f to be continuous assures that the integral $\int_a^x f$ exists. If f is discontinuous (e.g., blows up somewhere), A_f may have problems, too.

Why Is I Open? Working on an open interval I is a technical convenience. Doing so avoids troubles that might occur if the domain of f had gaps and endpoints.

In Other Symbols The last equation of the FTC is sometimes written as follows:

$$\frac{d}{dx}\left(\int_a^x f(t)\,dt\right) = f(x).$$

Which form is clearer?   This form avoids the symbol A_f but requires more careful unpacking. ◄

EXAMPLE 1   Let $f(x) = 2x\sin(x^2)$ and let $A_f(x) = \int_0^x f(t)\,dt$. Find a symbolic formula for A_f; interpret results graphically.

Solution   For the simple functions of the last section, we used elementary area formulas (for rectangles, triangles, trapezoids, and so on) to find explicit formulas for A_f. Here, simple formulas aren't enough. What, for instance, is the exact value of $A_f(2)$, shown shaded in the following figure?

Graph of $f(x) = 2x\sin(x^2)$; $A_f(2)$ is net shaded area

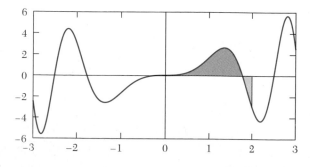

The FTC solves our problem. It says that A_f is *some* antiderivative of f; all we need do is find the right one.

Notice first that for any constant C, $F(x) = -\cos(x^2) + C$ is an antiderivative of $f(x) = 2x\sin(x^2)$. ▶ *Every* antiderivative of f—including A_f—has this form. ▶ Therefore, for some constant C,

Differentiate to see for yourself.

See the Fact that follows this example.

$$A_f(x) = -\cos(x^2) + C.$$

To choose the right value of C, we use the fact that $A_f(0) = 0$. ▶ In other words,

Why? Because $\int_0^0 f = 0$.

$$0 = A_f(0) = -\cos 0 + C \implies C = 1.$$

We've found our formula: $A_f(x) = -\cos(x^2) + 1$.

The following graphs agree; as it should, the A_f graph seems to describe the growth of signed area based at 0.

If $f(x) = 2x\sin(x^2)$,
then $A_f(x) = -\cos(x^2) + 1$

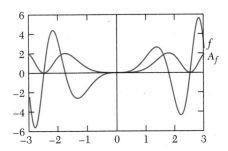

Notice, finally, that $A_f(2) = -\cos(4) + 1 \approx 1.65364$. ▶

■ *Is this result reasonable, given the general size of units?*

Finding All Antiderivatives of a Function

In Example 1 we claimed that any two antiderivatives of the function $f(x) = 2x\sin(x^2)$ differ only by an additive constant. This important principle deserves specific mention:

> **Fact** Suppose that $F'(x) = G'(x)$ for all x in an interval I. Then, for some constant C, $F(x) = G(x) + C$ for all x in I.

The Fact means, in practice, that finding *one* antiderivative for a function f on an interval I is tantamount to finding *all* antiderivatives: If $F(x)$ is *any* antiderivative, then every other antiderivative is of the form $F(x) + C$. (We gave more details on this Fact in Section 4.11.)

Continuous Functions Have Antiderivatives.   The fundamental theorem guarantees (among other things) that every continuous function f on I *has* an antiderivative on I—namely, the area function A_f.

For ordinary, well-behaved functions of elementary calculus, this is no great surprise. For naughtier functions, such as the one that follows, the situation is murkier.

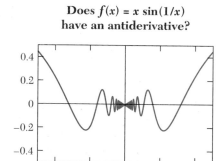

Does $f(x) = x \sin(1/x)$ have an antiderivative?

The fundamental theorem says that, naughty or not, f *has* an antiderivative. (*Finding* one in symbolic form may be difficult or impossible; the FTC offers no help there.)

Why the FTC Is Fundamental

The FTC deserves its high-sounding name for both theoretical and practical reasons. It's fundamental *theoretically* because it connects the two main concepts of the calculus: the derivative and the integral. Each is a sort of inverse of the other.

The FTC's *practical* consequences are at least as important; implicitly or explicitly, we'll use them again and again. Most important, the FTC leads to a practical method of calculating specific integrals. ◄

So far, we've relied entirely on elementary area formulas.

Using the FTC

Before proving the FTC, let's use it to prove another theorem, closely related to the FTC, but handier in computations.

Theorem 4 (Fundamental Theorem, Second Version)   Let f be continuous on $[a, b]$, and let F be *any* antiderivative of f. Then

$$\int_a^b f(x)\, dx = F(b) - F(a).$$

Proof.   This result follows readily from the original FTC. If F is any antiderivative of f, then $F(x)$ can differ from $A_f(x)$—another antiderivative of f, according to the FTC—only by some constant C. In other words, $F(x) = A_f(x) + C$. But then

$$F(b) - F(a) = \left(A_f(b) + C\right) - \left(A_f(a) + C\right) = A_f(b) - A_f(a)$$

$$= \int_a^b f - \int_a^a f = \int_a^b f,$$

just as claimed.   ∎

Calculating Areas with Antiderivatives

EXAMPLE 2   In Section 5.1 we claimed—but didn't show—that the shaded regions N and P have signed areas -0.25 and 0.25, respectively. Show it now.

Graph of $y = x^3$

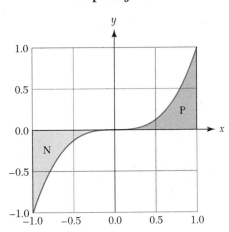

Solution   Let $f(x) = x^3$. The signed areas N and P are given by integrals:

$$\text{Signed area } N = \int_{-1}^0 f(x)\,dx; \qquad \text{Signed area } P = \int_0^1 f(x)\,dx.$$

Because $F(x) = x^4/4$ is an antiderivative of $f(x)$, Theorem 4 applies:

$$\int_{-1}^0 f(x)\,dx = F(0) - F(-1) = -\frac{1}{4}; \qquad \text{and} \qquad \int_0^1 f(x)\,dx = F(1) - F(0) = \frac{1}{4}$$

∎

Bracket Notation.   The **bracket notation** offers a convenient shorthand for calculating integrals. For any function F,

$$F(x)\Big]_a^b \qquad \text{means} \qquad F(b) - F(a).$$

Watch for brackets in the next example.

EXAMPLE 3   Evaluate $\int_0^1 x^2\,dx$ and $\int_0^1 x^{10}\,dx$. Interpret each integral as an area.

Solution   Theorem 4 says that it's enough to find *any* antiderivative, plug in endpoints, and subtract. We do just that:

$$\int_0^1 x^2\, dx = \left.\frac{x^3}{3}\right]_0^1 = \frac{1}{3}; \qquad \int_0^1 x^{10}\, dx = \left.\frac{x^{11}}{11}\right]_0^1 = \frac{1}{11}.$$

Each integral measures an area:

Areas under two curves: $y = x^{10}$ and $y = x^2$

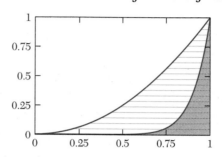

The lightly and darkly shaded areas represent, respectively, the integrals $\int_0^1 x^2\, dx = 1/3$ and $\int_0^1 x^{10}\, dx = 1/11$. The picture shows that the answers are numerically reasonable. ∎

EXAMPLE 4   Calculate the shaded area in the following picture.

Area under the graph of
$y = 1/x$ from 1 to b

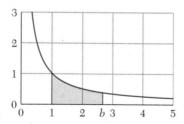

Which value of b makes the area 1?

Solution   The shaded area is the value of the integral $\int_1^b 1/x\, dx$. The function $\ln x$ is an antiderivative for $1/x$, so the FTC says that

$$\int_1^b \frac{1}{x}\, dx = \ln x\big]_1^b = \ln b - \ln 1 = \ln b.$$

If $b = e$ (as in the picture) then the area is $\ln e = 1$. ∎

From Rate to Amount

Restated slightly, the second version of the fundamental theorem reads like this:

Fact Let f be a well-behaved function on $[a, b]$, with derivative f'. Then

$$\int_a^b f'(t)\, dt = f(b) - f(a).$$

In words,

> *Integrating f' (the rate function) over $[a, b]$ gives the change in f (the amount function) over the same interval.*

This fact has important practical uses; with it, we can deduce *amount* information from *rate* information.

EXAMPLE 5   Demand for electric power varies more or less predictably over the course of a day. Drawing on experience, engineers in a mythical small town (a bedroom suburb, perhaps) use the formula

$$r(t) = 4 + \sin(0.263t + 4.7) + \cos(0.526t + 9.4)$$

to model the rate of power consumption, in megawatts, t hours after midnight. A graph of r follows. Is the model reasonable? What does the shaded area represent?

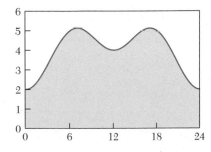

**Rate of power use
(megawatts) over 24 hours**

Solution   The graph looks believable (though the model is certainly over-simplified). Residential power demand peaks in the morning, declines during working hours, and rises again in the evening. The formula for r is complicated, but the ingredients—periodic functions and various constants—look right. ▶

The shaded area is the integral $\int_0^{24} r(t)\, dt$. To understand it in the present context, let $a(t)$ be the total *amount* of power consumed up to time t, starting at

Remember how additive and multiplicative constants affect periodic functions?

By definition, therefore,
a = 0 when t = 0.

midnight. ◄   Then $a'(t) = r(t)$; by the preceding Fact,

$$\int_0^{24} r(t)\, dt = a(24) - a(0) = a(24) = \text{total consumption, all day.}$$

With the "classic" FTC (and, from earlier information, a little help in finding an antiderivative) we can calculate this amount. It's easy to check (but harder to guess) that

$$4t - \frac{\cos(0.263t + 4.7)}{0.263} + \frac{\sin(0.526t + 9.4)}{0.526}$$

is an antiderivative for $r(t)$. The rest is straight calculation:

$$\int_0^{24} r(t)\, dt = 4t - \frac{\cos(0.263t + 4.7)}{0.263} + \frac{\sin(0.526t + 9.4)}{0.526} \Bigg]_0^{24}$$

$$\approx 95.781 \text{ megawatt hours.} \qquad\blacksquare$$

When *Not* to Use the FTC

Beautiful, too.

The FTC, useful ◄   as it is, won't solve all our integral problems. Evaluating integrals by antidifferentiation requires first finding a usable antiderivative. For a surprising number of integrands, even apparently simple ones, finding an antiderivative formula is hard or even impossible. None of the following functions, for instance, has an elementary antiderivative:

$$\sin(x^2); \qquad \frac{x}{\ln x}; \qquad 3 + \sin x + 0.3 \arcsin(\sin(7x)).$$

Nevertheless, all have perfectly sensible integrals over finite intervals. The following picture shows what $\int_0^{10} f$ means for the last of the preceding functions.

An FTC-resistant integral:
$$\int_0^{10} (3 + \sin x + 0.3 \arcsin(\sin(7x)))\, dx$$

Can you estimate it? We get about 30, *looking at the graph.*

With or without an antiderivative formula, the integral certainly exists. ◄   Soon we'll study techniques (that *don't* involve antiderivatives) for estimating integrals like the preceding one.

A Useful Notation for Antiderivatives—and a Caution

The **indefinite integral notation** for antiderivatives, as in

$$\int x^2\,dx = \frac{x^3}{3} + C \qquad \text{and} \qquad \int \cos x\,dx = \sin x + C$$

can be a convenient shorthand. It reflects, too, the fundamental connection between antiderivatives and integrals. A little care is necessary, however.

> **The Whole Family** $\int f(x)\,dx$ denotes a family of functions, one for each value of C.

> **Similar Signs, Different Meanings** The **definite integral** $\int_a^b f(x)\,dx$ and the **indefinite integral** $\int f(x)\,dx$ have very different meanings: The first is a *number*, and the second is a family of *functions*.

Proving the FTC

The FTC is clearly useful. By now, we hope, it's plausible. But why is it *true*?

The Idea of the Proof

Recall the setup: f is a continuous function, defined on an interval I containing $x = a$, and

$$A_f(x) = \int_a^x f(t)\,dt.$$

The theorem asserts that, for x in I,

$$A_f'(x) = \lim_{h \to 0} \frac{A_f(x+h) - A_f(x)}{h} = f(x).$$

We'll work directly with the difference quotient to prove this.

Notice first that $A_f(x+h) - A_f(x) = \int_a^{x+h} f - \int_a^x f = \int_x^{x+h} f$. The following pictures show why.

$A_f(x+h) - A_f(x) =$ the shaded area   **$A_f(x+h) - A_f(x)$ — a closer look**

        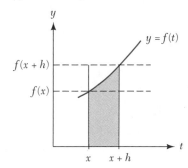

This second picture shows that, for small positive values of h,

$$A_f(x+h) - A_f(x) = \int_x^{x+h} f \approx f(x) \cdot h.$$

(The last quantity is the area of the shorter rectangle.) Thus, for h near 0,

$$\frac{A_f(x+h) - A_f(x)}{h} \approx \frac{f(x) \cdot h}{h} = f(x).$$

Therefore, as $h \to 0$, the difference quotient tends to $f(x)$, as desired. ∎

More on the Proof: Fine Print

A rigorous proof can be constructed from the preceding informal argument. The missing ingredient is a precise approach to the approximation $\int_x^{x+h} f \approx f(x) \cdot h$. It helps if f happens to be increasing near x—as it is in the preceding figure. In that case, ◄

Check each step!

$$f(x) \cdot h \le A_f(x+h) - A_f(x) = \text{shaded area} \le f(x+h) \cdot h,$$

so

$$f(x) \le \frac{A_f(x+h) - A_f(x)}{h} \le f(x+h).$$

As $h \to 0$, the quantities on both left and right tend to $f(x)$. So, therefore, must the difference quotient. (A similar idea works if f is not increasing, although some technical details differ.)

BASICS

1. Let $F(x) = \int_0^x f(t)\, dt$ where f is the function whose graph is shown below.

(a) Which of the following is the graph of F? Justify your answer.

Graph of f

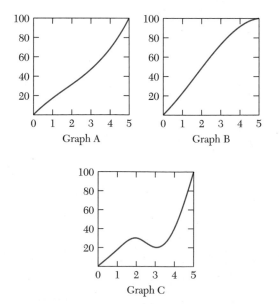

Graph A

Graph B

Graph C

(b) Suppose that $g(0) = 2$ and $g'(x) = f(x)$ for all x. Explain how the graphs of F and g are related.

2. Let $F(x) = \int_0^x f(t)\,dt$ where f is the function graphed below. (The graph of f is made up of straight lines and a semicircle.)

Graph of f

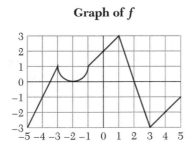

(a) Evaluate $F(-1)$, $F(0)$, and $F(2)$.
(b) Identify all the critical points of F in the interval $[-5, 5]$.
(c) Identify all the inflection points of F in the interval $[-5, 5]$.

(d) What is the average value of f over the interval $[-5, 5]$?
(e) Let $G(x) = \int_0^x F(t)\,dt$. On which subintervals of $[-5, 5]$, if any, is G concave upward?

3. Evaluate each of the following definite integrals using the Fundamental Theorem of Calculus.

(a) $\displaystyle\int_1^4 (x + x^{3/2})\,dx$

(b) $\displaystyle\int_0^\pi \cos x\,dx$

(c) $\displaystyle\int_{-2}^5 \frac{dx}{x + 3}$

(d) $\displaystyle\int_0^b x^2\,dx$

(e) $\displaystyle\int_1^b x^n\,dx \quad [n \neq -1]$

(f) $\displaystyle\int_2^{2.001} \frac{x^5}{1000}\,dx$

(g) $\displaystyle\int_0^{0.001} \frac{\cos x}{1000}\,dx$

(h) $\displaystyle\int_0^{\sqrt\pi} x \sin(x^2)\,dx$

FURTHER EXERCISES

4. Let $F(x) = \int_0^x f(t)\,dt$ where $f(t)$ is the function shown below:

Graph of f

(a) Does $F(x)$ have any local maxima within the interval $[0, 10]$? If so, where are they located?
(b) At what value of x does $F(x)$ attain its minimum value on the interval $[1, 10]$?
(c) On which subinterval(s) of $[0, 10]$, if any, is the graph of $F(x)$ concave up? Justify your answer.

5. Show that if $f(a) \neq 0$, then $\dfrac{d}{dx}\left[\displaystyle\int_a^x f(t)\,dt\right] \neq$
$\displaystyle\int_a^x \left[\dfrac{d}{dt} f(t)\right]\,dt.$

6. Let $g(x) = \int_1^x f(t)\,dt$ and $h(x) = \int_3^x f(t)\,dt$ where f is a function with the following properties:

(i) $f'(x) \geq 0$ when $0 \leq x \leq 3$ and $6 \leq x \leq 7$; $f'(x) \leq 0$ when $3 \leq x \leq 6$.
(ii) $f''(x) \leq 0$ when $0 \leq x \leq 5$; $f''(x) \geq 0$ when $5 \leq x \leq 7$.
(iii) f has the values shown below

x	0	1	2	3	4	5	6	7
$f(x)$	0	2	4	5	4	0	-2	0

(a) Sketch a graph of f.
(b) Evaluate $g'(4)$.
(c) Where is the graph of g concave down? Justify your answer.
(d) Find the value of x where $g(x)$ is maximum on the interval $[0, 7]$.
(e) Find the value of x where $h(x)$ is minimum on the interval $[0, 7]$.
(f) What is the average value of f' over the interval $[2, 5]$?

7. Let $g(u) = \int_1^u f(x)\,dx$ where f is the function graphed below.

Graph of f

(a) Evaluate $g'(4)$.
(b) Where in the interval $[0, 5]$ is g concave up? Justify your answer.
(c) How many roots does g have in the interval $[0, 5]$? Justify your answer.
(d) Rank the six numbers -1, 0, 1, $g(0)$, $g(2)$, and $g(4)$.
(e) Is the average value of g' over the interval $[0, 3]$ greater than 1? Justify your answer.

8. Suppose that g is a differentiable function, that the graph of g passes through the point $(5, 1)$, and that the derivative of g is the function sketched below. Let $G(x) = \int_0^x g'(t)\,dt$.

Graph of g'

(a) How are the graphs of g and G related? Explain.
(b) Explain why $g(x) = 1 + \int_5^x g'(t)\,dt$.
(c) How many solutions of the equation $g(x) = 0$ exist in the interval $[0, 10]$? Explain.
(d) How many solutions of the equation $G(x) = 0$ exist in the interval $[0, 10]$? Explain.

9. Suppose that $\int_0^x f(t)\,dt = 3x^2 + e^x - \cos x$. Find $f(2)$.

10. Find an equation of the line tangent to the graph of
$$F(x) = \int_1^x \sqrt[3]{t^2 + 7}\,dt \text{ at } x = 1.$$

11. Suppose that $f'(x) = \sqrt[3]{1 + x^2}$ and $f(1) = 0$.
(a) Explain why $f(x) = \int_1^x \sqrt[3]{1 + t^2}\,dt$.
(b) Express $\int_0^3 \sqrt[3]{1 + x^2}\,dx$ in terms of f.

12. Let $F(x) = \int_0^x \sqrt[3]{1 + t^2}\,dt$. Evaluate $\frac{d}{dx}\left[F(x^2)\right]$.

13. Let f be a function with the following properties:
(i) $f'(x) = ax^2 + bx$ (iii) $f''(1) = 18$
(ii) $f'(1) = 6$ (iv) $\int_1^2 f(x)\,dx = 18$
Find an algebraic expression for $f(x)$.

14. Let $f(x) = \begin{cases} -2, & \text{if } x < 0 \\ 1, & \text{if } x \geq 0 \end{cases}$ and
let $F(x) = \int_{-2}^x f(t)\,dt$.
(a) Evaluate $\int_{-1}^1 f(x)\,dx$.
(b) Sketch a graph of $F(x)$ and $f(x)$ on the same axes for x in $[-2, 2]$.
(c) Does $F'(1)$ exist? If so, evaluate it using the definition of the derivative. If not, explain why it doesn't exist.
(d) Does $F'(-1)$ exist? If so, evaluate it using the definition of the derivative. If not, explain why it doesn't exist.
(e) Explain why $F'(0)$ doesn't exist. Why doesn't this contradict the Fundamental Theorem of Calculus?

15. A company is planning to phase out a product because demand for it is declining. Demand for the product is currently 800 units/month and dropping by 10 units/month each month. To maintain customer and employee relations, the company has announced that it will continue to produce the product for one more year. At the present time, the company is producing the product at a rate of 900 units/month and 1680 units of the product are in inventory.
(a) Give expressions for the demand and production rates as functions of time t (take $t = 0$ as the present).
(b) Let $D(t)$ be the demand rate for the product at time t. Explain why $\int_0^t D(s)\,ds$ is the total demand for the product between now and time t.
(c) Give an expression for the inventory at time t. [**HINT:** The amount of the product in inventory is the difference between supply and demand at the end of t months.]
(d) The company would like to reduce production at a constant rate of R units/month, with R chosen so the product inventory is zero at the end of 12 months. Find a suitable R.

16. The function $\ln x$ is sometimes *defined* for $x > 0$ by integration:

$$\ln x = \int_1^x \frac{dt}{t}.$$

Assume $a > 0$ and $x > 0$. Derive the identity $\ln(ax) = \ln a + \ln x$ by carrying out and justifying each of the following steps:

(a) Show that $\left(\int_1^{ax} \frac{dt}{t} \right)' = \frac{1}{x}$. [**HINT:** Use the Chain Rule.]

(b) Explain why the result in part (a) implies that that $\ln(ax) = \ln x + C$.

(c) By choosing an appropriate x, show that the value of the constant C in part (b) is $\ln a$.

5.4 Approximating Sums: The Integral as a Limit

The integral $\int_a^b f$, as defined so far, is the signed ▶ area of the region bounded by the graph of f from $x = a$ to $x = b$. In the last section, we saw in the fundamental theorem of calculus the crucial inverse relationship between integrals and derivatives. Using the FTC, we can readily calculate any integral for which we can find an antiderivative.

Remember what "signed" means? It's important.

Despite their close relationship, integration and antidifferentiation are not the same thing. For many integrals (even simple-looking ones such as $\int_0^1 \sin(x^2)\,dx$), antidifferentiation is useless, because no usable antiderivative formula exists. Nevertheless, such integrals clearly exist, as graphs show.

In this section we take a new approach to the integral, ▶ defining it this time as a certain limit of "approximating sums." The limit definition offers new and useful perspectives on the integral and what it means. Equally important, it lends itself well—better than the FTC—to the problem of estimating integrals numerically.

The approach is new; the integral is the same!

This section resembles Section 2.5, where we first met the limit definition of the derivative. There, as here, the general problem was to translate an intuitively reasonable geometric concept ▶ into precise analytic language. For both derivative and integral, the idea of limit proves crucial; it links approximations to exact values.

There, the slope of the tangent line; here, the area under a graph.

Estimating Integrals by Approximating Sums

The formal definition of integral as a limit involves some formidable-seeming terminology and notation. The underlying ideas, however, are natural and straightforward, especially when expressed graphically.

To back up this claim, we'll approach the definition via a leisurely example, introducing terms and ideas as we go.

EXAMPLE 1 Part of the graph of $f(x) = x^3 - 3x^2 + 8$ follows; the shaded area represents the integral $\int_{0.5}^{3.5} f(x)\,dx$. ▶ Use the FTC to evaluate the integral *exactly*. Then use approximating sums to *estimate* the same integral.

First, estimate the area roughly, by eye alone. Write your answer in the margin.

What's the area?

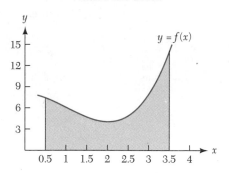

Check the steps; they're routine.

S o l u t i o n   With the FTC, an exact answer is easy to find by antidifferentiation. ◄

$$\int_{0.5}^{3.5} \left(x^3 - 3x^2 + 8 \right) dx = \left(\frac{x^4}{4} - x^3 + 8x \right)\Bigg]_{0.5}^{3.5} = \frac{75}{4} = 18.75.$$

How might we estimate this area? The following figures suggest four plausible strategies.

Left sum, 10 subdivisions

Right sum, 10 subdivisions

Midpoint sum, 10 subdivisions

Trapezoid sum, 6 subdivisions

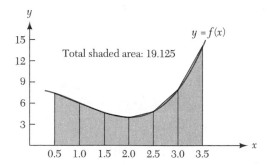

In each case we *approximated* the desired area with a sum of simpler areas—rectangles or trapezoids. (For trapezoids we used six, not ten, subdivisions in order to show more clearly the difference between the integral and the trapezoid estimate.) Adding up the areas of these simpler figures gives, in each case, a natural estimate to the desired area. ►

■ *Study the pictures; which estimate looks best? Which is best?*

Calculating Approximate Areas

How, exactly, did we calculate each of the preceding area estimates?

A close look at each picture gives the answer. Consider first L_{10}, the **left approximating sum with ten subdivisions**. The shaded area consists of ten rectangles; each has base 0.3 (i.e., $(3.5 - 0.5)/10$). The rectangles' *heights* vary: The height of each is the value of the function f at the left edge of the base—hence the name "left sum."

The second rectangle, for example, has the x-interval $[0.8, 1.1]$ for its base; its height is $f(0.8)$, i.e., the value of f at the *left* endpoint. Since $f(x) = x^3 - 3x^2 + 8$, $f(0.8) = 6.592$. Therefore, for the second rectangle,

$$\text{Area} = 0.3 \cdot 6.592 = 1.9776.$$

The *total* area of all ten left rectangles is therefore the sum

$$L_{10} = f(0.5) \cdot 0.3 + f(0.8) \cdot 0.3 + f(1.1) \cdot 0.3 + \cdots + f(3.2) \cdot 0.3$$

$$= 7.375 \cdot 0.3 + 6.592 \cdot 0.3 + 5.701 \cdot 0.3 + \cdots + 7.159 \cdot 0.3 + 10.048 \cdot 0.3$$

$$= 17.8725.$$

The **right approximating sum with ten subdivisions**, R_{10}, differs only slightly—the height of each rectangle is the value of f at the *right* endpoint of the base interval. ► Hence, the total area of all ten "right" rectangles is the sum

Think this sentence through carefully. Do you see how the picture says the same thing?

$$R_{10} = f(0.8) \cdot 0.3 + f(1.1) \cdot 0.3 + f(1.4) \cdot 0.3 + \cdots + f(3.5) \cdot 0.3$$

$$= 6.592 \cdot 0.3 + 5.701 \cdot 0.3 + \cdots + 10.048 \cdot 0.3 + 14.125 \cdot 0.3$$

$$= 19.8975.$$

For M_{10}, the **midpoint approximating sum with ten subdivisions**, heights are evaluated at the *midpoint* of each subinterval. Therefore,

$$M_{10} = f(0.65) \cdot 0.3 + f(0.95) \cdot 0.3 + f(1.25) \cdot 0.3 + \cdots + f(3.35) \cdot 0.3$$

$$= 7.007 \cdot 0.3 + 6.150 \cdot 0.3 + 5.226 \cdot 0.3 + \cdots + 11.928 \cdot 0.3$$

$$= 18.6825.$$

The use of a **trapezoid approximating sum** looks, geometrically, like a good idea. Calculating it, moreover, is almost as easy as calculating the rectangular sums. A key idea ► simplifies the work: The area of a trapezoid is the average of the areas of the two rectangles determined by its shorter and longer sides. From this

A picture makes the idea clear; draw one.

fact a useful observation follows:

> *The trapezoid approximation with n subdivisions is the average of the corresponding left and right approximations.*

The numerical details are left to you.

For *our* trapezoid sum (we'll call it T_6): ◄

$$T_6 = \frac{1}{2}(L_6 + R_6)$$

$$= \frac{1}{2}(17.4375 + 20.8125) = 19.125.$$

Sigma Notation; Partitions

Sums with many terms (the preceding sums, for instance) are tedious to write. **Sigma (Σ) notation** is more efficient than brute force. Even better, it shows clearly the similarities and the differences among summands.

EXAMPLE 2   Discuss and evaluate each expression:

$$\sum_{k=1}^{4} k; \qquad \sum_{j=1}^{4} j; \qquad \sum_{i=1}^{10} 3i.$$

Solution   By definition, $\sum_{k=1}^{4} k = 1 + 2 + 3 + 4$, or simply 10. Although the second sum is typographically slightly different, it means exactly the same thing: $\sum_{j=1}^{4} j = 1 + 2 + 3 + 4 = 10$. The moral: *The name of the **index variable** ◄ doesn't matter.*

First it was k, then j; i through n are all popular choices.

The third expression is more complicated, but the ideas are the same:

$$\sum_{i=1}^{10} 3i = 3 \cdot 1 + 3 \cdot 2 + 3 \cdot 3 + \cdots + 3 \cdot 9 + 3 \cdot 10$$

$$= 3 \cdot (1 + 2 + \cdots + 10) = 3 \cdot 55 = 165.$$

(Notice that the common factor can be pulled outside the sum: $\sum_{i=1}^{10} 3i = 3 \sum_{i=1}^{10} i$. This is the Σ-version of the distributive law.)   ■

EXAMPLE 3   Use sigma notation to rewrite the left, right, and midpoint approximating sums (L_{10}, R_{10}, and M_{10}) we already calculated.

Solution   All three approximating sums are based (literally!) on subdividing the interval [0.5, 3.5] into ten equal subintervals, each of length $\Delta x = 0.3$. Here are their endpoints, listed in order:

$$0.5 < 0.8 < 1.1 < 1.4 < \cdots < 3.2 < 3.5.$$

(This ordered set of 11 endpoints is called a **partition** of the x-interval [0.5, 3.5].) For convenience, let's name these endpoints, in the same order, $x_0 = 0.5$, $x_1 = 0.8, \ldots, x_{10} = 3.5$.

Now sigma notation is convenient:

$$L_{10} = \sum_{i=0}^{9} f(x_i)\,\Delta x = f(x_0)\Delta x + f(x_1)\Delta x + f(x_2)\Delta x + \cdots + f(x_9)\Delta x$$

$$= f(0.5) \cdot 0.3 + f(0.8) \cdot 0.3 + \cdots + f(3.2) \cdot 0.3;$$

$$R_{10} = \sum_{i=1}^{10} f(x_i)\,\Delta x = f(0.8) \cdot 0.3 + f(1.1) \cdot 0.3 + \cdots + f(3.5) \cdot 0.3;$$

$$M_{10} = \sum_{i=0}^{9} f\left(\frac{x_i + x_{i+1}}{2}\right)\Delta x = f(0.65) \cdot 0.3 + f(0.95) \cdot 0.3 + \cdots + f(3.35) \cdot 0.3.$$

The Σ form makes the approximating sums appear simple and compact. ▶

If we'd rather see numbers than symbols, we can always use the fact ▶ that $x_i = 0.5 + 0.3i$ to rewrite our sums. Here are two of them:

$$L_{10} = \sum_{i=0}^{9} f(x_i)\,\Delta x = 0.3 \cdot \sum_{i=0}^{9} f\left(0.5 + 0.3i\right);$$

$$M_{10} = \sum_{i=0}^{9} f\left(\frac{x_i + x_{i+1}}{2}\right)\Delta x = 0.3 \cdot \sum_{i=0}^{9} f\left(0.65 + 0.3i\right). \qquad ■$$

Why does the R_{10} sum run from $i = 1$ to $i = 10$, not from $i = 0$ to $i = 9$?

It is a fact. Do you see why?

Getting Better All the Time

It stands to reason that an approximating sum of any type should better approximate an integral as the number of subdivisions increases. The following pictures (still pertaining to $\int_{0.5}^{3.5} f(x)\,dx$) agree.

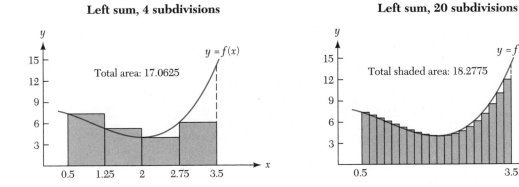

Left sum, 4 subdivisions

Total area: 17.0625

Left sum, 20 subdivisions

Total shaded area: 18.2775

Neither left approximation appears perfect, but the error committed with 20 subdivisions seems much less. Numerical values of various sums (all computed by machine, of course) show the same thing:

	Approximating sums for the integral $\int_{0.5}^{3.5} f(x)\,dx$			
n	**Left Sum**	**Right Sum**	**Midpoint Sum**	**Trapezoid Sum**
2	17.0625	27.1875	17.0625	22.125
4	17.0625	22.125	18.3281	19.5938
8	17.6953	20.2266	18.6445	18.9609
16	18.1699	19.4355	18.7236	18.8027
32	18.4468	19.0796	18.7434	18.7632
64	18.5951	18.9115	18.7484	18.7533
128	18.6717	18.8299	18.7496	18.7508
256	18.7107	18.7898	18.7499	18.7502

Riemann Sums and the Limit Definition of Integral

Riemann Sums and Their Ingredients

After the German mathematician G. F. B. Riemann (1826–66); he gave the first modern definition of the integral.

Left, right, and midpoint approximating sums are all special types of **Riemann sums**. ◀ Roughly speaking, any "rectangular" approximating sum is a Riemann sum. Riemann's idea of an approximating sum to an integral $\int_a^b f(x)\,dx$ is very liberal. He allows rectangles with unequal bases, and their heights can be determined anywhere—even randomly—within their respective subintervals.

Following, for instance, is the graphical version of a somewhat irregular Riemann sum for our by-now-familiar integral:

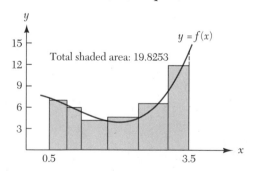

A Riemann sum, 6 (unequal) subdivisions

It may not be a *good* approximation, but it's a legitimate one.

We don't require such generality, but the formal definition of Riemann sum follows, for the record. The necessary ingredients are a function f and an interval $I = [a, b]$.

Definition Let I be partitioned into n subintervals by any $n + 1$ points

$$a = x_0 < x_1 < x_2 < \cdots < x_{n-1} < x_n = b;$$

let $\Delta x_i = x_i - x_{i-1}$ denote the width of the ith subinterval. Within each subinterval $[x_{i-1}, x_i]$, choose any point c_i. The sum

$$\sum_{i=1}^{n} f(c_i)\Delta x_i = f(c_1)\Delta x_1 + f(c_2)\Delta x_2 + \cdots + f(c_n)\Delta x_n$$

is a Riemann sum with n subdivisions for f on $[a, b]$.

Our left, right, and midpoint sums fit this definition with room to spare. Each is built from a **regular partition** of $[a, b]$ (i.e., one with subintervals of equal length) and some simple, consistent, scheme for choosing the **evaluation points** (or "sampling points") c_i. A *trapezoid* approximating sum, on the other hand, isn't quite a Riemann sum, since its approximating figures aren't rectangles. Nevertheless, trapezoidal sums do an excellent job of approximating the integral.

The Integral Defined as a Limit

Graphical intuition suggests that all of the approximating sums—L_n, R_n, M_n, and T_n—should approach the true signed area $\int_{0.5}^{3.5} f(x)\,dx$ as n increases. Numerical evidence ▶ argues irresistibly for a value somewhere near 18.75.

The limit definition of integral makes these ideas precise:

See the preceding table to dispel any lingering doubts.

Definition Let the function f be defined on the interval $I = [a, b]$. The **integral of f over I**, denoted $\int_a^b f(x)\,dx$, is the number to which all Riemann sums S_n tend as n tends to infinity and as the widths of all subdivisions tend to zero. In symbols:

$$\int_a^b f(x)\,dx = \lim_{n \to \infty} S_n = \lim_{n \to \infty} \sum_{i=1}^{n} f(c_i)\,\Delta x_i,$$

if the limit exists.

Here are some comments on the definition.

When Is a Function Integrable? A function for which the limit in the definition exists is called **integrable** on $[a, b]$. In practice, being integrable on $[a, b]$ is easy: Almost every calculus-style function that's defined and *bounded* (i.e., doesn't blow up or down) on $[a, b]$ is integrable. The surprise is the difficulty of finding a *non*integrable function. ▶

Well-Behaved Functions The limit in the definition, taken at face value, is a slippery customer. Understanding every ramification of permitting arbitrary partitions and sampling points, for example, can be tricky. Luckily, the matter is much simpler for the well-behaved functions (e.g., continuous functions) we typically encounter in calculus. For such functions,

Here's one: $f(x) = 1$ if x is rational; $f(x) = 0$ if x is irrational.

almost any respectable sort of approximating sum does what we'd expect—approaches the true value of the integral as n tends to infinity.

Notes on the dx Notation

The limit definition helps explain the otherwise mysterious dx in the notation $\int_a^b f(x)\,dx$. Consider, for example, the case of a *right* approximating sum with n equal subdivisions, each of length $(b-a)/n = \Delta x$. Then, by definition,

$$\int_a^b f(x)\,dx = \lim_{n\to\infty} \sum_{i=1}^n f(x_i)\,\Delta x.$$

We've seen this before:
$dy/dx = \lim_{\Delta x\to 0} \Delta y/\Delta x.$

Now the left side looks much like the right; dx on the left corresponds in a natural way to Δx on the right. ◄

Approximating Sums and Computers

Approximating sums offer simple, effective, and accurate approximations to almost any integral (simple for a computer or calculator, anyway). Calculation of approximating sums *by hand* is impractical except for simple integrands and few subdivisions. Calculation of approximating sums is the sort of "bureaucratic" task computers do best—it's simple, repetitive, and tedious.

BASICS

1. Let g be the function graphed below. Estimate the value of $\int_{-5}^5 g(x)\,dx$ by evaluating left, right, and midpoint sums each with 5 equal subintervals. Then draw sketches that illustrate these sums graphically.

(c) Estimate the value of $\int_1^7 f(x)\,dx$ using a right sum with three equal subintervals.

(d) Estimate the value of $\int_1^7 f(x)\,dx$ using a left sum with three equal subintervals.

(e) Estimate the value of $\int_0^8 f(x)\,dx$ using a right sum with four equal subintervals.

(f) Estimate the value of $\int_0^8 f(x)\,dx$ using a left sum with four equal subintervals.

Graph of g

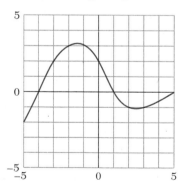

2. Let f be the function graphed to the right.

(a) Show that $\int_0^8 f(x)\,dx < 45$.

(b) Is $20 < \int_1^7 f(x)\,dx$? Justify your answer using a picture.

Graph of f

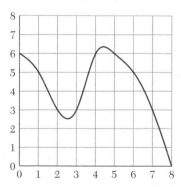

3. Let $I = \int_0^{20} x^2 \, dx$. Use sigma notation to write an expression for R_{10}, the right Riemann sum approximation to I with 10 equal subintervals.

4. At time t (measured in hours), $0 \le t \le 24$, a firm uses electricity at the rate of $E(t)$ kilowatts. The power company's rate schedule indicates that the cost per kilowatt-hour at time t is $c(t)$ dollars. Assume that both E and c are continuous functions.

   (a) Set up an N-term left sum that approximates the cost of the electricity consumed in a 24-hour period.

   (b) What definite integral is approximated by the sum you found in part (a)?

FURTHER EXERCISES

5. Let $I = \int_0^5 \sqrt[3]{2x} \, dx$. Use sigma notation to write the left, right, and midpoint approximations to I. (Assume that $N = 10$ equal subintervals are used.)

6. Let $I = \int_0^5 \sqrt{3x} \, dx$. Use sigma notation to write the left, right, and midpoint approximations to I. (Assume that N equal subintervals are used.)

7. Find a definite integral that is approximated by the right sum $\dfrac{2}{100} \sum_{k=1}^{100} \sin\left(\dfrac{2k}{100}\right)$. What is the value of this integral?

8. Find a definite integral that is approximated by the midpoint sum

$$\frac{2}{10} \sum_{k=1}^{40} \cos\left(\frac{2k-1}{10}\right).$$

What is the value of this integral? [**HINT**: Write out the first few terms of the sum.]

9. The following table contains values of a continuous function f at several values of x.

x	1	2	3	4	5	6
$f(x)$	0.14	0.21	0.28	0.36	0.44	0.54

   (a) Estimate $\int_2^5 f(x) \, dx$ using a left sum with 3 equal subintervals.

   (b) Repeat part (a) using the trapezoid rule.

   (c) Draw sketches that illustrate the computations you did in parts (a) and (b).

   (d) Show that $f(2) \cdot 2 + f(4) \cdot 3$ is a Riemann sum approximation to $\int_1^6 f(x) \, dx$. Justify your answer. [**HINT**: Draw a picture.]

10. The rate at which the world's oil is being consumed is increasing. Suppose that the rate (measured in billions of barrels per year) is given by the function $r(t)$, where t is measured in years and $t = 0$ is January 1, 1990.

    (a) Write a definite integral that represents the total quantity of oil used between the start of 1990 and the start of 1995.

    (b) Suppose that $r(t) = 32e^{0.05t}$. Find an approximate value for the definite integral from part (a) using a right sum with $n = 5$ equal subintervals.

    (c) Interpret each of the five terms in the sum from part (b) in terms of oil consumption.

    (d) Evaluate the definite integral from part (a) exactly using the FTC.

11. Evaluate $\lim_{n \to \infty} \frac{1}{n} \sum_{j=1}^n \left(\frac{j}{n}\right)^3$ by expressing it as a definite integral and then evaluating this integral using the Fundamental Theorem of Calculus.

12. Evaluate $\lim_{n \to \infty} \frac{2}{n} \sum_{j=1}^n \left(\frac{2j}{n}\right)^3$ by expressing it as a definite integral and then evaluating this integral using the Fundamental Theorem of Calculus.

13. (a) Write the sum

$$\frac{5}{2}(2.3)^2 + \frac{5}{6}(2.8)^2 + \frac{5}{12}(3.3)^2 + \frac{5}{20}(3.8)^2$$

   using sigma notation. [**HINT**: $2 = 1 \cdot 2$, $6 = 2 \cdot 3$, $12 = 3 \cdot 4$, and $20 = 4 \cdot 5$.]

   (b) Is the sum in part (a) a Riemann sum approximation to $\int_0^4 x^2 \, dx$? Explain.

14. For each of the following definite integrals, draw a sketch illustrating the approximations L_4, R_4, M_4, and T_4. Then compute each approximation and compare it with the exact value of the integral computed using the FTC. How could you have used the sketches to predict the results of these comparisons?

   (a) $\displaystyle\int_{-2}^2 x^2 \, dx$ (c) $\displaystyle\int_{-2}^2 (x+1) \, dx$

   (b) $\displaystyle\int_{-2}^2 x^3 \, dx$ (d) $\displaystyle\int_{-\pi/2}^{\pi/2} \sin x \, dx$

15. The following table gives speedometer readings at various times over a one-hour interval.

Speed Readings Over One Hour						
Time (min)	0	10	20	30	40	50	60
Speed (mph)	42	38	36	57	0	55	51

(a) Draw a plausible speed graph for the one-hour period.

(b) Estimate the total distance traveled using
  (i) A trapezoid approximating sum, 6 subdivisions.
  (ii) A left approximating sum, 6 subdivisions.
  (iii) A midpoint approximating sum, 3 subdivisions.
  Which answer is most convincing? Why?

(c) Draw a plausible distance graph for the one-hour period.

16. Find a definite integral for which $5\sum_{k=1}^{3} f(5k)$ is
  (a) A left Riemann sum with 3 equal subintervals.
  (b) A right Riemann sum with 3 equal subintervals.
  (c) A midpoint Riemann sum with 3 equal subintervals.

17. Find a definite integral for which $S = 2\big(f(2) + f(4) + f(6) + f(8)\big)$ is
  (a) A left Riemann sum with 4 equal subintervals.
  (b) A right Riemann sum with 4 equal subintervals.
  (c) A midpoint Riemann sum with 4 equal subintervals.

5.5 Approximating Sums: Interpretations and Applications

In the preceding section we defined the integral as a limit of approximating sums. In symbols,

$$\int_{a}^{b} f(x)\, dx = \lim_{n \to \infty} \sum_{i=1}^{n} f(c_i)\, \Delta x_i.$$

The limit involves some subtleties, but the basic idea of sums (left sums, right sums, midpoint sums, and so on) approximating an integral is natural and straightforward, especially if approached graphically.

In this section we examine approximating sums more closely. In the process we'll both see the integral in another light and apply the idea of approximating sums to several problems of measurement.

Different Views of Approximating Sums

Approximating sums are *discrete* (i.e., stepwise) approximations to integrals of *continuously* varying functions. This important relationship can be interpreted in several useful ways. We discuss three.

Approximating Sums and Simpler Areas

The simplest way to view approximating sums is geometric. Geometrically, approximating sums represent simpler versions (usually polygons) of curvy regions. ▶ The following shaded region, for example, might be called a **polygonal approximation** to the area from a to b under the graph of f.

Most of the pictures in this section and the last reflect this point of view.

A polygonal approximation to a curved area

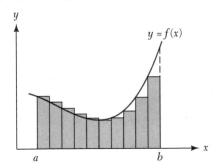

The total area of the polygonal region is the value of a left sum; it's evidently a reasonable approximation to $\int_a^b f$, the true area under the graph.

Approximating Sums and Simpler Functions

Another view of approximating sums for an integral $\int_a^b f$ focuses on the function f rather than on the area under graph. From this point of view, an approximating sum amounts to *replacement of the original function f with a new, simpler function*, one that's linear on each subinterval of the partition in question. In the following graph, for example, the function f of the preceding picture was replaced with a six-step linear approximation.

Replacing f with a piecewise-linear function

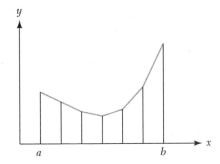

The area beneath is the value of the trapezoid approximating sum T_6.

Weighted Sums

A Riemann sum (left, right, midpoint, or any other kind) for $\int_a^b f$ has the form

$$\sum_{i=1}^{n} f(c_i)\,\Delta x_i = f(c_1)\Delta x_1 + f(c_2)\Delta x_2 + \cdots + f(c_n)\Delta x_n.$$

An ordinary average is another type of weighted sum.

In other words, an approximating sum is a **weighted sum** ◄ of values of f at various points c_i (the sampling points) in the interval $[a, b]$; the weights are the Δx_i. The exact choice of sampling points and the weights attached depend on the type of approximating sum chosen.

EXAMPLE 1   Let $I = \int_0^1 \sin x\,dx$. Let R_4 and M_4 be the midpoint and right approximating sums for I, each with four subdivisions. What are the sampling points? What are the weights? What are the weighted sums?

Convince yourself that the forms are correct.

Solution   Calculation of M_4 and R_4 is routine: ◄

$$R_4 = \sin(0.25) \cdot 0.25 + \sin(0.5) \cdot 0.25 +$$
$$\sin(0.75) \cdot 0.25 + \sin(1) \cdot 0.25 \approx 0.56248;$$
$$M_4 = \sin(0.125) \cdot 0.25 + \sin(0.375) \cdot 0.25 +$$
$$\sin(0.625) \cdot 0.25 + \sin(0.875) \cdot 0.25 \approx 0.460897.$$

The sampling points for R_4 are 0.25, 0.5, 0.75, and 1—the *right* endpoints of four equal subintervals. For M_4, the sampling points are 0.125, 0.375, 0.625, and 0.875, the midpoints of the same subintervals.

The weights in an approximating sum are the coefficients of the sample values $f(c_i)$. In the preceding sums, every sample value has weight $1/4$, the length of

Unequal subintervals would give unequal weights.

the subinterval from which it comes. ◄ This result seems just; there's no obvious reason to favor one value over another. The weighted sums are the numerical values of R_4 and M_4, shown earlier. ∎

Applications of Approximating Sums

Approximating sums are good for more than formally defining the integral. Indeed, compared to symbolic integrals, approximating sums are often simpler conceptually, handier in applications, and easier to calculate—especially for machines. For in-

Such integrals are common in practice.

tegrals that involve complicated formulas or no formulas at all, ◄ approximating sums may be the only practical resort.

Beyond Formulas—Integration with Tabular Data

Integration is important in the real world. Alas, real-world functions seldom come with handy formulas attached. The FTC, for all its other glories, is not much good without symbolic formulas on which to act. Approximating sums are the answer.

EXAMPLE 2   If $v(t)$ is the speed (in miles per hour) of a car at time t (in hours), then $\int_a^b v(t)\,dt =$ miles traveled from $t = a$ to $t = b$. To measure distance traveled from $t = 0$ to $t = 1$, all we need is the integral $\int_0^1 v(t)\,dt$.

Real cars don't adhere to explicit speed formulas. What we *can* observe, practically speaking, are numerical speed data such as the following ones. (Notice the irregular time intervals; real data sets often have gaps.)

Speed Readings over One Hour										
Time (min)	0	5	10	15	20	30	35	40	45	55	60
Speed (mph)	35	38	36	57	71	55	51	23	10	27	35

How far did the car travel in one hour?

Solution   Without more information (e.g., about speeds *between* the times measured), we'll never know exactly. But we can use approximating sums to estimate $\int_0^1 v(t)\,dt$ as well as possible.

These discrete speed data are mathematically compatible with many possible speed functions. The simplest possibility is drawn by "connecting the dots" (lower left): ▶

We converted time from minutes to hours.

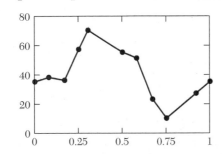

A possible speed function (hrs vs. mph)

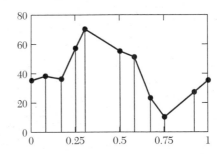

Estimating distance: total area = 40.375

The total bounded area (upper right) is easily computed using ten trapezoids. ▶ We get 40.375—a reasonable estimate for distance traveled over the hour.

The piecewise-linear speed function just plotted isn't the only possibility. *Any* curve through the known data points makes mathematical sense. ▶   The important

We'll spare you the arithmetic.

Some curves are physically more likely than others.

point is that the trapezoid estimate depends only on the 11 observed data, not on the particular curve drawn through them. The following picture illustrates this idea; it shows a new speed function ◀ but the same trapezoid estimate (40.375 miles) for distance.

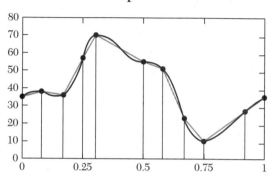

**New speed function (hrs vs mph),
same trapezoid area**

Areas in the Plane; Measuring by Slicing

We first defined the integral in terms of area bounded by the graph of one function. Not surprisingly, integrals can also be used to measure areas of more general plane areas, e.g., areas bounded by two or more graphs. Approximating sums show how and why.

EXAMPLE 3   Using an integral, measure the shaded area in the following picture. (It's bounded by the graphs of $y = \sin x$ and $y = \cos x$.)

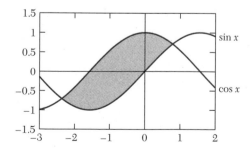

**Area between two curves:
$y = \sin x$ and $y = \cos x$**

Solution   The following picture suggests an approximate solution. We slice the area into vertical strips, approximate each strip with a rectangle, then add up the areas of the rectangles. Here's the picture for ten rectangles: ▶

The first rectangle has height 0.

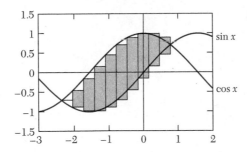

Approximating area with 10 "left-rule" rectangles

Look carefully:

- The curves intersect at $x = -3\pi/4$ and $x = \pi/4$. ▶ (Because $\cos x - \sin x \geq 0$ over the interval $[-3\pi/4, \pi/4]$, the integral gives the true shaded area desired, even though part of the area lies below the x-axis.)

Do you see why?

- The height of each rectangle is determined at its left edge, $x = x_i$; the height is $\cos(x_i) - \sin(x_i)$.

- The total area of all ten rectangles is L_{10}, the left sum with ten equal subdivisions for the function $\cos x - \sin x$ on the interval $[-3\pi/4, \pi/4]$. (It isn't necessary to solve the problem at hand, but calculating L_{10} is easy by machine; we get 2.80513.)

   Geometric intuition says that as $n \to \infty$, the area L_n tends to the exact area we want. By the limit definition of integral, L_n also tends to the integral in question. We conclude:

The area betwen the curves is the integral $\displaystyle\int_{-3\pi/4}^{\pi/4} (\cos x - \sin x)\, dx.$

Finding an exact answer is now easy, thanks to the FTC: ▶

But check the details.

$$\text{Area} = \int_{-3\pi/4}^{\pi/4} (\cos x - \sin x)\, dx$$

$$= \sin x + \cos x\Big]_{-3\pi/4}^{\pi/4}$$

$$= 2\sqrt{2} \approx 2.82843. \qquad \blacksquare$$

Two Basic Area Formulas

The simplest plane regions for area calculations are bounded above by the graph of one function, below by the x-axis, and on the left and right by vertical lines, as follows:

Area at its simplest

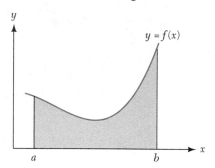

In this simplest case, the integral $\int_a^b f(x)\,dx$ measures the shaded area.

More interesting, but slightly more complicated, regions may be bounded by various types of curves—not necessarily all graphs of functions. Two common types follow. ◄

Some plane regions are of neither type shown. However, many regions, even quite complicated ones, are combinations of the two following types.

Region 1: top and bottom curves   **Region 2: left and right curves**

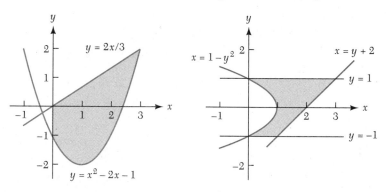

The following rules apply to finding areas such as the preceding ones.

Fact Let f and g be continuous functions.

Integrating in x Let R be the region bounded above by $y = g(x)$, below by $y = f(x)$, on the left by $x = a$, and on the right by $x = b$. Then R has area

$$\int_a^b \big(g(x) - f(x)\big)\,dx.$$

Integrating in y Let R be the region bounded on the right by $x = g(y)$, on the left by $x = f(y)$, below by $y = c$, and above by $y = d$. Then R has area

$$\int_c^d \big(g(y) - f(y)\big)\,dy.$$

For Region 2 in the preceding pictures, and other similar regions, integrating in y may be much simpler.

Using the Rules

Using the rules effectively may require a combination of graphical, algebraic, and symbolic operations. We illustrate by example.

E X A M P L E 4   Find the area of Region 1 in the preceding picture.

S o l u t i o n   The graphs *appear* to intersect at $x = 3$. Indeed they do: By the quadratic formula, ▶

Convince yourself of this!

$$\frac{2}{3}x = x^2 - 2x - 1 \iff x = 3 \text{ or } x = -\frac{1}{3}.$$

(This shows, in fact, that the two graphs intersect also at $x = -1/3$, to the left of the shaded region.) Now, by the first rule (and a close look at the picture):

$$\text{Area} = \int_0^3 \left(\tfrac{2}{3}x - (x^2 - 2x - 1)\right)\, dx.$$

An easy antiderivative calculation shows the result to be 6.   ■

E X A M P L E 5   Find the area of Region 2 in the preceding picture.

S o l u t i o n   Integration in x seems impractical, since defining upper and lower curves is troublesome. However, a close look at the figure shows that the second form of the preceding Fact applies, as follows:

$$\text{Area} = \int_{-1}^1 \left(y + 2 - (1 - y^2)\right)\, dy.$$

Another easy integration—this time in the variable y—shows that the area is 8/3.   ■

Any Way You Slice It

Some areas can be calculated by integration either in x or in y.

E X A M P L E 6   Find the area of the region R bounded by the curves $x = 0$, $y = 2$, and $y = e^x$.

Solution   Here's the picture:

Graphs of $y = e^x$, $y = 2$

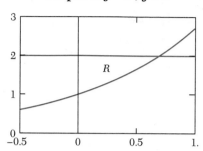

The two curves intersect where $e^x = 2$, i.e., at $x = \ln 2$. Therefore

$$\text{Area of } R = \int_0^{\ln 2} \left(2 - e^x \right) dx.$$

Why is this legitimate?

To find the same area in terms of y, we rewrite the equation $y = e^x$ in the equivalent form $x = \ln y$; ◄   the associated curve becomes the right boundary. Then

$$\text{Area of } R = \int_1^2 \ln y \, dy.$$

Do you recall this? If not, check it by differentiation.

Either integral is easy to calculate. For the second one (recalling that $y \ln y - y$ is an antiderivative of $\ln y$), ◄   we get

$$\text{Area} = y \ln y - y \Big]_1^2 = 2 \ln 2 - 1 \approx 0.3863. \qquad ■$$

BASICS

1. Let A be the region between the curves $y = x$ and $y = x^2$.
   (a) Draw a sketch of the region A.
   (b) Use 5 left rule rectangles to approximate the area of A.
      (Give a numerical answer.)
   (c) Draw a picture (similar to the one on p. 391) that illustrates the sum you computed in part (b). Use this picture to decide whether your answer overestimates or underestimates the area of A.
   (d) Find the area of A *exactly* by integration.

2. Find the area of the region bounded by the curves $y = \sin x$ and $y = \cos x$, and the lines $x = 1$ and $x = 3$ *exactly* (i.e., express your answer in terms of values of elementary functions rather than as a decimal number).

3. In this exercise, integration in x is used to find the area of Region 2 shown on page 392.
   (a) Find the area of the region bounded below by the x-axis, above by the line $y = 1$, on the right by the line

$x = y + 2$, and on the left by the curve $x = 1 - y^2$.
   (b) Find the area of the region bounded above by the x-axis, below by the line $y = -1$, on the right by the line $x = y + 2$, and on the left by the curve $x = 1 - y^2$.
   (c) Find the area of the region bounded by the lines $x = 1$, $x = y + 2$, and $y = 1$.
   (d) Use the results of parts (a), (b), and (c) to find the area of Region 2.

4. This exercise provides yet another approach to finding the area of Region 2 shown on page 392.
   (a) Find the area of the region bounded by the line $x = 0$ and the curve $x = 1 - y^2$.
   (b) Find the area of the region bounded by the lines $x = y + 2$, $y = -1$, and $x = 3$.
   (c) Find the area of the region bounded by the lines $x = 0$, $x = 3$, $y = 1$, and $y = -1$.
   (d) Use the results of parts (a), (b), and (c) to find the area of Region 2.

In Exercises 5–16, sketch the region bounded by the given curves and find the area of the region *exactly*.

5. $y = x^4$, $y = 1$

6. $y = x$, $y = x^3$

7. $y = x^2$, $y = x^3$

8. $y = x^2 - 1$, $y = x + 1$

9. $x = y^2 - 4$, $y = 2 - x$

10. $y = \sqrt{x}$, $y = 0$, $x = 4$

11. $y = \sqrt{x}$, $y = x^2$

12. $y = 9(4x + 5)^{-1}$, $y = 2 - x$

13. $y = 9(4x^2 + 5)^{-1}$, $y = 2 - x^2$

14. $y = 2 + \cos x$, $y = \sec^2 x$, $x = -\pi/4$, $x = \pi/6$

15. $y = e^x$, $y = 0$, $x = 0$, $x = 1$

16. $y = 2^x$, $y = 5^x$, $x = -1$, $x = 1$

5.6 Chapter Summary

This chapter introduced the **integral**. The derivative and the integral are the two most important concepts of calculus. Remarkably, they're closely related; the **fundamental theorem of calculus** describes how and why.

The Integral as Signed Area.   Section 5.1 introduced the integral graphically, as measuring **signed area**. ("Signed" means that area below the x-axis counts negatively.) The graphical viewpoint, though insufficient for *exact* calculations, explains clearly what the integral is and why some of its elementary properties hold.

The Area Function.   Given a function f and a starting point $x = a$, we defined the related **area function** A_f:

$$A_f(x) = \int_a^x f = \text{signed area bounded by } f \text{ from } a \text{ to } x.$$

Simple examples suggest that the area function is an **antiderivative** for f. This idea, suitably embellished, is the simplest form of the **fundamental theorem of calculus**.

The Fundamental Theorem.   Put more formally, the fundamental theorem says that if A_f is defined as we just stated, then

$$\frac{d}{dx}\left(A_f(x)\right) = f(x).$$

This result connects derivatives and integrals, the two main calculus ideas. For that reason, it deserves to be called *fundamental*.

The same idea, restated in another form, is called the **second version of the fundamental theorem**. If F is *any* antiderivative of f, then

$$\int_a^b f(x)\,dx = F(b) - F(a).$$

In other words,

To find $\int_a^b f$, find any antiderivative, plug in endpoints, and subtract.

The Integral as a Limit of Approximating Sums.   Like the derivative, the integral is defined rigorously as a limit of approximating sums. Section 5.4 described the procedure. Although symbolically messy, approximating sums make good graphical sense. Several varieties were introduced, compared, and calculated.

Approximating Sums and Applications of the Integral.   Thinking of the integral as a limit of sums suggests a variety of applications. Several were presented. The main point is that the integral is capable of various interpretations.

One application of the integral concerns rates and amounts:

If $f(t)$ tells the rate at which a quantity varies at time t, then $\int_a^b f$ tells the amount by which the same quantity changes over the interval from $t = a$ to $t = b$.

Another application of the integral is to calculate areas of plane regions defined by graphs of functions. Such regions come in many shapes and forms. Choosing the best integral to calculate a given area can be a challenging problem. We illustrated various techniques for doing so.

6

Finding Antiderivatives

6.1 Antiderivatives: The Idea

Integrals, Antiderivatives, and the FTC

This chapter is about antidifferentiation—finding, for a given function f, a new function F for which $F' = f$. The fundamental theorem of calculus ▶ (FTC) offers ample motivation for doing so. It says that if f is a continuous function and F is any antiderivative of f, then

See Section 5.3.

$$\int_a^b f(x)\,dx = F(b) - F(a).$$

Finding a nice ▶ antiderivative function F, in effect, translates an integral problem into something easier: plugging a and b into F. A simple example illustrates this principle and its importance.

Here, "nice" means "easy to evaluate."

EXAMPLE 1 Calculate $\displaystyle\int_0^\pi \sin x\,dx$. What does the answer mean graphically?

Solution Any function of the form $F(x) = -\cos x + C$ is an antiderivative of $f(x) = \sin x$. For simplicity, let's use $C = 0$. ▶ By the FTC,

Other values of C give the same answer.

$$\int_0^\pi \sin x\,dx = -\cos x \Big]_0^\pi = -\cos \pi + \cos 0 = 2.$$

Geometrically, the integral represents the area under one arch (from $x = 0$ to $x = \pi$) of the sine function. Surprisingly, this area measures exactly 2 square units. ■

Finding Antiderivatives

To use the FTC, we need first to find an antiderivative function F. Must there *be* an antiderivative? If so, how many antiderivatives are there?

It's reasonable to do so. If f is very badly behaved, $\int_a^b f$ may not even exist.

Existence.   If f is continuous (we'll assume so throughout this chapter) ◄ , the short answer—but not the whole answer—is yes: f does have an antiderivative. In fact, the other version of the FTC says that the area function

$$A_f(x) = \int_a^x f(t)\,dt$$

is an antiderivative of f.

It's nice to know in the abstract that every continuous function f has an antiderivative F. A better question for us, however, is whether f has an antiderivative function that we can use. To be useful for evaluating integrals, an antiderivative F must have a concrete *formula*, one that accepts numerical inputs and produces numerical outputs.

Most recently in Example 1. See also Section 5.3.

See the Fact on page 367.

How Many Antiderivatives?   As we've seen repeatedly, ◄ antiderivatives are not unique. Far from it: If $F(x)$ is an antiderivative of $f(x)$, then so is $F(x)+C$, for any constant C. Thus, a given function f has infinitely many different antiderivatives.

Not *very* different, however: We showed in Section 5.3 ◄ that any two antiderivatives can differ *only* by an additive constant on any domain interval. Finding *one* antiderivative is essentially tantamount to finding *all* antiderivatives.

Elementary Functions, Derivatives, and Antiderivatives

Functions with formulas are called **elementary**; they're built from the standard function **elements**—polynomials, trigonometric and exponential functions, and their inverses. ◄

Roughly speaking, elementary functions are those an inexpensive scientific calculator can handle.

Every elementary function has an elementary derivative. (This is just a pompous way of saying that differentiating a formula gives another formula.) The situation for antiderivatives is different: An antiderivative of an elementary function may or may not be elementary. Even deciding whether a function *has* an elementary antiderivative—let alone finding one—can be very difficult indeed. Thus, the generic problem of this chapter is as follows:

> Given an elementary function f, find an elementary function F such that $F' = f$.

Either by human or by machine. Maple's int *command doesn't always succeed, for example.*

It cannot always be solved. ◄   This problem is sometimes called **integration in closed form**; an elementary solution (if one exists) is called a **closed-form solution**.

Antidifferentiation in Closed Form: Art, Science, or Parlor Trick?

The uncertainty just discussed adds an element of intrigue to our problem, elevating it somewhat above the mundane level of routine calculation. (Differentiation, by contrast, *is* routine calculation.) Finding antiderivatives can be challenging, but the rewards are proportional to the effort.

We'll attack the antiderivative problem with a modest arsenal of antidifferentiation techniques. Like differentiation rules, antidifferentiation techniques can reduce a complicated problem to one or more simpler ones. But unlike differentiation rules, antidifferentiation rules don't always succeed; choosing the "right" antidifferentiation rule or rules combines art, science, intuition, and (sometimes) luck.

Notation and Terminology

We'll use several standard notations and technical terms in our tour of the antiderivative problem

Indefinite Integrals The expression

$$\int f(x)\, dx$$

is called an **indefinite integral**; f is the **integrand** and x is the **variable of integration.** The indefinite integral is a symbolic shorthand for the antiderivative problem. ▶ For example,

$$\int \cos x\, dx = \sin x + C$$

means that the antiderivatives of $\cos x$ are precisely the functions $\sin x + C$, where C, the **constant of integration**, may take any value.

One caveat is as follows. Interpreted scrupulously, the indefinite integral $\int f(x)\, dx$ denotes not just one function but many—the "family" of all possible antiderivatives of f. ▶ The distinction isn't always important in practice. Writing, say, $\int \cos x\, dx = \sin x$ seldom causes confusion.

Definite Integrals The expression

$$\int_a^b f(x)\, dx,$$

with endpoints definitely specified, is called a **definite integral**. ▶

The FTC links definite and indefinite integrals. In words,

The definite integral $\int_a^b f(x)\, dx$ equals the change $F(b) - F(a)$ from $x = a$ to $x = b$ in any indefinite integral.

> **Antidifferentiation by Machine.** Many high-powered mathematical computer programs attack the (hard) problem of elementary antidifferentiation. *Maple's* int command, for example, tries to find antiderivatives in closed form. For the good reasons already cited, any antiderivative program sometimes fails; a given problem may have no closed-form solution.
>
> Until recent times there was no reliable method of deciding, once and for all, whether a function has an elementary antiderivative. In the late 1960s, R. H. Risch discovered an algorithm that determines whether a function has an elementary antiderivative and, if it does, finds one. Although the Risch algorithm and its more recent variants can be implemented by computer, most modern software uses faster, less cumbersome methods.

Elementary Antiderivatives: Why Bother Looking?

Many useful functions do have elementary antiderivatives. We'll have great fun finding them. But why bother? The question is a serious one. As we'll soon see (starting in Chapter 7 of Volume 2), numerical methods let us compute many

The name and notation aren't perfect; integrals and antiderivatives aren't really the same thing.

For nice functions f, antiderivatives differ only by additive constants.

A definite integral has one value, not many.

definite integrals to high accuracy without fussing over antiderivatives. Following are three good reasons for our trouble.

Formulas for Success Elementary functions—those with formulas—are in many ways the simplest and most convenient possible functions. In applied mathematics, finding a formula that usefully models an interesting phenomenon represents the best imaginable outcome.

As a simple but important example, consider how an object moves along a straight line, starting from rest at time $t = 0$. If the object's acceleration a is constant (as in free fall), then the velocity function v is linear and the distance-traveled function s is quadratic. More precisely,

$$\text{If } a(t) = a, \text{ then } v(t) = at, \text{ and } s(t) = \frac{at^2}{2}.$$

Thus, a simple *formula* for acceleration leads in this case to almost equally simple *formulas* for velocity and position, its first two antiderivatives.

Not all applications work so smoothly, of course. Complex combinations of forces may produce complicated acceleration functions, perhaps without elementary antiderivatives. Nevertheless, the point remains: The practical and theoretical value of *having* antiderivative formulas amply justifies some effort in finding them.

Predicting Planetary Motion.   Planetary motion, the subject of **celestial mechanics**, has fascinated observers throughout history. The search for formulas to describe and predict planetary motion has continued for many thousands of years.

Nicolaus Copernicus, Johannes Kepler, and Galileo Galilei, working in the 16th and early 17th centuries, began the task of describing planetary and gravitational phenomena in modern mathematical terms. Isaac Newton, in his *Principia* (1687), used the tools of calculus to extend and unify earlier work. Newton showed, for instance, that his inverse-square law for gravitation (gravity varies inversely with the square of the distance) implies Kepler's first law—that the orbit of a planet is an ellipse with the sun as one focus.

Exact Answers Evaluation of integrals in closed form gives exact answers. Numerical results, no matter how accurate, are only approximate. The difference can be important. For example, the fact that

$$\int_{-1}^{1} \frac{2}{1 + x^2}\, dx \approx 3.141592614$$

is easy to show numerically. It's not very exciting, however; one number is much like another. But the fact that

$$\int_{-1}^{1} \frac{2}{1 + x^2}\, dx = 2 \arctan x \Big]_{-1}^{1} = \pi,$$

which depends on antidifferentiation, is an interesting surprise—how in the world does π arise?

Integrals with Parameters: Many for the Price of One Many useful integrals involve parameters—constants whose values we don't want to specify in advance. Consider, for example, the integral

$$\int_{0}^{\pi} \sin(kx)\, dx,$$

for k a nonzero integer. ▶ Without a specific value for k, numerical methods are helpless. ▶ Antidifferentiation solves the problem for all values of k at one blow: ▶

$$\int_0^\pi \sin(kx)\,dx = -\frac{\cos(kx)}{k}\Big]_0^\pi = \frac{1 - \cos(k\pi)}{k}.$$

If $k = 0$, the problem is trivial.

Think about it. Do you see why?

Check details.

Thus, the parity of k determines the value of the integral: It's zero if k is even, $2/k$ if k is odd.

Derivatives and Antiderivatives: Basic Strategies

Antidifferentiation is differentiation in reverse. The expert antidifferentiator, therefore, has a ready mastery of the easier skill of differentiation and remembers this rule:

> *Antidifferentiation is hard; differentiation is easy. Therefore, check antiderivatives by differentiation.*

EXAMPLE 2 Is $\int e^x \sin x \, dx = e^x \cos x + C$?

Solution No. Here's why: $(e^x \cos x)' = e^x \cos x - e^x \sin x \neq e^x \sin x$. ■

Caveat. Some care is necessary. When checking antiderivatives, one should realize that some functions (especially trigonometric functions) can be written in more than one way.

EXAMPLE 3 Is $\int \sin(2x)\,dx = \sin^2(x) + C$?

Solution Let's check. By the chain rule,

$$\left(\sin^2(x)\right)' = 2\sin x \cos x.$$

We hoped for $\sin(2x)$, so we're tempted to answer no. The right answer, however, is yes. One of the many trigonometric identities says that

$$2\sin x \cos x = \sin(2x).$$

We stand corrected. ■

Basic Antiderivative Formulas

Differentiation techniques (e.g., the product and quotient rules) start with a few known derivatives; combining them properly produces new derivatives.

Antidifferentiation, too, requires a basis of known formulas to which others are reduced. We'll use the following ones.

Basic Antiderivative Formulas

$$\int x^k \, dx \;=\; \frac{x^{k+1}}{k+1} + C \quad \text{(if } k \neq -1\text{)}$$

$$\int \frac{1}{x} \, dx \;=\; \ln|x| + C$$

$$\int e^x \, dx \;=\; e^x + C$$

$$\int a^x \, dx \;=\; \frac{a^x}{\ln a} + C \quad \text{(if } a \neq 1\text{)}$$

$$\int \sin x \, dx \;=\; -\cos x + C$$

$$\int \cos x \, dx \;=\; \sin x + C$$

$$\int \sec^2 x \, dx \;=\; \tan x + C$$

$$\int \frac{dx}{\sqrt{1-x^2}} \;=\; \arcsin x + C$$

$$\int \frac{dx}{1+x^2} \;=\; \arctan x + C$$

Notice:

Why They're True   Antiderivative rules are nothing more than derivative rules in reverse. The rule $\int \cos x \, dx = \sin x + C$, for instance, means simply that

$$(\sin x + C)' = \cos x + 0 = \cos x.$$

A Subtlety: Logs and Absolute Values   The formula $\int \frac{1}{x} \, dx = \ln|x| + C$ has an unexpected ingredient: the absolute value. After all, $\ln x$ itself is an antiderivative for $1/x$—why bother with an absolute value?

The answer has to do with domains. The function $f(x) = 1/x$ is defined for all $x \neq 0$; ideally, an antiderivative should have the same domain. Unfortunately, the unadorned logarithm function $F(x) = \ln x$ accepts only *positive* inputs x. Use of $G(x) = \ln|x|$ solves this problem neatly. This G has both the right domain (positive and negative numbers) and the right antiderivative. The following graphs give convincing evidence. ◄

For more formal arguments, see the exercises at the end of this section.

Graph of $y = 1/x$

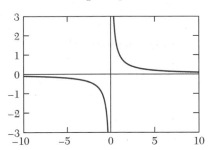

Graph of $y = \ln |x|$

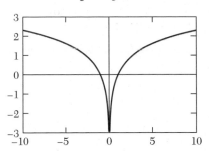

Combining Antiderivatives: Sums and Constant Multiples

Every derivative rule, read backwards, says something about antiderivatives. For sums and constant multiples, the rules are especially simple. ▶ If f and g are continuous functions and k is any constant, then

We'll read the chain rule backwards in the next section.

$$\int \big(f(x) + g(x)\big)\, dx = \int f(x)\, dx + \int g(x)\, dx;$$

$$\int k \cdot f(x)\, dx = k \cdot \int f(x)\, dx.$$

EXAMPLE 4   Find $\displaystyle\int (3\cos x - 2e^x)\, dx$.

Solution   Use the rules and the basic formulas:

$$\int (3\cos x - 2e^x)\, dx = \int 3\cos x\, dx - \int 2e^x\, dx$$

$$= 3\int \cos x\, dx - 2\int e^x\, dx$$

$$= 3\sin x - 2e^x + C.$$

The answer, as usual, is easy to check by differentiation.   ■

Guess and Check

Educated guessing is a perfectly respectable strategy for antidifferentiation. Because differentiation is simple, most guesses are easily checked. Incorrect guesses, moreover, are often easy to repair. The next two examples show what we mean.

EXAMPLE 5   Find $\int \cos(2x)\, dx$ by guessing.

Solution   Since $\int \cos x\, dx = \sin x + C$, we might—just guessing—try $\sin(2x)$ as an antiderivative. Differentiation shows that we missed (but by only a factor of 2).

$$\big(\sin(2x)\big)' = \cos(2x) \cdot 2 \neq \cos(2x).$$

A minor alteration is all that's needed:

$$\left(\frac{\sin(2x)}{2}\right)' = \cos(2x) \implies \int \cos(2x)\, dx = \frac{\sin(2x)}{2} + C.$$ ■

E X A M P L E 6   Find $\int 6/(1 + 4x^2)\,dx$ by creative guessing.

Check this; use the chain rule carefully.

S o l u t i o n   The integrand's form suggests an arctangent, perhaps with some constants attached, as in $F(x) = A\arctan(Bx)$. Differentiation gives ◄

$$F'(x) = \big(A\arctan(Bx)\big)' = \frac{A}{1 + (Bx)^2} \cdot B = \frac{AB}{1 + B^2 x^2}.$$

Our desired antiderivative therefore has $AB = 6$ and $B^2 = 4$. Setting $A = 3$ and $B = 2$ does nicely; therefore, $F(x) = 3\arctan(2x)$ is a suitable antiderivative.   ■

BASICS

In Exercises 1–12, evaluate the antiderivative. Check your answer by differentiation.

1. $\int (3x^5 + 4x^{-2})\,dx$

2. $\int \dfrac{dx}{3x}$

3. $\int \dfrac{dx}{4\sqrt{1 - x^2}}$

4. $\int \dfrac{3}{x^2 + 1}\,dx$

5. $\int 3e^{4x}\,dx$

6. $\int \big(2\sin(3x) - 4\cos(5x)\big)\,dx$

7. $\int 4\sec^2(3x)\,dx$

8. $\int \big(1 + \sqrt{x}\big)^2\,dx$

9. $\int (x + 1)^2 \sqrt[3]{x}\,dx$

10. $\int \dfrac{(3 - x)^2}{x}$

11. $\int \dfrac{(1 - x)^3}{\sqrt{x}}$

12. $\int e^x(1 - e^x)\,dx$

In Exercises 13–15, determine whether the antiderivative formula is true or false by differentiating the right side.

13. $\int e^{\sin x}\,dx = e^{-\cos x} + C$

14. $\int \dfrac{dx}{1 + 4x^2} = \arctan(2x) + C$

15. $\int (1 + \tan^2 x)\,dx = \tan x + C$

FURTHER EXERCISES

Some antiderivatives are difficult to *find* from scratch, but may be *checked* easily by differentiation. In Exercises 16–22, verify the antiderivative formula by differentiation.

16. $\int \ln x\,dx = x\ln x - x + C$

17. $\int \arctan x\,dx = x\arctan x - \dfrac{1}{2}\ln(1 + x^2) + C$

18. $\int \dfrac{dx}{x^2 + a^2} = \dfrac{1}{a}\arctan\left(\dfrac{x}{a}\right) + C$

19. $\int \tan x\,dx = \ln|\sec x| + C$

　　[**HINT:** Recall that $\big(\ln|x|\big)' = 1/x$ for all $x \neq 0$.]

20. $\int \sec x\,dx = \ln|\sec x + \tan x| + C$

21. $\int \dfrac{dx}{1 - x^2} = \dfrac{1}{2}\ln\left|\dfrac{1 + x}{1 - x}\right| + C$

　　[**HINT:** $\ln\left|\dfrac{1 + x}{1 - x}\right| = \ln|1 + x| - \ln|1 - x|.$]

22. $\int \arcsin x\,dx = x\arcsin x + \sqrt{1 - x^2} + C$

23. (a) Use the trigonometric identity $2\sin^2 x = 1 - \cos(2x)$ to evaluate $\int \sin^2 x\,dx$.

　　(b) Explain why $\displaystyle\int_0^{2\pi} \sqrt{1 - \cos(2x)}\,dx \neq \sqrt{2}\int_0^{2\pi} \sin x\,dx.$

Some antiderivatives—trigonometric ones, especially—can be written in more than one way. In Exercises 24–26, verify the formula using the trigonometric identities $\sin^2 x + \cos^2 x = 1$, $\sin(2x) = 2\cos x \sin x$, and $\cos(2x) = 2\cos^2 x - 1$.

24. $\displaystyle\int 2\sin x \cos x \, dx = \sin^2 x + C = -\cos^2 x + C$

25. $\displaystyle\int \cos(2x) \, dx = \frac{1}{2}\sin(2x) + C = \cos x \sin x + C$

26. $\displaystyle\int \cos^2 x \, dx = \frac{x}{2} + \frac{1}{2}\cos x \sin x + C$
$$= \frac{x}{2} + \frac{1}{4}\sin(2x) + C$$

In Exercises 27–36 evaluate the antiderivative. Check your answers by differentiation.

27. $\displaystyle\int \frac{x^2}{1+x^2}\, dx \quad \left[\textbf{HINT: } \frac{x^2}{1+x^2} = 1 - \frac{1}{1+x^2}.\right]$

28. $\displaystyle\int \frac{4x^2 + 3x + 2}{x+1}\, dx$
$\left[\textbf{HINT: } \dfrac{4x^2 + 3x + 2}{x+1} = 4x - 1 + \dfrac{3}{x+1}.\right]$

29. $\displaystyle\int \frac{x-1}{x+1}\, dx \quad [\textbf{HINT: } x - 1 = (x+1) - 2.]$

30. $\displaystyle\int \frac{dx}{\sqrt{x-1} + \sqrt{x+1}}$
$\left[\textbf{HINT: } \left(\sqrt{x-1} + \sqrt{x+1}\right)\left(\sqrt{x-1} - \sqrt{x+1}\right) = -2.\right]$

31. $\displaystyle\int \tan^2 x \, dx$
$\left[\textbf{HINT: } \tan^2 x = \dfrac{\sin^2 x}{\cos^2 x} = \dfrac{1 - \cos^2 x}{\cos^2 x} = \sec^2 x - 1.\right]$

32. $\displaystyle\int \frac{dx}{1+\sin x}\, dx$
$[\textbf{HINT: } (1 + \sin x)(1 - \sin x) = 1 - \sin^2 x = \cos^2 x.]$

33. $\displaystyle\int \frac{dx}{1 + 9x^2} \quad [\textbf{HINT: } \text{See Example 6.}]$

34. $\displaystyle\int \frac{6}{9 + x^2}\, dx \quad [\textbf{HINT: } 9 + x^2 = 9\left(1 + (x/3)^2\right).]$

35. $\displaystyle\int \frac{2}{\sqrt{1 - 9x^2}}\, dx$

36. $\displaystyle\int \frac{8}{\sqrt{4 - x^2}}\, dx$

6.2 Antidifferentiation by Substitution

The simplest and most important antidifferentiation technique is called, variously, **direct substitution**, **u-substitution**, ▶ or just plain **substitution**.

Antidifferentiation by any method is nothing more than differentiation in reverse. Substitution, in particular, can be understood as the chain rule running backward. We'll explain the connection more precisely later in this section. In the meantime, watch the chain rule appear again and again as we check our answers.

For obscure reasons, the letter u is almost invariably used in this setting. We'll follow convention.

Substitution: How it Works

We'll use examples ▶ to see both *how* substitution works and *that* it works.

Read them all carefully; they're the heart of this section.

EXAMPLE 1 Find $\int 2x\,\cos(x^2)\,dx$. (i.e., antidifferentiate $2x\,\cos(x^2)$).

Solution Let $u = u(x) = x^2$; then $du/dx = 2x$, so $du = 2x\,dx$. (Whoa—can du and dx legally be separated this way? The short answer is yes, but we'll return to such questions later.) Substituting everything into the original integral gives

$$\int \cos(x^2)\, 2x \, dx = \int \cos u \, du.$$

The last integral is simpler than the first. The rest is routine. We first antidifferentiate, then substitute back for u:

$$\int \cos u \, du = \sin u + C = \sin(x^2) + C.$$

Watch the chain rule in action.

Did it work? An easy differentiation shows that it did: ◄

$$\left(\sin(x^2) + C\right)' = \cos(x^2) \cdot 2x,$$

as we wanted.   ■

Separating du and dx might seem suspicious, for example.

That was easy, quick, and possibly dirty. ◄   Notice, though, that the method—dirty or not—did work. It produced an answer we could easily check.

EXAMPLE 2   Find $\int \dfrac{x}{1+x^2}\,dx$.

Solution   Let $u(x) = 1 + x^2$; then $du/dx = 2x$ and $du = 2x\,dx$. Now we substitute:

$$\int \frac{x}{1+x^2}\,dx = \frac{1}{2}\int \frac{2x}{1+x^2}\,dx = \frac{1}{2}\int \frac{du}{u}.$$

The first step was for convenience; multiplying by 2 inside the integral sign "made" the desired du. (We compensated for this by dividing by 2 outside.)

In its new, simpler form, the antiderivative problem is easier:

$$\frac{1}{2}\int \frac{du}{u} = \frac{1}{2}\ln|u| + C = \frac{1}{2}\ln(1+x^2) + C.$$

Checking the answer is another chain-rule exercise:

$$\left(\frac{1}{2}\ln(1+x^2) + C\right)' = \frac{1}{2}\cdot\frac{2x}{1+x^2} = \frac{x}{1+x^2}.$$ ■

The Idea of u-Substitution: Trading One Integral for Another

As the preceding examples show, a successful *u*-substitution transforms one indefinite integral into another, the second being simpler than the first. The process has three steps:

Substitute   Judiciously choose a function $u = u(x)$, and write $du = u'(x)\,dx$. Then substitute both u and du into the original integral, $\int f(x)\,dx$, to produce a new one of the form $\int g(u)\,du$.

Antidifferentiate   Solve $\int g(u)\,du$ as an antidifferentiation problem in u; i.e., find a function $G(u)$ for which $G'(u) = g(u)$.

Resubstitute   Substitute *back* to eliminate u. The result, $F(x) = G\big(u(x)\big)$, is an antiderivative of f; so is any function of the form $F(x) + C$.

Examples: A Substitution Sampler

This makes success in substitution especially satisfying.

Substitution is simple enough in theory. In practice, however, finding the right substitution is as much art as science. ◄   In both realms, practice (false starts and all) makes perfect. The following examples illustrate some of the possibilities and some useful tricks of the trade.

Choosing *u*: Wise and Foolish Choices.   The first step—choosing *u* properly—is the tricky part. Substitution works best when (as in the preceding examples) both

u and du appear conveniently in the original integrand. Morever, the new integral should be simpler or more familiar than the old.

EXAMPLE 3 Find $\int \sin^3 x \cos x \, dx$.

Solution If $u = \sin x$, then $du = \cos x \, dx$. With u and du staring at us, the situation looks promising. The rest is routine:

$$\int \sin^3 x \cos x \, dx = \int u^3 \, du = \frac{u^4}{4} + C = \frac{\sin^4 x}{4} + C. \qquad \blacksquare$$

EXAMPLE 4 Find $\int \dfrac{x}{1 + x^4} \, dx$.

Solution No obvious substitution suggests itself. We might try $u = 1 + x^4$, but then $du = 4x^3 \, dx$, and nothing similar appears in the given integrand. A better choice, it turns out, is $u = x^2$. Then $du = 2x \, dx$, and ▶

Watch the 2's carefully.

$$\int \frac{x}{1 + x^4} \, dx = \frac{1}{2} \int \frac{2x}{1 + x^4} \, dx = \frac{1}{2} \int \frac{du}{1 + u^2}.$$

We've made progress at last. The last antiderivative is standard:

$$\frac{1}{2} \int \frac{du}{1 + u^2} = \frac{1}{2} \arctan u + C = \frac{1}{2} \arctan(x^2) + C.$$

Checking the answer involves, as usual, the chain rule. ▶ $\qquad \blacksquare$ *The details are left to you.*

Traveling Constants. Multiplicative constants can move freely (and legally) in and out of integrals. We exploit this freedom in the next example, as we did in Example 2.

EXAMPLE 5 Find $\int \dfrac{3x}{5x^2 + 7} \, dx$.

Solution The numerator is almost (except for multiplicative constants) the derivative of the denominator; this brings $\int du/u$ to mind. If we set $u = 5x^2 + 7$, then $du = 10x \, dx$, and ▶

Watch the constants carefully.

$$\begin{aligned}
\int \frac{3x}{5x^2 + 7} \, dx &= \frac{3}{10} \int \frac{10x}{5x^2 + 7} \, dx \\
&= \frac{3}{10} \int \frac{du}{u} \\
&= \frac{3}{10} \ln |u| + C \\
&= \frac{3}{10} \ln (5x^2 + 7) + C.
\end{aligned}$$

Differentiation shows that we're right:

$$\left(\frac{3}{10} \ln (5x^2 + 7) + C \right)' = \frac{3}{10} \frac{10x}{5x^2 + 7} = \frac{3x}{5x^2 + 7}. \qquad \blacksquare$$

Inverse Substitutions: Writing x and dx in Terms of u and du

In the examples so far, we've written u and du in terms of x and dx. Sometimes it's simpler to write x and dx in terms of u and du. The next two examples illustrate this strategy.

EXAMPLE 6   Find $\displaystyle\int \frac{x}{\sqrt{2x+3}}\, dx$.

Convince yourself.

Solution   We start as usual. If $u = \sqrt{2x+3}$, then ◄ $du = dx/\sqrt{2x+3}$. Substitution gives

$$\int \frac{x}{\sqrt{2x+3}}\, dx = \int x\, \frac{dx}{\sqrt{2x+3}} = \int x\, du.$$

So far, so good, but what about that remaining x?

The trick is to write x in terms of u, as follows:

$$u = \sqrt{2x+3} \implies u^2 = 2x + 3 \implies \frac{u^2 - 3}{2} = x.$$

Now we're on our way:

$$\int x\, du = \int \frac{u^2 - 3}{2}\, du$$

$$= \frac{u^3}{6} - \frac{3u}{2} + C$$

$$= \frac{(2x+3)^{3/2}}{6} - \frac{3\sqrt{2x+3}}{2} + C.$$

For variety, let's redo the problem a bit differently, still using $u = \sqrt{2x+3}$. Notice first that

$$\frac{u^2 - 3}{2} = x \implies \frac{2u}{2}\, du = dx \implies u\, du = dx.$$

Finally, substitution for x, dx, and $\sqrt{2x+3}$ gives

$$\int \frac{x}{\sqrt{2x+3}}\, dx = \int \frac{u^2 - 3}{2u}\, u\, du = \int \frac{u^2 - 3}{2}\, du,$$

just as before.   ■

EXAMPLE 7   Find $\displaystyle\int \frac{dx}{1 + \sqrt{x}}$.

This is not the only possibility: $u = 1 + \sqrt{x}$ also works.

Solution   Let $u = \sqrt{x}$. ◄   Then easy calculations give

$$x = u^2 \qquad \text{and} \qquad dx = 2u\, du.$$

Substituting for x and dx gives

$$\int \frac{dx}{1 + \sqrt{x}} = \int \frac{2u}{1 + u}\, du.$$

The last integral is certainly simpler than the first. Another substitution makes it simpler still. If $v = 1 + u$, then $dv = du$ and $u = v - 1$, so ▶

Check the algebra.

$$\int \frac{2u}{1+u}\, du = \int \frac{2(v-1)}{v}\, dv = 2\int \left(1 - \frac{1}{v}\right) dv = 2(v - \ln|v|) + C.$$

Finally, we substitute back for x:

$$2(v - \ln|v|) + C = 2(1 + u - \ln|1 + u|) + C = 2 + 2\sqrt{x} - 2\ln|1 + \sqrt{x}| + C. \quad ∎$$

Substitution in Definite Integrals

The next examples show two ways of evaluating *definite* integrals by substitution.

EXAMPLE 8   Find $\displaystyle\int_0^{\sqrt{\pi/2}} 2x\, \cos(x^2)\, dx$.

Solution   (**Antidifferentiate in x; Plug in Endpoints**)   We write $f(x) = 2x\, \cos(x^2)$. If $F(x)$ is any antiderivative of $f(x)$, then (by the FTC) $F(\sqrt{\pi/2}) - F(0)$ is our answer.

We already found a suitable F: Using the substitution $u = x^2$, $du = 2x\, dx$, we saw that $F(x) = \sin(x^2)$ is an antiderivative of f. ▶   Thus,

We need only one antiderivative, so we set $C = 0$.

$$\int_0^{\sqrt{\pi/2}} 2x\, \cos(x^2)\, dx = \sin(x^2)\Bigg]_{x=0}^{x=\sqrt{\pi/2}} = \sin(\pi/2) - \sin 0 = 1. \quad ∎$$

EXAMPLE 9   Find $\int_0^{\sqrt{\pi/2}} 2x\, \cos(x^2)\, dx$ again.

Solution   (**Substitute; Create a New Definite Integral**)   As before, let $u = x^2$ and $du = 2x\, dx$. At the endpoints $x = 0$ and $x = \sqrt{\pi/2}$, $u = 0$ and $u = \pi/2$, respectively. Therefore, substitution for u, du, *and the endpoints* gives

$$\int_0^{\sqrt{\pi/2}} 2x\, \cos(x^2)\, dx = \int_0^{\pi/2} \cos u\, du.$$

The last integral can be calculated just as it stands: ▶

No need to substitute x back in.

$$\int_0^{\pi/2} \cos u\, du = \sin u\Bigg]_0^{\pi/2} = \sin(\pi/2) - \sin 0 = 1,$$

as before.   ∎

Look Again.   Notice carefully how we used substitution differently in the last two examples. In Example 8 we used substitution only as a temporary aid to find an antiderivative of the given integrand, with respect to the original variable x. In Example 9 we used substitution to transform the original definite integral into an entirely new definite integral, with new limits and a new integrand. The final result was, the same in each case.

Substitution in General: Why It Works

In each preceding example, substitution worked—it led successfully to a correct antiderivative. To see *why* substitution works, we'll describe it in general language.

For any indefinite integral $\int f(x)\,dx$, substitution means finding a function $u = u(x)$ such that $f(x)$ can be rewritten in the form

$$f(x) = g\big(u(x)\big)\,u'(x)$$

for some new function g. To say that substitution works is to say that

$$\int f(x)\,dx = \int g(u)\,du,$$

i.e., that if G is a function for which $G'(u) = g(u)$, then $G\big(u(x)\big)$ is an antiderivative of f.

The chain rule guarantees that $G\big(u(x)\big)$ has the desired property:

$$\frac{d}{dx}\left(G\big(u(x)\big) \right) = G'\big(u(x)\big) \cdot u'(x) = g\big(u(x)\big) \cdot u'(x) = f(x),$$

so $G\big(u(x)\big)$ is indeed the antiderivative we seek.

Substitution in *definite* integrals works for essentially the same reason. The fact that it does is sometimes known as the **change-of-variables theorem:**

Theorem 1 Let f, u, and g be continuous functions such that, for all x in $[a, b]$,

$$f(x) = g\big(u(x)\big) \cdot u'(x).$$

Then

$$\int_a^b f(x)\,dx = \int_{u(a)}^{u(b)} g(u)\,du.$$

Proof.   Let G be an antiderivative of g. By the FTC,

$$\int_{u(a)}^{u(b)} g(u)\,du = G\big(u(b)\big) - G\big(u(a)\big).$$

As we showed before (using the chain rule), $G\big(u(x)\big)$ is an antiderivative for $f(x)$. Therefore,

$$\int_a^b f(x)\,dx = G\big(u(b)\big) - G\big(u(a)\big),$$

so the two integrals are equal.

Differentials: du, dx, and All That

The symbols du and dx are called the **differentials** of u and x, respectively. If $u = u(x)$, then du and dx are related in the expected way: $du = u'(x)\,dx$.

A rigorous (and rather subtle) theory of differentials exists in higher mathematics; we won't need it in this book. For us, the differential in an integral expression (dx in $\int f(x)\,dx$, for instance) is mainly an aid to memory: It reminds us of the variable of integration. In substitution problems, trading dx's for du's (or vice versa) is a useful bookkeeping device; it helps keep us honest with respect to Theorem 1.

BASICS

In Exercises 1–10, evaluate the antiderivative using the indicated substitution, then check your answer by differentiation.

1. $\int (4x+3)^{-3}\,dx; \quad u = 4x+3$

2. $\int x\sqrt{1+x^2}\,dx; \quad u = 1+x^2$

3. $\int e^{\sin x}\cos x\,dx; \quad u = \sin x$

4. $\int \dfrac{(\ln x)^3}{x}\,dx; \quad u = \ln x$

5. $\int \dfrac{\arctan x}{1+x^2}\,dx; \quad u = \arctan x$

6. $\int \dfrac{\sin\left(\sqrt{x}\right)}{\sqrt{x}}\,dx; \quad u = \sqrt{x}$

7. $\int \dfrac{e^{1/x}}{x^2}\,dx; \quad u = 1/x$

8. $\int x^3\sqrt{9-x^2}\,dx; \quad u = 9-x^2$

9. $\int \dfrac{e^x}{1+e^{2x}}\,dx; \quad u = e^x$

10. $\int \dfrac{x}{1+x}\,dx; \quad u = 1+x$

In Exercises 11–14, find real numbers a and b so that equality holds, then evaluate the definite integral.

11. $\displaystyle\int_{-2}^{1} \dfrac{x}{1+x^4}\,dx = \dfrac{1}{2}\int_{a}^{b}\dfrac{1}{1+u^2}\,du$

12. $\displaystyle\int_{-\sqrt{\pi/2}}^{\sqrt{\pi}} x\cos(3x^2)\,dx = \dfrac{1}{6}\int_{a}^{b}\cos u\,du$

13. $\displaystyle\int_{0}^{3} \dfrac{x}{(2x^2+1)^3}\,dx = \dfrac{1}{4}\int_{a}^{b} u^{-3}\,du$

14. $\displaystyle\int_{1}^{2} x^2 e^{x^3/4}\,dx = \dfrac{4}{3}\int_{a}^{b} e^u\,du$

FURTHER EXERCISES

15. Example 4 says that $\int \dfrac{x}{1+x^4}\,dx = \dfrac{1}{2}\arctan(x^2) + C$. Verify this by differentiation.

16. Example 6 says that $\int \dfrac{x}{\sqrt{2x+3}}\,dx = \dfrac{(2x+3)^{3/2}}{6} - \dfrac{3\sqrt{2x+3}}{2} + C$. Verify this by differentiation.

17. (a) Use the substitution $u = 1 + \sqrt{x}$ to find the antiderivative $\int \dfrac{dx}{1+\sqrt{x}}$.
    (b) The expression obtained in part (a) and the expression for this antiderivative given in Example 7 are not identical. Show that both are correct.

18. Let $a > 0$ be a constant. Derive the antiderivative formula $\int \dfrac{dx}{\sqrt{a^2-x^2}} = \arcsin(x/a) + C$ from the antiderivative formula $\int \dfrac{dx}{\sqrt{1-x^2}} = \arcsin x + C$.

19. Let a and $b > 0$ be constants. Use the substitution $u = a + b/x$ to find the antiderivative $\int \dfrac{dx}{ax^2+bx}$.

20. Show that $\int \dfrac{dx}{1+e^x} = x - \ln(1+e^x) + C$.
    [**HINT:** $\dfrac{1}{u(1+u)} = \dfrac{1}{u} - \dfrac{1}{1+u}$.]

21. Suppose that $\int_0^{12} g(x)\,dx = \pi$. Evaluate $\int_0^4 g(3x)\,dx$.

22. Let $I = \int \sec^2 x \tan x\,dx$.
    (a) Use the substitution $u = \sec x$ to show that $I = \frac{1}{2}\sec^2 x + C$.
    (b) Use the substitution $u = \tan x$ to show that $I = \frac{1}{2}\tan^2 x + C$.
    (c) The expressions for I in parts (a) and (b) look different, but both are correct. Explain this apparent paradox.

In Exercises 23–69, evaluate the antiderivative, then check your answer by differentiation.

23. $\int \cos(2x+3)\,dx$

24. $\int \sin(2-3x)\,dx$

25. $\int (3x-2)^4\,dx$

26. $\int x\cos(1-x^2)\,dx$

27. $\int \dfrac{2x^3}{1+x^4}\,dx$

28. $\int x(3x+2)^4\,dx$

29. $\int \dfrac{dx}{1-2x}$

30. $\int \sqrt{3x-2}\,dx$

31. $\int x\sqrt{3-2x}\,dx$

32. $\int \dfrac{\ln x}{x}\,dx$

33. $\int \dfrac{\sqrt{1+1/x}}{x^2}\,dx$

34. $\int xe^{x^2}\,dx$

35. $\displaystyle\int x^3(x^4-1)^2\,dx$

36. $\displaystyle\int \frac{x^3}{1+x^2}\,dx$

37. $\displaystyle\int \frac{x^2}{1+x^6}\,dx$

38. $\displaystyle\int \frac{x}{\sqrt{1-x^4}}\,dx$

39. $\displaystyle\int x^2\sqrt{4x^3+5}\,dx$

40. $\displaystyle\int \frac{x}{\sqrt{1+x^2}}\,dx$

41. $\displaystyle\int \frac{x+4}{x^2+1}\,dx$

42. $\displaystyle\int x(1-x^2)^{15}\,dx$

43. $\displaystyle\int \frac{2x+3}{\left(x^2+3x+5\right)^4}\,dx$

44. $\displaystyle\int (x+2)(x^2+4x+5)^6\,dx$

45. $\displaystyle\int \frac{x+1}{\sqrt[3]{3x^2+6x+5}}\,dx$

46. $\displaystyle\int \frac{e^{2x}}{\left(1+e^{2x}\right)^3}\,dx$

47. $\displaystyle\int \frac{e^x}{(2e^x+3)^2}\,dx$

48. $\displaystyle\int \frac{e^{\sqrt{x}}}{\sqrt{x}}\,dx$

49. $\displaystyle\int \frac{2x+3}{(x+1)^2}\,dx$

50. $\displaystyle\int \frac{x^2}{x-3}\,dx$

51. $\displaystyle\int \tan x\,dx$

52. $\displaystyle\int \sec x\tan x\,dx$

53. $\displaystyle\int \frac{\arcsin x}{\sqrt{1-x^2}}\,dx$

54. $\displaystyle\int \sec x\tan x\sqrt{1+\sec x}\,dx$

55. $\displaystyle\int \frac{5x}{3x^2+4}\,dx$

56. $\displaystyle\int \frac{\cos x}{1+\sin^2 x}\,dx$

57. $\displaystyle\int \frac{\sec^2 x}{\sqrt{1-\tan^2 x}}\,dx$

58. $\displaystyle\int \tan^2 x\csc x\,dx$

59. $\displaystyle\int x\tan(x^2)\,dx$

60. $\displaystyle\int x^2\sec^2(x^3)\,dx$

61. $\displaystyle\int \frac{\cos x}{\sin^4 x}\,dx$

62. $\displaystyle\int x^4\sqrt[3]{x^5+6}\,dx$

63. $\displaystyle\int \ln(\cos x)\tan x\,dx$

64. $\displaystyle\int \frac{\left(1+\sqrt{x}\right)^3}{\sqrt{x}}\,dx$

65. $\displaystyle\int \frac{dx}{\sqrt{x}\left(\sqrt{x}+2\right)^3}$

66. $\displaystyle\int x^3\sqrt{x^2+2}\,dx$

67. $\displaystyle\int \sqrt{1+\sqrt{x}}\,dx$

68. $\displaystyle\int \frac{e^x}{e^{2x}+2e^x+1}\,dx$

69. $\displaystyle\int \frac{e^{\tan x}}{1-\sin^2 x}\,dx$

In Exercises 70–75, evaluate the definite integral.

70. $\displaystyle\int_0^2 \frac{x}{\left(1+x^2\right)^3}\,dx$

71. $\displaystyle\int_{-19}^8 \sqrt[3]{8-x}\,dx$

72. $\displaystyle\int_1^e \frac{\sin(\ln x)}{x}\,dx$

73. $\displaystyle\int_e^{4e} \frac{dx}{x\sqrt{\ln x}}$

74. $\displaystyle\int_0^\pi \sin^3 x\cos x\,dx$

75. $\displaystyle\int_{-\pi/2}^\pi e^{\cos x}\sin x\,dx$

76. Evaluate $\displaystyle\int_0^1 x\sqrt{1-x^4}\,dx$.   [**HINT:** Use a substitution to relate the given integral to one that can be evaluated geometrically.]

77. Suppose that the function f is continuous on the interval $[-1, 1]$. Use the substitution $u = \pi - x$ to show that

$$\int_0^\pi xf(\sin x)\,dx = \frac{\pi}{2}\int_0^\pi f(\sin x)\,dx.$$

78. Suppose that the function f is continuous on the interval $[a, b]$. Show that

$$\int_a^b f(x)\,dx = \int_a^b f(a+b-x)\,dx.$$

79. Use the substitution $u = 1 - x$ to show that

$$\int_0^1 x^n(1-x)^m\,dx = \int_0^1 x^m(1-x)^n\,dx.$$

The substitution $u^n = ax + b$ can sometimes be used to find antiderivatives of expressions involving the form $\sqrt[n]{ax + b}$. Use this substitution to evaluate each of the following antiderivative in Exercises 80–81.

80. $\displaystyle\int x\sqrt{2x + 1}\, dx$

81. $\displaystyle\int \frac{dx}{\sqrt{x} + \sqrt[3]{x}}$

82. Find the flaw in the following "proof" that
$$I = \int_{-1}^{1} \frac{dx}{x^2 + 1} = 0:$$

$$I = \int_{-1}^{1} \frac{dx}{x^2 + 1} = \int_{-1}^{1} \frac{x^{-2}}{1 + x^{-2}}\, dx = -\int_{-1}^{1} \frac{du}{1 + u^2} = -I$$

so $I = 0$.

83. Find the flaw in the following "proof" that $\int_0^\pi \sqrt{1 - \sin x}\, dx = 0$:

$$\int \sqrt{1 - \sin x}\, dx = \int \frac{\sqrt{1 - \sin^2 x}}{\sqrt{1 + \sin x}}\, dx$$

$$= \int \frac{\cos x}{\sqrt{1 + \sin x}}\, dx$$

$$\to \int \frac{du}{\sqrt{u}}$$

$$= 2\sqrt{u} + C \to 2\sqrt{1 + \sin x} + C.$$

Thus,

$$\int_0^\pi \sqrt{1 - \sin x}\, dx = 2\sqrt{1 + \sin \pi} - 2\sqrt{1 + \sin 0} = 0.$$

6.3 Integral Aids: Tables and Computers

Although finding symbolic antiderivatives from scratch can be difficult, there are excellent educational (and recreational) reasons for doing so: Finding the right substitution or other method may reveal satisfying order and pattern in apparent symbolic chaos.

Whatever its educational and aesthetic value, the process of symbolic antidifferentiation is difficult, tricky, and laborious enough that, in practice, calculus users freely resort to integral tables, computer software, or anything else that comes to hand.

Like all power tools, tables and mathematical software require some care. In this section we illustrate some of their possibilities and pitfalls.

Integral Tables

Standard scientific reference works contain extensive integral tables. ▶ The *CRC Handbook of Chemistry and Physics*, 48th edition, for instance, lists nearly 600 different antiderivative formulas on more than 40 pages. One of them reads as follows: ▶

"Antiderivative table" would be a better name.

We added the absolute value signs.

F a c t (Formula 403)

$$\int \frac{dx}{a + be^{px}} = \frac{x}{a} - \frac{1}{ap} \ln|a + be^{px}|$$

Notice the missing (but understood) constant of integration; supplying one every time would be typographically tiresome.

Parameters and Pattern Matching

Integral tables use parameters—letters other than the variable of integration—to stand for numerical constants that can vary from problem to problem. With this device, a single integral formula can solve an entire family of similar problems. The next few examples concern the parameters in the preceding formula.

EXAMPLE 1   Verify Formula 403 by differentiation.

Watch the algebra carefully.

Solution   We differentiate the right side with respect to x. All other letters are constants. ◄

$$\left(\frac{x}{a} - \frac{1}{ap} \ln|a + be^{px}| \right)' = \frac{1}{a} - \frac{1}{ap} \frac{bpe^{px}}{a + be^{px}}$$

$$= \frac{1}{a} - \frac{1}{a} \frac{be^{px} + a - a}{a + be^{px}}$$

$$= \frac{1}{a} - \frac{1}{a} \left(1 - \frac{a}{a + be^{px}} \right)$$

$$= \frac{1}{a + be^{px}}.$$

That's what we wanted; the *CRC Handbook* got it right.   ■

> **Getting It Right.**   Not every equation in published integral tables is correct. Integral tables change over time as new antiderivative formulas come into favor and old ones fall into disuse. With each new version of an integral table, new typographical errors may creep in; some errors have lain undiscovered for years.
>
> Standard integral tables are sometimes used as benchmarks to check mathematical software programs for accuracy. This process sometimes uncovers errors—not in the software but in the tables. Some integral tables have been found to have error rates in the double-digit percentage range.

EXAMPLE 2   Use Formula 403 to find $\int \dfrac{5}{3 - 2e^{-x}} \, dx$.

Solution   Writing

$$\int \frac{5}{3 - 2e^{-x}} \, dx = 5 \int \frac{dx}{3 - 2e^{-x}}$$

Don't forget the 5 out front.

makes the parameter values directly apparent: $a = 3$, $b = -2$, $p = -1$. ◄   The result is

$$\int \frac{5}{3 - 2e^{-x}} \, dx = 5 \left(\frac{x}{3} + \frac{1}{3} \ln|3 - 2e^{-x}| \right) + C.$$ ■

Making the Shoe Fit

Antiderivatives that don't appear to fit an integral template can sometimes be made to do so. A *u*-substitution, some symbolic algebra, or a combination of both often effects this Cinderella-like transformation. ▶

Sometimes the shoe just won't fit. Remember the Cinderella story?

EXAMPLE 3   Find $I = \int \dfrac{\cos x}{5 + 2e^{3\sin x}}\, dx.$

Solution   As it stands the integral resembles Formula 403 only vaguely. However, substituting $u = \sin x$ and $du = \cos x\, dx$ reveals the familiar pattern:

$$\int \frac{\cos x}{5 + 2e^{3\sin x}}\, dx = \int \frac{du}{5 + 2e^{3u}}.$$

Setting $a = 5$, $b = 2$, and $p = 3$ completes the problem:

$$I = \frac{u}{5} - \frac{\ln\left|5 + 2e^{3u}\right|}{15} + C = \frac{\sin x}{5} - \frac{\ln\left|5 + 2e^{3\sin x}\right|}{15} + C. \qquad \blacksquare$$

EXAMPLE 4   Find $I = \int \dfrac{e^x}{3e^{-x} + 2e^x}\, dx.$

Solution   A touch of exponential algebra ▶ puts the integrand into the necessary form:

Multiplying top and bottom by e^{-x}.

$$\frac{e^x}{3e^{-x} + 2e^x} = \frac{e^x e^{-x}}{e^{-x}\left(3e^{-x} + 2e^x\right)} = \frac{1}{2 + 3e^{-2x}}.$$

Now we're ready for Formula 403, with $a = 2$, $b = 3$, $p = -2$:

$$I = \int \frac{dx}{2 + 3e^{-2x}} = \frac{x}{2} + \frac{1}{4}\ln\left|2 + 3e^{-2x}\right| + C. \qquad \blacksquare$$

Completing the Square

Quadratic expressions (of the form $ax^2 + bx + c$) appear often in antiderivative problems. Completing the square sometimes helps fit such integrands into an integral table's template. We illustrate with two examples.

EXAMPLE 5   Find $I = \int \dfrac{dx}{x^2 + 4x + 5}.$

Solution   Rummaging through an integral table (such as the one at the end of this book) turns up two likely possibilities:

$$\int \frac{dx}{x^2 + a^2} = \frac{1}{a}\arctan\left(\frac{x}{a}\right); \qquad \int \frac{dx}{x^2 - a^2} = \frac{1}{2a}\ln\left|\frac{x-a}{x+a}\right|.$$

Which one, if either, applies?

To decide, we'll complete the square in the denominator:

$$x^2 + 4x + 5 = (x^2 + 4x + 4) + (5 - 4) = (x + 2)^2 + 1.$$

This, in turn, suggests the substitution $u = x + 2$, $du = dx$. Here's the result, it's easily checked by differentiation.

$$\int \frac{dx}{x^2 + 4x + 5} = \int \frac{du}{u^2 + 1} = \arctan u + C = \arctan(x + 2) + C. \qquad \blacksquare$$

EXAMPLE 6   Find $I = \int \frac{dx}{x^2 + 4x + 3}$.

Solution   The same two possibilities suggest themselves as in Example 5. Again we complete the square:

$$x^2 + 4x + 3 = (x^2 + 4x + 4) + (3 - 4) = (x + 2)^2 - 1.$$

The rest follows easily. Substitution of $u = x + 2$ and $du = dx$ gives

$$\int \frac{dx}{x^2 + 4x + 3} = \int \frac{du}{u^2 - 1} = \frac{1}{2} \ln \left| \frac{u - 1}{u + 1} \right| + C = \frac{1}{2} \ln \left| \frac{x + 1}{x + 3} \right| + C. \qquad \blacksquare$$

A Reality Check: Do the Graphs Look Right?

Good sense dictates that results obtained by one method be tested by another, from time to time. Symbolic antiderivatives, especially, deserve such attention. Functions, after all, are more than purely formal expressions; they're graphical and numerical objects as well. If symbolic operations really make sense, they should stand up to graphical scrutiny. In other words, symbolically produced antiderivatives should "look right." Now and then, we'll check to make sure that they do.

Examples 5 and 6 invite such a look. Their symbolic results,

$$\int \frac{dx}{x^2 + 4x + 5} = \arctan(x + 2); \qquad \int \frac{dx}{x^2 + 4x + 3} = \frac{1}{2} \ln \left| \frac{x + 1}{x + 3} \right|,$$

are a bit mysterious. Although the integrands are almost identical, the antiderivatives look very different. Is there some mistake?

Let's approach this graphically. Here, first, are graphs of the two integrands:

Graphs of $f_1(x) = 1/(x^2 + 4x + 5)$
and $f_2(x) = 1/(x^2 + 4x + 3)$

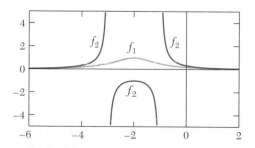

Now the mystery begins to dissolve. Despite their typographical similarity, the two integrands are quite different: f_1 behaves tamely, while f_2 has vertical asymptotes at $x = -3$ and $x = -1$. As $x \to \pm\infty$, on the other hand, both f_1 and f_2 tend to zero.

Plots of the antiderivatives F_1 and F_2 show that they, too, behave as they should:

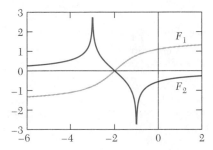

Graphs of $F_1(x) = \arctan(x + 2)$
and $F_2(x) = 0.5 \ln|(x + 1)/(x + 3)|$

Notice the following features.

Asymptotes Like f_1, its antiderivative F_1 is tamely behaved. Like f_2, the antiderivative F_2 has vertical asymptotes at $x = -3$ and $x = -1$, but with different directions.

Long-Run Behavior Both F_1 and F_2 appear to tend toward the horizontal as $x \to \pm\infty$. This is as it should be: Their respective derivatives both tend to zero in the long run.

Reduction Formulas: One Antiderivative in Terms of Another

Some formulas in integral tables have antiderivatives on *both* sides of an equation. Here's another entry from the *CRC Handbook*:

Fact (Formula 311) If $n \neq 1$, then

$$\int \tan^n x \, dx = \frac{\tan^{n-1} x}{n - 1} - \int \tan^{n-2} x \, dx.$$

Applying such a **reduction formula**—repeatedly, perhaps—transforms one integral problem into another, with the second being easier than the first.

EXAMPLE 7   Find $\displaystyle\int \tan^4 x \, dx$.

Solution   Formula 311, with $n = 4$, says that

$$\int \tan^4 x \, dx = \frac{\tan^3 x}{3} - \int \tan^2 x \, dx.$$

The last integral may itself appear in an integral table. If so, we're finished. If not, we can apply Formula 311 again (with $n = 2$, this time).

$$\int \tan^2 x \, dx = \frac{\tan x}{1} - \int \tan^0 x \, dx = \tan x - x.$$

Combining the results,

$$\int \tan^4 x \, dx = \frac{\tan^3 x}{3} - \tan x + x + C. \qquad \blacksquare$$

Antidifferentiation by Machine

Old mathematical handbooks contain vast tables of trigonometric, exponential, and logarithm function values. Practically speaking, scientific calculators have rendered such tables obsolete.

To a lesser (but significant) degree, sophisticated mathematical software effectively replaces integral tables. *Derive*, *Maple*, *Mathematica*, and other programs "know" not just integral formulas but—much more challenging—how and when to apply them.

Symbolic Camouflage: Different Forms of the Same Result

Equivalent mathematical expressions can appear in radically different symbolic forms. Trigonometric expressions are especially tricky; the many "trigonometric identities" offer almost unlimited possibilities for disguise.

Computer symbolic operations often produce variant or unexpected forms of symbolic results. (What's convenient for a computer may be baffling to you and me—and vice versa.)

> **EXAMPLE 8**   According to one integral table,
>
> $$\int \sin^2 x \, dx = \frac{x}{2} - \frac{\sin(2x)}{4}.$$

Maple says:

```
> int( sin(x)^2, x );

              - 1/2 cos(x) sin(x) + 1/2 x
```

Which is right?

Sometimes it's much harder.
See for yourself.

> **Solution**   Both are right (we forgive *Maple* the omitted constant of integration). This time the reason is simple. ◀ Stirring the trigonometric identity $\sin(2x) = 2 \sin x \cos x$ into the first formula readily produces the second. ◀ $\blacksquare$

When Nothing Works

Sometimes nothing seems to work. Symbolic methods, integral tables, algebraic tricks, and software may all fail to find an antiderivative in elementary form. Two reasons are possible: Either we haven't (or a computer hasn't) searched cleverly enough, or there *is* no elementary antiderivative. As we remarked at the beginning of this chapter, many elementary functions, even simple ones, don't have (elementary antiderivatives, (i.e., antiderivatives with formulas). None of these does, for instance:

$$e^{x^2}, \qquad \sin(x^2), \qquad \frac{\sin x}{x}, \qquad \frac{x}{\ln x}.$$

Like all continuous functions, these have antiderivatives, but not elementary ones. Notice how *Maple* handles one of them:

```
> int( sin(x)/x, x );
```

$$Si(x)$$

Unable to find an elementary answer, *Maple* returned the function *Si*, which is *defined* to be an antiderivative of $\sin(x)/x$. In effect, *Maple* punted.

The Bright Side.   Even when symbolic methods fail, all is not lost. Numerical methods, such as those we'll study in the next chapter, often succeed. For example, numerical methods will nicely handle *definite* integrals of all the preceding "problematic" functions.

BASICS

Evaluate the integrals in Exercises 1–31 using a table of integrals or a software package. Making simple *u*-substitutions or completing the square *first* may help.

1. $\displaystyle\int \frac{dx}{3 + 2e^{5x}}$

2. $\displaystyle\int \frac{dx}{x(2x + 3)}$

3. $\displaystyle\int \frac{dx}{x^2(3 - x)}$

4. $\displaystyle\int_e^{4e} x^2 \ln x \, dx$

5. $\displaystyle\int \tan^3(5x) \, dx$

6. $\displaystyle\int \frac{dx}{x^2\sqrt{2x + 1}}$

7. $\displaystyle\int x \sin(2x) \, dx$

8. $\displaystyle\int x^2 e^{3x} \, dx$

9. $\displaystyle\int e^{2x} \cos(3x) \, dx$

10. $\displaystyle\int \frac{2x + 3}{4x + 5} \, dx$

11. $\displaystyle\int \frac{dx}{4 - x^2}$

12. $\displaystyle\int x^2\sqrt{1 - 3x} \, dx$

13. $\displaystyle\int \frac{4x + 5}{(2x + 3)^2} \, dx$

14. $\displaystyle\int \frac{dx}{x\sqrt{3x - 2}}$

15. $\displaystyle\int \frac{dx}{4x^2 - 1}$

16. $\displaystyle\int \frac{dx}{(4x^2 - 9)^2}$

17. $\displaystyle\int \frac{x + 2}{2 + x^2} \, dx$

18. $\displaystyle\int \frac{3}{\sqrt{6x + x^2}} \, dx$

19. $\displaystyle\int \frac{5}{4x^2 + 20x + 16} \, dx$

20. $\displaystyle\int \frac{dx}{\sqrt{x^2 + 2x + 26}}$

21. $\displaystyle\int x^3 \cos(x^2) \, dx$

22. $\displaystyle\int \frac{x}{\sqrt{x^2 + 4x + 3}}\, dx$

23. $\displaystyle\int \frac{dx}{(x^2 + 3x + 2)^2}$

24. $\displaystyle\int x^2 (\cos x + 3 \sin x)\, dx$

25. $\displaystyle\int \frac{e^x}{e^{2x} - 2e^x + 5}\, dx$

26. $\displaystyle\int \cos x \sin x \sin^2 (2 \cos^2 x + 1)\, dx$

27. $\displaystyle\int \sqrt{x^2 + 4x + 1}\, dx$

28. $\displaystyle\int \frac{\cos x}{3 \sin^2 x - 11 \sin x - 4}\, dx$

29. $\displaystyle\int \frac{\cos x \sin x}{(\cos x - 4)(3 \cos x + 1)}\, dx$

30. $\displaystyle\int \frac{e^{2x}}{\sqrt{e^{2x} - e^x + 1}}\, dx$

31. $\displaystyle\int x \sin(3x + 4)\, dx$

Real Numbers and the Coordinate Plane

The real numbers might seem, at first glance, almost *too* familiar. What is there to say? Surprisingly, a great deal—far more than a calculus course could cover.

Here's just one hint at the subtleties: It's been known since Pythagoras (*ca.* 500 B.C.E.) that some numbers, such as $\sqrt{2}$, are **irrational**—that is, they can't be written as *ratios* of integers. But how thickly scattered are the irrationals among the reals? Are "most" real numbers rational or irrational? The answers are remarkable, even mysterious: "Most" real numbers are irrational; a dart thrown randomly at a number line will almost always hit an irrational. Yet plenty of rational numbers exist; in any interval, no matter how small, there are *infinitely many* rational numbers! Resolving these tantalizing mysteries must wait for a later mathematics course. For us, a brief, brisk, selective review of the properties of real numbers will suffice.

> **Newton Managed.** Isaac Newton himself did quite well without a fully rigorous development of the real numbers. The first mathematically satisfactory definition of the real numbers was given by Richard Dedekind, over 130 years after Newton's death.

The Real Number Line

The real numbers are really no more *real*, in an existential or philosophical sense, than the integers, the rationals, or even the complex numbers. The best reason for the adjective "real" may be that the set of real numbers can be thought of naturally as an unbroken, unending line. Once an origin and a unit of distance are fixed, every point has its own number and every number its own point. Here is the usual picture:

Order.   The line shows nicely how the real numbers are ordered: *As we move to the right on the line, real numbers increase.* This is important, so we restate it as an inequality:

$$a > b \text{ if and only if } a \text{ is to the right of } b.$$

At the risk of redundancy, we'll say it one last time: For real numbers, "greater than" means "to the right of." Thus, for example, $-3 > -4$, even though -4 has greater *magnitude* than -3.

Subsets of the Reals, Subsets of the Line

The set of real numbers, like any infinite set, has a gigantic infinity of possible subsets. Some subsets have fiendishly complicated structures. ("Fractals," for example, are sets that are irregular or ragged at any scale. The most spectacular fractal sets are usually shown in two dimensions, but many fractals have one-dimensional forms.) In calculus, luckily, most subsets of interest are easily understood.

Set Notation: Brackets and More Brackets.   Sets and subsets pop up everywhere in mathematics. By using set notation efficiently, we can avoid tedious and sometimes confusing verbal descriptions.

In calculus, for instance, we might ask:

Let $f(x) = x^3 - x$. For which real numbers x is $f(x) < 0$?

Plot the curve, say for $-2 \le x \le 2$, to see for yourself.

Plotting $y = f(x)$ makes the answer—the **solution set**—apparent. ◄   The solution set, expressed (a bit long-windedly) in words, is

the set of real numbers x such that either x is less than -1 or x lies between 0 and 1.

That's a mouthful, even for this relatively simple problem. In set notation, the same answer is much more concise: $\{x \mid x < -1 \text{ or } 0 < x < 1\}$. Read aloud, the symbols say exactly the same thing as the words above. In particular, the vertical line | means "such that." It's understood here, moreover, that x represents a real number.

One caution: Three types of brackets ($\{\}$, $()$, and $[\,]$) are commonly used to describe sets. Each type has a different meaning; watch carefully for them below.

Calculus inevitably requires that some mathematical language (closed, open, infinite, bounded, etc.) be learned, and then used carefully. The payoff in efficiency can be great.

Intervals.   Intervals are "pieces" chopped from the real line. They are the simplest and most useful subsets of the reals. Intervals come in various styles: They can be bounded or infinitely long, and they may or may not include their endpoints. (**Closed intervals** contain their endpoints, **open intervals** don't.) ◄   The following table of examples illustrates the full taxonomy. It illustrates, too, how intervals can be defined by one or more inequalities:

Taxonomy of Intervals		
Interval	**Description**	**Inequality Form**
$(-2, 3)$	an *open* interval	$\{x \mid -2 < x < 3\}$
$[a, b]$	a *closed* interval	$\{x \mid a \leq x \leq b\}$
$[2, b)$	a half-open interval	$\{x \mid 2 \leq x < b\}$
$(5, \infty)$	an infinite interval	$\{x \mid x > 5\}$
$(-\infty, \infty)$	another infinite interval	

More Subsets of $\mathbb{R}$.   Below are more examples of subsets of the reals. Notice carefully our use of various mathematical notations, especially $\mathbb{R}$, $\mathbb{Q}$, and $\mathbb{Z}$ for the sets of reals, rationals, and integers, and the $\cup$ (union) and $\cap$ (intersection) symbols for combining sets.   These notations are standard in mathematical writing; we will use them often.

Subsets of the Reals		
Set	**Explanation**	**Alternative Notation(s)**
$\mathbb{R}$	the real numbers	$(-\infty, \infty)$
$\mathbb{N}$	the natural numbers	$\{1, 2, 3, \ldots\}$
$\mathbb{Z}$	the integers	$\{0, \pm 1, \pm 2, \pm 3, \ldots\}$
$\mathbb{Q}$	the rationals	
$\{0, 2, 5\}$	a finite set	$\{x \mid x = 0 \text{ or } x = 2 \text{ or } x = 5\}$ or $\{x \mid x(x-2)(x-5) = 0\}$
$(1, 4) \cup (5, 6)$	union of two intervals	$\{x \mid 1 < x < 4 \text{ or } 5 < x < 6\}$
$(1, 6) \cap [4, 7]$	intersection of two intervals	$\{x \mid 1 < x < 6 \text{ and } 4 \leq x \leq 7\}$ or $\{x \mid 4 \leq x < 6\}$ or $[4, 6)$

Pictures of Subsets.   Uncomplicated subsets of the reals are easy to "draw" on the real line. Intervals are nicest of all: They appear as line segments, including some, all, or none of their endpoints. Errors of approximation are inevitable ▶ in *all* pictures, but for most sets they do no harm. Several straightforward examples appear below; a *closed* dot indicates an *included* point; an *open* dot indicates an *excluded* point.

For example, because ideal "points" on a line, unlike pencil points, have zero thickness. Errors of omission are also inevitable; no picture can show everything.

Drawing Sets	
Set	**Picture**
$(-2, 3)$	
$[3, \infty)$	
$(-3, 2]$	
$\mathbb{Z}$	
$[-4, -2) \cup (1, 3)$	

The Symbol ∞: Not a Number

Every real number corresponds to a unique point on the real line. The symbol ∞ does not; ∞ *is not a real number*.

What *is* ∞, if not a number? That's a thorny question indeed, in a league with "What is truth?" A fully rigorous answer is far beyond the scope of this book. A good enough answer for us (and for most working mathematicians, most of the time) is that the symbol ∞ represents the idea of unboundedness, or endlessness. When we say that the real line has *infinite* length, for instance, we mean that the line goes on forever.

The symbol ∞ is a useful, convenient, and suggestive shorthand for the idea of "without bound." Use it sparingly and with care; ∞ is *not* a number and seldom behaves like one. In particular, "algebra" with the infinity symbol takes special care. ◄

For more on "algebra with ∞," see Section 2.7.

Absolute Value, Distance, and Inequalities

The idea of **absolute value** (or **magnitude**) should be familiar from long experience. Its use in simple expressions like these—

$$|-3| = 3; \qquad |\sin x| \le 1; \qquad \sqrt{x^2} = |x|$$

will cause no trouble. What absolute value means is usually clear in context, but sometimes a formal definition is helpful. Here it is:

$$|x| = \begin{cases} x & \text{if } x \ge 0 \\ -x & \text{if } x < 0 \end{cases}$$

If you learn nothing else about absolute values, learn these facts.

Familiar as all this is, absolute values can be sticky in practice—especially if inequalities are involved. The following two tricks of the trade often help: ◄

Perhaps we should say the distance between a and b, since the distance in both directions is the same. The equation $|a - b| = |b - a|$ says the same thing.

Interpreting Absolute Value as Distance For *any* numbers a and b, $|a - b|$ can (and should!) be thought of as the *distance from a to b*. ◄ Thinking of absolute value as distance helps us visualize algebraic expressions geometrically. For example, the absolute value equation

$$|x + 2| = 3$$

isn't hard to solve algebraically, but the algebraic solution doesn't give us much insight. Distance brings the situation to life. Because

$$|x + 2| = |x - (-2)| = \text{distance from } x \text{ to } -2,$$

the equation $|x + 2| = 3$ says that *x lies 3 units away from* -2. Clearly, $x = -5$ and $x = 1$ are the only solutions of the equation.

Trading Absolute Values for Inequalities: Two for One   In calculus, inequalities often involve absolute values. Here's one:

$$|x + 2| < 3$$

Solving such inequalities ▶ requires care. Here's the key idea:

> *One absolute value inequality corresponds to two ordinary inequalities.*

To solve an inequality involving a variable x means to find the solution set and describe it concisely.

The inequality above is equivalent to these two:

$$-3 < x + 2 \quad \text{and} \quad x + 2 < 3$$

or, more succinctly, $-3 < x + 2 < 3$. Now it's easy to find the solution set: $-5 < x < 1$. In other words, the inequality $|x + 2| < 3$ holds for those x that lie in the interval $(-5, 1)$, i.e., the set of points that lie within 3 units of -2.

More generally, for *any* quantities x and b:

$$|x| < b \qquad \text{if and only if} \qquad -b < x < b$$

$$|x| > b \qquad \text{if and only if} \qquad -b > x \text{ or } x > b$$

(Similar rules hold for $\leq$ and $\geq$ inequalities.)

EXAMPLE 1   Any open or closed finite interval can be described in terms of its **midpoint** and **radius**; absolute value inequalities help. The open interval $(-3, 5)$, for instance, has midpoint $x = 1$ and radius 4 because the distance from every point in the interval to 1 is less than 4. Therefore, $(-3, 5)$ corresponds naturally to the inequality $|x - 1| < 4$. In formal language:

$$(-3, 5) = \{ x \mid |x - 1| < 4 \}$$ ∎

EXAMPLE 2   Solve the inequality $|2x + 3| \leq 5$.

Solution   By the second fact above, the inequality

$$|2x + 3| \leq 5$$

is equivalent to

$$-5 \leq 2x + 3 \leq 5.$$

Now the usual algebraic rules for inequalities apply:

$$-5 \leq 2x + 3 \leq 5 \iff -8 \leq 2x \leq 2 \iff -4 \leq x \leq 1.$$

(The symbol $\iff$ means "if and only if"; it indicates that each side implies the other.) We conclude that the inequality holds if and only if x lies in the interval $[-4, 1]$.   ∎

EXAMPLE 3   Solve $|x - 2| > |x + 3|$.

Solution   Having absolute values on *both sides* of the inequality complicates things. A workable—but tedious—strategy is to look separately at three cases: $x < -3$, $-3 \le x < 2$, and $x \ge 2$.

Convince yourself of this—it's easy but important.

Draw a picture!

An easier approach is to interpret both sides as distance. If we do, the given inequality says: ◄ *x is farther from 2 than from* −3. Now the solution set is easy to visualize; it's everything to the *left* of −1/2, the point halfway between 2 and −3. ◄ In interval notation, the solution set is $(-\infty, -1/2)$.   ■

Most absolute value inequalities arise in richer contexts. Here is one in its natural habitat.

EXAMPLE 4   Solve the inequality $|x - \sin(x)| < 0.1$.

Solution   How might such an inequality arise? Why does an answer matter? Consider these graphs of two functions:

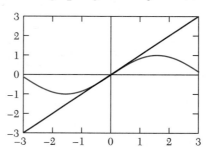

The picture shows that $\sin(x) \approx x$ if $x \approx 0$. This approximation has practical value, but only if we know something about *how* closely x approximates $\sin(x)$. Thus we might ask: For which values of x do x and $\sin(x)$ differ (in either direction) by less than 0.1? That's exactly what our original inequality, $|x - \sin(x)| < 0.1$, asks.

We'll solve our inequality graphically. Here's a piece of the graph of $y = x - \sin(x)$:

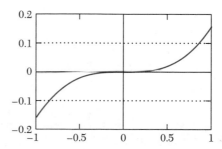

For what values of x does the inequality $-0.1 < x - \sin(x) < 0.1$ hold? The graph shows ▶ that this occurs for $-0.8 < x < 0.8$.

Approximately, of course.

Further graphical insight is available. Here's a closer look at the two original graphs:

Two graphs: $y = x$ and $y = \sin(x)$

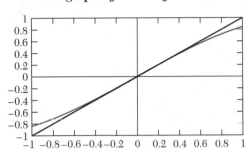

In this setting the inequality $|x - \sin(x)| < 0.1$ holds if and only if the graphs of $y = x$ and $y = \sin(x)$ are less than 0.1 unit apart, vertically. As we've already seen graphically, this occurs for $-0.8 < x < 0.8$. ▶   ■

The numbers are approximate.

The Coordinate Plane

The real numbers correspond to points on a *line*. In much the same way, pairs of real numbers correspond to points in the **coordinate plane**.

Coordinates and Why They Matter

The idea of attaching numerical coordinates to points in the plane is credited to the French philosopher René Descartes (1596–1650). It is no exaggeration to describe Descartes's idea, natural as it seems in hindsight, as one of the most powerful in intellectual history.

Descartes's idea is powerful because it links geometry with algebra and analysis. Geometry is the study of figures—lines, curves, circles, etc.—in the plane. Algebra studies operations—addition, multiplication, etc.—on numbers and on functions. Analysis is harder to define, but it includes the main ideas of calculus, such as function, derivative, and integral.

Before Descartes, geometry was relatively isolated from other branches of mathematics; geometers used compasses and other drawing tools to construct and measure plane figures. Figures *not* based on circles and lines were harder to describe and manipulate. The parabola, for instance, was described geometrically as the set of points equidistant from a fixed point and a fixed line. Properties of the parabola (such as the **reflection property**: All rays parallel to the axis of symmetry "bounce" to the same point, the **focus**) were hard to understand and prove with purely geometric tools and ways of thinking.

The great advantage of Cartesian coordinates is that they allow *algebraic* descriptions of geometric figures in the plane. The algebraic equation $y = x^2$, for example, describes a parabola in a far simpler and, for many purposes, more convenient fashion than its geometric properties.

When geometric figures are described algebraically the powerful tools and ideas of algebra and analysis become available. The connection, moreover, goes both ways: Geometric intuition sheds light on algebraic and analytic problems.

This link is the essence of the field of mathematics known as **analytic geometry**. In its most familiar form, analytic geometry studies properties of lines, circles, ellipses, parabolas, and hyperbolas. Collectively, these figures are called **conic sections**, because each can be thought of, geometrically, as a slice (i.e., "section") of a cone. (The cone in question is hollow, and may be "double," as in the case of a hyperbola.)

The key idea for us is that the conic sections can be described *algebraically*, as graphs of quadratic equations. Thus, properties of quadratic equations are, from another point of view, properties of conic sections, and vice versa. Knowledge of either tells us something of the other.

Throughout calculus, linking geometric, algebraic, and analytic ideas is crucial to understanding; it is the most important theme of this book.

Coordinate Systems

A **rectangular coordinate system** in the plane has three parts: two perpendicular axes, with scales, and an origin at the point where they intersect. By convention, the x-axis is horizontal and numbered from left to right; the y-axis is vertical and numbered from bottom to top. (These are *only* conventions. "Tilted" axes, other numbering schemes, and other variable names are all possible.)

With a coordinate system in place, every point's coordinates are defined. Here is the familiar picture; several points have coordinates labeled:

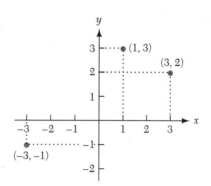

A coordinate system defines a one-to-one correspondence between points in the plane and pairs of real numbers: Each point corresponds uniquely to its two coordinates.

Cartesian Addresses.   The coordinates of a point are, in a sense, the point's "address"— they locate the point in the plane. A point's Cartesian address, moreover, gives explicit directions for getting there: To reach (a, b), from the origin, go a units "east" and b units "north." This is more than can be said for typical street addresses—123 Elm Street, for instance. (Some cities, of course, do use Cartesian-style numerical addresses.)

Calculating with Coordinates: Distance and Midpoint

Using Cartesian coordinates, many questions about points in the plane can be reduced to straightforward, direct calculations.

The Midpoint Formula.   Recall the situation in *one* dimension. If x_1 and x_2 are
points on the real *line*, their midpoint, m, is just their average: $m = (x_1 + x_2)/2$.

A similar situation holds in the plane. To find the midpoint, M, of the segment
joining points $P = (x_1, y_1)$ and $Q = (x_2, y_2)$, we average their x- and y-coordinates
separately:

$$M = \left(\frac{x_1 + x_2}{2}, \frac{y_1 + y_2}{2} \right)$$

A picture makes everything plausible:

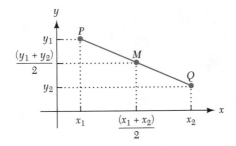

In particular, the coordinates of M are, respectively, midpoints of intervals on the
two axes.

Distance.   On the line the situation is simple: The distance between x_1 and x_2
is $|x_2 - x_1|$. The distance formula in the plane, like the midpoint formula, derives
from the one-dimensional situation.

What's the distance between the points $P = (x_1, y_1)$ and $Q = (x_2, y_2)$? Here's
the picture:

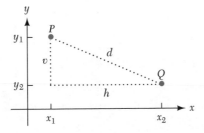

As the picture shows, the distance from P to Q (labeled d) is the hypotenuse of
a right triangle, with sides (labeled h and v) parallel to the axes. We'll use the
Pythagorean theorem to find d, but first we need values for h and v. By the *one-
dimensional* formula:

$$h = |x_2 - x_1| \qquad \text{and} \qquad v = |y_2 - y_1|.$$

Notice where (and why!) the absolute value signs conveniently vanished.

The distance formula in the plane now follows easily: ◄

$$d = \sqrt{h^2 + v^2} = \sqrt{|x_2 - x_1|^2 + |y_2 - y_1|^2} = \sqrt{(x_2 - x_1)^2 + (y_2 - y_1)^2}$$

The last form is standard. To summarize:

Fact The distance from $P = (x_1, y_1)$ to $Q = (x_2, y_2)$ is given by $d(P, Q) = \sqrt{(x_2 - x_1)^2 + (y_2 - y_1)^2}$.

Elementary properties of distance in the plane follow from this formula:

The Same Distance Either Way. Intuition says that the distance from P to Q should be equal to the distance from Q to P. All is well (but check the equation carefully to see why):

$$d(P, Q) = \sqrt{(x_2 - x_1)^2 + (y_2 - y_1)^2}$$
$$= \sqrt{(x_1 - x_2)^2 + (y_1 - y_2)^2}$$
$$= d(Q, P).$$

Finding the Midpoint. It stands to reason that the midpoint M between $P = (x_1, y_1)$ and $Q = (x_2, y_2)$ should be equidistant from both. An ugly computation (check the details, as an algebra exercise!) shows that this is indeed so. Squaring everything in sight helps a little:

$$d(P, M)^2 = \left(\frac{x_1 + x_2}{2} - x_1\right)^2 + \left(\frac{y_1 + y_2}{2} - y_1\right)^2$$
$$= \frac{(x_2 - x_1)^2}{4} + \frac{(y_2 - y_1)^2}{4}$$
$$= \frac{(x_1 - x_2)^2}{4} + \frac{(y_1 - y_2)^2}{4}$$
$$= \left(\frac{x_1 + x_2}{2} - x_2\right)^2 + \left(\frac{y_1 + y_2}{2} - y_2\right)^2$$
$$= d(M, Q)^2.$$

Using the Distance and Midpoint Formulas

Simply finding distances and midpoints is neither very edifying nor, fortunately, very common in calculus. Such calculations usually arise in more interesting contexts.

Circles. A circle is, by definition, the set of points at some fixed distance, say r, from a center, say $C = (a, b)$. The distance formula puts all this conveniently and simply, in algebraic form:

The point $P = (x, y)$ lies on the circle if and only if $d(P, C) = \sqrt{(x - a)^2 + (y - b)^2} = r$.

It's equivalent to say that the circle has equation $(x - a)^2 + (y - b)^2 = r^2$.

EXAMPLE 5   Find an algebraic equation for the circle of radius 5, centered at $C = (3, 2)$.

Solution   A point $P = (x, y)$ lies on this circle if and only if $d(P, C) = 5$. Thus,

$$\sqrt{(x-3)^2 + (y-2)^2} = 5$$

is an algebraic equation for this circle. We're done; we've found a perfectly acceptable algebraic equation. For anyone squeamish about square roots, there are other, equivalent, forms: ▶

The $\iff$ symbol means "if and only if."

$$\sqrt{(x-3)^2 + (y-2)^2} = 5 \iff (x-3)^2 + (y-2)^2 = 25$$
$$\iff x^2 + y^2 - 6x - 4y - 12 = 0$$
$$\iff x^2 + y^2 - 6x - 4y = 12 \qquad ■$$

EXAMPLE 6   Is $x^2 + 4x + y^2 - 2y = 10$ the equation of a circle? If so, which one?

Solution   Let's **complete the square** in both x and y:

$$x^2 + 4x + y^2 - 2y = 10 \iff (x^2 + 4x + 4) + (y^2 - 2y + 1) = 10 + 4 + 1$$
$$\iff (x+2)^2 + (y-1)^2 = 15$$

The answer can now be read off: The equation describes a circle of radius $\sqrt{15}$, centered at $(-2, 1)$. ■

EXAMPLE 7   It's well known among those who know such things that *the segment joining the midpoints of two sides of a triangle is parallel to the third side, and half its length.* Classical geometers would have proved this "constructively," with drawing tools. With coordinates, the proof is an easy calculation.

The picture follows. We arranged things to simplify calculations, but convince yourself that we did nothing illegal:

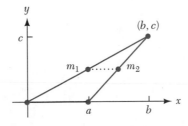

To prove our first claim—that the segment joining m_1 and m_2 is horizontal—it's enough to show that m_1 and m_2 have the same "height." By the midpoint formula,

$$m_1 = (b/2, c/2) \qquad \text{and} \qquad m_2 = \big((a+b)/2, c/2\big).$$

Thus m_1 and m_2 have the same y-coordinate, as claimed. We leave the claim about lengths as an exercise. ■

BASICS

1. Express each of the following intervals using absolute value inequalities.
   (a) $(0, 10)$ (d) $(2, 8)$
   (b) $[-\pi, \pi]$ (e) $[-3, 7]$
   (c) $(-2, 4)$ (f) $[-10, -2]$

2. Express each of the following sets using absolute value inequalities.
   (a) $\{x \mid x < -3 \;\; \text{or} \;\; x > 3\}$
   (b) $\{x \mid x < -5 \;\; \text{or} \;\; x > 3\}$
   (c) $\{x \mid -7 < x \;\; \text{and} \;\; x < 7\}$
   (d) $\{x \mid -7 < x \;\; \text{and} \;\; x < 5\}$

3. Express each of the following sets using interval notation.
   (a) $\{x \mid -10 < x \le 15\}$ (e) $\{x \mid |x - 2| \le 5\}$
   (b) $\{x \mid -6 \le x < 4\}$ (f) $\{x \mid |x + 3| \le 4\}$
   (c) $\{x \mid -13 < x < 17\}$ (g) $\{x \mid |x - a| < b\}$
   (d) $\{x \mid |x| \le 3\}$ (h) $\{x \mid |x - a| < a\}$

4. Find the distance between each of the following pairs of points.
   (a) $(0, 0)$ and $(2, 3)$ (c) $(-4, 2)$ and $(-7, -4)$
   (b) $(1, 2)$ and $(3, 4)$ (d) $(-5, -2)$ and $(2, 5)$

5. Explain why
$$\sqrt{|x_2 - x_1|^2 + |y_2 - y_1|^2} = \sqrt{(x_2 - x_1)^2 + (y_2 - y_1)^2}.$$

6. Find the points in the xy-plane that are at distance 13 from the point $(1, 2)$ and have y-coordinate equal to -3. [**HINT**: Draw a picture.]

7. Find an algebraic equation for the circle of radius 4 centered at $(-3, 7)$.

8. Is $x^2 + y^2 - 6x - 8y = -9$ the equation of a circle? If so, which one?

9. Show that $3x^2 + 3y^2 + 4y = 7$ is the equation of a circle. Find the radius and center of this circle.

10. Show that the line segment joining the midpoint of two sides of a triangle is half the length of the third side.

11. Explain why each of the following statements is true for every real number r such that $3 < r < 7$.
    (a) $|r| < 7$ (c) $|r - 5| < 2$
    (b) $|r| > 3$

12. Suppose that s is a real number and that $-2 \le s \le 1$.
    (a) Must $|s| \le 1$ be true? Justify your answer.
    (b) Must $|s| \le 2$ be true? Justify your answer.
    (c) Must $|s| < 10$ be true? Justify your answer.

13. Which of the following statements is true for every real number x such that $-3 \le x \le 11$? Justify your answers.
    (a) $-3 \le x \le 9$ (c) $-7 \le x \le -4$
    (b) $2 \le x \le 10$ (d) $5.9999 \le x \le 6.0001$

14. Suppose that $L \le x \le U$. Show that $\left| x - \frac{1}{2}(L + U) \right| \le \frac{1}{2}(U - L)$. [**HINT**: Draw a picture.]

15. Suppose that $|x - 3| \le 0.005$ and $|y - 2| \le 0.003$. Find real numbers A and B such that $|(x + y) - A| \le B$. [**HINT**: Draw a picture.]

16. Express each of the following English sentences in mathematical notation using absolute values.
    (a) The temperature outside is within $5°$ (Fahrenheit) of freezing.
    (b) The temperature in that room is closer to $100°$F than to freezing.
    (c) The high temperature today is closer to yesterday's high temperature than yesterday's high was to the high temperature two days ago.

Lines and Linear Functions

Manipulating lines and linear functions is important throughout calculus. In this appendix we review the rudiments of lines: algebra, definitions, and techniques.

Slope

Slope measures a line's *steepness* (i.e., the rate at which a line rises or falls). Before reviewing the formal definition, study the picture below. Several lines appear, each labeled ▶ with its slope.

The letter m almost invariably stands for slope. Perhaps from the French monter—to climb.

Lines with various slopes

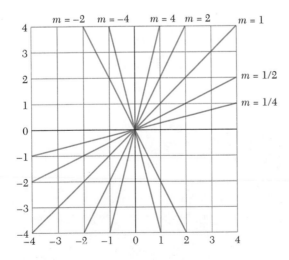

The picture illustrates the general situation:

> *Lines with positive slope rise (from left to right); lines with negative slope fall. The steeper the line, the larger the absolute value of its slope.*

> **Definition** The **slope** of a line is the ratio
> $$\frac{y_2 - y_1}{x_2 - x_1} = \frac{\Delta y}{\Delta x}$$
> where (x_1, y_1) and (x_2, y_2) are any two points on the line.

Notice:

Same Slope Everywhere A line has the same slope everywhere. *Any* two points can be used to calculate the slope. *Order* doesn't matter either— the slope from P to Q is the same as the slope from Q to P. In algebraic language, ◀

Keeping the same order in numerator and denominator does matter, of course.

$$\frac{y_2 - y_1}{x_2 - x_1} = \frac{y_1 - y_2}{x_1 - x_2}.$$

Parallel and Perpendicular Lines with equal slope are parallel. The slopes of perpendicular lines are negative reciprocals (i.e., their product is -1.)

Horizontal and Vertical Horizontal lines have slope 0; slope is undefined for *vertical* lines. The definition shows why.

How Steep? Scale Effects It's easy to see that the line $y = x$ has slope 1. How steep the line *looks*, on the other hand, depends strongly on the size of units (i.e., the **scales**) on the x- and y-axes. To see the effect of scales, consider these three graphs of the line $y = x$:

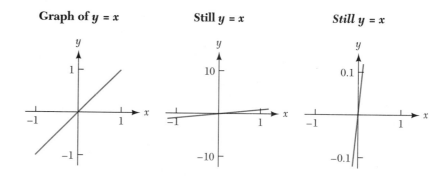

If—but only if—identical scales are used on both axes, the line $y = x$ makes a 45° angle with the x-axis.

Rise and Run The formula for slope is sometimes written as $\Delta y / \Delta x$, or as *rise/run*. ◀ The Δ-notation is a convenient shorthand. Invariably, Δ*anything* refers to a change in anything; it's usually calculated as the difference between a final value and an initial value. *Which* values these are must be made clear from context. For lines, things are simple: Δx is a *run*, or horizontal change; Δy is a *rise*, or vertical change. Either or both can be negative.

Convince yourself that these alternative versions really say the same thing.

Slope as a Rate Slope is defined as the ratio $\Delta y / \Delta x$. Slope, therefore, is a rate of change—the rate of change of y with respect to x. The straightness of a line means that this rate is constant.

Lines, Equations, and Functions

A line is completely determined by two bits of information. It's enough to know either (1) two points on the line, or (2) just one point and the slope. With either type of information, several simple algebraic descriptions of a line are possible. Here are two:

> **Definition** (Point-Slope Form) The line through (a, b) with slope m has equation $y = m(x - a) + b$.
>
> (Slope-Intercept Form) The line with slope m and y-intercept b has equation $y = mx + b$.

EXAMPLE 1   Consider the lines

ℓ_1: through $(0, 1)$ and $(2, 5)$;

ℓ_2: through $(-1, 3)$, with slope -2.

Describe each in point-slope form and in slope-intercept form.

Solution   The line ℓ_1 has slope $(5 - 1)/(2 - 0) = 2$. Now we know both the slope and (at least) one point on each line. Hence, in point-slope form:

$$\ell_1 : y = 2(x - 2) + 5 \quad \text{and} \quad \ell_2 : y = -2(x + 1) + 3$$

Now for slope-intercept forms. By inspection, ▶ ℓ_1 has slope 2 and y-intercept 1. Thus, ℓ_1 has slope-intercept equation $y = 2x + 1$. How about ℓ_2? Its slope is -2, so the slope-intercept equation looks like $y = -2x + b$. But what's b? Well, since ℓ_2 passes through $(-1, 3)$, those coordinates must satisfy the equation. In other words,

$$3 = -2 \cdot (-1) + b = 2 + b.$$

Clearly, $b = 1$. Therefore, the slope-intercept equation is $y = -2x + 1$.   ■

"By inspection" is the mathematician's grandiloquent way of saying "just by looking."

> **Unique or Not?**   The slope-intercept form of a line is *unique*—there's only one way to write a line in slope-intercept form. Moreover, the slope-intercept form works for any line that *has* slope (i.e., for any line that isn't vertical). Notice that the point-slope form of a line is not unique—*any* point on the line can appear in the point-slope equation. Since, for example, $(10, 21)$ lies on ℓ_1, $y = 2(x - 10) + 21$ is another point-slope form for ℓ_1.

EXAMPLE 2   Write an equation in slope-intercept form for the line ℓ through two given points (a, b) and (c, d). Where does ℓ intersect each axis?

Solution   If $a = c$, then ℓ is vertical; its equation is just $x = a$.

If $a \neq c$, ℓ has slope $(d - b)/(c - a)$, and the point-slope form applies:

$$y = \left(\frac{d - b}{c - a}\right)(x - a) + b.$$

More algebra yields this (slightly clumsy) equation for ℓ in slope-intercept form:

$$y = \left(\frac{d - b}{c - a}\right)x + \left(b - a\left(\frac{d - b}{c - a}\right)\right).$$

Check it, please.

The y-intercept can now be read off; it's everything after the $+$ sign. To find the x-intercept, we set $y = 0$ and solve for x. The result: ◄

$$x = a - b\left(\frac{c-a}{d-b}\right).$$ ■

EXAMPLE 3   How many lines do the equations

$$y + 2x = 6, \qquad y = -2(x+2) + 6, \qquad \text{and} \qquad y = -2(x-5) - 8$$

represent? Why?

Solution   Putting everything in slope-intercept form gives

$$y = -2x + 6, \qquad y = -2x + 2, \qquad y = -2x + 2.$$

This shows that the *three* equations represent *two* different lines.   ■

In the next example, lines appear in a typical calculus-style setting.

EXAMPLE 4   Find an equation in point-slope form for the line ℓ that meets the graph of $y = \sin x$ at $x = 1.3$ and at $x = 1.5$. (A line that "cuts" a graph at two prescribed points is called a **secant line**.) ◄   Here's a picture:

From the Latin, secare: to cut.

A secant line to the sine graph

Solution   First let's find the slope of ℓ. Any two points on ℓ will do. As the graph shows, ℓ passes through $(1.3, \sin(1.3))$ and $(1.5, \sin(1.5))$. The slope of ℓ is, therefore,

$$m = \frac{\sin 1.5 - \sin 1.3}{1.5 - 1.3} \approx \frac{0.9975 - 0.9636}{1.5 - 1.3} = 0.1695\,.$$

Hence an equation for ℓ is

$$y = m(x - 1.3) + \sin(1.3) \approx 0.1695(x - 1.3) + 0.9636\,.$$ ■

Linear Functions

The line with equation $y = 5x + 2$ can be thought of in various ways: as a geometric figure, as an algebraic equation, or—more typically in calculus—as the function $f(x) = 5x + 2$. Functions whose graphs are lines are called, naturally enough, **linear** ▶ **functions**. Such functions have an especially simple algebraic form:

Unfortunately, the word "linear" is used in more than one way in mathematics; confusion can result.

> **Definition** A linear function is one whose rule can be written in the form
>
> $$f(x) = Ax + B$$
>
> where A and B are constants.

A crucial property of linear functions, from our point of view, is that they "grow" at *constant* rates. In geometric terms, graphs of linear functions are straight. The numerical interpretation is also important.

EXAMPLE 5   The graph of the linear function $f(x) = 5x + 2$ is a line with slope 5. Interpret this slope *numerically*, as a rate of change.

Solution   Here's a sampler of x- and y-values:

Sample Values: $y = f(x) = 5x + 2$											
x	1.2	1.3	1.4	1.5	1.6	...	−3.4	−3.3	−3.2	−3.1	−3.0
y	8.0	8.5	9.0	9.5	10.0	...	−15.0	−14.5	−14.0	−13.5	−13.0

Notice what the numbers say: Whenever x increases by 0.1, y increases by 0.5 (i.e., 5 times as much). In symbols: $\Delta y = 5 \Delta x$. ▶   In words:

The Δ symbol denotes a change, or increment, in a variable. We'll use it repeatedly.

> *Along the line $y = 5x + 2$, y increases 5 times as fast as x.*
>
> This rate, moreover, is *constant*—it applies for all values of x.   ■

EXAMPLE 6   Its curved graph shows that the sine function is *not* linear. How do numerical values show the same thing?

Solution   Here's another sampler, this time of values of the sine function:

Sample Values: $y = \sin x$										
x	1.2	1.3	1.4	1.5	1.6	...	−3.4	−3.3	−3.2	−3.1
y	0.932	0.964	0.985	0.997	1.00	...	0.256	0.158	0.0584	−0.0416

The nonlinearity of the sine function shows clearly in the numbers. This time ▶ there's no regular "growth" pattern. Increasing x by 0.1 produces varying increases— or even decreases!—in y.   ■

As opposed to the previous table.

Lines and Linear Functions in Calculus

Lines are important throughout calculus—perhaps surprisingly so, given that most functions of interest aren't linear.

An idea we'll return to repeatedly.

We'll return repeatedly to this subtle but important idea. Rough synonyms for "locally straight" include "almost linear" and "locally linear."

To choose a domain point at random.

Local Straightness.   The key idea, ◄   and the link to linear functions, is that a typical calculus function is "locally straight," i.e., it "looks straight at small scale" near every point of its domain. ◄   The next (important) example explores what local straightness means.

EXAMPLE 7   In what *graphical* sense is the sine function "locally straight" near $x = 1.4$? ◄   *Which* line does the sine graph resemble near $x = 1.4$? What does all this mean *numerically*?

Solution   "Locally straight" means that the sine graph looks straight near the point $(1.4, \sin 1.4)$ if viewed in a small enough window—i.e., after enough "zooming in." One view of the graph appears in Example 4. Here are two successively closer views: ◄

Zooming is easy with a graphing calculator. As always, a graph's appearance depends on the x- and y-scales.

A closer view: $y = \sin x$

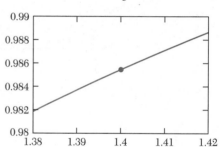

Closer still: $y = \sin x$

The last view makes our point: the sine graph is all but indistinguishable from a straight line.

To estimate slope accurately, read values from the graph at $x = 1.38$ and $x = 1.42$.

Allowing for some estimation.

Which straight line?   Because our graph passes through $(1.4, \sin 1.4)$ with slope ≈ 0.17, ◄   the line we want has equation ◄

$$y = \ell(x) = \sin 1.4 + 0.17(x - 1.4) = 0.17x + 0.7474.$$

What all this means is that $\sin x$, although *not* a linear function, behaves similarly to $\ell(x)$ for x near 1.4. Numerical data support this conclusion:

				If $x \approx 1.4$, $\sin x \approx \ell(x)$					
x	1.380	1.385	1.390	1.395	1.400	1.405	1.410	1.415	1.420
$\sin x$	0.9819	0.9828	0.9837	0.9846	0.9854	0.9863	0.9871	0.9879	0.9887
$\ell(x)$	0.9820	0.9829	0.9837	0.9846	0.9854	0.9863	0.9871	0.9880	0.9888

■

The next example illustrates an entirely different use of linear functions.

EXAMPLE 8   Are hardness of water and central nervous system (CNS) disorders in infants related, as some health researchers suspect? Below are some relevant data; hardness is measured in parts per million, the disorder rate in malformations per 1000 births. [These data come from *Biostatistics: A Foundation for Analysis in the Health Sciences*, 2nd ed., Wayne W. Daniel (Wiley, 1978).]

Hard Water: Good or Bad?										
Hardness	50	25	15	75	100	150	180	250	275	220
Disorder Rate	7.2	8.1	11.1	9.3	9.4	5.0	5.8	3.3	3.6	4.8
Hardness	160	50	45	60	100	155	200	240	40	65
Disorder Rate	6.3	12.5	15.0	6.5	8.0	10.0	5.3	4.9	7.2	11.9

Is there pattern in the data? It's hard to tell from the table. Let's plot the data:

What *does* the picture show? The data don't lie on any obvious curve or line, but they suggest a general tendency—with harder water, the CNS disorder rate seems to decrease.

Can we describe this situation mathematically? One possibility is to "fit" a reasonable line to the data, perhaps something like this:

The picture raises obvious questions: How was the line chosen? Is it a "good" fit? Might another line fit better? What does the line mean about CNS disorders? Does *any* line really fit this situation?

We chose the line entirely by appearance—it just *looked* right. Having confessed this, let's find an equation for the line. The y-intercept, 12, can be read directly from the graph. Reading from upper left to lower right shows a rise of -11, a run of 300, and hence a slope of $-11/300 \approx -0.0367$. A suitable equation, therefore, is

$$y = -\frac{11}{300}x + 12.$$

What good is such an equation? If $x = 125$, then

$$y = -\frac{11}{300} \cdot 125 + 12 \approx 7.4.$$

This means, in context, that in an area with water hardness 125 ppm we'd predict about 7.4 CNS disorders per 1000 births.

How trustworthy is such a prediction? The answer depends both on the philosophy and on the mathematics of mathematical modeling. ∎

BASICS

1. A line L passes through the points $(1, 2)$ and $(-2, 0)$.

   (a) Find an equation for the line L in slope-intercept form.

   (b) Does the line L pass through the point $(\pi, 2\pi)$? Explain your answer.

2. (a) Find an equation of the line through $(2, 1)$ that has slope 2/3.

   (b) Find an equation of the line through $(2, 1)$ that is perpendicular to the line in part (a).

3. Find equations in slope-intercept form for

   (a) the line that is parallel to the x-axis and passes through $(-3, 2)$.

   (b) the line that is parallel to the y-axis and passes through $(2, 4)$.

   (c) the line that is parallel to the line $y = x$ and passes through $(-3, 1)$.

   (d) the line through $(-3, 1)$ that is perpendicular to the line $y = 5x + 7$.

4. Find the x- and y-intercepts of the following lines

   (a) $3x + 4y = -12$

   (b) $y = m(x - a) + b$

5. For each condition below, find *all* values of k, if any, for which the line $2x + ky = -4k$ satisfies the condition.

   (a) The line has slope 3.

   (b) The line has slope $m \neq 0$.

   (c) The line passes through the point $(0, 4)$.

   (d) The line is horizontal.

   (e) The line is vertical.

6. Consider the curve shown below.

For each of the following pairs of points, find the slope of the secant line passing through the points and an equation of this line. [**HINT:** Draw the secant lines on this page.]

   (a) **A, B**   (c) **B, D**

   (b) **A, C**   (d) **C, D**

7. The graph of the cosine function is "locally straight" near $x = 1$. Which line does the cosine graph resemble near $x = 1$?

8. The graph $y = \sqrt{x}$ is "locally straight" near $x = 4$. Which line does this graph resemble near $x = 4$?

9. The graph $y = x^3$ is "locally straight" near $x = 2$. Which line does this graph resemble near $x = 2$?

10. The graph $y = 1/x$ is "locally straight" near $x = 1$. Which line does this graph resemble near $x = 1$?

Polynomial Algebra: A Brisk Review

Polynomials

Polynomials are among the simplest and most useful calculus functions. In this appendix we offer a brisk review of the *precalculus* of polynomials: basic ideas, terminology, techniques, and algebra. First, the definition:

> **Definition** A **polynomial** in x is an expression of the form
> $$a_0 + a_1x + a_2x^2 + a_3x^3 + \cdots + a_nx^n,$$
> that is, a sum of *constant* multiples of *nonnegative integer* powers of x.

EXAMPLE 1   Each of the following expressions
$$5x + 2, \qquad -x + 5x^{17}, \qquad \text{and} \qquad (x + 3)(\pi x^{17} + 7)$$
is a polynomial in x. ▶ (*Linear* functions, in particular, are polynomials—among the simplest ones.) None of

Use algebra to write the last example in standard form.

$$3 - \frac{1}{x}, \qquad \sqrt{x}, \qquad \text{and} \qquad \frac{x + 3}{\pi x^{17} + 7}$$

is a polynomial. ■

Basic Facts, Terms, and Techniques.   To help recall basic polynomial properties and concepts, here is a short glossary of polynomial language:

Degree The **degree** of a polynomial is its largest power. Constant polynomials, for instance, have degree zero. The polynomials $x^3 + 5x^2 + 1$, $-x + 5x^{17}$, and $(x + 3)(\pi x^{17} + 7)$ have degree 3, 17, and 18, respectively.

Polynomials of small degree have useful, suggestive nicknames. ▶ Polynomial functions of degree one are called **linear**, those of degree two **quadratic**, or **parabolic**. The naming sequence continues: cubic, quartic, quintic, ...

Learn and use such names; they're part of mathematical culture.

Factoring and Expanding To **factor** a polynomial is to write it as a product of lower-degree polynomials. To **expand** is to multiply out. The following table shows polynomials in both **factored form** and **expanded form**:

Polynomials in Two Forms	
Expanded Form	**Factored Form**
$x^2 + 3x + 2$	$(x + 1)(x + 2)$
$x^4 - 1$	$(x + 1)(x - 1)(x^2 + 1)$
$x^4 + 6x^3 + 11x^2 + 6x$	$x(x + 1)(x + 2)(x + 3)$
$x^4 + 6x^3 + 11x^2 + 6x$	$x(x + 1)(x + 2)(x + 3)$
$x^2 + 2ax + a^2$	$(x + a)^2$
$x^3 + 3ax^2 + 3a^2x + a^3$	$(x + a)^3$
$x^3 + 3ax^2 + 3a^2x + a^3$	$(x + a)^3$
$x^3 - a^3$	$(x - a)(x^2 + ax + a^2)$

The idea of root makes sense for any function f. The number r is a root of f if $f(r) = 0$. Finding roots of arbitrary functions is an important, and sometimes difficult, problem. We'll return to it.

Roots A number r is a **root** of a polynomial ◀ p if $p(r) = 0$. Graphically speaking, real roots and x-intercepts are the same thing.

Graphs show nicely the connections among real roots, intercepts, and factors. See the following graph for the polynomial expression $y = x^5 - x = x(x + 1)(x - 1)(x^2 + 1)$.

Graph of $y = x^5 - x$

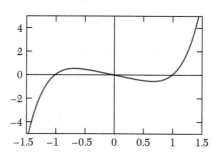

Notice, in particular, the *number* of real roots. Each linear factor contributes one root, in the obvious place. The remaining quadratic factor, $x^2 + 1$, is always positive, so it contributes no real roots.

Roots and Factors Roots and factors have an important algebraic relationship:

$(x - r)$ *is a factor of the polynomial $p(x)$ if and only if r is a root of p.*

The connection between roots and factors leads to a simple strategy for factoring polynomials: Somehow find (graphically, by guessing—whatever works) a root; then divide out the corresponding linear factor. ▶ It's easy to see, for instance, that $x = a$ is a root of the polynomial $p(x) = x^3 - a^3$. It follows that $(x - a)$ is a factor of $p(x)$. ▶

How Many Roots? A polynomial of degree n can have at most n real roots. Geometrically, this means that the graph of a polynomial of degree n crosses the x-axis at most n times. ▶

As we've seen, some polynomials of degree n have fewer than n real roots; *any* number up to n is possible. Want a polynomial of degree 43 with 27 real roots? Here's one (of many):

$$(x - 1)(x - 2) \cdots (x - 27)(x^2 + 1)^8 .$$

Can Every Polynomial Be Factored? No. For linear polynomials there's nothing to do; they're already factored. Factoring a quadratic polynomial p is either impossible or easy, depending on whether p has real roots. The **quadratic formula**

$$r = \frac{-b \pm \sqrt{b^2 - 4ac}}{2a}$$

finds the roots—if any exist—of the quadratic function $ax^2 + bx + c$.

Polynomials of degree three and higher are stickier. For the record, abstract theory (well beyond this course) says that *every* polynomial of degree three or higher has at least some lower-degree factors. In other words, every nonconstant polynomial can be written as a product of *linear* and *quadratic* factors; finding these factors, however, can be hard.

Completing the Square It's sometimes convenient to rewrite a quadratic polynomial in the general form $a(x + b)^2 + c$; doing so is called **completing the square**. The following example illustrates the technique.

> *This strategy seldom works in real life. In calculus textbooks, it seldom fails—the problems are rigged to make it so.*
>
> *See the last line of the table on page 442.*
>
> *Here's one consequence: The sine function isn't a polynomial.*

EXAMPLE 2   Let $p(x) = 2x^2 + 8x + 1$. Complete the square; use the result to find all roots of p.

Solution   We want to write $p(x)$ in the form $p(x) = 2(x + b)^2 + c$. Here goes—follow each algebra step closely:

$$2x^2 + 8x + 1 = 2(x^2 + 4x) + 1 = 2(x^2 + 4x + 4) + 1 - 8 = 2(x + 2)^2 - 7.$$

The third equation explains the name. Adding 4 to $x^2 + 4x$ "completed" the perfect square $(x + 2)^2$. ▶

The roots of p can now be read off:

$$p(x) = 0 \iff 2(x + 2)^2 - 7 = 0 \iff 2(x + 2)^2 = 7 \iff (x + 2)^2 = \frac{7}{2}$$

$$\iff x + 2 = \pm\sqrt{7/2} \iff x = -2 \pm \sqrt{7/2}.$$

(It's easy to check ▶ that the quadratic formula gives the same roots.)   ■

> *We "paid" for doing so by subtracting 8 at the end.*
>
> *Do so!*

Polynomials as Functions. The polynomial *expression* $2x^3 + 5x$ corresponds naturally to the *function* $p(x) = 2x^3 + 5x$. *Any* function defined by a polynomial expression is called a **polynomial function**. In practice the technical distinction between polynomial and polynomial function is seldom important. We'll use the terms interchangeably.

Domain and Range.   Because polynomials involve only multiplication and addition, every polynomial function has the same domain: all real numbers.

Polynomials' ranges vary drastically. The range of $p(x) = 3$ is $\{3\}$, a set with just one member. By contrast, the range of $q(x) = x^2 + 3$ is $[3, \infty)$, an infinite interval. The range of $r(x) = x^3 - 7x^2$ may not be obvious from the formula. Here's a graph:

Graph of $y = x^3 - 7x^2$

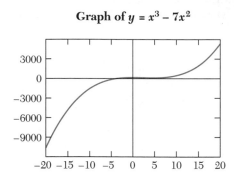

The graph shows how r behaves in the long run. It "blows up" (tends to ∞) as $x \to +\infty$ and "blows down" (tends to $-\infty$) as $x \to -\infty$. Therefore, apparently, r "hits" all real values; its range is $(-\infty, \infty)$.

Polynomial-Speak.   Polynomials have their own specialized vocabulary. Watch carefully in the following discussion for such words as **root**, **degree**, **factor**, **expand**, and **quadratic formula**.

EXAMPLE 3   Find the roots of the fourth-degree polynomial $p(x) = x^4 + 6x^3 + 11x^2 + 6x$; interpret results both algebraically and graphically.

Solution   In its expanded form, the formula for $p(x)$ is unhelpful. But the factored form reveals much more:

$$p(x) = x^4 + 6x^3 + 11x^2 + 6x = x(x + 1)(x + 2)(x + 3).$$

Now it's clear, algebraically, that p has *four* real roots—one where each of the four factors is zero. The following picture tells the same story:

Graph of $y = x(x + 1)(x + 2)(x + 3)$

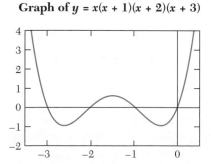

Whether a factored or expanded form of a polynomial is "better" depends on what we're doing. Here the factored form was preferable; the expanded form, on the other hand, may be handier for such symbolic operations as differentiation and integration. ▶ ■   *As we'll see!*

Some polynomials have *fewer* real roots than their degrees allow.

EXAMPLE 4 · Consider the cubic polynomial polynomial $p(x) = x^3 - 2x^2 + x - 2$. Factor p as much as possible; find all of p's real roots.

Solution   Here's a graph of p:

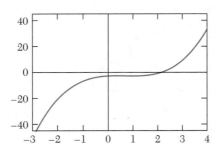

Graph of $y = x^3 - 2x^2 + x - 2$

The graph suggests that 2 is a root; a simple check confirms that, indeed, $p(2) = 0$. Therefore $(x - 2)$ is a factor of p. Indeed, long dividing $(x - 2)$ into $x^3 - 2x + x - 2$ shows ▶ that

Check the equality by multiplying out the product on the right.

$$p(x) = x^3 - 2x^2 + x - 2 = (x - 2)(x^2 + 1).$$

Is further factorization possible? No. Additional factors, if any, would come from $x^2 + 1$. But $x^2 + 1$ has no real roots ▶ and hence no linear factors.   *Why not?*

Note that the factored form $p(x) = (x - 2)(x^2 + 1)$ tells us something new: $p(x) = 0$ *only if $x = 2$*. In other words, 2 is the only real root. No graph could tell us this, because we can never see all of any polynomial's graph. ■

Why Polynomials Matter

Polynomial functions are simple but surprisingly useful. They lurk in calculus problems ranging from the sublime:

> *Show that among all rectangles of given perimeter, the square encloses the most area.*

to the ridiculous:

> *Design the most spacious possible rectangular pigpen, using 100 feet of fence. One side abuts a river . . . .*

A subtler but equally important use of polynomials is in *approximating* less convenient, nonpolynomial functions. For example, the following graphs show that, for x near 0, the polynomial function $x - x^3/6$ does a good job of approximating the sine function, which *isn't* a polynomial. ◄

Can you tell which is which?

Two graphs: $y = x - x^3/6$, $y = \sin(x)$

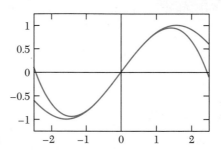

This is worth knowing. It suggests how a calculator might compute numerical values for nonalgebraic functions, such as the sine function.

Rational Functions

Is this really obvious? Think about it; try some examples.

Adding or multiplying two polynomials creates a new polynomial. ◄   *Dividing* two polynomials doesn't (usually) give another polynomial; it gives a **rational function**. Formally:

> **Definition** A rational function is a function that can be written as the quotient, or "ratio," of two polynomials.

EXAMPLE 5   Each function below is rational:

$$f(x) = \frac{1}{x} \qquad\qquad g(x) = \frac{x^2}{x^2 + 3}$$

$$h(x) = \frac{x^7 + x^6 + x^5 - x^4}{x + 1} \qquad k(x) = \frac{2}{x} + \frac{3}{x + 1}$$

The function $m(x) = \sqrt{x} = x^{1/2}$ is not rational because it involves a *fractional* power of x.

The function k may look suspicious. It's certainly the *sum* of two rational functions, but is k itself rational? The answer is yes, because we can rewrite $k(x)$

in the form:

$$k(x) = \frac{2}{x} + \frac{3}{x+1} = \frac{2(x+1)}{x(x+1)} + \frac{3x}{x(x+1)} = \frac{5x+2}{x(x+1)}.$$ ■

This computation illustrates a more general fact:

Fact Let r and s be rational functions. The sum, difference, product, and quotient of r and s are all rational functions—each can be written as the ratio of polynomials.

Graphs of Rational Functions

Rational functions ▶ have an important new feature: the possibility of **horizontal** and **vertical asymptotes**. Asymptotes are straight lines toward which a graph "tends."   *Unlike polynomials.*

EXAMPLE 6   Consider the rational function

$$r(x) = \frac{x^2}{x^2 - 1} = \frac{x^2}{(x-1)(x+1)}$$

and its graph:

Graph of $r(x) = x^2/(x^2 - 1)$

Technicalities aside, it's evident graphically that this function has three asymptotes: the *vertical* lines $x = -1$ and $x = 1$, and the *horizontal* line $y = 1$.

In Section 1.5, page 54, we show these facts carefully, relating them to infinite limits and limits at infinity.   ■

BASICS

1. Let p, q, r, and t be the polynomials

$$p(x) = 2x^3 + 5x^2 - x$$

$$q(x) = 4x^5 - 6x^3$$

$$r(x) = 5x^{12} - 123x^8 + 47$$

$$t(x) = 17x^{123} - 26x^{19} + 71x^5$$

What is the degree of each polynomial?

2. Find a polynomial of degree 5 with $x = -2$, $x = 1$, and $x = 3$ as its *only* real roots. Sketch its graph. (There are many possible answers.)

3. The following graphs show (i) a quadratic polynomial with two real roots, (ii) a quadratic polynomial with no real roots, (iii) a cubic polynomial with only one real root, and (iv) a cubic polynomial with three real roots.
   (a) Which graph is which? Label the graphs (i)–(iv).
   (b) Find an equation for the quadratic polynomial with two real roots.
   (c) Find an equation for the cubic polynomial with three real roots.
   (d) The cubic polynomial with only one real root can be written in the form $p(x) = (x + r)(x^2 + s)$. Find the values of r and s.
   (e) Find an equation for the quadratic polynomial with no real roots.

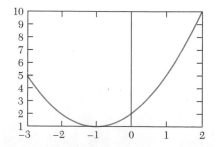

4. Let a be any constant. The table on page 442 gives the formula $x^3 - a^3 = (x - a)(x^2 + ax + a^2)$.
   (a) Verify the formula by multiplying out the right side.
   (b) If $a \neq 0$, then the second factor (i.e., $x^2 + ax + a^2$) cannot be factored any further. Show this either by completing the square or by using the quadratic formula.
   (c) Use the table on page 442 to find a factored formula for $x^3 + b^3$. [**HINT**: Let $a = -b$.]

5. The table on page 442 gives the formula $(x + a)^3 = x^3 + 3ax^2 + 3a^2x + a^3$.
   (a) Verify the formula by multiplying out the left side.
   (b) Let $p(x) = x^3 + 3x^2 + 3x + 1$. Check that $x = -1$ is a root of $p(x)$.
   (c) Since $x = -1$ is a root of $p(x) = x^3 + 3x^2 + 3x + 1$, $(x + 1)$ must be a factor of $p(x)$. Divide $p(x)$ by $(x + 1)$ to get a quadratic polynomial.
   (d) Factor $p(x) = x^3 + 3x^2 + 3x + 1$ as a product of linear factors.

6. Match each of the following functions

    (i) $f(x) = (x-1)(x+2)$

    (ii) $g(x) = \dfrac{x+3}{(x-1)^2}$

    (iii) $h(x) = \dfrac{2x^2 + 3x - 1}{x^2 - 2x - 8}$

to one of the following graphs. Then determine all values of x for which each function is positive.

Real-World Calculus: From Words to Mathematics

Translation.   Calculus is more than a brilliant theoretical invention. It is also a practical tool—among the most powerful ever discovered—for posing and solving important real-world problems. Real-world problems seldom arrive as neat mathematical packages. More common are untidy bundles of questions, assumptions, information, and (sometimes) disinformation. Important parts may be missing; useless extras may be included.

The first (and sometimes hardest) step in tackling such problems is to translate them, as concisely as possible, into mathematical language. Since calculus operates mainly on functions, "translation" in calculus usually means interpreting a problem in *functional language.*

To understand real phenomena—the Consumer Price Index, gravitational forces, population growth, planetary motion, and so on—we must be able to *predict* their behavior. Useful predictions, moreover, must be *quantitative*, not merely qualitative. Mathematics is the route to quantitative results.

Different areas of mathematics are best suited to describe and predict different sorts of phenomena. Describing phenomena of change is the special genius of calculus; calculus offers both language in which to describe changing quantities, and rules with which to predict their behavior.

Story Problems: You May Feel a Little Discomfort . . . .   Mathematics has been compared (in jest, of course) to dentistry. "Story problems," in particular, sometimes elicit emotions normally associated with the whine of a high-speed drill.

Such fear and loathing can be alleviated—without pharmaceutical intervention. This is not to deny the difficulty. Translation of any sort is expert work, combining both technical and "literary" ◄ expertise.

In our case, mathematical.

Learn the Words by Heart.   The essential point is that progress is possible. Indeed, the brevity, bluntness, and precision of mathematical language make mathematical translation simpler and more straightforward than it would otherwise be. For the same reason, mastering basic mathematical vocabulary and "grammar" is especially important.   Calculus has its own recurring key words and phrases— **proportional**, **parameter**, **domain**, **maximum**, **increasing**, and many others. An important goal of this appendix is to review several important mathematical terms and techniques and to illustrate their use in calculus-style applications.

EXAMPLE 1   As a rocket rises, the force earth's gravity exerts on it decreases. This *qualitative* observation, although perfectly true, is utterly useless in practice.

It doesn't take a rocket scientist to see why. Without more information, we can't compute anything: how much fuel to load, how much thrust is needed, how long anything will take.

Newton's law of universal gravitation ▶   describes gravity's effect *quantitatively*.   It says that for a 1000-lb rocket d miles from the center of the earth, the approximate force (in pounds) of earth's gravity is given by the function

$$F(d) = \frac{1.6 \times 10^{10}}{d^2}.$$

We discuss it more fully elsewhere. See p. 458.

With the formula, predictions are possible. If, say, $d = 20{,}000$ miles, ▶   then the force of gravity is only

$$F(20{,}000) = \frac{1.6 \times 10^{10}}{20{,}000^2} = 40 \text{ pounds.} \blacksquare$$

A reasonable distance for a communications satellite.

Garden-Variety Optimization

Many real-life problems boil down to finding the largest or smallest value of some function. Finding the right function hidden in the verbiage may be the main challenge.

EXAMPLE 2   Gardener Alpha plans a rectangular lettuce patch; 100 feet of rabbit-proof fencing is available. What are the best dimensions for Alpha's lettuce patch?

Could "best" possibly mean anything else? Probably so. For real lettuce patches, "best" would involve harvesting convenience, ease of watering, etc.

Translating Alpha's Problem.   "Best" can mean different things; we'll take it to mean "largest in area." ▶   The problem, then, is to find the largest rectangle with perimeter 100 feet.

The answer—a square—isn't surprising; translation corroborates our intuition. Thus, let x and y represent the lengths of the "vertical" and "horizontal" sides of the garden:

Alpha's garden

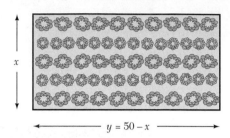

$y = 50 - x$

Equivalently, that
$y = 50 - x$.

(That 100 feet of fence is available says, algebraically, that $100 = 2x + 2y$.) ◄   The crucial point is that the area A, given by

$$A = \text{length} \times \text{width} = xy = x(50 - x),$$

is a function of the variable x.

The rule $A(x) = x(50 - x)$ makes *algebraic* sense for any real number x. In context, though, both x and $50 - x$ must be nonnegative, because each represents a physical length. Thus the "feasible" input set, or domain, is the x-interval $[0, 50]$.

We've arrived at an optimization problem:

Find the maximum value, and where it occurs, of the function

$$A(x) = x(50 - x)$$

on the interval $[0, 50]$.

S o l u t i o n   The area A is a quadratic function of x, so its graph is (part of) a parabola:

Graph of $A(x) = x(50 - x)$

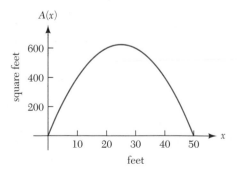

The graph suggests that the maximum value of A occurs at $x = 25$, where the area is $A(25) = 625$ square feet. Graphical guesses are often imperfect, but this time we're right: symmetry guarantees that the vertex of a parabola lies *exactly* midway between the two roots.

In the original language of the problem: *Alpha's lettuce patch should be square, 25 feet on a side, with area 625 square feet.* ■

EXAMPLE 3   Beta, too, has 100 feet of fence. Beta's barn will form one side of a rectangular patch. (The barn is long enough ▶   so that its length isn't a constraint.) Which dimensions are best now?

Over 100 feet, say.

Solution   This time, the answer *isn't* obvious; posing the problem mathematically will help us solve it.

Another "generic" picture applies:

Beta's garden

As before, area is a quadratic function of the side we called x: ▶

$$A(x) = xy = x(100 - 2x) = 100x - 2x^2 = 2x(50 - x).$$

Convince yourself that $y = 100 - 2x$.

The feasible domain is (again) the interval $[0, 50]$. ▶   In functional language, the problem is to *maximize the function A over its domain.*

Because x and y represent physical lengths.

The area function A is again quadratic, so its maximum value occurs midway between the two roots, at $x = 25$. The graph of A agrees:

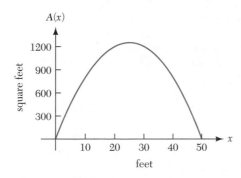

Graph of $A(x) = 2x(50 - x)$

In barnyard language: *The largest feasible garden has dimensions 25′ × 50′ and area A(25) = 1250 square feet.* ■

E X A M P L E 4   Gardener Gamma plans to fence exactly 1000 square feet of land. *Any* rectangular shape will do. For convenience in watering, one side will lie along a (straight) river. How much fencing will Gamma need for the other three sides?

Notice the new twist: Alpha and Beta aimed to *maximize area, for a fixed perimeter.* Gamma wants to *minimize perimeter, for a fixed area.* Optimization problems often come in such pairs. ◀

Paired optimization problems are known as "duals." This problem could be called a fencing dual.

S o l u t i o n   A familiar picture applies:

Gamma's garden

Using it, let's find our function. Let x and y be the lengths shown, and let L denote the total length of the fence. By assumption, $xy = 1000$, so

$$L = 2x + y = 2x + \frac{1000}{x}.$$

There's our function.

The Domain of *L*: A Subtlety.   What is L's "natural" domain? Technically, *any* positive number is a feasible input—a fence a mile long and 2.27 inches wide makes mathematical sense. Since no such ridiculously oblong design could possibly be most economical, we can safely ignore very large and very small positive inputs x. Given the general size of the numbers, the domain $5 \le x \le 50$ leaves plenty of room. Here, finally, is our problem:

Find the minimum value, and where it occurs, of the function $L(x) = 2x + 1000/x$ on the interval $[5, 50]$.

Solving the problem without calculus is hard, so we'll work approximately, with graphics. Here's a look:

Graph of $L(x) = 2x + 1000/x$

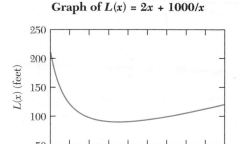

The graph shows that L attains its minimum value (about 90 feet) around $x = 22$. ► In agricultural terms: *Gamma needs 90 feet of fence; the garden should have approximate dimensions* $22.5' \times 45'$. ■

With a bit of calculus, we can find exact answers:
$x = \sqrt{500} \approx 22.3607$ *and*
$L(\sqrt{500}) \approx 89.4427$.

Parameters

We assumed earlier that Alpha and Beta each had 100 feet of fencing; our answers, naturally, reflected that assumption. But was the number 100 essential in either problem? Hardly. With 200 feet of fence we'd *still* design a square garden for Alpha, and a rectangular one, with proportions 1:2, for Beta.

Such "accidental" aspects of a problem are conveniently handled using **parameters**—symbols that stand for *any* constant value. The next example illustrates a simple use of parameters.

EXAMPLE 5   Delta and Epsilon, horticulturally incompatible, agree to "divorce" their gardens as shown:

Dividing the estate

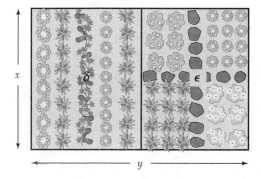

They have L feet of fence. What shape is best?

Notice two aspects of the problem:

Sorry; it just worked out that way.

- In this problem, the parameter is perimeter. ◄

- Before proceeding, *guess* an answer. Should the total garden be square? Vertically oblong? Horizontally oblong?

Let x and y be as shown. As before, the area A satisfies the equation $A = xy$. The garden's peculiar design dictates that the total length of fence needed is $L = 3x + 2y$, so ◄

Solve this equation for y to get $y = (L - 3x)/2$.

$$A(x) = xy = \frac{x(L - 3x)}{2}.$$

In context, both x and $L - 3x$ must be nonnegative, so $0 \leq x \leq L/3$ must hold. Our problem, then, is to maximize A for x in the interval $[0, L/3]$.

The presence of the parameter L now complicates life slightly. We'll show two possible strategies.

Maximizing A with Algebra.   For *any* nonnegative number L the function

$$A(x) = \frac{x(L - 3x)}{2}$$

is quadratic in x. Its graph is a parabola, opening *downward*. Symmetry guarantees that the vertex of a parabola lies midway between its two x-intercepts. From the formula above it's clear that $A(x) = 0$ if $x = 0$ or $x = L/3$. Therefore, the largest value of A occurs halfway between, at $x = L/6$.

Maximizing A with Graphics.   Another strategy is to plot $A(x)$ for *several* reasonable values of the parameter, and look for a pattern:

Graphs of A for various lengths L

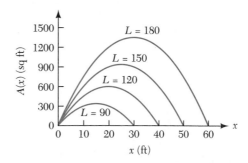

The picture shows that for each value of L, the maximum area occurs at $x = L/6$, the *midpoint* of the domain $[0, L/3]$.

Conclusion.   The most spacious double garden uses $L/6$ feet of fence for each of the three "vertical" runs. The remaining $L/2$ feet is divided equally for the two "horizontal" runs. ▶   The result, drawn proportionately to scale, looks like this:

Again, half the fence runs in each direction. Is this coincidence, or is something happening here?

Biggest gardens with perimeter _L_

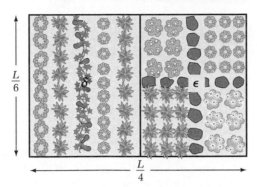

Taking Leave of Our Fences.   Fencing problems, although unglamorous, illustrate nicely the standard and important process of defining a pertinent function on a sensible domain, finding maximum or minimum values, and then interpreting the results.

Proportionality

Proportionality and inverse proportionality are relationships between varying quantities:

Definition Let _A_ and _B_ be quantities which vary together. _A_ is **proportional** to _B_ if, for some nonzero constant _k_, the equation

$$A = k\,B$$

holds for _all_ values of _A_ and _B_. _A_ and _B_ are **inversely proportional** if

$$A = \frac{k}{B} \,.$$

EXAMPLE 6   The formula ▶

$$C = \pi\,d$$

says that the circumference of a circle is proportional to the diameter, with **proportionality constant** π. The formula

$$A = \pi\,r^2$$

shows that the area of a circle is proportional to the _square_ of the radius, but not to the radius itself.   ■

Although familiar, this formula is not trivial. That the same constant, π, appears here and in the area formula is a remarkable fact.

Gravity—Applying Proportionality

It's well known that objects weigh less on the moon than on Earth. How much less? Why?

Our words—Newton wrote in Latin.

Newton's law of universal gravitation describes the relationships among quantities involved in gravitation: ◄

The attractive force of gravity exerted on one object by another is proportional to the mass of each, and inversely proportional to the square of the distance between them. ◄

In each case, all other quantities are held constant.

In algebraic form, Newton's law says that

$$F = k\frac{M_1 M_2}{d^2},$$

For a sphere the center of gravity is the ordinary (geometric) center.

where F represents gravitational *force*, M_1 and M_2 are the objects' masses, d is the distance between the objects' centers of gravity, ◄ and k is a constant of proportionality. The numerical value of k depends on the units chosen. In the metric system (using kilograms for mass, meters for distance, and newtons for force), $k \approx 6.7 \times 10^{-11}$.

EXAMPLE 7   How much less does an object weigh on the moon than on Earth? What would the same object weigh on Jupiter?

Solution   An object's weight is defined as the force gravity exerts on it. According to Newton's law, the "Earth-weight" F_e of an object with mass M, at the surface of Earth, is

$$F_e = k\frac{M M_e}{R_e^{\,2}},$$

where M_e is Earth's *mass*, and R_e is Earth's *radius*. Similarly, the same object's "moon-weight" and "Jupiter-weight," are, respectively,

$$F_m = k\frac{M M_m}{R_m^{\,2}} \qquad \text{and} \qquad F_j = k\frac{M M_j}{R_j^{\,2}},$$

where M_m and M_j are the masses, and R_m and R_j are the radii, of the moon and Jupiter, respectively.

Our problem is about the ratios F_m/F_e and F_j/F_e. By easy algebra,

$$\frac{F_m}{F_e} = \frac{M_m R_e^{\,2}}{M_e R_m^{\,2}} \qquad \text{and} \qquad \frac{F_j}{F_e} = \frac{M_j R_e^{\,2}}{M_e R_j^{\,2}}.$$

Remarkably enough, every constant above is known—physics and astronomy textbooks have tables like this one:

Celestial Constants		
Object	**Mass (kg)**	**Radius (km)**
Moon	7.35×10^{22}	1740
Earth	5.98×10^{24}	6378
Jupiter	1.90×10^{27}	71398

Now only simple arithmetic remains:

$$\frac{F_m}{F_e} \approx 0.165; \qquad \frac{F_j}{F_e} \approx 2.535.$$

In human terms, an astronaut weighing 75 kg (about 165 pounds) on Earth weighs only $0.165 \cdot 75 \approx 12.38$ kg (about 27 pounds) on the moon. On Jupiter, the same astronaut weighs a staggering 190.13 kg (over 418 pounds).   ∎

Mass or Weight?   Physicists distinguish between an object's *mass* and its *weight*. An object's weight can vary (as our example shows) but its mass is the same everywhere. In the metric system, the standard unit of force (and therefore of weight) is the **newton**. At the surface of the earth an object with mass 1 kilogram has weight 9.8 newtons. (Actually, since Earth is not perfectly spherical, an object's weight is slightly different at different points on Earth's surface.)

BASICS

1. Zeta plans to fence off a rectangular field along the bank of a river; no fence is needed along the river. Fencing costs $2 per foot for the ends and $3 per foot for the side parallel to the river. Zeta's budget is $900.

   (a) Let x be the length of the side of the field that lies along the river. Explain why the formula $A(x) = x(225 - 0.75x)$ describes the area of Zeta's field.

   (b) What is the domain of A? Why?

   (c) What are the dimensions of Zeta's *largest* field?

   (d) If given twice as much money, can Zeta enclose twice as much area? Why or why not?

2. Water is flowing into a cubical tank. Each side of the tank is ten feet.

   (a) Write a formula for V, the volume of water in the tank, as a function d, the depth of water in the tank.

   (b) What is the domain of V? What is the range of V?

   (c) Write a formula for d as a function of V.

3. Guido's pizza is delivered in boxes formed from 50-cm by 100-cm rectangular pieces of cardboard. The boxes are made by cutting out six squares of equal size, three from each of the 100-cm edges (one square at each corner, and one in the middle of the edge), then folding in the obvious way.

   (a) Find the volume of the box as a function of the length of the side of the square cutouts.

   (b) How large should the cutouts be to maximize the volume of the box?

4. Suppose that A and B are *directly* proportional, but A and C are *inversely* proportional. Suppose, further, that if $A = 1$, then $B = 2$ and $C = 3$.

   (a) What are the values of B and C if $A = 10$?

   (b) Are B and C directly proportional? Inversely proportional? Explain.

   (c) Suppose that $B = 10$. Find values for A and C.

5. Is the volume of a sphere proportional to its radius? Explain.

6. The weight of an object is inversely proportional to the square of its distance from the center of Earth. A satellite weighs 200 pounds when 10,000 miles from the center of Earth.

   (a) Express the weight of the satellite as a function of d, its distance from the center of Earth.

   (b) At what distance from the center of Earth does the satellite weigh 100 pounds?

7. An open box with a capacity of 36,000 cubic inches is needed. The box must be twice as long as it is wide. The material out of which the box is to be constructed costs $0.10 per square foot. Find the cost of the material for the box as a function of the width of the box.

8. It's 9:45 P.M. Your snowmobile is out of gas and you are 3 miles due south of a major highway in a field. The nearest service station on the highway is 6 miles east of your position. You can walk 4 miles per hour on roads, but only 3 miles per hour through snowy fields.

   Express the time to reach the service station as a function of the distance r you walk along the road. [**HINT:** You could walk straight through the fields, or partway through the fields, then on the road.]

Algebra of Exponentials

Exponential functions are among the most important in both the theory and the applications of calculus. Indeed the exponential function $f(x) = e^x$ ◄ has been called the most important function in all of mathematics. To understand and use such functions effectively requires a basic familiarity with exponential algebra.

Exponential Expressions

An **exponential expression** is any algebraic expression that contains exponents. Each of the following expressions is exponential:

$$2^x, \quad 3^{x+h}, \quad b^3, \quad 2^{x^2}, \quad \frac{2^x}{3^x}, \quad \left(\frac{2}{3}\right)^x, \quad \left(1+\frac{1}{n}\right)^n.$$

In any expression of the form b^x, b is the **base** and x is the **exponent**.

The Natural Exponential Function

The base-e exponential function, defined by $f(x) = e^x$, is called the **natural exponential function** or simply **the exponential function**. (The definite article in the latter name is slightly unfortunate. Expressions with *other* bases—2^x, 10^x, and so on—also deserve the name "exponential.") Various notations are used for the natural exponential function:

$$\exp(x), \qquad \exp x, \qquad \text{and} \qquad e^x$$

all mean the same thing.

Basics of Exponential Algebra

The basic algebraic rules for exponential expressions are few and simple:

$$b^x b^y = b^{x+y} \tag{E.1}$$

$$\frac{b^x}{b^y} = b^{x-y} \tag{E.2}$$

$$\left(b^x\right)^y = b^{xy} \tag{E.3}$$

$$a^x b^x = (ab)^x \tag{E.4}$$

$$b^0 = 1 \tag{E.5}$$

$$b^{-x} = \frac{1}{b^x} \tag{E.6}$$

In each case, b can be any *positive* number; x and y can take any real values. ▶

We'll see below why $b > 0$ matters.

EXAMPLE 1   Explain Rule (E.1): ▶ $b^x b^y = b^{x+y}$, assuming that x and y are positive integers.

The "addition rule."

Solution   If x and y are positive integers, then

$$b^x = \overbrace{b \cdot b \cdot b \cdots b}^{x\text{-times}} \quad \text{and} \quad b^y = \overbrace{b \cdot b \cdot b \cdots b}^{y\text{-times}}.$$

Therefore,

$$b^x b^y = \overbrace{b \cdot b \cdot b \cdots b}^{x} \cdot \overbrace{b \cdot b \cdot b \cdots b}^{y} = \overbrace{b \cdot b \cdot b \cdots b}^{x+y} = b^{x+y}. \quad \blacksquare$$

EXAMPLE 2   Make sense of the "zeroth power" rule: $b^0 = 1$.

Solution   Rule(E.5) follows directly from the addition rule: If b is any positive number, then

$$b \cdot b^0 = b^1 \cdot b^0 = b^{1+0} = b^1 = b.$$

Thus $b \cdot b^0 = b$; dividing through by b gives $b^0 = 1$. ∎

EXAMPLE 3   Rule (E.6) tells how to switch the sign of an exponent. It helps with calculations like this:

$$\left(\frac{1}{2}\right)^{-17} = \frac{1}{2^{-17}} = 2^{17}. \quad \blacksquare$$

EXAMPLE 4   Derive Rule (E.6) from the other rules.

Solution   Use Rules (E.1) and (E.5) to convince yourself of each equality below:

$$b^{-x} \cdot b^x = b^{x-x} = b^0 = 1.$$

Dividing the first and last quantities by b^x gives Rule (E.6). ∎

This example is algebraically tricky, but it shows one way exponential algebra is really used in calculus.

E X A M P L E 5 Let k be any constant, and define the function f by $f(x) = 3^{kx}$. Show ◀ that

$$\frac{f(x+h) - f(x)}{h} = f(x) \cdot \frac{f(h) - f(0)}{h}.$$

S o l u t i o n Watch carefully—we'll use several of the rules cited previously:

$$\frac{f(x+h) - f(x)}{h} = \frac{3^{k(x+h)} - 3^{kx}}{h} = \frac{3^{kx+kh} - 3^{kx}}{h}$$

$$= \frac{3^{kx} \cdot 3^{kh} - 3^{kx}}{h} = 3^{kx} \cdot \frac{3^{kh} - 1}{h}$$

$$= 3^{kx} \cdot \frac{3^{kh} - 3^0}{h} = f(x) \cdot \frac{f(h) - f(0)}{h}. \qquad \blacksquare$$

Thinking of Positive Powers: Why $b > 0$ Is Best

Exponential expressions with negative bases can cause trouble. To see how, consider the expression $(-8)^x$ for various values of x:

$$(-8)^2 \qquad (-8)^{-2} \qquad (-8)^{1/2} \qquad (-8)^{1/3} \qquad (-8)^{3/2} \qquad (-8)^{4/3}$$

The first two make good sense:

$$(-8)^2 = 64 \qquad \text{and} \qquad (-8)^{-2} = \frac{1}{64}.$$

What is this value?

The third, $\sqrt{-8}$, isn't a real number, but the fourth, $(-8)^{1/3}$, has a real value. ◀ The fifth expression is meaningless (it involves $\sqrt{-8}$) but the sixth makes good sense:

$$(-8)^{4/3} = (-8) \cdot (-8)^{1/3} = (-8) \cdot (-2) = 16.$$

Faced with such difficulties, we retreat. In calculus, we'll take the base b of every exponential expression b^x to be positive.

Transcendental Meditations (Optional)

The *exponential* function $f(x) = 2^x$ differs from the *power* function $g(x) = x^2$ far more than typography would suggest. One important difference is that the expression x^2 is an algebraic formula (i.e., an explicit recipe for computing output values). If, say, $x = 2.3$, it's easy, even without a calculator, to compute the output: $2.3^2 = 5.29$.

Do you see why? Try it.

A curious word. It suggests that exponential functions somehow float above, or transcend, the mundane world of algebra.

The expression 2^x, by contrast, is not algebraic. Finding the output value $2^{2.3}$ by hand is almost impossible. ◀ For this reason, exponential functions (and other "nonalgebraic" functions) are referred to as **transcendental**. ◀ We explore some ramifications of the nonalgebraic character of exponential functions that follow.

Fractional and Irrational Exponents: Abuse of Powers?

What does b^x really mean if the exponent x isn't an integer? What if x is irrational?

EXAMPLE 6   Find (or estimate) decimal values for

$$2^{1/3}, \qquad 1.6^{11/13}, \qquad \text{and} \qquad \pi^{-2/3}.$$

Solution $2^{1/3}$ is the *cube root* of 2. According to any scientific calculator,

$$\sqrt[3]{2} \approx 1.25992104989.$$

For *any* integer $n > 0$,

$$b^{1/n} = \sqrt[n]{b}.$$

The connection between roots ▶ and powers suggests how to estimate $1.6^{11/13}$ numerically:

$$1.6^{11/13} = \left(1.6^{1/13}\right)^{11} \approx 1.03681563373^{11} \approx 1.4883905652.$$

Is it obvious that every positive number b has a positive nth root? How would one estimate, say, the 11th root of 13?

Similarly,

$$\pi^{-2/3} = \frac{1}{\pi^{2/3}} = \frac{1}{\sqrt[3]{\pi^2}} \approx \frac{1}{\sqrt[3]{9.869604401089}}$$

$$\approx \frac{1}{2.14502939711} \approx 0.466194077036.$$ ■

So far so good: The examples show how to handle *rational* exponents. How about *irrational* exponents?

EXAMPLE 7   Estimate the value of $\pi^{\sqrt{2}}$.

Solution   The "obvious" approach—to approximate everything in sight with rational numbers—works: ▶

$$\pi^{\sqrt{2}} \approx 3.1415^{1.4142} \approx 5.0472.$$ ■

Decimal numbers are rational, aren't they?

Fair Questions.   Does all this approximation really work? Is it really OK to approximate *both* b and x with convenient decimal numbers before computing b^x? The short answer to all of these good questions is "yes." In mathematical language, the fortunate fact is that

$$\text{if } a \approx b \text{ and } x \approx y, \text{ then } b^x \approx a^y.$$

Reasonable as this seems, a rigorous proof is subtle; we'll omit it.

BASICS

1. Evaluate 64^x if x equals
   (a) 0
   (b) 1
   (c) 1/2
   (d) $-1/2$
   (e) 2/3
   (f) 5/6

2. Express each of the following numbers as 4^x for a suitable exponent x.
   (a) 16
   (b) 2
   (c) 8
   (d) $1/\sqrt{2}$
   (e) $4\sqrt{2}$
   (f) 16^{35}

3. Let a be a positive number. Rewrite each of the following expressions in the form a^x.
   (a) $(\sqrt{a})^2$
   (b) $1/a^2$
   (c) 1
   (d) a^3/a^4
   (e) $(a^3/a^5)^{10}$
   (f) $a/\sqrt[3]{a}$

4. Here's further evidence for the fact that $b^0 = 1$:
   $$b^0 \cdot b^0 = b^{0+0} = b^0.$$
   Therefore, b^0 *is its own square.* Does it follow that b^0 must be 1? Explain.

5. Let f be the function $f(x) = 5^{2x}$. Show that
   $$\frac{f(x+h) - f(x)}{h} = f(x) \cdot \frac{f(h) - f(0)}{h}.$$

In Exercises 6–8, mimic the arguments given for the "addition" rule in Example 1 to show that the rule is "obvious" for positive integer powers.

6. Rule (E.2)   8. Rule (E.4)
7. Rule (E.3)

9. Let $g(x) = Ae^{0.2x}$ where A is a real number.
   (a) Suppose that $g(0) = 5$. Find A.
   (b) Suppose that $g(1) = 5$. Find A.
   (c) Suppose that $g(-1) = 5$. Find A.

10. Describe how the graph of each of the following functions is geometrically related to the graph of $y = 2^x$. [**EXAMPLE:** The graph of $y = 2^{-x}$ is the reflection of the graph of $y = 2^x$ about the y-axis.]
   (a) $(1/2)^x$
   (b) $(1/2)^{-x}$
   (c) -2^{-x}
   (d) 2^{x-1}
   (e) $2 - 2^x$
   (f) $2 \cdot 2^{x-1}$

11. Find an equation for the graph that results from each of the following operations on the graph of $y = e^x$.
   (a) Reflection about the x-axis.
   (b) Reflection about the y-axis.
   (c) Reflection about the line $y = 4$.
   (d) Reflection about the line $x = -1$.
   (e) Reflection about the y-axis, then a reflection of the resulting graph about the x-axis.

12. (a) Explain how to obtain the graph of $y = 2e^{x+3}$ from the graph of $y = e^x$ using *only* vertical stretching.
   (b) Explain how to obtain the graph of $y = 2e^{x+3}$ from the graph of $y = e^x$ using *only* horizontal translation.
   (c) Can the graph of $y = 2e^{4x+3}$ be obtained from the graph of $y = e^x$ using only the operations of vertical stretching and horizontal translation? Explain.

Algebra of Logarithms

Logarithm functions (i.e., functions defined by logarithmic expressions) are important in both the theory and the applications of calculus. To understand and use such functions effectively requires a basic familiarity with the algebra of such expressions.

Logarithms and Exponentials: How They're Related

Logarithmic expressions like these—

$$\log_2 x, \qquad \log_{10} t, \qquad \log_e z, \qquad \ln w, \qquad \text{and} \qquad \log_b x$$

may be mystifying at first glance. What does "$\log_b$" *mean*? What is the role of b? What is "ln?" Some mystification is natural: Like "sin," "cos," and "exp," "$\log_b$" is the *name* of a function—not a recipe for computation.

Demystifying logarithms means, first and foremost, understanding how they're related to exponentials. For us, this relation *defines* the idea of logarithm.

> **Definition** For any suitable **base** b and positive number x, the expression $\log_b x$ is defined by the condition
> $$y = \log_b x \iff x = b^y.$$
> (NOTE: "Suitable" means $b > 0$ and $b \neq 1$.)

EXAMPLE 1 Use the definition to simplify the expressions

$$\log_2 8, \qquad \log_{10} 100, \qquad \log_e e^{25}, \qquad \text{and} \qquad \log_b \frac{1}{b^3}.$$

Solution   By definition,

$$y = \log_2 8 \iff 8 = 2^y.$$

It's clear from the second equation that y must be 3; hence $\log_2 8 = 3$. By similar reasoning,

$$y = \log_{10} 100 \iff 10^y = 100 \iff y = 2;$$

$$y = \log_e e^{25} \iff e^y = e^{25} \iff y = 25;$$

$$y = \log_b \frac{1}{b^3} \iff b^y = b^{-3} \iff y = -3. \quad \blacksquare$$

Logarithm and Exponential Functions as Inverses

The relation

$$y = \log_b x \iff x = b^y$$

means that the logarithm function $\log_b x$ and the exponential function b^x are *inverses*. For any two functions f and g to be inverses means that each function "undoes" the other. In other words, the equations

$$f\big(g(x)\big) = x \quad \text{and} \quad g\big(f(x)\big) = x$$

hold for all sensible inputs x. This general fact, applied specifically to the logarithm and exponential functions with base b, looks like this:

$$\log_b b^x = x \quad \text{and} \quad b^{\log_b x} = x.$$

For more on domains, see Section 1.5.

These equations hold for all x in the respective domains of $\log_b$ and b^x. ◄

Basics of Logarithmic Algebra

A few algebraic rules for logarithmic expressions go a long way. Here are four to know:

$$\log_b xy = \log_b x + \log_b y \qquad (F.1)$$

$$\log_b x/y = \log_b x - \log_b y \qquad (F.2)$$

$$\log_b x^r = r \log_b x \qquad (F.3)$$

$$\log_b 1/x = -\log_b x \qquad (F.4)$$

(Note: $b > 0$ and $b \neq 1$.)

Using the Rules

The rules simplify computations that involve logarithms—and even some that don't seem to.

EXAMPLE 2   Just before it died, your calculator said that $\log_{10} 2 \approx 0.301$ and $\log_{10} 3 \approx 0.477$. You *really* need to know

$$\log_{10} 6, \qquad \log_{10} 1/18, \qquad \log_{10} 2000, \qquad \text{and} \qquad \log_{10} 0.003 .$$

What can you do?

Solution   With the rules there's no problem: ▶

$$\log_{10} 6 = \log(3 \cdot 2) = \log_{10} 3 + \log_{10} 2 \approx 0.301 + 0.477 = 0.778$$

$$\log_{10} 1/18 = -\log_{10}(2 \cdot 3^2) = -\log_{10} 2 - 2\log_{10} 3 \approx -1.255$$

$$\log_{10} 2000 = \log_{10} 1000 + \log_{10} 2 = 3 + \log_{10} 2 \approx 3.301$$

$$\log_{10} 0.003 = \log_{10} 0.001 + \log_{10} 3 = -3 + \log_{10} 3 \approx -2.523 \qquad ∎$$

But watch closely, and check the results with your own (live) calculator.

EXAMPLE 3   Solve (for x) each of the equations

$$200 \cdot 1.05^x = 375 \qquad \text{and} \qquad \ln x + \ln x^2 = 5 .$$

Solution   For the first equation we'll take the natural logarithm of both sides, and then use Rules (F.1) and (F.3):

$$200 \cdot 1.05^x = 375 \iff \ln 200 + \ln(1.05^x) = \ln 375$$

$$\iff x \ln 1.05 = \ln 375 - \ln 200$$

$$\iff x = \frac{\ln 375 - \ln 200}{\ln 1.05} \approx 12.8839$$

To solve the second equation we'll first use Rules (F.1) (in "reverse") and (F.3), and then apply an exponential function to undo the logarithm. Notice first that

$$\ln x + \ln x^2 = \ln x^3 = 3 \ln x .$$

Therefore,

$$\ln x + \ln x^2 = 5 \iff 3 \ln x = 5$$

$$\iff \ln x = 5/3$$

$$\iff x = e^{5/3} \approx 5.2945 . \qquad ∎$$

Why the Rules Hold: Translating between Logarithms and Exponentials

The familiar properties of logarithms listed above are, in a sense, disguised versions of similar rules for exponentials. Properties of logarithms can be "translated" into properties of exponentials, and vice versa.

EXAMPLE 4   Explain, in terms of exponentials, Rule (F.1):

$$\log_b xy = \log_b x + \log_b y .$$

Solution   What does Rule (F.1) really say? The equation $\log_b xy = z$—where z is *any* real number—means that $b^z = xy$. Therefore Rule (F.1) really says that

$$b^{\log_b x + \log_b y} = xy.$$

Now exponential algebra comes in:

$$b^{\log_b x + \log_b y} = b^{\log_b x} \cdot b^{\log_b y} = xy,$$

Why does the last equation hold? Because logarithms and exponentials are inverses.

just as we wanted. ◄

Which Base?

The general ideas of logarithm and exponential functions are the same for *any* base b. In practice, though, either $b = 10$ or $b = e \approx 2.71828$ is almost invariably used. Logarithms with these bases are known as the **common logarithm** and the **natural logarithm**, respectively. The natural log function, usually written as $\ln x$, is for many good reasons ◄ the favorite in calculus. We'll use base e almost exclusively.

Several good reasons appear in Section 3.3.

Moving Around the Bases: Two Key Identities

To exploit the advantages of base e—for both logarithm and exponential functions—we'll need to move easily from *any* base b to the special base e. Two key identities tell how. For any base b and input x,

$$b^x = e^{x \cdot \ln(b)};$$

$$\log_b(x) = \frac{\ln(x)}{\ln(b)}.$$

Why do these equations hold? Both follow from the fact that logarithm and exponential functions are inverses, and from basic properties of logarithms. For the first:

$$b^x = \left(e^{\ln(b)}\right)^x = e^{x \cdot \ln(b)},$$

Why does $b = e^{\ln(b)}$?

as claimed. ◄ The second identity is similar:

$$x = b^{\log_b x} \implies \ln(x) = \ln\left(b^{\log_b x}\right) = \log_b x \, \ln b$$

$$\implies \log_b(x) = \frac{\ln(x)}{\ln(b)}.$$

EXAMPLE 5 Rewrite the exponential function 3^x in base e.

Solution From the first identity above,

$$3^x = e^{x \ln(3)} \approx e^{1.09861228867x}.$$

EXAMPLE 6 Rewrite $\log_5 x$ in terms of the *natural* logarithm.

Solution From the second identity above,

$$\log_5 x = \frac{\ln x}{\ln 5} \approx \frac{\ln x}{1.60943791243}.$$

BASICS

1. Translate each of the following equations into the language of logarithms.
   (a) $8^{2/3} = 4$
   (b) $10^3 = 1000$
   (c) $10^{-4} = 0.0001$
   (d) $(1/2)^2 = 4$

2. Translate each of the following equations into the language of exponentials.
   (a) $\log_2 7 = x$
   (b) $\log_5 1/25 = -2$
   (c) $\log_{10} 1000 = 3$
   (d) $\log_{64} 128 = 7/6$
   (e) $\log_{1/2} (1/4) = 2$

3. Evaluate each of the following expressions *exactly*.
   (a) $2^{\log_2 16}$
   (b) $2^{\log_2 1/2}$
   (c) $10^{\log_{10} 7}$
   (d) $\log_6 \sqrt{6}$
   (e) $\log_5 5^3$
   (f) $\log_3 (1/27)$

4. If $\log_4 A = 2.1$, evaluate each of the following expressions *exactly*.
   (a) $\log_4 A^2$
   (b) $\log_4 (1/A)$
   (c) $\log_4 16A$
   (d) $\log_2 A$
   (e) $\log_{16} A$

5. Interpret Rules (F.2) and (F.3) in exponential language.

6. Explain why $\ln(x^2 - x - 2) = \ln(x - 2) + \ln(x + 1)$.

7. (a) Explain how to obtain the graph of $y = 2 + \ln(3x)$ from the graph of $y = \ln x$ using *only* vertical translation.
   (b) Explain how to obtain the graph of $y = 2 + \ln(3x)$ from the graph of $y = \ln x$ using *only* horizontal stretching.
   (c) Can the graph of $y = 2 + \ln(3x + 4)$ be obtained from the graph of $y = \ln x$ using *only* the operations of vertical translation and horizontal stretching? Explain.

8. Let $f(x) = 20e^{0.1x}$.
   (a) Find an x for which $f(x) = 20$.
   (b) Find an x for which $f(x) = 40$.
   (c) Find an x for which $f(x) = 80$.
   (d) Find an x for which $f(x) = 10$.

9. Let $h(x) = 20e^{kx}$ where k is a real number.
   (a) Suppose that $h(1) = 20$. Find k.
   (b) Suppose that $h(1) = 40$. Find k.
   (c) Suppose that $h(1) = 10$. Find k.

10. Let $j(x) = Ae^{kx}$ where A and k are real numbers.
    (a) Suppose that $j(0) = 5$ and $j(3) = 10$. Find A and k.
    (b) Suppose that $j(1) = 5$ and $j(3) = 10$. Find A and k.

11. In each part below, plot *both* graphs on the same axes. Explain what you see, referring to properties of exponential and logarithm functions.
    (a) $y = 2^x$, $y = 4^{x/2}$
    (b) $y = \ln(\exp(x^2))$, $y = x^2$
    (c) $y = \ln((\exp x)^2)$, $y = 2x$
    (d) $y = \exp(\ln x)$, $y = x$
    (e) $y = \exp(2 \ln x)$, $y = x^2$
    (f) $y = \exp(\frac{1}{2} \ln x)$, $y = \sqrt{x}$

Trigonometric Functions

The six trigonometric functions are introduced and defined in Chapter 1, Section 5. Start there for definitions and basic properties. In this appendix we collect a variety of information, in more detail, on trigonometric functions, their properties, their graphs, and their importance in calculus.

Trigonometric Function Notation

And a few others.

Trigonometric functions ◄ are often typeset without the customary $f(x)$-style parentheses. It's standard to write

$$\sin t, \qquad \cos t, \qquad \text{and} \qquad \tan t$$

rather than

$$\sin(t), \qquad \cos(t), \qquad \text{and} \qquad \tan(t).$$

But not slavishly.

We'll usually ◄ follow this convention, although a bit reluctantly—parentheses remind us that *functions* are involved. Whenever confusion seems possible (e.g., with the expression $\sin(t + 1)$) or we want to emphasize the functional point of view, we'll use parentheses.

Other quirks of trigonometric notation involve *powers* of trigonometric functions. It's conventional, for instance, to write

Other powers are handled similarly—e.g., $\sin^{17} x$ means $(\sin x)^{17}$.

$$\sin^2 t + \cos^2 t = 1 \qquad \text{instead of} \qquad \left(\sin(t)\right)^2 + \left(\cos(t)\right)^2 = 1,$$

although both mean the same thing. ◄ Neither of these equations should *ever*, of course, be confused with

$$\sin t^2 + \cos t^2 = 1 \quad \text{or} \quad \sin(t^2) + \cos(t^2) = 1,$$

which are equivalent to each other, but *false*.

Sines, Cosines, Angles, and Triangles

Any real number is a legitimate input to the sine or cosine function. We've interpreted such an input as the *signed length of an arc*. (For more on this interpretation, see the definitions on page 62.) Thus $\cos(2) \approx -0.42$, because -0.42 is (approximately) the x-coordinate of the point 2 *units of distance* counterclockwise from $(1, 0)$ on the unit circle. (The same units of distance are used both on the circle and on the axes.) Here's the picture for $\cos(2)$: ▶

Convince yourself that all the quantities shown are plausible.

Why cos(2) ≈ –0.42

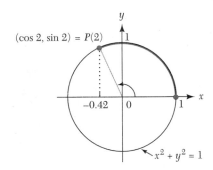

Sometimes we prefer to interpret inputs to the sine and cosine functions as *angles*, not arclengths. Any arc on the unit circle determines an obvious angle: the one between the two line segments joining the origin to the ends of the arc. (See the picture above.) From this point of view, $\cos 2$ is the cosine of the *angle* determined by an arc of length 2.

From Radians to Degrees and Back.   Two standard units (**radians** and **degrees**) are used to measure angles. Radian measure will prove far more convenient for us; we'll use it consistently. ▶   To convert between radians and degrees, use these formulas:

So should you. So should your calculator. Lock it in radian mode and throw away the key.

Radians ⇄ Degrees	
radian measure	$= \dfrac{\pi}{180} \times$ degree measure
degree measure	$= \dfrac{180}{\pi} \times$ radian measure

(Both follow from the fact that 2π radians $= 360°$.) In particular, the angle shown is

$$2 \text{ radians} = 360/\pi \approx 114.6°\,.$$

Right Triangles.   "Trigonometry" refers, etymologically, to *measuring triangles*. Our "circular" viewpoint on trigonometric functions is easily linked to right triangles; doing so sometimes aids intuition.

If $P(t)$ lies in the first quadrant, i.e., if $0 \le t \le \pi/2$, then $P(t)$ naturally determines, as shown below, a right triangle with unit hypotenuse, one of whose angles measures t radians:

Triangles and trigonometric functions

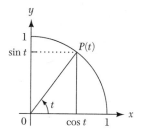

It's easy to see that this triangle has the "right" trigonometric properties. For example:

$$\sin t = \frac{\text{opposite}}{\text{hypotenuse}} = \frac{y}{1} = y; \qquad \cos t = \frac{\text{adjacent}}{\text{hypotenuse}} = \frac{x}{1} = x.$$

Famous Sines and Cosines

If it isn't, review the definitions to convince yourself. It may also help to review the definition of "evident."

We've already violated our prohibition against degree measure! Maybe we should have said $\pi/6$–$\pi/3$–$\pi/2$ and $\pi/4$–$\pi/4$–$\pi/2$.

It's evident ◄   from the definitions of the sine and cosine functions that $\sin 0 = 0$ and $\cos(\pi/2) = 0$. Other important sines and cosines, such as $\sin(\pi/4) = \sqrt{2}/2$ and $\cos(\pi/3) = 1/2$ are well known but less obvious. All the well-known values can be computed as ratios of sides of 30–60–90 and 45–45–90 right triangles. ◄   Know these important values:

Celebrated Sines and Cosines					
t	0	$\pi/6$	$\pi/4$	$\pi/3$	$\pi/2$
$\cos t$	1	$\sqrt{3}/2$	$\sqrt{2}/2$	1/2	0
$\sin t$	0	1/2	$\sqrt{2}/2$	$\sqrt{3}/2$	1

In daily practice one needs to know sines and cosines of *all* integer multiples of $\pi/4$ and $\pi/6$. This is easier than it seems. Using the values tabulated above for angles in the first quadrant we can use the symmetry of the unit circle to find sines and cosines of many other angles.

EXAMPLE 1   Use symmetry to find $\sin(7\pi/6)$ and $\cos(7\pi/6)$.

Solution   The following diagram

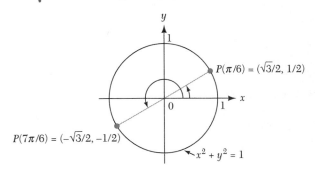

Exploiting symmetry

shows that

$$\sin\left(\frac{7\pi}{6}\right) = -\sin\left(\frac{\pi}{6}\right) = -\frac{1}{2} \quad \text{and} \quad \cos\left(\frac{7\pi}{6}\right) = -\cos\left(\frac{\pi}{6}\right) = -\frac{\sqrt{3}}{2}. \quad ■$$

Obscure Sines and Cosines

For most inputs, precise values of the trigonometric functions are far from obvious. For example, what are $\sin(1)$ and $\cos(13)$? We could estimate values graphically using a unit circle, but, as it turns out, there is *no* simple algebraic formula for computing *exact* values of trigonometric functions. Fortunately, calculus offers methods of *approximating* values of these functions.

Other Trigonometric Functions

Four additional trigonometric functions are defined in terms of the sine and cosine functions.

Definition For real numbers t,

$$\tan t = \frac{\sin t}{\cos t} \qquad \cot t = \frac{\cos t}{\sin t}$$

$$\sec t = \frac{1}{\cos t} \qquad \csc t = \frac{1}{\sin t}$$

Graphs of all these functions appear on page 64, in Section 1.5.

> **How Many Trigonometric Functions?**   Cosine, sine, tangent, secant, cosecant, and cotangent make six. From another point of view, there are only two—sine and cosine—from which all the others are derived.
>
> If we were serious about this numbers game, we could count at least *ten* trigonometric functions: The familiar six and their (justifiably) obscure relatives *exsecant, versine, coversine,* and *haversine.* Taking the opposite tack, we might argue for only *one* trigonometric function, say cosine, from which sine, and hence all the others, can be derived.
>
> In the end, the number of trigonometric functions is a matter more philosophical than mathematical. What's important is that, whatever their number, *all* trigonometric functions are closely related to each other and to the unit circle.

More on the Tangent Function

Among the "other" trigonometric functions, the tangent function arises often enough to deserve special attention. To start, here's a close look at its graph. Notice all the asymptotes, shown dotted:

Graph of $y = \tan x$

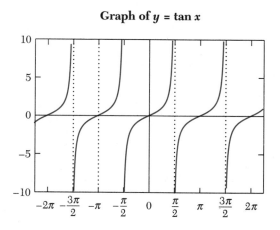

Many properties of the tangent function follow easily from its definition:

Domain and Range   The tangent function accepts all real inputs t *except* those for which $\cos t = 0$ (i.e., $t = \pm\pi/2, \pm 3\pi/2, \pm 5\pi/2$, and so on). The domain of the tangent function, therefore, consists of all real numbers *other* than these obvious offenders. The graph reflects these troubles. The range is the full set of real numbers. ◄

The graph shows this, too.

Periodicity   The tangent function (like the sine and cosine functions) repeats itself on intervals of length 2π. Here's why:

$$\tan(t + 2\pi) = \frac{\sin(t + 2\pi)}{\cos(t + 2\pi)} = \frac{\sin t}{\cos t} = \tan t .$$

Look at it. Do you agree?

The graph reveals the same property. ◄

More on Periodicity—A Surprise   Look again. The tangent function's graph repeats its basic shape on intervals of length π—*twice as often* as the computation above suggests. In other words, the tangent function appears to be π-periodic. Is something wrong?

   No. Every π-periodic function is *also* 2π-periodic, automatically. The tangent function repeats itself on intervals of length π for the following reason:

$$\tan(t + \pi) = \frac{\sin(t + \pi)}{\cos(t + \pi)} = \frac{-\sin t}{-\cos t} = \tan t.$$

The second equality is the key. It holds because $\sin(t + \pi) = -\sin t$ and $\cos(t + \pi) = -\cos t$. ▶

Are these equalities obvious? If not, draw the points $P(t)$ and $P(t + \pi)$ on the unit circle. How are they related?

Trigonometric Identities

Equations like

$$\sin^2 t + \cos^2 t = 1, \qquad \sin(t + \pi/2) = \cos t, \qquad \text{and} \quad \tan^2 t + 1 = \sec^2 t,$$

that relate various trigonometric functions to each other are known as **trigonometric identities**. A great many such relations exist—not surprisingly, because all six trigonometric functions spring from the same unit circle idea.

   Some trigonometric identities, like the first one above, follow directly from the definitions. Others are far subtler. Memorizing every trigonometric identity is out of the question. It's wiser to memorize a few truly basic identities. ▶

The three above enjoy this exalted status.

   Other—less obvious but still useful—trigonometric identities exist; one should know *that* they exist and *where* to find them.   Below are several such second-rank identities.

F a c t (Addition Formulas)

$$\sin(s + t) = \sin s \cos t + \cos s \sin t$$

$$\cos(s + t) = \cos s \cos t - \sin s \sin t$$

(**Double-Angle Formulas**)

$$\sin(2t) = 2 \sin t \cos t \qquad \cos(2t) = \cos^2 t - \sin^2 t$$

$$\sin^2 t = \frac{1}{2}\big(1 - \cos(2t)\big) \qquad \cos^2 t = \frac{1}{2}\big(1 + \cos(2t)\big)$$

The double-angle formulas are junior-grade versions of the addition formulas; the former are readily derived from the latter.

E X A M P L E 2   Derive the formula for $\cos(2t)$.

S o l u t i o n   Using the addition formula,

$$\cos(2t) = \cos(t + t) = \cos t \cos t - \sin t \sin t = \cos^2 t - \sin^2 t. \qquad ■$$

BASICS

1. The angle drawn within the unit circle shown below measures 1.2 radians. The tick marks around the circumference of the circle indicate tenths of a radian.

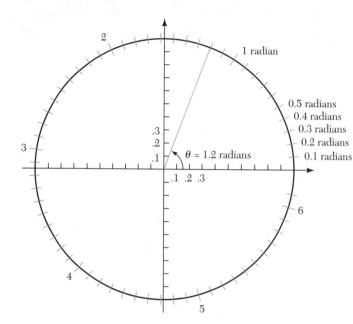

Use this diagram to estimate each of the following values:
   (a) $\cos(1.2)$ (d) $\cos(0.2)$
   (b) $\sin(0.5)$ (e) $\sin(4.5)$
   (c) $\tan(2.8)$ (f) $\cos(13)$

2. Use the diagram above to estimate each of the following. (Then check your answers with a calculator.)
   (a) Two values of x such that $\sin x = 1/2$.
   (b) Two values of x such that $\tan x = 5$.

3. Let $P(t)$ be the point t units counterclockwise from $(1, 0)$ on the unit circle. Explain in words:
   (a) How $P(t)$ and $P(t + \pi)$ are related.
   (b) Why $\cos(t + \pi) = -\cos(t)$.
   (c) Why $\sin(t + \pi) = -\sin(t)$.
   (d) Why $\tan(t + \pi) = \tan(t)$.

4. Find the domain of each of the following functions.
   (a) $\cot x$ (b) $\sec x$

5. Graphs of $y = \sin x$ and $y = \cos x$ are shown below.
   (a) Which graph is which? Explain.
   (b) It's a fact that $\cos x = \sin(x + \pi/2) \approx \sin(x + 1.57)$. How do the graphs reflect this fact?
   (c) For which values of x is $\tan x = 0$? How can you tell from the graphs above?
   (d) For which values of x is $\tan x = 1$? How can you tell from the graphs above?

(e) Draw the graphs of $y = \sin x$ and $y = 3 \sin x$ over the interval $[-2\pi, 2\pi]$ on the same axes.

Graphs of $y = \sin x$ and $y = \cos x$

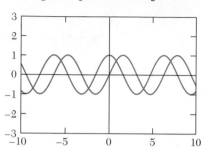

6. Suppose that $\sin a = b$ and $\cos c = 0.3$. Evaluate each of the following expressions. (Your answers should involve only real numbers and the constant b.)
   (a) $\sin(-a)$ (d) $\sin c$
   (b) $-\sin a$ (e) $\sin^2 a$
   (c) $\cos(-c)$ (f) $\sin(2a)$

7. Evaluate each of the following expressions *exactly*. (No decimal approximations allowed!)
   (a) $\cos(5\pi/6)$ (e) $\sec(5\pi/3)$
   (b) $\sin(2\pi/3)$ (f) $\csc(7\pi/4)$
   (c) $\tan(3\pi/4)$ (g) $\sin(257\pi/3)$
   (d) $\cot(4\pi/3)$ (h) $\tan(-2000\pi/3)$

8. Use the addition formulas for sine and cosine to evaluate each of the following expressions *exactly*.
   (a) $\cos\left(\dfrac{5\pi}{12}\right) = \cos\left(\dfrac{\pi}{4} + \dfrac{\pi}{6}\right)$
   (b) $\sin\left(\dfrac{19\pi}{12}\right) = \sin\left(\dfrac{3\pi}{4} + \dfrac{5\pi}{6}\right)$
   (c) $\sin\left(\dfrac{\pi}{12}\right) = \sin\left(\dfrac{\pi}{4} - \dfrac{\pi}{6}\right)$
   (d) $\tan\left(\dfrac{7\pi}{12}\right) = \tan\left(\dfrac{5\pi}{4} - \dfrac{2\pi}{3}\right)$

9. Which of the following pairs of expressions define the same function?
   (a) $\sec^2 x$ and $1 + \tan^2 x$
   (b) $\cos^2 x$ and $1 - \sin^2 x$
   (c) $\cos x$ and $\sin(x - \pi/2)$
   (d) $\sin x$ and $\cos(x + \pi/2)$

10. Find subtraction formulas for sine and cosine similar to the addition formulas.
    (a) $\sin(s - t) = \text{???}$ (b) $\cos(s - t) = \text{???}$

11. Derive the formula $\sin(2t) = 2 \sin t \cos t$.

12. Derive the formula $\cos^2 t = \dfrac{1}{2}\left(1 + \cos(2t)\right)$.

Selected Proofs

This appendix contains careful proofs of various results, more or less in order of subtlety. The proofs are intended to be read "as needed," as particular questions arise. Alternatively, some readers may enjoy reading proofs for their own sake, as an introduction to the formal ideas and methods of analytic mathematics.

A Challenging Limit

Many limits that arise in calculus courses can be evaluated either by "inspection" (i.e., at a glance) or with elementary algebra. All of the following limits are of these types:

$$\lim_{x \to \infty} \frac{1}{x} = 0; \qquad \lim_{x \to 3} (x + 2) = 5; \qquad \lim_{x \to 2} \frac{x^2 - 4}{x - 2} = 4.$$

Only the last one requires any argument, and simple algebra is all that's needed:

$$\lim_{x \to 2} \frac{x^2 - 4}{x - 2} = \lim_{x \to 2} \frac{(x - 2)(x + 2)}{x - 2} = \lim_{x \to 2} (x + 2) = 4.$$

The important ▶ limit

$$\lim_{t \to 0} \frac{\sin t}{t} = 1$$

One of the 2 or 3 most important in calculus.

is different. Because $\sin t$ is *not* an algebraic expression, the usual proof tricks—factoring, expansion, and so on—are useless. A rigorous proof must take another tack.

Before beginning a rigorous proof, we remark that overwhelming circumstantial evidence—in graphical and numerical forms—that $\sin t / t \to 1$ is available. (See, e.g., Example 4, page 152.) Still, pictures and numbers aren't proofs. On the contrary, what a computer or calculator "understands" about the sine function probably depends implicitly on some knowledge of the very limit in question.

A Proof.   Because $\sin t$ and t are both *odd* functions of t, their ratio is *even*. Thus it's enough to consider only positive values of t, i.e., to show that

$$\lim_{t \to 0^+} \frac{\sin t}{t} = 1.$$

477

The following unit-circle diagram describes the situation:

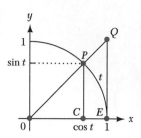

The diagram supports three key claims:

i. $\sin t \leq t$; (the segment CP is shorter than the arc EP)

ii. $t \leq \sin t + 1 - \cos t$; (the arc EP is shorter than EC+CP—the route traveling *west*, then *north*)

iii. as $t \to 0^+$, $\dfrac{1 - \cos t}{t} \to 0$.

Claim (iii) requires some support. Here's an an airtight proof, based on (i):

$$\frac{1 - \cos t}{t} = \frac{1 - \cos t}{t} \cdot \frac{1 + \cos t}{1 + \cos t} = \frac{1 - \cos^2 t}{t(1 + \cos t)} = \frac{\sin^2 t}{t(1 + \cos t)}$$

$$= \frac{\sin t}{t} \cdot \frac{\sin t}{1 + \cos t} \leq \frac{\sin t}{1 + \cos t}.$$

Now it follows that

$$0 \leq \frac{1 - \cos t}{t} \leq \frac{\sin t}{1 + \cos t}.$$

As $t \to 0^+$, the right side tends to 0. By the squeeze principle, (iii) holds as claimed.

To finish our proof we'll use the squeeze principle again. From (i) and (ii) we have

$$t - 1 + \cos t \leq \sin t \leq t.$$

Dividing through by t gives

$$\frac{t - 1 + \cos t}{t} = 1 - \frac{1 - \cos t}{t} \leq \frac{\sin t}{t} \leq 1.$$

Taking limits gives

$$1 - \lim_{t \to 0^+} \frac{1 - \cos t}{t} \leq \lim_{t \to 0^+} \frac{\sin t}{t} \leq 1.$$

From (iii), the first limit is 1; we're done. ∎

Another Proof. Another "squeeze" proof of the same limit can be based on areas. The same picture shows the following area relation:

$$\text{area(OCP)} \leq \text{area(OEP)} \leq \text{area(OEQ)}.$$

Here OCP and OEQ are triangles; OEP is a circular sector. Calculating these three areas gives

$$\text{area}(OCP) = \frac{\text{base} \times \text{height}}{2} = \frac{\cos t \, \sin t}{2};$$

$$\text{area}(OEP) = \pi \times \text{fraction of circle} = \pi \times \frac{t}{2\pi} = \frac{t}{2};$$

$$\text{area}(OEQ) = \frac{\text{base} \times \text{height}}{2} = \frac{\tan t}{2}.$$

Thus,

$$\frac{\cos t \, \sin t}{2} \leq \frac{t}{2} \leq \frac{\tan t}{2}.$$

If $t > 0$, then multiplying through by $2/\sin t$ ▶ gives   *A positive number!*

$$\cos t \leq \frac{t}{\sin t} \leq \frac{1}{\cos t}.$$

As $t \to 0^+$ the left and right quantities tend to 1; so, therefore, does the middle. ∎

Proving the Product Rule

Recall the product rule:

Fact (**The Product Rule**) Suppose that u and v are differentiable at x. If $p(x) = u(x) \cdot v(x)$, then p is differentiable at x, and

$$p'(x) = \big(u(x) \cdot v(x)\big)' = u'(x) \cdot v(x) + u(x) \cdot v'(x).$$

In Section 3.5 we argued the case informally. Below is a formal proof; notice that it shows—rather than assumes—that the product uv is differentiable.

Proof.   Let u and v be differentiable functions. By definition,

$$(uv)'(x) = \lim_{h \to 0} \frac{u(x+h)v(x+h) - u(x)v(x)}{h}.$$

We need to show that this limit exists, and equals $u'(x)v(x) + u(x)v'(x)$.

An algebraic trick—subtracting and adding the same quantity in the numerator—does the job: ▶   *Check each step.*

$$\begin{aligned}
(uv)' &= \lim_{h \to 0} \frac{u(x+h)v(x+h) - u(x)v(x)}{h} \\
&= \lim_{h \to 0} \frac{u(x+h)v(x+h) - u(x)v(x+h) + u(x)v(x+h) - u(x)v(x)}{h} \\
&= \lim_{h \to 0} \frac{u(x+h)v(x+h) - u(x)v(x+h)}{h} + \lim_{h \to 0} \frac{u(x)v(x+h) - u(x)v(x)}{h} \\
&= \lim_{h \to 0} \frac{u(x+h) - u(x)}{h} \cdot \lim_{h \to 0} v(x+h) + u(x) \cdot \lim_{h \to 0} \frac{v(x+h) - v(x)}{h} \\
&= u'(x)v(x) + u(x)v'(x).
\end{aligned}$$

We're done. ∎

An $\epsilon - \delta$ Limit Proof

Recall the $\epsilon - \delta$ definition of limit:

> **Definition** Suppose that for every positive number ϵ, no matter how small, there is a positive number δ so that
>
> $$|f(x) - L| < \epsilon \qquad \text{whenever} \qquad 0 < |x - a| < \delta.$$
>
> Then
>
> $$\lim_{x \to a} f(x) = L.$$

Other limit properties, in the same spirit, appear in Theorem 1, page 168.

Recall, too, this plausible-seeming property: ◄

> **Fact** Suppose that
>
> $$\lim_{x \to a} f(x) = L \qquad \text{and} \qquad \lim_{x \to a} g(x) = M,$$
>
> where L and M are finite numbers. Then
>
> $$\lim_{x \to a} [f(x) + g(x)] = L + M.$$

A rigorous proof must use the rigorous definition. Here's how:

Proof. We need to show that for any number $\epsilon > 0$ there's a corresponding number $\delta > 0$ such that

$$0 < |x - a| < \delta \implies |(f(x) + g(x)) - (f(a) + g(a))| < \epsilon.$$

Let $\epsilon > 0$ be given. To find the desired δ, we'll first set $\epsilon_1 = \epsilon_2 = \epsilon/2$. Since ϵ_1 and ϵ_2 are positive, there exist, by our assumption, other positive numbers δ_1 and δ_2 such that

$$0 < |x - a| < \delta_1 \implies |f(x) - f(a)| < \epsilon/2;$$

$$0 < |x - a| < \delta_2 \implies |g(x) - g(a)| < \epsilon/2.$$

If we now set δ to be the smaller of δ_1 and δ_2, then we have what we want:

$$0 < |x - a| < \delta \implies 0 < |x - a| < \delta_1 \qquad \text{and} \qquad 0 < |x - a| < \delta_2,$$

so

$$|(f(x) + g(x)) - (f(a) + g(a))| \le |f(x) - f(a)| + |g(x) - g(a)|$$

$$< \epsilon/2 + \epsilon/2$$

$$= \epsilon. \qquad \blacksquare$$

A Graphical Glossary of Functions

Graphical reasoning—representing functions and their derivatives graphically—is an invaluable skill in calculus. This appendix offers a brief "atlas" of representatives of the most important classes of calculus functions. Each function is shown together with its derivative.

Graph of $f(x) = x$

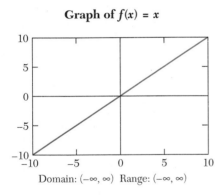

Domain: $(-\infty, \infty)$ Range: $(-\infty, \infty)$

Graph of $f'(x) = 1$

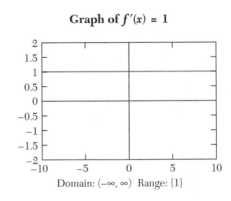

Domain: $(-\infty, \infty)$ Range: $\{1\}$

Notes. The derivative of a linear function is a constant function.

Graph of $f(x) = x^2$

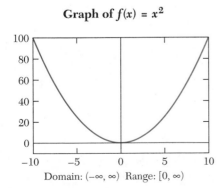

Domain: $(-\infty, \infty)$ Range: $[0, \infty)$

Graph of $f'(x) = 2x$

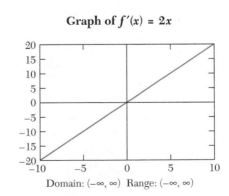

Domain: $(-\infty, \infty)$ Range: $(-\infty, \infty)$

Notes.   The derivative of a quadratic function is a linear function. The quadratic function's vertex corresponds to a root of the derivative.

Graph of $f(x) = \dfrac{x}{x^2 - 9}$

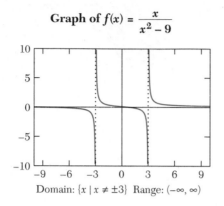

Domain: $\{x \mid x \neq \pm 3\}$ Range: $(-\infty, \infty)$

Graph of $f'(x) = -\dfrac{x^2 - 9}{(x^2 - 9)^2}$

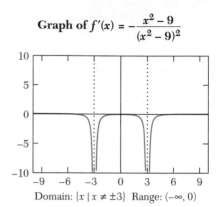

Domain: $\{x \mid x \neq \pm 3\}$ Range: $(-\infty, 0)$

Notes.   The derivative of a rational function is another rational function. Notice the behavior of asymptotes: If f has a *vertical* asymptote at $x = a$, then so does f'. If f has a *horizontal* asymptote at $y = b$, then f' has a horizontal asymptote at $y = 0$.

Graph of $f(x) = \sin(x)$

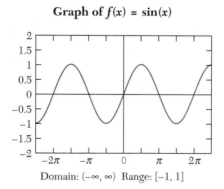

Domain: $(-\infty, \infty)$ Range: $[-1, 1]$

Graph of $f'(x) = \cos(x)$

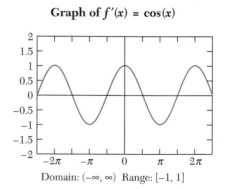

Domain: $(-\infty, \infty)$ Range: $[-1, 1]$

Notes.   The sine function is 2π-periodic; so, therefore, is its derivative. Both the sine function and its derivative oscillate between -1 and 1. At the origin, the sine graph has slope 1.

Graph of $f(x) = \cos(x)$

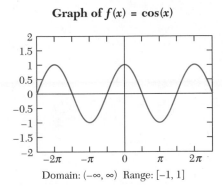

Domain: $(-\infty, \infty)$ Range: $[-1, 1]$

Graph of $f'(x) = -\sin(x)$

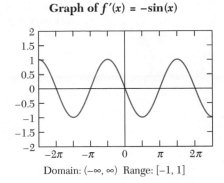

Domain: $(-\infty, \infty)$ Range: $[-1, 1]$

Notes.   The cosine function is 2π-periodic; so, therefore, is its derivative. Both the sine function and its derivative oscillate between -1 and 1. At the origin, the cosine graph is horizontal.

Graph of $f(x) = \tan(x)$

Domain: $\{x \mid x \neq \pi \pm k\pi/2\}$ Range: $(-\infty, \infty)$

Graph of $f'(x) = \sec^2(x)$

Domain: $\{x \mid x \neq \pi \pm k\pi/2\}$ Range: $[1, \infty)$

Notes.   The tangent function is π-periodic; so is its derivative. Unlike the sine and cosine functions, the tangent function and its derivative have vertical asymptotes. At the origin, the tangent graph has slope 1.

Graphs of $f(x) = e^x$ and $g(x) = 2^x$

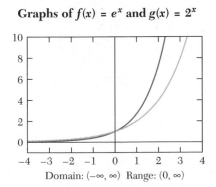

Domain: $(-\infty, \infty)$ Range: $(0, \infty)$

Graphs of $f'(x) = e^x$ and $g'(x) = 2^x \ln 2$

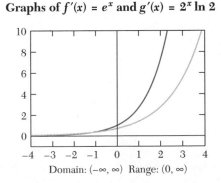

Domain: $(-\infty, \infty)$ Range: $(0, \infty)$

Notes. The natural exponential function $y = e^x$ is its own derivative. Other exponential functions, such as $y = 2^x$, are *proportional* to their own derivatives. The $y = e^x$ graph has slope 1 at $x = 0$; the $y = 2^x$ graph has slope $\ln 2$ at $x = 0$. All exponential functions have horizontal asymptotes; so do their derivatives.

Graph of $f(x) = \ln x$

Domain: $(0, \infty)$ Range: $(-\infty, \infty)$

Graph of $f'(x) = 1/x$

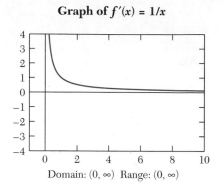

Domain: $(0, \infty)$ Range: $(0, \infty)$

Notes.   The natural logarithm function $y = \ln x$ is the inverse function to $y = e^x$. It is defined only for positive inputs. The graph of $y = \ln x$ has slope 1 at $x = 1$.

Index

Table of Integrals

Basic Forms

1. $\int x^n dx = \dfrac{x^{n+1}}{n+1}, n \neq -1$

2. $\int \dfrac{dx}{x} = \ln |x|$

3. $\int e^x dx = e^x$

4. $\int b^x dx = \dfrac{1}{\ln b} b^x$

5. $\int \sin x \, dx = -\cos x$

6. $\int \cos x \, dx = \sin x$

7. $\int \tan x \, dx = \ln |\sec x|$
$= -\ln |\cos x|$

8. $\int \cot x \, dx = \ln |\sin x|$
$= -\ln |\csc x|$

9. $\int \sec^2 x \, dx = \tan x$

10. $\int \csc^2 x \, dx = -\cot x$

11. $\int \sec x \tan x \, dx = \sec x$

12. $\int \csc x \cot x \, dx = -\csc x$

13. $\int \dfrac{dx}{x^2 + a^2} = \dfrac{1}{a}\arctan\left(\dfrac{x}{a}\right),$ $a \neq 0$

14. $\int \dfrac{dx}{x^2 - a^2} = \dfrac{1}{2a}\ln\left|\dfrac{x-a}{x+a}\right|$

15. $\int \dfrac{dx}{\sqrt{a^2 - x^2}} = \arcsin\left(\dfrac{x}{a}\right),$ $a > 0$

16. $\int \ln x \, dx = x(\ln x - 1)$

17. $\int \sec x \, dx = \ln |\sec x + \tan x| = \ln \left|\tan\left(\dfrac{x}{2} + \dfrac{\pi}{4}\right)\right|$

18. $\int \csc x \, dx = \ln |\csc x - \cot x| = \ln \left|\tan\left(\dfrac{x}{2}\right)\right|$

Expressions Containing $ax + b$

19. $\int (ax+b)^n dx = \dfrac{(ax+b)^{n+1}}{a(n+1)}, n \neq -1$

20. $\int \dfrac{dx}{ax+b} = \dfrac{1}{a}\ln |ax+b|$

21. $\int \dfrac{x}{ax+b}dx = \dfrac{x}{a} - \dfrac{b}{a^2}\ln |ax+b|$

22. $\int \dfrac{x}{(ax+b)^2}dx = \dfrac{b}{a^2(ax+b)} + \dfrac{1}{a^2}\ln |ax+b|$

23. $\int \dfrac{dx}{x(ax+b)} = \dfrac{1}{b}\ln \left|\dfrac{x}{ax+b}\right|$

24. $\int \dfrac{dx}{x^2(ax+b)} = -\dfrac{1}{bx} + \dfrac{a}{b^2}\ln \left|\dfrac{ax+b}{x}\right|$

25. $\int \sqrt{ax+b} \, dx = \dfrac{2}{3a}\sqrt{(ax+b)^3}$

26. $\int x\sqrt{ax+b} \, dx = \dfrac{2(3ax-2b)}{15a^2}\sqrt{(ax+b)^3}$

27. $\int \dfrac{dx}{\sqrt{ax+b}} = \dfrac{2\sqrt{ax+b}}{a}$

28. $\int \dfrac{dx}{x\sqrt{ax+b}} = \dfrac{1}{\sqrt{b}}\ln \left|\dfrac{\sqrt{ax+b}-\sqrt{b}}{\sqrt{ax+b}+\sqrt{b}}\right|, b > 0$

29. $\int \dfrac{dx}{x\sqrt{ax-b}} = \dfrac{2}{\sqrt{b}}\arctan\sqrt{\dfrac{ax-b}{b}}, b > 0$

Expressions Containing $ax^2 + c$, $x^2 \pm p^2$, and $p^2 - x^2, p > 0$

30. $\int \dfrac{dx}{p^2 - x^2} = \dfrac{1}{2p}\ln \left|\dfrac{p+x}{p-x}\right|$

31. $\int \dfrac{dx}{ax^2 + c} = \dfrac{1}{\sqrt{ac}}\arctan\left(x\sqrt{\dfrac{a}{c}}\right), a > 0, c > 0$

32. $\int \dfrac{dx}{ax^2 - c} = \dfrac{1}{2\sqrt{ac}}\ln \left|\dfrac{x\sqrt{a}-\sqrt{c}}{x\sqrt{a}+\sqrt{c}}\right|, a > 0, c > 0$

33. $\int \dfrac{dx}{(ax^2+c)^n} = \dfrac{1}{2(n-1)c}\dfrac{x}{(ax^2+c)^{n-1}} + \dfrac{2n-3}{2(n-1)c}\int \dfrac{dx}{(ax^2+c)^{n-1}}, n > 1$

34. $\int x(ax^2+c)^n \, dx = \dfrac{1}{2a}\dfrac{(ax^2+c)^{n+1}}{n+1}, n \neq -1$

35. $\int \dfrac{x}{ax^2+c}dx = \dfrac{1}{2a}\ln |ax^2+c|$

36. $\int \sqrt{x^2 \pm p^2} \, dx = \dfrac{1}{2}\left(x\sqrt{x^2 \pm p^2} \pm p^2 \ln \left|x + \sqrt{x^2 \pm p^2}\right|\right)$

37. $\int \sqrt{p^2 - x^2} \, dx = \dfrac{1}{2}\left(x\sqrt{p^2 - x^2} + p^2\arcsin\left(\dfrac{x}{p}\right)\right), p > 0$

38. $\int \dfrac{dx}{\sqrt{x^2 \pm p^2}} = \ln \left|x + \sqrt{x^2 \pm p^2}\right|$

Expressions Containing Trigonometric Functions

39. $\int \sin^2(ax)\,dx = \dfrac{x}{2} - \dfrac{\sin(2ax)}{4a}$

40. $\int \sin^n(ax)\,dx = -\dfrac{\sin^{n-1}(ax)\cos(ax)}{na} +$

$\dfrac{n-1}{n}\int \sin^{n-2}(ax)\,dx, n > 0$

41. $\int \cos^2(ax)\,dx = \dfrac{x}{2} + \dfrac{\sin(2ax)}{4a}$

42. $\int \cos^n(ax)\,dx = \dfrac{\cos^{n-1}(ax)\sin(ax)}{na} +$

$\dfrac{n-1}{n}\int \cos^{n-2}(ax)\,dx$

43. $\int \sin(ax)\cos(bx)\,dx = -\dfrac{\cos((a-b)x)}{2(a-b)} -$

$\dfrac{\cos((a+b)x)}{2(a+b)}, a^2 \neq b^2$

44. $\int \sin(ax)\sin(bx)\,dx = \dfrac{\sin((a-b)x)}{2(a-b)} -$

$\dfrac{\sin((a+b)x)}{2(a+b)}, a^2 \neq b^2$

45. $\int \cos(ax)\cos(bx)\,dx = \dfrac{\sin((a-b)x)}{2(a-b)} +$

$\dfrac{\sin((a+b)x)}{2(a+b)}, a^2 \neq b^2$

46. $\int x \sin(ax)\,dx = \dfrac{1}{a^2}\sin(ax) - \dfrac{x}{a}\cos(ax)$

47. $\int x \cos(ax)\,dx = \dfrac{1}{a^2}\cos(ax) + \dfrac{x}{a}\sin(ax)$

48. $\int x^n \sin(ax)\,dx = -\dfrac{x^n}{a}\cos(ax) +$

$\dfrac{n}{a}\int x^{n-1}\cos(ax)\,dx, n > 0$

49. $\int x^n \cos(ax)\,dx = \dfrac{x^n}{a}\sin(ax) -$

$\dfrac{n}{a}\int x^{n-1}\sin(ax)\,dx, n > 0$

50. $\int \tan^n(ax)\,dx = \dfrac{\tan^{n-1}(ax)}{a(n-1)} -$

$\int \tan^{n-2}(ax)\,dx, n \neq 1$

51. $\int \sec^n(ax)\,dx = \dfrac{\sec^{n-2}(ax)\tan(ax)}{a(n-1)} + \dfrac{n-2}{n-1}\int \sec^{n-2}(ax)\,dx, n \neq 1$

Expressions Containing Exponential and Logarithm Functions

52. $\int xe^{ax}\,dx = \dfrac{e^{ax}}{a^2}(ax-1)$

53. $\int x^n e^{ax}\,dx = \dfrac{1}{a}x^n e^{ax} - \dfrac{n}{a}\int x^{n-1}e^{ax}\,dx, n > 0$

54. $\int e^{ax}\sin(bx)\,dx = \dfrac{e^{ax}}{a^2+b^2}(a\sin(bx) - b\cos(bx))$

55. $\int e^{ax}\cos(bx)\,dx = \dfrac{e^{ax}}{a^2+b2}(a\cos(bx) + b\sin(bx))$

56. $\int x^n \ln(ax)\,dx = x^{n+1}\left(\dfrac{\ln(ax)}{n+1} - \dfrac{1}{(n+1)^2}\right),$

$n \neq -1$

57. $\int (\ln x)^n\,dx = x(\ln x)^n - n\int (\ln x)^{n-1}\,dx$

58. $\int \dfrac{dx}{a+be^{px}} = \dfrac{x}{a} - \dfrac{1}{ap}\ln|a+be^{px}|$

Expressions Containing Inverse Trigonometric Functions

59. $\int \arcsin(ax)\,dx = x\arcsin(ax) + \dfrac{1}{a}\sqrt{1-a^2x^2}$

60. $\int \arccos(ax)\,dx = x\arccos(ax) - \dfrac{1}{a}\sqrt{1-a^2x^2}$

61. $\int \arctan(ax)\,dx = x\arctan(ax) - \dfrac{1}{2a}\ln(1+a^2x^2)$

62. $\int \operatorname{arccot}(ax)\,dx = x\operatorname{arccot}(ax) + \dfrac{1}{2a}\ln(1+a^2x^2)$

63. $\int \operatorname{arccsc}(ax)\,dx = x\operatorname{arccsc}(ax) + \dfrac{1}{a}\ln\left|ax + \sqrt{a^2x^2-1}\right|$

64. $\int \operatorname{arcsec}(ax)\,dx = x\operatorname{arcsec}(ax) - \dfrac{1}{a}\ln\left|ax + \sqrt{a^2x^2-1}\right|$